Recombinant DNA and Biotechnology

A Guide for Teachers

Recombinant DNA and Biotechnology

A Guide for Teachers

Helen Kreuzer, Ph.D.
Biotechnology Department
Carolina Biological Supply Company
Burlington, North Carolina

Adrianne Massey, Ph.D.
Education and Training Program
North Carolina Biotechnology Center
Research Triangle Park, North Carolina

ASM Press
Washington, D.C.

ISBN 1-55581-101-9

10 9 8 7 6 5 4 3 2 1

Support and funding for the preparation of these materials were provided by the
State of North Carolina through the North Carolina Biotechnology Center, which
support is gratefully acknowledged.

Cover and interior design: Susan Brown Schmidler
Illustrations: Network Graphics and Impressions Book and Journal Services, Inc.

Cover figure: Genetic scientists with plasmids in the background (VU/©K. G. Murti)

To the memory of Paul W. Bryant

About the Authors

Helen Kreuzer describes herself as cross-trained in both science and education. She has degrees from the University of Alabama in chemistry, microbiology, and education and received her Ph.D. from Duke University in molecular genetics. During the 3 years that she coordinated the North Carolina Biotechnology Center's teacher education program, she developed many of the teaching activities presented in this book. She has also worked in academic research and served on the faculty of a 4-year liberal arts college, where she taught genetics and molecular biology to undergraduates. Dr. Kreuzer joined the staff of the Carolina Biological Supply Company in 1995, where she is developing new molecular biology teaching materials, including wet labs, models, and videos. She has taught biotechnology workshops to high school teachers and college faculty at sites across the country.

Adrianne Massey is the Director of the North Carolina Biotechnology Center's award-winning Education and Training Program. She received her Ph.D. in zoology from North Carolina State University and has taught biology to undergraduate and graduate students at North Carolina State University and the University of Georgia. She has conducted research in evolutionary genetics and developed interactive exhibits and related educational materials for science and technology centers. At the North Carolina Biotechnology Center, she has developed extensive curriculum materials and videos for educators in high schools, colleges, and technical training centers. Dr. Massey has also been a frequent speaker on both the science and public policy issues of biotechnology.

Note to Readers

This book, *Recombinant DNA and Biotechnology: A Guide for Teachers,* contains four parts: part I (*Laying the Foundation*), part II (*Classroom Activities*), part III (*Societal Issues*), and part IV (*Appendixes*). In addition to the material intended only for teachers, the *Student Activity* pages in parts II and III contain material for students as well.

The accompanying book, *Recombinant DNA and Biotechnology: A Guide for Students,* contains part I mentioned above in its entirety but only the *Student Activity* pages mentioned above for parts II and III. Thus, the page numbering for the teacher and student guides is different from part II on.

Finally, the teacher guide contains 12 appendixes. The student guide has only one appendix, Appendix A (Templates), which corresponds to Appendix K in the teacher guide. The reader will need to remember that a reference in the *Student Activity* pages to "Appendix A" is referring to the student guide appendix of templates and not to Appendix A in the teacher guide. In those cases, the reader needs to refer to Appendix K in the teacher guide.

Contents

Foreword

This book, *Recombinant DNA and Biotechnology: A Guide for Teachers,* by Helen Kreuzer and Adrianne Massey and the accompanying book for students come out at a propitious time, a time when a new energy is infusing our nation. Today we are poised to make dramatic progress in improving our educational institutions with thousands of volunteer scientists promoting a new type of science education as a catalyst that can reinvigorate the excitement of learning in all grades.

In 1989, our nation's 50 governors, seriously concerned about our future, promulgated a set of national goals. One of these was to provide the world's finest science and math education to our young people by the year 2000. Recognizing that their goals would be difficult, if not impossible, to achieve without a roadmap to guide us, the governors also called for the development of a set of national education standards. In 1991, the challenge of leading the effort to produce the standards for science education was assigned to the National Research Council, the operating arm of the National Academies of Sciences and Engineering. The development process proceeded by means of a series of widely circulated drafts that were reviewed by thousands of scientists, teachers, science educators, and others. The result is the National Science Education Standards, a consensus document that was published in late 1995. This document is available in book form from the National Academy Press, as well as electronically on the World Wide Web (www.nas.edu).

The National Science Education Standards provide a tool that should allow all interested people in the United States to work together in an effective and consistent way to improve the education that we provide to our young people. Nothing is more important for the future of this nation or for the future of our children and grandchildren.

What we seek are large numbers of students graduating from our high schools with dramatically improved skills—not just the most academically gifted of our young people but all of them, regardless of their initial interests, abilities, or cultural backgrounds. Today, there are very few rewarding jobs for those who do not go on to college. Our manufacturing industries are finding it very difficult to find qualified employees to interview, despite a large excess of applicants. Without productive employment, large numbers of our high school graduates are becoming disillusioned with their prospects for an enjoyable life.

Unfortunately, many aspects of our present educational system will have to change in concert to achieve the vision of education described in the Standards. It is for this reason that only one-half of the Standards document deals with content standards—what every child should know and be able to do by the end of the 4th, 8th, and 12th grades. Other parts of the document deal with standards for teaching, for the professional development of teachers, for assessments, for school programs, and for the broad educational system.

Americans are not known for their patience. Consider, for example, our insistence on measuring the performance of business enterprises through quarterly reports of earnings or on measuring the performance of schools by scores on standardized tests that are given annually. The consensus vision of science education presented in the Standards should allow us to give all of the players—school systems, teachers, parents, business, universities, and policy makers—the patience they will need to maintain a coherent strategy long enough to be rewarded with dramatic results.

Recombinant DNA and Biotechnology: A Guide for Teachers by Helen Kreuzer and Adrianne Massey and the accompanying book for students support the Standards in at least three ways:

1. The books provide a series of hands-on, inquiry-based exercises that allow students to work through many of the fundamental concepts and

processes that underlie the revolution in molecular biology and biotechnology. As the Standards stress, "learning science is something students do, not something that is done to them." Some of the exercises, those that involve paper cutouts and intellectual problem solving, can be done in any classroom. Others will require access to special equipment that scientists and school districts should help to provide.

2. The books provide connections to many issues that students will read about in newspapers and hear about on television. Because the science taught is current, with clear relevance to their lives, students will immediately see the purpose for learning it. And because we do not know all of the answers to the questions that will be raised, students will acquire the scientific reasoning skills that will be important to them as citizens, long after they graduate. As the Standards stress, "scientific literacy implies that a person can identify scientific issues underlying national and local decisions and express positions that are scientifically and technologically informed."

3. The books represent a joint effort of scientists and outstanding science teachers, each with a unique contribution to make. The authors point out how much they have learned in preparing these books from the teachers who have helped them. New sets of permanent alliances have also formed, as local scientists and teachers have come to appreciate and respect each other's expertise. As the Standards stress, "Scientists must understand the vision of science education in the Standards and their role in achieving the vision."

In summary, the National Science Education Standards represent not an end but a beginning—a beginning of a rejuvenation of our educational system that will require a persistent, patient effort over at least 10 years. These books, and many like them in other areas of science, will be needed to fill the terrible gap that now exists in the available curricular choices for students, especially in our high schools but also in our colleges. I applaud the collaboration of scientists and teachers that has produced these fine books and the American Society for Microbiology for publishing them, thereby providing an example of what scientific societies can and should do in the precollege and college teaching arenas.

Bruce Alberts
President, National Academy of Sciences
Washington, D.C.

Preface

Teaching biology today is no simple task. The pace at which the science is advancing is exhilarating but also exhausting. It seems that every day, newspapers announce a new and exciting finding about life at the molecular level. Educators interested in keeping up with this explosion of knowledge face a daunting task.

Why has the pace of biological discovery quickened so much in the last decade? Because findings in the previous decade furnished biologists with a bag full of new research tools—molecular techniques that provide answers to old questions and create a whole set of new questions. These techniques are commonly grouped under the inclusive heading "biotechnology." Scientists are using the tools of biotechnology to tackle problems in all branches of biology, from cellular and molecular biology to ecology and evolution.

Biotechnology provides us with more than research tools for scientists. It touches all of us in many ways, for we are using the diverse technologies that constitute biotechnology to produce an even broader array of products. Thanks to monoclonal antibody technology, we have more than 100 new tests for diagnosing plant, animal, and human diseases. In bioprocessing technology, we are exploiting the biochemical machinery of microbes to degrade environmental pollutants and synthesize enzymes useful in the textile, food processing, and paper industries. Genetic engineering has corrected genetic disorders in children and made possible the development of crops resistant to insects, diseases, and environmental stresses.

Even though biotechnology includes all of these technologies and more, in this text we focus only on technologies based on DNA and the DNA-RNA-protein interconnection. To provide a conceptual background for understanding the technologies and their applications, we also include information on some of the science fundamental to DNA-based technologies.

In writing this book, we wanted to provide not only background information to help educators stay abreast of advances in biology but also activities that teach students the science of molecular genetics and its application to biotechnology. Our experience working with educators in North Carolina has taught us that both types of instructional materials are essential for effecting the inclusion of biotechnology in biology courses. Most educators began teaching before the molecular biology explosion occurred. As a result, updating educators on scientific findings is a requisite first step in enabling them to bring newfound knowledge into their classrooms. On the other hand, without clear, easy-to-follow activities that respect the constraints under which most educators must operate, successful transmission of these new understandings to students is unlikely.

The first part of this book, *Laying the Foundation*, provides basic information on topics covered by the activities in the remainder of the book. In chapter 1, we introduce the entire scope of biotechnology before narrowing our focus to DNA-based technologies for the remainder of the book. Chapter 2 provides a historical and conceptual context for the detailed discussions of molecular biology and DNA-based technologies given in chapters 3 and 4. These four chapters contain much more information than most biology instructors will ever use in the classroom. They are meant to supply a broad grounding in a complex subject. We hope readers will refer to *Laying the Foundation* much as they would a biology textbook.

Parts II and III, *Classroom Activities* and *Societal Issues,* resemble a laboratory manual or workbook with activities appropriate for audiences ranging from middle school students to upper-level undergraduates. The introduction to each activity contains germane background information not provided in *Laying the Foundation.*

Part II, *Classroom Activities,* provides wet and dry laboratory activities for teaching both the basic science of molecular genetics and the hands-on techniques of DNA-based technologies. Activities are presented in lesson plan format. Included in the lesson plans are teaching objectives, lists of required materials, preparation procedures, teaching tips, and answers to student questions. In this edition we have also included the material provided in the student edition: additional background information and directions for guiding students through the activity.

Some of the activities utilize paper models or other objects to mimic molecular phenomena or laboratory techniques. The value of these activities may not be apparent to some postsecondary biology instructors, particularly those who have chosen a career in scientific research. The activities may seem to be superfluous games intended to keep students busy. We are sympathetic to that point of view because it used to be ours. Originally, we believed that precise words and clear pictures would allow almost anyone to grasp the concepts contained in this book. However, after leaving an academic research environment, teaching a wide range of audiences of various ages and educational levels, and learning about teaching and learning from North Carolina teachers, we realized that the ability to think abstractly and learn through visual and verbal information is rarer than we had previously thought. Activities that seemed superfluous to us originally—mimicking protein synthesis by stringing beads together, acting out transcription and translation, or splicing genes by cutting and pasting pieces of construction paper—we now accept as crucial components of learning. For many people, irrespective of age and educational level, molecular events can be understood only through manipulating models or acting out cellular and molecular events. No matter how apt our words or lucid our graphics, we often do not see that telltale sign of understanding—the light that appears in a student's eyes—until we have that student *do* something.

Part II is divided into four sections, each of which stresses a concept central to understanding biological systems and our manipulation of them.

- The first group of activities focuses primarily on basic molecular biology, specifically the structure and function of DNA. The pervasive relationship between structure and function in the natural world is a basic unifying principle that should be integrated into every part of any biology course.

- In the second group of activities, *Manipulation and Analysis of DNA,* we shift the focus from basic science to some of the hands-on techniques commonly used in DNA-based technologies.

- The activities in the third group teach both the basic science of the methods microbes use for transferring genetic information and the laboratory techniques that exploit these natural means of gene transfer.

- The fourth group of activities emphasizes the importance of genetic variation to both biological organisms and biotechnology. Like the relationship between structure and function, the fundamental importance of genetic variation in biological systems is a scientific concept that should be woven into the whole biology course. We begin the section with an activity designed to show a primary means of generating genetic variation in eukaryotes—sexual reproduction—and then concentrate on genetic variation in one eukaryotic species, our own. We describe methods of looking directly at variation in DNA molecules to determine paternity, investigate crimes, and diagnose genetic disorders.

Science and technology don't exist in a social vacuum. They exert a powerful influence on society, and society, in turn, alters how science is conducted and how technologies are developed. Because biotechnology has generated wide public debate about a number of important issues, we would be remiss if we described only the scientific foundations and technological applications of biotechnology and excluded its attendant societal issues.

Part III, *Societal Issues,* includes both background information and classroom activities useful in teaching students how to approach the difficult issues that biotechnology, or any technology, spawns. All too often, discussions about controversial social issues are merely emotional exchanges of opinions that may have nothing to do with facts. These unproductive debates contribute to rather than alleviate confusion. Teaching students how to objectively analyze the complex relationships among science, technology, and society is one of the greatest contributions an educator can make to society.

In *Societal Issues,* we offer tools for rationally analyzing and discussing a number of difficult issues. We begin by asking readers to consider the complex relationship and interdependence of scientific understanding, technological development, and societal structure. In chapters 30 and 31, we introduce the science of risk assessment and risk-benefit analysis, and we provide suggestions for guiding debate on the risks and benefits of biotechnology. Chapters 32 to 34 move to more difficult questions about the bioethical ramifications of biotechnology. To facilitate productive discussion, we provide a field-tested model for guiding thoughtful consideration of bioethical dilemmas. Finally, in chapter 35, we describe career options in biotechnology, highlighting different types of jobs and giving their educational requirements.

The final segment of the book, Part IV, provides appendixes that contain overhead masters, biosafety information, recipes, information on teaching resources, and other practical information.

In this book, we want not only to provide information and activities but also to share our understanding and appreciation of biology with educators. Our understanding comes from years spent in an academic environment. During that time, we were fortunate enough to be able to immerse ourselves in biology. It takes years of a single-minded pursuit of knowledge for knowledge to be transformed into understanding. With that understanding comes a very deep respect and reverence for the workings of the natural world. If we convey just a fraction of that reverence to our readers, then we will have succeeded.

Dr. Helen Kreuzer
and Dr. Adrianne Massey

Acknowledgments

Many people contribute directly and indirectly to the production of a book. Ours is no exception. Without the technical, emotional, and financial support of others, publication of this book would not have been possible. We are indebted to all who have shared their time, insights, expertise, and talents with us.

First and foremost we thank the teachers throughout North Carolina who have worked with us over the years and assisted us in a variety of ways during the production of the book. Their enthusiasm for learning new and difficult material, determination to introduce this exciting science to their classes, willingness to share so generously with their colleagues, and unflagging devotion to their students have inspired and reinforced us. We feel blessed to have had the opportunity to work with all of them, but those who were particularly helpful in the production and field-testing of this book and who deserve special thanks include Sherri Andrews, Leslie Brinson, Nancy Evans, Marilyn Garner, Britt Hammond, Bobbie Hinson, Marlene Jacoby, Elizabeth Ruc, Thea Sinclair, and Brian Wood.

We appreciate the North Carolina scientists who reviewed various portions of the book for conceptual clarity and scientific accuracy. In particular, we thank Drs. Harold Coble and Fred Gould of North Carolina State University, Dr. Michael Hudson of the University of North Carolina at Charlotte, Drs. Ken Kreuzer and Jeremy Sugarman of the Duke University Medical Center, and Dr. Scott Shore of the North Carolina Department of Agriculture.

Thanks are also due to the following reviewers who critically read portions of the manuscript for ASM Press and provided many helpful suggestions: Bobbie Hinson, Providence Day School, Charlotte, N.C.; Carolyn Britt Hammond, Eastern Guilford High School, Gibsonville, N.C.; Sheila Gilligan, Convent of the Sacred Heart, New York, N.Y.; Toby Mogollon Horn, Thomas Jefferson High School for Science and Technology, Alexandria, Va.; Jim Amara, Minuteman Regional Vocational Technical School District, Lexington, Mass.; Judy Brown, Maryland Biotechnology Institute; David F. Betsch, Bryant College, Smithfield, R.I.; Brian Shmaefsky, Kingwood College, Kingwood, Tex.; Janet Glaser, University of Illinois, Urbana, Ill.; Tom Zinnen, University of Wisconsin Biotechnology Center, Madison, Wis.; Joy A. Macmillan, Madison Area Technical College, Madison, Wis.; Walter R. Fehr, Iowa State University, Ames, Iowa; Rebecca Ross, VPI, Blacksburg, Va.; David Schlessinger, Washington University, St. Louis, Mo.; Marshall Bloom, Rocky Mountain Biology Laboratory, Hamilton, Mont.; and Robert I. Krasner, Providence College, Providence, R.I. Any shortcomings in the book are, of course, the responsibility of the authors.

A special note of thanks to Stanley Falkow at Stanford University for his support and Bruce Alberts, President of the National Academy of Sciences, for his encouragement and willingness to write the foreword for our book.

Current and former North Carolina Biotechnology Center employees assisted in various ways during the production of the text. Karyn Hede George, Dr. Kathleen Kennedy, Peggy Nelsen, Mary Sox, and Barry Teater improved the verbal and visual quality of the book immeasurably. Special thanks to Dr. Charles Hamner, President of the Center, for his wisdom, understanding, and support of our work.

Our sincere thanks to the North Carolina General Assembly for its foresight in establishing the Center and its continued support of our educational programs and to the citizens of North Carolina whose tax dollars provide funding for the Center.

Some lesson plans published by the National Association of Biology Teachers and the North Carolina Biotechnology Center in *A Sourcebook of Biotechnology Activities* and original classroom activities by Drs. Sherri Andrews, Edward Baptist, Lynn Elwell, Kathleen Kennedy, and Robert Thompson were modified for inclusion in this book.

Moral support during stressful times was provided by Lindsay Gould, Marty Groder, Kathleen Kennedy, and Ken Kreuzer. Thanks, y'all.

Susan Birch, our production editor, gently, but persistently, pushed us to the completion of the book. The text has benefited from the skills of our copy editor, Marie Smith. Their conscientious oversight improved the quality of the book immensely.

Finally, thanks to Patrick Fitzgerald, Director of ASM Press, for his encouragement and willingness to take a risk and try something new.

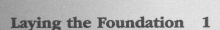

PART I

Laying the Foundation

*Laying the foundation contains background
information for conceptual grounding in
both the science fundamental to DNA-based
technologies and the technologies themselves.
We introduce the broad scope of biotechnology
before narrowing our focus to the historical
development and scientific underpinnings of the
biotechnologies that are DNA based.*

An Overview of Biotechnology

Introduction

When one reads or hears the word "biotechnology," the word "revolution" is often close behind. The combination of words is appropriate in many ways, for advancements in biotechnology will revolutionize major aspects of our lives and our relationship with the natural world.

In the field of human health, biotechnology will bring new ways to diagnose, treat, and prevent diseases. In agriculture, every aspect, from the seed placed in the ground to the food that appears on our tables, will be affected. Biotechnology is often touted as an environmental savior of sorts, for it will offer new, cleaner energy sources, methods for cleaning up environmental contamination, and products and processes that are more environmentally benign than some we have used before.

Even though all of us are certain that biotechnology is important, most of us are not sure we know exactly what biotechnology is. Such confusion is understandable, for "biotechnology" is an ambiguous term. To make matters even worse, it is used differently by different people. So what is biotechnology anyway?

Biotechnology Definitions

Defining "biotechnology" is actually very easy. Break it into its root words, "bio" and "technology," and you have the following definition:

Biotechnology: the use of living organisms to solve problems or make useful products.

After reading that definition, you may question the appropriateness of the phrase "biotechnology revolution." We domesticated plants and animals 10,000 years ago (Fig. 1.1). For thousands of years we have used microbes such as yeasts and bacteria to make useful food products like bread, wine, cheese, and yogurt. Virtually all antibiotics come from microbes, as do the enzymes we use in manufacturing processes as diverse as making high-fructose corn syrup and stone-washed jeans. In agriculture, we have used microbes since the 19th century to control insect pests and have inoculated the soil with nitrogen-fixing bacteria to improve crop yield. Microbes have been used extensively in sewage treatment for decades. Certain vaccines are based on the use of live, but weakened, viruses.

So why are people talking about a biotechnology revolution? Why go to the trouble to coin a new phrase for activities we have engaged in for ages?

The answer is this. During the 1960s and 1970s, our knowledge of cellular and molecular biology reached the point that we could begin to manipulate organisms at those levels. Manipulating organisms to our advantage is not new. What is new is *how* we are manipulating them.

Before, we used whole organisms; now, we use their cells and molecules. Before, we manipulated organisms and had little or no understanding of the mechanisms underlying the manipulations. Our manipulations were hit or miss ventures. Now, we understand our manipulations at the most basic level, the molecular level. As a result, we can predict what effect our manipulations will have, and we can direct the change we want with great specificity.

So "biotechnology" in the new sense of the word can be defined as follows:

"New" biotechnology: the use of cells and biological molecules to solve problems or make useful products.

Biological molecules are the large macromolecules unique to living organisms that are usually composed of repeating subunits (see Table 1.1). The biological molecules most often utilized in biotechnology today are **nucleic acids,** such as **DNA** and **RNA,** and **proteins.** As our understanding of the roles other biological molecules play in cellular structure and activities

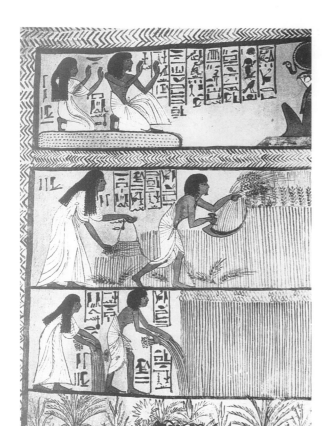

Figure 1.1 Painting depicting Egyptian agriculturists 6,000 years ago. (Photograph courtesy of the Bettmann Archive.)

Table 1.1 The four classes of biological molecules

Biological molecule (polymer)	Repeating subunit (monomer)
Protein	Amino acid
Carbohydrate	Simple sugar
Nucleic acid	Nucleotide
Lipid[a]	

[a]Lipids are a heterogeneous group of compounds that are grouped together because they are soluble in organic solvents and insoluble in water. Lipids exhibit a great deal of variety in chemical structure, size and complexity, so they are not as tidy a class to describe as classes of other biological molecules are. Listing a single repeating subunit is not possible. The most useful fact about the biochemistry of lipids is that they are primarily hydrocarbons, that is, chains of carbon molecules bound to hydrogens.

increases, we are finding, and no doubt will continue to find, new ways to use these molecules to our advantage.

Biotechnology: A Collection of Technologies

Some of the confusion surrounding the word "biotechnology" could eliminated by simply changing the singular noun to its plural form, "biotechnologies," because biotechnology is not a singular entity. Instead, biotechnology is a collection of technologies, all of which utilize cells and biological molecules (Fig. 1.2).

Developing technologies that use cells and biological molecules in place of whole multicellular organisms allows us to capitalize on a critical aspect of life at the cellular and molecular level: the extraordinary specificity of interactions. Because of this specificity, the tools and techniques of biotechnology are quite precise and are tailored to operate in known, predictable ways.

Any one of the biotechnologies can produce a wide variety of products; any one product may have diverse applications. For example, xanthan gum, currently produced by bacteria through bioprocessing technology, can be used to thicken salad dressings and jams or to clean residual oil from oil wells. A genetically engineered bacterium could synthesize a protein-digesting enzyme dissolving blood clots or unstopping drains.

The Technologies and Their Uses

What are some of these technologies that use cells and biological molecules, and how are we using these technologies?

Monoclonal antibody technology

Monoclonal antibody (MCAb) technology uses cells of the immune system that make proteins called **antibodies.** Your immune system is composed of a number of cell types that work together to locate and destroy substances that invade your body. Each cell type has a specific task. Some of these cell types are able to distinguish self from nonself. They recognize, on a molecular level, what is you, and therefore belongs inside of you, and what isn't you and should be eliminated. One of these very smart cells, the **B lymphocyte,** responds to invaders by producing antibodies that bind to the foreign substance with extraordinary specificity. We are now harnessing the ability of B lymphocytes to make these very specific antibodies.

Because of their specificity, MCAb are powerful tools for detection, quantification, and localization. Measurements based on MCAb are faster, more accurate, and more sensitive because of this specificity.

Diagnostic and Therapeutic Uses of MCAb
The substances that MCAb detect, quantify, and localize are remarkably varied and are limited only by the substance's ability to trigger the production of antibodies. Home pregnancy kits use an MCAb that binds

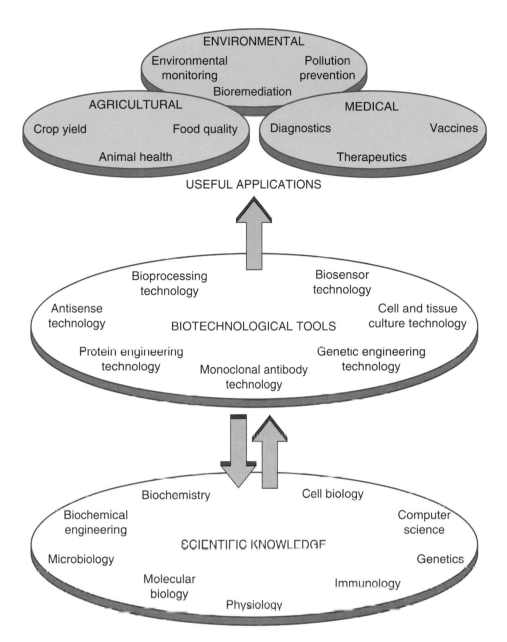

Figure 1.2 Synthesis of scientific and technical knowledge from many academic disciplines has produced the set of enabling technologies we call biotechnology. Any one technology will be applied to a number of industries to produce an even broader array of products.

to a hormone produced by the placenta (Fig. 1.3). MCAb are currently being used to diagnose a number of infectious diseases such as strep throat and gonorrhea. Because cancer cells differ biochemically from normal cells, we can make MCAb that detect cancers by binding selectively to tumor cells.

Ultimately, we hope to use the tumor antibody not just to diagnose cancers but also to treat many types of cancer by tagging a radioisotope or toxin to the tumor antibody. Because of the specificity of MCAb, the tumor antibodies will bypass normal cells and de-

liver the radioisotope or chemotherapeutic agent directly to the tumor.

In addition to diagnosing diseases in humans, MCAb are being used to detect plant and animal diseases, food contaminants, and environmental pollutants.

Bioprocessing technology

Bioprocessing technology uses living cells or components of their biochemical machinery doing what they normally do: synthesizing products, breaking

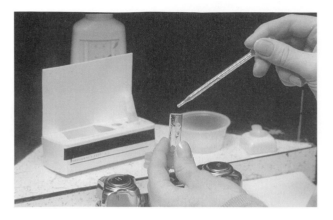

Figure 1.3 A pregnancy test based on the use of monoclonal antibodies (photograph courtesy of Custom Medical Stock Photo/©1993 Amethyst).

down substances, and releasing energy. The living cells most frequently used are one-celled microorganisms such as bacteria and yeasts; the cellular components most often used are proteins called **enzymes.**

Enzymes are essential for life. They catalyze all cellular biochemical reactions, most of which would occur much too slowly to support life. Through enzyme-catalyzed reactions, organisms break down large organic molecules to obtain energy and then use the breakdown products as raw materials or building blocks for synthesizing new molecules. We are using microbes and their enzymes for our own synthesis, degradation, and energy production needs (Fig. 1.4).

Fermentation

The oldest and most familiar bioprocess is microbial **fermentation**. Originally, the microbial fermentation products we used were derived from the series of enzyme-catalyzed reactions microbes use to break down glucose. In the process of metabolizing glucose

Figure 1.4 Industrial fermenter (photograph courtesy of Science VU-UFCSIM).

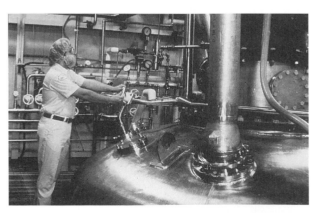

to acquire energy, microbes synthesize by-products we can use: carbon dioxide for leavening bread, ethanol for brewing wine and beer, lactic acid for making yogurt, and acetic acid (vinegar) for pickling foods (Fig. 1.5).

Now we have extended our use of the rich biochemical machinery of microbes beyond the metabolic pathway for glucose breakdown. We use microbial fermentation to synthesize an extraordinary array of products, including antibiotics, amino acids, hormones, vitamins, industrial solvents, pesticides, food processing aids, pigments, enzymes, enzyme inhibitors, and pharmaceuticals.

Biodegradation

Microbes and the enzymes they use to break down organic molecules are helping us clean up certain environmental problems: oil spills, toxic waste sites, and leakage from underground storage tanks. The use of microbial populations to clean up pollution is known as **bioremediation.** Probably the best-known example of bioremediation is the use of oil-eating bacteria to clean up oil spills such as the *Exxon Valdez* spill in Alaska's Prince William Sound in 1989 and spills in Iraq after the 1991 Gulf War (Fig. 1.6).

In the future we may be able to use sewage and agricultural refuse as energy sources by exploiting microbes that degrade these organic compounds and, in the process, release energy.

Cell and tissue culture technology

Cell and tissue culture technology is the growing of cells or tissues in appropriate nutrients in laboratory containers.

Plant Cell Culture

Plant cell and tissue culture is an essential aspect of plant biotechnology. The centrality of cell and tissue culture to plant biotechnology stems from a property unique to plant cells, their totipotency, or the potential to generate an entire multicellular plant from a single differentiated cell (Fig. 1.7B).

We genetically engineer plants at the level of the single cell. When a leaf cell is genetically engineered to contain a useful trait such as resistance to insect pests, that cell must develop into a whole plant if it is to be useful to farmers. This regeneration is accomplished through cell and tissue culture.

Animal Cell Culture

Plant cell culture is not the only type of cell culture being applied to agriculture. Using insect cell culture

A

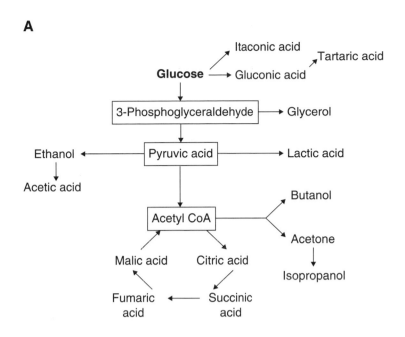

B

Organic chemical	Microbial source(s)	Industrial uses
Ethanol	*Saccharomyces*	Industrial solvent, fuel, beverages
Acetic acid	*Acetobacter*	Industrial solvent, rubber, plastics, food acidulant (vinegar)
Citric acid	*Aspergillus*	Food, pharmaceuticals, cosmetics, detergents
Gluconic acid	*Aspergillus*	Pharmaceuticals, food, detergent
Glycerol	*Saccharomyces*	Solvent, sweetener, printing, cosmetics, soaps, antifreezes
Isopropanol	*Clostridium*	Industrial solvent, cosmetic preparations, antifreeze, inks
Acetone	*Clostridium*	Industrial solvent, intermediate for many organic chemicals
Lactic acid	*Lactobacillus, Streptococcus*	Food acidulant, fruit juice, soft drinks, dyeing, leather treatment, pharmaceuticals, plastic
Butanol	*Clostridium*	Industrial solvent, intermediate for many organic chemicals
Fumaric acid	*Rhizopus*	Intermediate for synthetic resins, dyeing, acidulant, antioxidant
Succinic acid	*Rhizopus*	Manufacture of lacquers, dyes, and esters for perfumes
Malic acid	*Aspergillus*	Acidulant
Tartaric acid	*Acetobacter*	Acidulant, tanning, commercial esters for lacquers, printing
Itaconic acid	*Aspergillus*	Textiles, paper manufacture, paint

Figure 1.5 (A) Useful metabolic products of glucose breakdown in various microorganisms. (B) Chemicals currently produced by microbial fermentation of glucose and their industrial applications.

to grow viruses that infect insects should enable us to broaden the application of viruses and **baculoviruses** as biological control agents. Mammalian cell culture is also being used in livestock breeding.

The medical community uses animal cell culture to study such topics as the safety and efficacy of pharmaceutical compounds, the molecular mechanism of viral infection and replication, the toxicity of compounds, and basic cell biochemistry. Combining two biotechnologies, mammalian cell culture and bioprocessing technology, provides a mechanism for producing large quantities of cellular compounds. Advances in mammalian cell culture now permit scientists to grow most human cell types, introducing

the prospect of eventually providing appropriate tissue substitutes when large areas of functional tissue are lost to disease or accidents (Fig. 1.7A).

Biosensor technology

Biosensor technology represents the joining of molecular biology and microelectronics. A **biosensor** is a detecting device composed of a biological substance linked to a transducer (Fig. 1.8). The biological substance might be a microbe, a single cell from a multicellular animal, or a cellular component such as an enzyme or an antibody. Biosensors allow us to measure substances that occur at extremely low concentrations.

1. An Overview of Biotechnology • 7

Figure 1.6 Exxon Valdez oil spill, Alaska. Rocks on the beach before (right) and after (left) bioremediation. (Photograph courtesy of the U.S. Environmental Protection Agency.)

How do biosensors work? When the substance we want to measure collides with the biological detector, the transducer produces a tiny electrical current. This electrical signal is proportional to the concentration of the substance.

Biosensors will be used in ways as varied as measuring the freshness and safety of food, monitoring industrial processes in real time with immediate feedback for process control, and detecting minute quantities of substances in blood. By coupling a glucose biosensor to an insulin infusion pump, the correct blood concentration of glucose could be maintained at all times in diabetics.

Genetic engineering technology

Genetic engineering technology is often referred to as **recombinant DNA technology.** Recombinant DNA is made by the joining or recombining of genetic material from two different sources. In nature, genetic material is constantly recombining. Each of the following is just one of many ways nature joins genetic material from two sources:

- when **crossing over** occurs between **homologous** maternal and paternal chromosomes during gamete formation

- when egg and sperm fuse during fertilization

- when bacteria exchange genetic material through **conjugation, transformation,** and **transduction**

In each of these examples of natural recombination, when genetic material from two different sources is combined, the result is increased genetic variation. The genetic variation that exists in nature has provided the raw material for evolutionary change driven by natural selection or artificial selection imposed by humans.

A

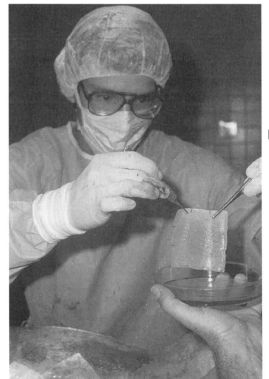

B

***Figure 1.*7** (A) Human tissue culture. Surgeon with sheet of skin cells (photograph courtesy of VU/©SIU). (B) Stages of plant cell/tissue culture from callus to plantlet.

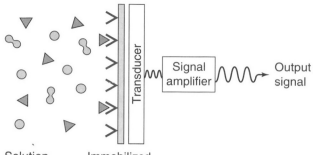

Solution containing substance to be detected (▷)

Immobilized biological recognition element (>)

Figure 1.8 Schematic drawing of a simple biosensor. A biosensor is a detecting device that utilizes an immobilized biological sensing element and a transducer to produce an electrical signal proportional to the concentration of substance to be detected.

Using Existing Genetic Variation in Selective Breeding

As soon as humans domesticated organisms, we began to use selective breeding to alter their genetic makeup to suit our needs. Certain individuals in a population had traits, and therefore genes, we valued, and we chose these individuals to serve as parents for the next generation. By selecting certain genetic variants from a population and excluding others, we intentionally directed the recombining of genetic material. As a result, we radically changed the genetic makeup of the organisms we domesticated. If you question the extent to which human-imposed ("artificial") selection has altered the genetic makeup of organisms, see the photograph showing today's corn next to its presumed ancestor, teosinte (Fig. 1.9).

So existing genetic variation has been a valuable natural resource that humans have exploited for cen-

Figure 1.9 Teosinte, the presumed ancestor of corn, next to a modern variety of field corn.

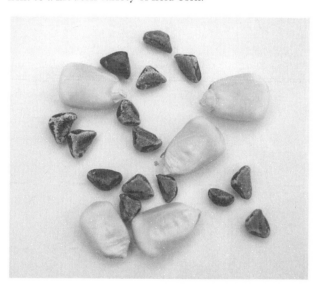

turies. The tools and knowledge we need to make selective breeding more predictable and more precise have been continually evolving. Genetic engineering is the next step in that continuum.

Generating Novel Genetic Variation with Genetic Engineering

The term "recombinant DNA technology," or genetic engineering, refers to the precise molecular techniques that join specific segments of DNA molecules from different sources. We recombine DNA by using enzymes (**restriction endonucleases**) designed to cut and join DNA in predictable ways. To ferry the DNA into the target organism, we usually use bacteria and viruses that transport DNA in nature, or we use their DNA molecules.

Therefore, in addition to directing the recombining of genetic material through the intentional joining of eggs and sperm (pollen in plants) in selective breeding, we can now recombine genetic material with greater precision by working at the molecular level.

Selective Breeding versus Genetic Engineering

Many scientists view genetic engineering as simply an extension of selective breeding, because both techniques join genetic material from different sources to create organisms that possess useful new traits. However, even though genetic engineering and selective breeding bear a fundamental resemblance to one another, they also differ in important ways (Table 1.2 and Fig. 1.10).

In genetic engineering, we move single genes whose functions we know from one organism to another, while in selective breeding, sets of genes of unknown function are transferred. By increasing the precision and certainty of our genetic manipulations, the risk of producing organisms with unexpected traits de-

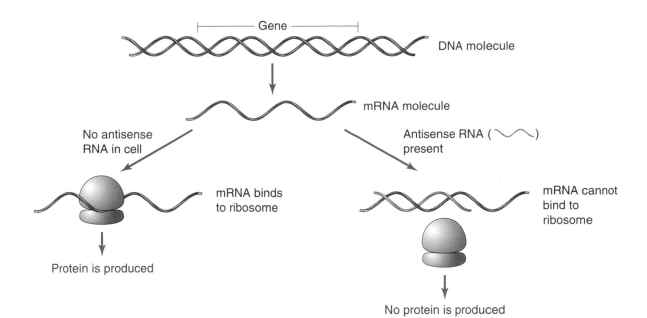

Figure 1.12 Schematic representation of antisense technology. In this example, the antisense oligonucleotide is an RNA molecule that blocks protein production by preventing the binding of mRNA to the ribosome.

would be beneficial, antisense technology provides a valuable approach to the problem. An obvious example is a situation in which you want to block the production of a protein product. Currently, researchers are using this technology to slow food spoilage, control viral diseases, and treat cancers that have a genetic basis.

Metabolic Engineering

A less obvious but very exciting use of antisense technology will be in **metabolic engineering.** Many compounds in nature that have great commercial applicability are not proteins. For example, many of the compounds produced by plants to deter insect feeding could be useful as crop protectants but are not proteins. We could increase their production in crop plants by using antisense technology to block the production of enzymes in certain pathways, thus rerouting the plant's metabolism toward these compounds.

On the other hand, we might want to decrease the production of a substance that is not a protein. An excellent example is cholesterol. Given the right antisense molecule, perhaps we could significantly decrease production of cholesterol by blocking key enzymatic steps in its synthesis.

To visualize how metabolic engineering might work, refer to Fig. 1.5A, which depicts some of the valuable products that result from the sequential enzymatic breakdown of a single starting substance, glucose. None of these useful products is a protein, but all have commercial value. If you wanted to maximize the production of isopropanol in a microbial fermentation process, which steps, and therefore enzymes, would you block with antisense technology?

The Applications of Biotechnology

In the coming years, the diverse collection of technologies labeled "biotechnology" will yield an even broader array of products. Most of the commercial applications of biotechnology will be in three markets: human health care, agriculture, and environmental management.

Medical biotechnology

Biotechnology has already provided us with quicker diagnostic tests, therapeutic compounds with fewer side effects, and safer vaccines.

Diagnostics

We can now detect many diseases and medical conditions more quickly and with greater accuracy because of the sensitivity of new diagnostic tools developed through biotechnology. The time required to diagnose strep throat and gonorrhea has dropped from days to minutes. Certain cancers are now diagnosed by simply taking a blood sample, thus eliminating the need for invasive and costly surgery. Scientists are also making remarkable progress in identifying and sequencing genes. These advances will greatly assist doctors in diagnosing hereditary diseases, many of which we cannot detect currently.

The tools and techniques of biotechnology that are making progress in disease diagnosis possible include MCAb technology, **DNA probes, restriction fragment length polymorphisms (RFLPs),** and the **polymerase chain reaction (PCR).** More information about DNA probes, RFLPs, and PCR is provided in chapter 4 and in Classroom Activities, chapters 18 and 24.

Therapeutics

Biotechnology will provide improved versions of today's therapeutic regimes as well as treatments that would not be possible without these new techniques. Here are just a few examples of the novel therapeutic advances biotechnology now makes feasible.

Pharmaceuticals from plants. Many plants produce compounds with human therapeutic value. For years, we have used a chemical derived from foxglove for treating heart conditions. More recently, a chemical extracted from Madagascar periwinkle has proven to be a successful therapeutic agent for treating leukemia.

Growing these small herbaceous plants requires little space and time. But what about others? Some therapeutic compounds come from large trees or endangered species. How can we acquire enough of these therapeutic compounds to make their commercialization feasible?

We are circumventing these constraints with cell and tissue culture. For example, plant cell and tissue culture is being used to produce a naturally occurring plant compound, **taxol,** recently found to be effective in treating ovarian and breast cancers. Without plant tissue culture, the supply of taxol would be so limited that its potential as a chemotherapeutic agent would never be realized (Fig. 1.13).

Endogenous therapeutic agents. The human body produces many of its own therapeutic compounds, and many of them are proteins. As proteins, they are prime candidates for possible production by genetically engineered bacteria. Such production would provide quantities that would allow us to better analyze their functions and make their commercialization economically feasible. As we increase our understanding of these and other endogenous therapeutic agents, we will be able to capitalize more on the body's innate healing ability. Following are exam-

Figure 1.13 Stripping bark of the Pacific yew (*Taxus brevifolia*) for taxol extraction (photograph courtesy of VU/©Peter K. Ziminski).

ples of some endogenous therapeutic compounds and applications.

- **Interleukin-2** activates **T-cell** responses. It may be effective in treating certain cancers and is currently produced by genetically engineered bacteria.

- **Erythropoietin** regulates red blood cell (erythrocyte) production and appears to be successful in treating severe anemia.

- **Tissue plasminogen activator** dissolves blood clots. It is now being produced by genetically engineered bacteria.

- **Growth factors** are small proteins such as epidermal growth factor, nerve growth factor, fibroblast growth factor, and macrophage colony-stimulating growth factor. As we learn more about their natural functions, the roles they may play in treating diseases will become more apparent.

Designer drugs. Using principles of protein engineering and computer molecular modeling, a scientist may be able to design effective therapeutic compounds before ever stepping into a laboratory.

Replacement therapies. Many disease states result from defective genes that result in the total lack or inadequate production of substances the body normally produces. Following are some examples of compounds being produced (thanks to genetic engineering and bioprocessing technologies) in sufficient quantity to provide a reliable source of replacement therapies.

- **Factor VIII** is a protein involved in the blood-clotting process. Some hemophiliacs lack this protein.

- **Insulin** is a protein hormone that regulates blood glucose levels by affecting cellular uptake of glucose. Diabetes results from an inadequate supply of insulin.

- **Growth hormone** is also a protein hormone. Dwarfism results from insufficient production of this hormone.

Gene therapy. The ability to isolate and clone specific genes gives us the power to do far more than simply replace missing or dysfunctional proteins. Through gene therapy, we will be able to treat genetic disorders by giving patients functional genes in place of defective ones.

In the near future, the only candidates for correction by gene therapy will be certain hereditary diseases: those that are controlled by one gene. Of those diseases, only a few are amenable to correction via gene therapy.

A primary candidate for gene therapy is adenosine deaminase (ADA) deficiency, or the "bubble boy" disease. This disorder results from a mutation in the gene that codes for the enzyme ADA. When ADA is not produced, intracellular toxins build up. Cell types particularly susceptible to these toxins are the B and T cells of the immune system. Successful gene therapy experiments have already been conducted on children who suffer from ADA deficiency (Fig. 1.14). For a partial list of gene therapy trials conducted over the last few years, see Table 1.3.

Figure 1.14 Both of these children were born with a rare genetic defect that results in a defective immune system. The child on the left, the "bubble boy," was born prior to advances in biotechnology (photograph courtesy of James DeLeon, Jr., Children's Hospital, Texas Medical Center, Houston, Texas). The girl on the right received gene therapy treatment and, as a result, leads a relatively normal, healthy life (photograph courtesy of Jennifer Coate, March of Dimes).

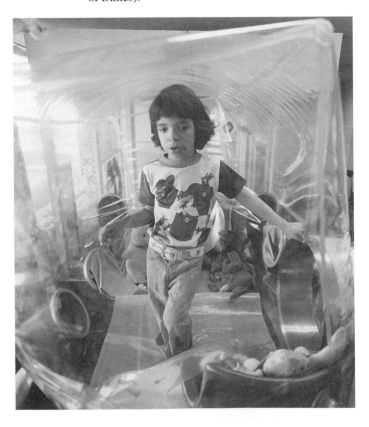

Table 1.3 Partial list of gene therapy trials conducted since 1990[a]

Disease	Gene inserted[b]
ADA deficiency	ADA
Advanced cancer	Tumor necrosis factor
Advanced cancer	Interleukin-2
Hypercholesterolemia	HDL receptor
Ovarian cancer	Thymidine kinase
Neuroblastoma	Interleukin-2
Hemophilia	Factor VIII
Brain tumor	Interleukin-2
Cystic fibrosis	Cystic fibrosis transmembrane regulator
Malignant melanoma	Gamma interferon
Lung cancer	Antisense ras/p53
Kidney cancer	G-CSF
Gaucher's disease	Glucocerebrosidase
AIDS	HIV coat protein

[a]For more information on some of these disorders and gene therapy, see chapters 28 and 33.

[b]In most cases, the protein product of the missing or malfunctioning gene is given here. These products are primarily enzymes, endogenous therapeutic proteins, or membrane-bound proteins that function as receptors or transport molecules. Abbreviations: HDL, high-density lipid; G-CSF, granulocyte colony-stimulating factor; HIV, human immunodeficiency virus.

Cancer therapies. In addition to chemotherapeutic agents derived from plants, antisense technology to block cancer-causing genes, and MCAb that selectively deliver toxins to tumors, biotechnology research tools have permitted progress on another front that offers potential routes for treating cancer: understanding the genetic basis of cell growth and differentiation.

When genes involved in certain critical events of cell growth and development mutate, they become **oncogenes,** or tumor-producing genes. If we could find a way to block expression of these genes, we might be able to inactivate them. Antisense technology may provide us with this blocking ability.

All of us normally have genes called tumor suppressor genes. When these genes function correctly, they act as cell growth suppressors. When both copies of these genes become inactive or are absent, the genes then act as oncogenes. A type of eye cancer, retinoblastoma, results from abnormal copies of tumor suppressor genes. By introducing normal copies of these genes into tumor cells, the tumor may regress. (See chapter 28 for more information on the genetics of cancer.)

Immunosuppressive therapies. Without a responsive immune system, we would face death as a real possibility every day. Sometimes, however, our vigilant immune system works against us. In organ transplant rejections and autoimmune diseases, suppressing our immune system would be in our best interest. Currently, we are testing the feasibility of using MCAb to accomplish this suppression. Here's how it works.

When exposed to foreign tissue in an organ transplant, the T cells of the immune system go to work to rid the body of this nonself component. We have produced an MCAb that binds to a protein found on the surface of T cells. The MCAb takes the T cells out of commission by binding to this protein.

Patients injected with this MCAb show significantly less transplant rejection than those given an immunosuppressive drug such as cyclosporin. By using a specific MCAb in place of an immunosuppressive drug, we selectively knock out only one aspect of the immune response, the T cells, leaving other aspects intact. Because immunosuppressive drugs suppress all immune function, they leave organ transplant patients vulnerable to infection. Once again, because of the extraordinary specificity of this technique, we can zero in on the problem and cause few, if any, side effects.

In autoimmune diseases, such as rheumatoid arthritis and multiple sclerosis, our immune system turns against our own tissue. No one knows why our immune system sometimes loses the ability to discern self from nonself. These diseases are characterized by progressive degeneration of the tissues the immune system is attacking. Given the success of MCAb in treating organ transplant rejection, researchers believe a similar tactic might slow the progression of autoimmune diseases.

Vaccines

There is much truth to the adage "an ounce of prevention is worth a pound of cure." The best way to battle diseases is not to develop new therapeutics but to prevent the diseases. Through biotechnology, we are developing better preventative agents by improving on a practice that has been with us since the 19th century: vaccination. The effectiveness of vaccination as a method of preventing infectious diseases is best illustrated by a single example. The disease smallpox has been eliminated from human populations by a worldwide vaccination effort.

Vaccines we have produced to prevent smallpox and other diseases (polio, diphtheria, tetanus, measles) are based on the use of either killed viruses or live but weakened viruses. When vaccinated with such a non-virulent virus, your body produces antibodies to that organism, but you don't get the disease. If you are exposed to that virus again, your body has a ready supply of antibodies to defend itself. Vaccines are analogous to the "threat of war" that incites us to build up a supply of weapons. For the most part, vaccines cause no serious problems, but they do have side effects: allergic reactions, aches and pains, and fever. In a very few individuals, the vaccine has caused the disease it was intended to prevent.

A second problem with this method of vaccination is vaccine production. Growing large amounts of some human pathogenic viruses outside of the human body is not easy. Viruses are quasi-living organisms; they need the biochemical machinery of a living cell to reproduce.

Using genetic engineering, scientists are making vaccines that contain only that part of the disease-causing organism (pathogen) that triggers the production of antibodies. Usually one or a few proteins on the surface of the pathogen trigger the production of antibodies. By isolating the gene for the pathogen's cell surface protein(s) and inserting it into a bacterium such as *Escherichia coli,* bioprocess scientists can produce large quantities of this protein to serve as the vaccine. When the protein is injected, the body produces antibodies that can recognize the pathogen but suffers none of the adverse side effects that sometimes go hand in hand with being vaccinated. Using these new techniques of biotechnology, scientists have developed vaccines against the life-threatening diseases hepatitis B and meningitis.

Medical Research Tools

One way that biotechnology will help us prevent diseases is less obvious than the ready example of vaccinations. For medical researchers, some of the most important outcomes of advances in biotechnology are not commercial products but the powerful research tools biotechnology provides.

The technologies of biotechnology discussed earlier are not simply techniques for producing vaccines, pharmaceuticals, and diagnostic kits. As research tools, these technologies have given us a much more valuable product: a deeper understanding of biological systems. They have provided us with new answers to old questions and have prompted us to ask questions we never would have dreamed of asking. Learning more about healthy biological processes and what goes wrong when they fail will enable us to develop even better diagnostics, therapeutics, and preventative agents.

Agricultural biotechnology

The opportunities biotechnology will create for agriculture are as impressive and extensive as those for human health. In the near future, we will witness progress in both improving the quality and yields of our agricultural products and decreasing production costs. A related aspect of agriculture that will profit from the tools of biotechnology is food processing.

Plant Agriculture

Because plants are genetically complex, plant agricultural biotechnology lagged behind medical advances in biotechnology. Another, perhaps equally important factor that may explain the different rates of progress is that over the years, animal research has received much more federal funding than plant research. In the last 10 years, we have made remarkable progress in plant biotechnology, largely because of improvements in two fundamental techniques of biotechnology: genetic engineering and plant cell and tissue culture. In the 1980s, around the time we were already using genetically engineered bacteria to produce human insulin, we discovered a way to genetically engineer plants by using recombinant methods. To insert novel genes into plants, scientists exploited a genetic engineer that occurs in nature, **Agrobacterium tumefaciens.** A common soil bacterium, *A. tumefaciens* infects plants by injecting a portion of its DNA into plant cells (Fig. 1.15). Since the mid-1980s, plant geneticists have used *A. tumefaciens* to genetically engineer more than a dozen important agricultural crops (see chapter 22).

The traits agricultural scientists are incorporating into our crops through genetic engineering are the same traits we have incorporated into crops through selec-

Figure 1.15 Scanning electron micrograph of *Agrobacterium tumefaciens* infecting plant cell (photograph courtesy of Ann Matthysse, Biology Department, University of North Carolina, Chapel Hill, N.C.).

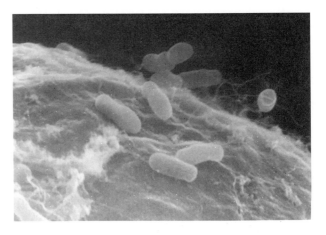

tive breeding: improved nutritional content; delayed ripening; resistance to diseases caused by bacteria, fungi, and viruses; better taste; the ability to withstand harsh environmental conditions such as freezes and droughts; greater nitrogen fixation capabilities; and resistance to pests such as insects, weeds, and nematodes.

Biocontrol and biotechnology. Just as biotechnology is allowing us to make better use of the natural therapeutic compounds our bodies produce, it is also providing us with more opportunities to work with nature in plant agriculture. To deter crop pests, we will be able to rely more heavily on biological methods of pest control. To encourage increased productivity, we may be able to exploit a number of positive symbiotic relationships found in nature.

Biological control, or biocontrol, as it is often called, is the suppression of pests and diseases through the use of biological agents. For example, a virus may be used to control an insect pest, or a fungus may deter the growth of a weed (Fig. 1.16).

Biological control has been used in agricultural systems in the United States since the 1800s, when we began using bacteria and a type of virus called a baculovirus to control pests, but the potential for biological control has been limited by the constraints that exist in nature. It is no simple task to find a specific microbe that kills a specific insect pest but not others, a certain fungus that causes diseases in weeds but not in crop plants, or an insect predator that preys only on "bad" and not "good" insects.

Figure 1.16 Parasitic wasps are effective biocontrol agents. Female wasps deposit eggs inside caterpillars where larvae develop and feed. After larvae emerge, they form the white, ovoid pupal cases attached to the caterpillar's body. (Photograph courtesy of VU/©W. Mike Howell.)

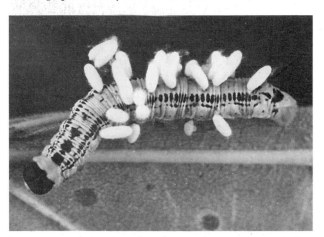

On the other hand, sometimes the unique characteristics of an organism or the ecological relationship between an organism and a crop would make the organism an excellent biocontrol agent, but it lacks the "ammunition." A perfect example of this situation is provided by the bacterium that lives in the stalks of corn plants and was discussed earlier under genetic engineering and Bt.

The flexibility that comes from genetic engineering removes those constraints. If we find a gene for insect resistance in an oak tree, we can incorporate it into our crops. If a very rare mold that occurs only in showers in homes in Australia has a gene that codes for a protein that can kill weeds, we can isolate that gene, engineer it into bacteria, and have the bacteria pump out massive quantities of the herbicidal protein.

Exploiting cooperative relationships in nature. In addition to capitalizing on nature's negative interactions—predation and parasitism—to control pests, we might also use existing positive relationships that are important to plant nutrition. One example is the symbiosis between plants in the bean family and certain nitrogen-fixing bacteria (Fig. 1.17). By providing the crop plant with a usable form of nitrogen, the bacteria encourage plant growth. Scientists are working to understand the genetic basis of this symbiotic relationship so that we can give nitrogen-fixing capabilities to crops other than legumes.

A second and less well-known example of a positive symbiotic relationship we might exploit for increasing crop production involves fungi and plant roots and is known as a **mycorrhiza.** In mycorrhizal associations, fungi extract nutrients from the soil and make them available to plants.

At this time, plant genetic engineers (or any genetic engineers, for that matter) are limited to one- or two-gene traits, those controlled by one or two genes. Most of the agronomically valuable traits are **multigenic;** that is, they are controlled by many genes. One example is nitrogen fixation, which is controlled by at least 15 different genes. So while the idea of having our major crops obtain useful nitrogen from the air via a bacterium that lives on their roots is an appealing one, it is unrealistic in the near future.

Animal Agriculture
The field of animal agriculture will develop as biotechnology provides new ways to improve animal health and increase productivity. As is true of human health, improvements in animal health will come from advances in diagnostics, therapeutics, and vaccines. We have already seen progress in these areas as a result of biotechnology.

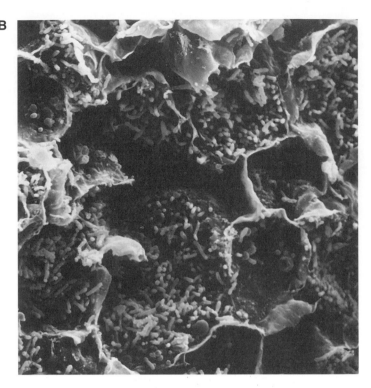

Figure 1.17 (A) Alfalfa nitrogen-fixing nodules (photograph courtesy of VU/©C.P. Vance); (B) alfalfa nodule cells containing actively nitrogen-fixing bacteroids. ×640. (Photograph courtesy of VU/©C.P. Vance.)

Animal health. Farmers are using MCAb on their livestock for early detection of pregnant females and more accurate diagnosis of certain diseases. The same principles discussed under human health apply here: because of their specificity, these diagnostics are faster, more accurate, and more sensitive than traditional diagnostics. Diagnosing a disease sooner and with greater accuracy means that appropriate therapy can be chosen and started sooner, thus decreasing the spread of the disease. Brucellosis in cattle and trichinosis in pigs are a few of the economically important diseases we can now diagnose thanks to advances in biotechnology.

Another parallel exists in the type of therapies biotechnology is providing for both humans and animals. Genetic engineering is being used to produce sufficient quantities of the endogenous therapeutic proteins animals produce. Cattle naturally make **interferon** and interleukin-2, proteins produced by the immune system to fight viruses. The genes for these proteins have been cloned and placed into bacteria for mass production. Injections of interferon and interleukin-2 are helping decrease the incidence of shipping fever in cattle, a disease that costs the beef industry more than $250 million a year. These proteins are also proving effective against equine influenza.

Great progress has also been made in disease prevention. Techniques being used to improve human vaccines are finding their way into veterinary medicine as well. Vaccines are now available for a number of diseases and parasites, including pseudorabies, foot-and-mouth disease, coccidiosis in poultry, tapeworms, ticks, and tick-borne diseases.

Increasing yields. Farmers are always interested in improving the productivity of agricultural animals. Their goal is to obtain either the same output (milk, eggs, meat) with less input (food) or increased output with the same input. One method we are using to increase the productivity of our livestock is a variation on the theme of selective breeding. We first choose those individuals that possess desirable traits. Then, instead of breeding the animals, we simply remove their gametes (eggs and sperm) and allow fertilization to occur in a laboratory dish. This in vitro fertilization is followed by embryo culture, a form of mammalian cell culture in which the fertilized egg develops into an embryo. At some stage early in development, the embryo is taken from the laboratory dish and implanted into a female of the same species but not necessarily of the same breed. This is known as **embryo transplant.** Using this method, a farmer can improve the genetic makeup of the herd more quickly than by simply relying on a single female who produces one calf per year.

Scientists have also tested ways to increase production efficiency by applying the same principle to some very different animals: cows, pigs, and fish. The principle involves increasing the level of endogenous growth hormone found in these animals. Protein synthesis in all animals is dependent upon growth hormone. As a result, growth hormone in cows, which is also known as **bovine somatotropin (BST),** is important in growth and milk production. By increasing a cow's blood level of growth hormone, farmers can increase milk output or growth rate through stimulating protein synthesis.

Scientists have isolated the gene for BST and inserted it into bacteria that now produce BST through microbial fermentation. The BST is extracted from the fermentation tanks, purified, and injected into dairy cows, causing a 10 to 20% increase in milk production.

Although the Food and Drug Administration (FDA) has determined that milk taken from cows injected with BST contains the same amount of BST as the milk of uninjected cows, some public concern about the safety of BST still exists. A second factor that may make the commercialization of BST difficult is based on economics, not safety. The country's dairy farmers already produce more milk than we need, so American consumers have asked a very valid question: Do we need a substance that increases milk production?

On the other hand, the product **porcine somatotropin (PST)** has applications that are both economically valuable and attractive to consumers. Injections of PST, the growth hormone in pigs that is the functional equivalent of BST, cause a dramatic increase in protein and a decrease in the amount of pork fat (Fig. 1.18).

Recently, scientists at Auburn University conducted an experiment using transgenic carp that may prove important for a growing segment of the agricultural market, fish-farming, or aquaculture, as it is sometimes called. Auburn scientists introduced the gene that codes for the production of growth hormone in rainbow trout into carp eggs to boost the natural production of growth hormone. Scientists are hopeful that by genetically engineering farm-raised fish in this way, the fish will reach a marketable size sooner than fish with lower levels of growth hormone.

Food Processing

An aspect of agriculture that people often forget is food processing: what happens to the food when it leaves the farm. The food processing industry will be affected by a variety of developments in biotechnology. Some will involve the food product itself, but just as many, if not more, will be directed toward improving food additives.

Traditionally, microbes have been essential to the food processing industry not only for the role they play in the production of fermented foods but also as a rich source of food additives and enzymes used in food processing (Table 1.4). Through advances in genetic engineering and bioprocessing technology, their importance to the food industry will only increase in the future.

Product quality. In plant agriculture we are altering crops to have more desirable processing qualities. The Campbell Soup Co. now uses a tomato for its soup that was derived from a biotechnology technique, **somaclonal variant selection,** that operates on the same principles as selective breeding. The new tomato contains 30% less water and is processed into soup with greater efficiency. A 0.5% increase in the solid content is worth $35 million to the U.S. processed-tomato industry.

The first product of plant genetic engineering approved by the FDA was a tomato that is allowed to ripen on the vine instead of being picked while it is green. Calgene, the company that commercialized the Flavr Savr tomato, developed this tomato by biochemically separating the ripening process from the spoiling process. These processes usually go hand in hand,

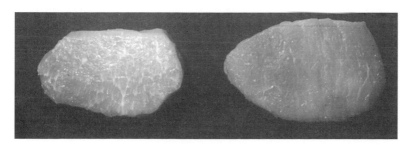

Figure 1.18 Because porcine somatotropin (PST) decreases fat deposition in pigs, consumers will have access to leaner cuts of pork. (Left) Boneless pork chop control; (right) boneless pork chop from pig treated with PST. (Photograph courtesy of VU/©Ken Prusa.)

Table 1.4 Microbial fermentation is essential to the production of foods, food additives, and enzymes used in food processing

Foods

Cheese	Vinegar
Yogurt	Bologna
Buttermilk	Salami
Sour cream	Tofu
Soy sauce, tamari	Miso
Bread and other baked goods	Sauerkraut
Wine, beer	Pickles
Cider	Olives

Food additives

Acidulants (citric and lactic acids)

Amino acids (glutamine, lysine)

Vitamins (β-carotene, riboflavin)

Flavor enhancers (monosodium glutamate)

Thickeners (xanthan gum)

Stabilizers (dextran)

Flavors (methyl salicylate [wintergreen], benzaldehyde [almond])

Enzymes

Proteinase (Gouda and Edam cheeses)

Peptidase (cheddar cheese)

Lipase (blue cheese)

Amylase (high-fructose corn syrup)

Invertase (soft-centered candies)

Pectinase (fruit juices)

but in fact, some different enzymes and therefore different genes are involved. Using antisense technology, Calgene blocked a gene that codes for one of the enzymes involved in spoiling while leaving the ripening enzymatic pathways intact.

Food additives are substances used to increase nutritional value, retard spoilage, change consistency, and enhance flavor. The compounds food processors use as food additives are substances nature has provided and are usually of plant or microbial origin, such as xanthan gum and guar gum, which are produced by microbes.

Through genetic engineering, food processors will be able to produce many compounds that could serve as food additives but that now occur in scant supply or are found in microbes that are difficult to maintain in fermentation systems. We will also be able to capitalize on the extraordinary diversity of the microbial world and obtain new enzymes that will prove important in food processing.

Food safety. The most important advance in food processing, however, will be in food safety. MCAb,

biosensors, and DNA probes are being developed that will be used to determine harmful bacteria such as *Salmonella* species and *Clostridium botulinum.* Once again, these tests will be quicker and more sensitive to low levels of contamination than previous tests because of the increased specificity of molecular techniques.

Environmental biotechnology

Few people doubt that we are in the midst of an environmental crisis. The air, soil, and water of the Earth are contaminated as a result of human activities. We are depleting the Earth's nonrenewable resources and generating massive amounts of wastes that don't degrade. Many people look with hope to biotechnology to bail us out of some of our environmental problems.

People believe that in addition to solving environmental problems, biotechnology will cause fewer environmental problems than previous physical or chemical technologies. This optimism is based on the use of biological, renewable resources in place of chemical, nonrenewable ones and the greater specificity, precision, and predictability that characterize biologically based technologies. Because we will be working with the biology of organisms in more specific and targeted ways, we should be able to develop technological solutions that generate fewer side effects and have fewer or less severe unintended consequences.

In addition, just as biotechnology is providing us with new tools for diagnosing health problems and detecting harmful contaminants in food, it is yielding new methods of monitoring environmental conditions and detecting environmental pollutants.

Cleaning Up Pollution through Bioremediation

Using biotechnology to treat pollution problems is not a new idea. Communities have depended on complex populations of naturally occurring microbes for sewage treatment for many years. Microbes help purify water by breaking down solid organic wastes before the water is recycled. But solid organic wastes are not the only type of pollutants that need to be removed from our water supplies. More and more, we are finding that our water is contaminated by chemical pollutants; more and more, we are turning to microbes for help in removing pollutants both from water and from soil (Fig. 1.19).

Why microbes? Over the billions of years they have been on Earth, microbial populations have adapted to every imaginable environment. No matter how harsh

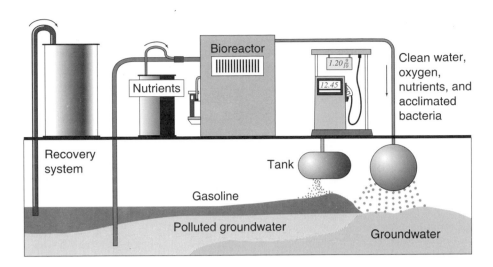

Figure 1.19 Gasoline from an underground storage tank seeps through the soil to the water table. After the leak is stopped, the free-floating gasoline is pumped out to a recovery tank, and polluted groundwater is pumped into a bioreactor with oxygen, nutrients, and hungry microbes. After the microbes eat the gasoline, the mixture of clean water, nutrients, and microbes is pumped back into the ground so that more of the pollutant can be degraded (From *Carolina Genes,* 1994, North Carolina Biotechnology Center Education and Training Program, Research Triangle Park, N.C.)

the habitat, some microbe has found a way to make a living there. We have found microbes in hot springs that are the source of the geysers in Yellowstone, thousands of feet underwater in hydrothermal vents, in salty seas and lakes, and living off inorganic materials such as copper sulfide in copper mines.

Life in unusual habitats makes for unique biochemical machinery, so the range of compounds microbes can degrade is enormous. We are capitalizing on this biochemical diversity to repair a number of pollution problems. We are using a fungus to clean up a noxious substance discharged by the papermaking industry. Indigenous microbes that contribute to the natural cycling of metals, such as mercury, offer prospects for removing heavy metals from water. Other naturally occurring microbes that live on toxic waste dumps are degrading wastes such as polychlorobiphenyls to harmless compounds.

Not all pollution problems that threaten human health result from human activities. During the hot summer months in Australia, aquatic cyanobacterial populations increase rapidly and secrete carcinogenic chemical compounds into the water supply. To counter these population blooms, copper sulfate, which kills the cyanobacteria but doesn't break down the toxins, is dumped into the water. As you might expect, microbes that coexist with the cyanobacteria have evolved a better solution to the problem. Recently, scientists discovered a bacterium that produces three enzymes that sequentially break down

cyanobacterial toxins until they are harmless. Conveniently for scientists, the three genes that code for the enzymes are clustered together in a single functional unit, an **operon.**

In these examples, naturally occurring organisms are performing the cleaning duties. Using only naturally occurring microbes can be somewhat limiting, however. Microbes that degrade hazardous wastes in the clay soils of North Carolina may not work in the silty soils of the Mississippi delta. Genetic engineering gives us the flexibility we need to maximize our use of the biochemical capabilities of microbes and at the same time circumvent problems such as habitat specificity.

If useful enzymes are discovered in microbes that can survive only in certain habitats, the gene that codes for that enzyme may be movable into microbes known to prosper in other habitats. For example, we have discovered a number of species of aquatic microbes that degrade some of the 200 naturally occurring halogenated hydrocarbons found in the ocean. Some of our primary soil pollutants belong to this class of chemical compounds. If we isolate the genes that code for the enzymes that degrade these compounds, we could insert these genes into soil microorganisms. These genetically engineered microorganisms might then be able to clean up hazardous waste sites.

William Reilly, the former head of the Environmental Protection Agency (EPA), the World Wildlife Fund, and

The Conservation Foundation, is enthusiastic about the extraordinary potential of bioremediation. Not only is it a natural process, but it is also inexpensive and effective and requires little energy input. After the *Exxon Valdez* oil spill, Reilly created a new office at EPA to investigate and develop bioremedial solutions to some of our environmental problems.

Preventing Environmental Problems

If an ounce of prevention is worth a pound of cure in human health, surely the same can be said for environmental health. Rather than cleaning up environmental problems, wouldn't it be better if we did not create them? Biotechnology is opening up a number of avenues for preventing environmental problems.

Biotechnology and renewable resources. The source of some of our pollution problems is the use of petroleum-based chemicals in many of our industrial manufacturing processes. The wastes produced by chemically based manufacturing processes are not easy to degrade. In addition, petroleum products are nonrenewable resources.

In the future, more companies will look to abundant and renewable organic matter such as agricultural refuse and municipal wastes or plentiful biological molecules such as starch and cellulose as the raw materials for manufacturing processes. Other chemicals used in the production process, such as the solvents listed in Fig. 1.5B, will increasingly be of biological origin. Biological molecules are easy to degrade, their by-products are nontoxic, and they are renewable. In short, through biotechnology we will develop manufacturing processes that make better sense both economically and environmentally.

Biodegradable plastic through biotechnology. Another environmental problem we have heard a good deal about lately is solid waste disposal. There is little room left in our landfills. We are about to be buried under our own trash.

Many communities have initiated recycling programs to help lessen this problem. Another way to lessen the trash load would be to use biodegradable plastic. Some recent advances by researchers in England and Michigan make a truly biodegradable plastic feasible. Guess who we should thank for help on this front? Another microbe.

In England, scientists discovered a soil bacterium that produces a compound, **polyhydroxybutyrate (PHB)** that can be used to make biodegradable plastic. They tried to produce PHB through microbial fermentation, but extracting PHB from the bacterium is very

difficult and costly. Scientists in Michigan have now cloned the gene responsible for producing PHB in the bacterium and inserted the gene into plants, from which extraction is no problem.

Alternative energy sources. Using alternative energy sources could also help prevent environmental problems. With so many sources of renewable energy on this planet, we could easily decrease our reliance on fossil fuels. Many of our air pollution problems result from burning these products to provide energy to heat our homes, cool our offices, and drive our cars.

If we could develop systems for generating energy that rely on biological rather than chemical energy sources, we could lessen air pollution. One country, Brazil, has already made steps in that direction. Instead of using gas in their cars, many Brazilians are using alcohol. They are using microbial fermentation to convert the starch in plants such as sugar cane and sugar beets to alcohol.

Another biological energy source that is virtually untapped is sewage. Solid wastes from domestic and agricultural sources are degraded by many bacteria, some of which produce methane gas as a by-product of this degradation. If we could capture that methane gas, we could put it to work for us.

Monitoring the Environment

The techniques of biotechnology are providing us with novel methods for diagnosing environmental problems and assessing normal environmental conditions so that we will be more informed environmental stewards in the future.

A North Carolina company, EnSys, Inc., has developed a method of detecting harmful organic pollutants such as polychlorobiphenyls in the soil by using MCAb (Fig. 1.20). Not only is this method cheaper than the current laboratory methods, which require large and expensive instruments, but the MCAb test kit is also portable. Rather than gathering soil samples and sending them to a laboratory for analysis, scientists can measure the level of contamination on site and know the results immediately.

Societal Issues

You often hear it said that advances in biotechnology will give rise to problems, issues, and concerns humans have never before faced. But what exactly are they? Are the societal issues said to be rooted in biotechnology unique to the new powers biotechnology gives us, or are they the old, unresolved, some-

Figure 1.20 EnSys, Inc., diagnostic test for measuring environmental pollutants in the soil (photograph courtesy of EnSys, Inc.).

times unacknowledged problems precipitated by earlier technologies?

Many of the issues attributed to the development of biotechnology have actually been with us in slightly different forms for many years, but we haven't been paying attention. We have passively enjoyed the fruits of technology without considering its implications. Now we recognize that the short-term gratification technology provides is sometimes accompanied by long-term costs.

Biotechnology is being born in this environment. As a result, it is developing under exceptionally watchful eyes on a stage flooded with light. We are focusing more attention on this technology at an earlier developmental stage than we did with any of the previous technologies, and much of the attention is negative. Why is biotechnology receiving so much negative attention? Is the negativity deserved, or is biotechnology paying the price for the mistakes of previous technologies?

We can trace some of the negativity to an antitechnology sentiment that is not uncommon in today's society. Our sentiments about some technologies seem to have swung from unmindful acceptance of the benefits to cynical disillusionment with the costs. Unfortunately, evaluating technology is not an either-or issue. Forming a balanced and fair opinion of technology depends on looking at its costs and benefits side by side.

Considering the benefits of technology

Take a minute and try to imagine the world in the absence of any technology. That is very difficult, for technology is woven tightly into the fabric of our lives and ourselves. The essence of our species seems inseparable from technology. We cannot objectively analyze technology and our relationship to it any more than a fish can step back and observe water. Nonetheless, try and imagine a world without any technology.

Would you really want to go back to the "good old days" of no running water, no electricity, sewage in the streets, and an expected life span of 40 years? These conditions existed very recently in the United States and continue to exist in many parts of the world. According to the World Bank, 1 billion of the 3 billion people living in underdeveloped countries cannot count on finding food or shelter from one day to the next. Every year, 14 million children under the age of 4 years die from starvation or starvation-related diseases.

As recently as 100 years ago, almost 50% of us had no choice but to live on a farm. That number of farm workers was required to provide enough food to feed our population. As a result, most children had no schooling beyond the first few years because they were needed as laborers. Now all of our food is produced by less than 2% of the population. We can trace that remarkable change to the development of new farm technologies that allowed machines and chemicals to take the place of people. The people who used to work on farms now sell insurance, play professional sports, watch professional sports, and write books on biotechnology. Much of the growth of this country's economy is linked to the movement of people from the farm and into the work force. How many of you truly want to get rid of machines and chemicals and go back to the "good old days" of hard manual labor and no schooling? Some of you might, but most of you probably don't.

Look around the room you are in. What is "natural," and what is a product of technology? How often do you really come face to face with nature? Not often, because humans rejected the idea of living with nature long ago.

In evaluating any technology, we must, first and foremost, be honest with ourselves. We depend on technology, and few of us could survive without it.

Considering the costs of technology

Most of the negative sentiment surrounding technology stems from the perception that technology is destroying the environment. Technology has a profound effect on our relationship with the natural world. That, after all, is its intent. We use technology to insulate us from the often harsh realities of nature.

But does technology have to be incompatible with nature? Not necessarily. If we look at technology as engineering solutions to the problems of life on Earth, then technology exists in nature. Other species have evolved technologies that are harmonious with nature. The most obvious example is the conversion of solar radiation into chemical energy (glucose) by plants. This describes an engineering solution to a problem all living organisms must face: obtaining energy.

Biologists call the technological achievements of other organisms "adaptations." The forms of adaptations that have evolved have been shaped by the indifferent force of natural selection. Natural selection cares nothing about politics and economics, so these factors have neither driven nor shaped the technological solutions evolution has designed. So maybe it's not technology per se but the technological solutions we have chosen that have had detrimental environmental effects.

Biotechnology has the potential to offer solutions more compatible with nature because it allows us to solve biological problems in biological ways. Knowing what we do now, perhaps we will aim for applications of biotechnology that are energy, resource, and capital efficient as well as minimally polluting.

However, even if we do develop technologies that are more environmentally benign, that will not rectify the relationship between technological advance and environmental health. No matter what ingenious engineering solutions we observe in other organisms, only one species has been able to use technology to increase its population size beyond ecologically sustainable limits, and that's us. Even if we, like other organisms, develop technologies that are harmonious with nature, if our population continues to increase at its current rate, we will only increase the strain we are placing on the planet and ourselves.

This, in fact, is the way that advances in biotechnology will probably contribute to our current environmental problems. Biotechnology will give us what we want: lives that are long and healthy. As appealing as that is to us as individuals, the result will be more people living on an already overcrowded planet. As a species we should remember the adage taught in the myth of Midas and the Golden Touch: Be careful what you wish for—you may get it.

Technology alters our relationship with nature in ways even more subtle than allowing an increase in population size. It has insulated us from nature, and increasingly from one another, so effectively that we have forgotten we are part of the natural world. We are no longer conscious of the very real biological ties we have with other species and with each other.

We are in the midst of an environmental crisis and must begin to change our ways if we and countless other species are to survive. Loss of this sense of connectedness with the natural world ultimately makes the tasks ahead of us even more difficult. The less we feel we are a part of nature, the more we feel we can play outside its rules. The less we know about nature, the harder it is for us to respect it. Without respect for nature and each other, will we be able to make the necessary sacrifices to balance our wants and needs with the needs of other species?

Our evolutionary heritage has provided us with a conscious mind. As thinking beings, we question and explore and then use our discoveries to control our environment and change things to our liking. For the most part, we have directed these abilities toward things *outside* of ourselves. That's what science and technology are all about. Turning inward and questioning, exploring, controlling, and changing ourselves does not come quite as easily to us. Now we must stretch the minds that evolution has provided and learn the more difficult behaviors of questioning our sense of values and our place in the scheme of things, exploring ourselves and our motives, and accepting responsibility for our actions.

We can do better than muddling through problems we create, trying to clean up our mess as we go along. We know tomorrow's problems can be predicted, in part, from our past and our present. Rather than simply react, we can anticipate and plan. More than any technology, biotechnology places this challenge before us. We will not meet this challenge by relying solely on the cleverness of our science. This technology raises deeper philosophical questions that science can't help answer. For these we need wisdom.

How can we act wisely in directing biotechnology? Neither blind acceptance nor wholesale rejection qualifies as acting wisely. By attempting to completely block the development of biotechnology, we ultimately deny ourselves possible choices. What wisdom is there in limiting our options? On the other hand, ignoring the implications of this technology and passively allowing it to develop on its own will more than likely guarantee that we will do no better with this technology than we have with others. Instead, it is in our best interest to attend to biotechnology's development by consciously making choices based on critical evaluation of real issues. How do we do this?

Analyzing the issues of biotechnology

First of all, we must place biotechnology in context. Remember, biotechnology is one of many technologies we have developed. Medicine, agriculture, and energy production are age-old attempts to control our environment with technology. Biotechnology now places at our disposal a new set of technologies for shaping the world. The tools may be different, but the goals remain the same: to improve human health and alter our environment so that our lives are as long and easy as possible.

In evaluating the societal issues raised by biotechnology, placing biotechnology in the historical context of a continuum of technological change can help us

- understand the changes the technology will bring and determine whether the changes are significant;
- assess the uses of this technology compared with those of past activities;
- separate real problems from remote possibilities.

Second, in analyzing the issues raised by biotechnology, we should consider each issue separately. Analyzing each issue apart from the others makes the job a lot easier and also may reveal any hidden—and perhaps incorrect—assumptions.

Once we have focused on an issue, we should ask ourselves whether this issue is new or unique to biotechnology. Does the concern or issue derive from biotechnology's novel powers that provide capabilities we never before have had, or do we currently engage in practices that raise similar or even identical issues? If the practice and issues it raises are unique to biotechnology, do all aspects of the issue raise concerns? If not, specifically define those areas that trigger concern, and focus your attention on them.

You may determine that the practices that elicit your concerns are not at all unique to biotechnology but are accepted as part of life. Does their commonness mean they are not worth analyzing? Absolutely not. Just because we *are* doing something does not mean we *should* be doing it. For example, you may be concerned that we are attempting to improve food through genetic engineering, which allows us to move genes between different species. Below, you will learn that much of the food we consume on a daily basis has been derived by interbreeding different species. After learning this, you may still believe we should not move genes between species, whether we accomplish this through genetic engineering or crossbreeding. In other words, what we are doing, not how we are doing it, poses a problem for you.

Here are a few examples of issues said to be derived from advances in agricultural and medical biotechnology and suggested ways of analyzing them. We will begin with agricultural biotechnology because many of the applications of agricultural biotechnology, and therefore the resulting issues, are identical to today's agricultural practices and issues. Beginning on ground that is somewhat familiar will allow us to get comfortable with the method of analyzing complex issues before we move into a more difficult area, that of medical biotechnology. As we progress through these issues, notice that most of the concerns expressed about biotechnology revolve around one particular type of biotechnology, gene-based biotechnology.

Agricultural Biotechnology
ISSUE: **Environmental Introductions of Genetically Engineered Organisms**

Concern: Many people have expressed concerns about the environmental risks associated with introducing genetically engineered organisms into our agricultural systems.

Their concerns focus primarily on

- the transfer of the gene(s) to other organisms;
- the creation of "exotic species" that will, like kudzu, displace indigenous species.

Are these concerns new or unique to biotechnology? No. We have been introducing genetically altered organisms into the environment for thousands of years. If we decide that introducing genetically engineered organisms into the environment should not be allowed, we will need to reassess the way we practice agriculture today.

We have changed the genetic makeup of all of the crops we plant and the livestock we raise. Every year, seed companies introduce many new varieties that differ genetically from those they've sold before. Not only have we released genetically altered plants and animals on a worldwide scale for centuries, but we have also conducted innumerable releases of incalculable numbers of microbes. For over 100 years, we have used microbes in agriculture for biological control and nitrogen fixation. Food processing and fermentation involve releasing genetically modified microbes into the environment. In fact, every time you throw away yogurt or sour cream, you introduce thousands of genetically altered microbes into the environment.

All of the genetically modified organisms we have released, whether plant, animal, or microbe, differ from genetically engineered organisms only in the methods used to alter their genetic makeup. Traditionally, we have used selective breeding and mutagenesis, methods of genetic alteration that are less specific than genetic engineering.

Our methods of genetically altering organisms may have changed, but the traits we have tried to incorporate into organisms in the past will, for the most part, be those we will incorporate through genetic engineering. As a result, we have a large body of experience to draw on in assessing the environmental and public health risks of releases of genetically engineered organisms.

However, a few of the genetically engineered traits we are incorporating will be novel. As described earlier, we are genetically engineering crops such as tobacco to serve as sources of pharmaceutical compounds. Because we have no previous experience with this type of crop, we have no body of experience to use for predictive purposes. In the future, novel traits introduced through genetic engineering must be assessed thoroughly to determine whether or not they pose a risk to human health or the environment.

As of June 1995, the U.S. Department of Agriculture has approved more than 2,000 environmental introductions of genetically engineered organisms into agricultural systems, and there have been no adverse environmental consequences. A detailed discussion of environmental introductions of genetically engineered organisms is provided later. For now, suffice it to say that our extensive experience releasing genetically altered organisms serves as a rich information base from which to discuss this issue.

ISSUE: Crossing Genetic Boundaries Established by Nature

Concern: People have expressed concern that genetic engineering allows us to cross the normal reproductive barriers that exist in nature.

Is this ability new or unique to biotechnology? No. When we selectively bred many agricultural species, both plant and animal, we forced fertilizations that would never have occurred without human intervention. Many of the crosses we forced involved organisms belonging to different species but the same genus. A few, however, occurred between organisms of different genera. So we have already crossed genetic barriers that would not have been crossed in nature.

If the issue is that we should not force crosses between organisms that would not breed in nature, what if the two organisms are in the same species? Sometimes, for a variety of reasons, members of the same species will not mate or will not produce viable

offspring. Is it acceptable for us to intervene in these cases? If it is, why?

If we decide that crossbreeding organisms in different species is acceptable but we object to moving genes between species through genetic engineering, we must try to delineate our concerns precisely. What is the heart of the issue? Is it the method, genetic engineering, that causes the concern? Why? Is it that we shouldn't move genes between organisms that are "too different" from each other? How will we define "too different"? Where will we draw the line, and why will we draw it there?

Does it matter that, left to their own devices in the absence of human intervention, organisms do not respect the genetic boundaries we have erected to delineate species? Some don't. This natural exchange of genetic material between species is called horizontal gene transfer, and it is more common than you might imagine.

Most of us are familiar with the ready transfer of genes between bacteria of different species. That is one of the factors responsible for the rapid spread of resistance to antibiotics in the microbial world. But bacteria are not the only organisms that exchange genetic material between species. So do many other species, and they are not "lowly" organisms like bacteria. In fact, primate ecologists have described natural hybrids between two different monkey species. In plants, outcrossing between different species has been crucial to the evolution of the higher plants, the angiosperms.

The evolutionary distance over which gene transfer occurs in nature can be very great. According to molecular evidence, it seems that viruses have transferred genetic material between distantly related insect and mammalian species. Similar evidence shows that a parasitic mite has transferred genes between two *Drosophila* species and kept a few *Drosophila* genes for itself.

Of course, significant obstacles block the frequent transfer of genetic information across species boundaries. Blocking interspecific gene flow is much more prevalent than permitting it. Nonetheless, it is important for us to realize that the issue of natural versus unnatural gene transfer is not as clear cut as we might like to think.

ISSUE: The Evolution of Resistance to Pesticides

Concern: People have expressed concern that pests will evolve resistance to the pesticidal products of biotechnology just as they have evolved resistance to other chemical pesticides. In particular, people are concerned that insects will evolve resistance to Bt and its toxic protein.

Is this evolutionary response new or unique to biotechnology? No. The evolution of resistance is not a problem specific to biotechnology. We can expect all pests—insects, weeds, fungi, nematodes, bacteria, viruses—to evolve resistance to *any* control measure we use against them. Weeds have evolved to physically resemble crop plants to avoid being pulled up by hand. Insect pests have lengthened their diapause from 1 year to 2 to escape control by crop rotation and have even skirted biological control mechanisms.

If we continue to try and control insects, weeds, and disease, then there can be no doubt that certain pests will evolve resistance to certain pesticidal products produced through biotechnology. No matter what methods we use to kill pests, they will do their best to evolve resistance. We are powerless to stop evolution or change the mechanism through which it acts—natural selection. In order to stop the evolution of resistance to control mechanisms, we would have to stop trying to control our crop pests.

We can, however, slow evolution down greatly by decreasing the selective pressure we place on pest populations. We should use what we've learned about retarding the rate of the evolution of resistance to maximize the efficiency of all products of biotechnology.

Some say that evolution of resistance to Bt is a unique problem because Bt is a rare, natural resource that shouldn't be squandered. We have discovered very few organisms selectively toxic to pests but not to beneficial organisms.

Again, this problem and the solution are not specific to biotechnology. Knowing what we do about the extraordinary adaptability of pests, we should use all products judiciously so that we can maximize their effective life span. In other words, we should use all products designed to control problem organisms—insecticides, fungicides, herbicides, and even antibiotics—in ways that lessen the selective pressure we place on a population from any single control measure.

ISSUE: The Safety of Genetically Engineered Foods

Concern: People have expressed concern over the safety of foods we have genetically engineered.

There is something new and unique to biotechnology here, but it is only a portion of the concerns that have been expressed.

By now you know that we have genetically altered everything we consume except for nondomesticated organisms like the wild fish we catch or the blackberries we gather. None of our food has escaped our genetic tinkering via selective breeding, a way of changing an organism's genetic makeup that is less precise than genetic engineering. So the overarching concern about the safety of genetically engineered foods compared to our current foods is misplaced.

The aspect of public concern that is appropriate relates to food allergens. A handful of foods are known to be common allergens: soybeans, peanuts, wheat, eggs, and shellfish. These foodstuffs are so common in our diet that people who are allergic to them become aware of their allergies early on. They then avoid those foods. We now have the ability to move genes from common allergens to foods not typically seen as allergenic. If we are interested in transferring certain traits in an allergenic food, we might move the gene that codes for the protein that triggers the allergic reaction into a nonallergenic food. People with food allergies should and will be alerted to that fact. In its 1993 announcement regarding the regulation of genetically engineered foods, the FDA announced that it will require labeling when a gene from a known allergen has been moved to a nonallergen.

But what about unknown allergens or foods that are allergenic in very rare cases? Should the government require labeling for a genetically engineered product that might cause an allergic reaction in 0.001% of the population? Nongenetically engineered products currently on the market might cause allergic reactions in some people. Should they be labeled? When kiwi fruits were first introduced into grocery stores in the United States, many people learned they were allergic to kiwi fruits only after buying and eating them. Should the fruits have carried warning labels? What should the labels have said?

Medical Biotechnology

The most difficult issues raised by biotechnology deal with medical biotechnology and bioethics. Here are just a few examples. For a more thorough treatment of these issues, see chapters 32 to 34.

ISSUE: Genetic Testing

Concerns: People have voiced a number of interrelated and complex concerns associated with genetic testing. For simplicity's sake and so that the discussion will lead to increased understanding, each concern will be considered in turn.

ISSUE: Genetic Screening for Inherited Disorders

Concern: Some believe that our ability to identify genetic defects will cause more harm than good by placing individuals in the middle of psychological, emotional, and social quandaries that defy easy answers.

They fear that people with information about personal genetic disorders will be psychologically harmed by having this knowledge. They believe that knowing you have a genetic disorder that causes an incurable disease, such as Huntington's chorea, would be emotionally devastating. Others say that knowing you are pregnant with a child suffering from a genetic disorder would force parents into a psychologically and emotionally difficult decision regarding abortion.

What about this issue is new or unique to biotechnology? First, let's consider what's not new, because people and physicians grapple with many of the same issues today.

If a routine physical examination reveals that you have an incurable cancer, do you expect your physician to share that information with you? Probably so. But just a few decades ago, physicians did not tell their patients they had cancer because they thought not knowing was best for the patient's emotional well-being. The tension between withholding and disclosing information is not new to medicine and varies with the disease, the physician, the patient, and the prevailing practice. It is not specific to new genetic testing capabilities.

The same can be said for prenatal testing for inherited disorders. Today, many obstetricians routinely advise women over 35 to have amniocentesis or chorionic villus sampling to test fetuses for genetic disorders. Biochemical tests on fetal cells can detect more than 200 metabolic genetic disorders and a number of chromosomal abnormalities. A few of these disorders can be treated once the baby is born. For example, the course of the inherited disorder phenylketonuria can be changed by placing the infant on a special diet early in life. The mental retardation associated with phenylketonuria will not result if the individual maintains a diet low in the amino acid phenylalanine.

However, most of the genetic disorders these tests can detect are incurable. Their progress cannot be subverted by any known treatments. Parents today must decide what to do with that information. If we are concerned about the emotional trauma caused by sharing fetal genetic information with parents, we should reassess our current practices.

Here's what is new about genetic screening and the issues it creates. The first issue relates to methodology: how the genetic basis of a disorder is discovered

and how screening methods for this disorder are developed.

Before knowing the exact location of a gene that causes a disorder, scientists first develop "markers" that are associated with the defective genes. The process of developing markers depends on acquiring blood samples from many members of a family and subjecting their DNA to analysis. As a part of this discovery process, a person's genetic makeup becomes public. (See chapter 4 for more information about finding genes.)

For an individual, a dilemma is posed between the right to "genetic privacy" and a responsibility to members of the family, both nuclear and extended. Where would you draw the line balancing your need for privacy with their need to know?

The second new issue raised by widespread availability of genetic testing is more a question of a change in degree than a true novelty. We have had the power to conduct fetal testing for genetic disorders for many years. Parents have had to face some very difficult decisions as a result, because most of those tests have been for incurable diseases or serious medical problems that leave little or no room for human intervention.

In the future, we will be able to detect many more defective genes than those that cause incurable illnesses. Some genetic defects will directly cause serious problems that can be reversed with appropriate treatment (hemophilia). Other defective genes will simply indicate a propensity to develop a certain disease (emphysema). Still others will cause defects that are in no way life threatening (color blindness).

Where will we draw the line when defining a genetic defect and determining how to handle it? When is a defect truly a defect? At what point will eliminating genetic defects gradually inch toward selecting desired characteristics? Will each couple be free to make decisions regarding the fate of offspring carrying a defective gene?

Currently, genetic alteration in human germ cells (sperm and eggs) is not allowed. If the government ever changes its position on germ line gene changes, the door for altering the species' gene pool will be opened. Will we do our best to decrease genetic diversity? We may praise genetic diversity in the abstract, but historically, our behaviors are not consonant with that viewpoint. If we decrease genetic diversity in the human gene pool, will we ultimately be limiting our species' capacity to evolve?

You are probably asking why we see these issues as potential problems for individuals seeking information on the genetic makeup of their offspring. Can't people avoid these difficult decisions by simply not subjecting themselves to genetic testing? After all, no one requires them to take the tests, do they?

If we look at history, we see that technologies that are supposed to increase our options often increase our obligations instead. A choice at one time may become a requirement later on. For example, most parents probably would be surprised to learn that their newborn babies had been tested for as many as nine inherited disorders immediately after delivery. Such tests are required by law in most states. So although genetic testing is an option today, tomorrow we may have no choice but to be tested.

ISSUE: The Meaning of Genetic Information

Concern: What will this information really tell us?

People object to the large-scale use of genetic testing because, by nature, the test results are often ambiguous. The presence of certain genetic defects does not guarantee that an individual will actually develop the disease. In many cases, the presence of a defective gene means only that the person is a carrier. In other, more ambiguous cases, the presence of a defective gene means only that you have a genetic predisposition to that disease. In other words, it tells you that you are at risk.

But how does this differ from current practices? Every time we have our blood pressure taken or our cholesterol measured, we are assessing risk factors. When we describe our family's medical history to a physician, we are providing data on genetic predisposition. Is the information gained through genetic testing any more ambiguous than the information physicians already gather?

Both bodies of information are ambiguous, but while most people are prepared to accept the ambiguity of blood pressure readings and cholesterol measurements, they do not know that genetic testing data can be equally ambiguous. The nature and power of genes are thoroughly misunderstood by the public. An aura of predestination surrounds a gene. People believe that if you have a gene for a given trait, then all is determined. Nothing is left to chance.

Biologists know that this is not true. The path from genotype to phenotype is an indirect one filled with opportunities for myriad factors other than a single gene to exert an influence (Fig. 1.21). The final expression of a gene is affected by many variables: the

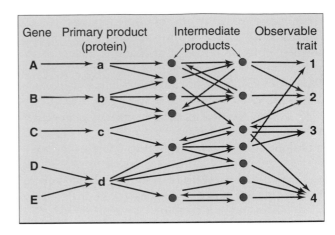

Figure 1.21 This diagram illustrates the complex relationship between genes and observable traits. As you can see, the path from the gene to its primary protein product to visible traits is not as straightforward and linear as many people assume. One gene can affect a number of traits, and any one trait may be the result of interactions between many genes. In addition, gene products have feedback effects on the activities of other genes. Although the diagram is complex, it is an oversimplification of the actual processes. Consider how complicated the diagram would be if we listed all intermediate products and regulatory steps between the gene and the trait, and all the environmental influences that affect the observable trait.

allelic forms of the gene in question, other genes, and environmental influences.

These facts are often ignored when people are assessing the meaning of genetic data. Members of the public who are in positions of power might assume that genetic information is much more meaningful than it really is. Such unenlightened viewpoints can be particularly pernicious when held by individuals in a position to exercise control over our lives or by institutions and organizations concerned with issues other than humanitarianism. This leads us to the final issue related to genetic testing.

ISSUE: Genetic Privacy

Concern: Who will have access to this information and for what purposes?

The final concern involves the confidentiality of the results of the genetic tests and the interpretation of these results. People fear that the results will be used unjustly by employers to withhold jobs and promotions and by insurance companies to deny coverage.

What is unique about this concern? Again, let's start by describing what's not new, because a precedent for some of these concerns already exists. Many American companies go to great lengths to screen their employees. At least 50% of the employers in the United States require that job applicants take medical examinations. In those examinations, many of the tests conducted are indirect measures, that is, phenotypic manifestations, of genetically based traits. As of 1982, 23 of 366 companies surveyed by the Office of Technology Assessment had begun to use genetic tests in screening applicants (another 57 expressed an interest in using them).

Companies are interested in the health of workers for reasons related to decreasing costs.

- Time lost due to sickness or disability costs money.
- Hiring healthy people allows corporations to hold down health care costs.
- By screening workers susceptible to certain occupational hazards, companies protect themselves against lawsuits.

Currently, insurance companies do not require genetic testing, but historically, insurers have used medical technologies to identify preexisting conditions as a basis for denying coverage or raising premiums. They have also excluded people from coverage because their family medical history put them in a high-risk category.

Even though these concerns about privacy and the use of genetic information are not specific to new genetic technologies, other concerns can be traced to these newer technologies. In general, concerns specifically related to genetic technologies are derived from the true nature of genes and people's misunderstandings about their nature as described previously.

When insurance companies require medical examinations to determine the health of their applicants, the physicians measure physiological and morphological correlates of health and disease. That is, if the person has begun to develop cardiovascular disease, the physician will be able to discern the problem through blood tests, electrocardiograms, and blood pressure measurements. The same is true of a lung disease such as emphysema. Measurable factors such as lung capacity and force of exhalation indicate that the applicant-patient is developing or has developed the disease.

Genetic information is unlike this type of data. Genes foretell the future. They might allow physicians to predict the possible appearance of cardiovascular or lung disease in the absence of symptoms. Will these telltale markers then be classified as determinants of preexisting conditions, thus allowing insurance coverage to be denied?

A second problem regarding the nature of genetic information is linked to the general public misunder-

standing of what genes are, what they do, and what genetic information means. If employers and insurance companies view genes as final arbiters of traits rather than one of many variables that determine the final product, any genetic defect may be seen as a preexisting condition.

This leads us to another important and difficult question that must be answered. What exactly will be considered a defect? There is extraordinary genetic variation for most traits in all populations. Which variants of which traits will be labeled normal and which will be labeled defective? Geneticists estimate that probably everyone has 5 to 10 genes that could cause an illness of some sort. As a result, all of us could be considered genetically defective. How could all of us be uninsurable and unemployable? Again, where will we draw the line, and who will draw it?

Summary

Biotechnology is a set of very flexible and powerful tools that offers great potential for improving human health, increasing the quality and yield of our agricultural products, and mending our relationship with the environment.

Because biotechnology is in the earliest stages of development, we are in a position to fashion these tools to meet our needs and to use them to achieve goals deemed desirable by society. By carefully overseeing the development of biotechnology, we are better able to ensure that the decisions shaping its future are informed by past mistakes and current understandings. We have an opportunity to avoid the mistakes of the past, to act a little wiser, and to use what we have learned.

However, careful oversight of technological development sometimes becomes excessive vigilance that encourages stagnation. Moving cautiously may mean we don't move at all. Should we ask biotechnology to pay the price for previous technologies by scrutinizing it so thoroughly that we stifle its development and effectively limit our options?

An essential advantage of biotechnology over other technologies is that it is based on biology. It can work with the biology of organisms in very specific, predictable ways to solve biological problems or make products. Of all the technologies developed so far, biotechnology has the potential to be more compatible with life on this planet.

Today, many people are antitechnology because of our past mistakes. They equate technology with problems. Unless we are prepared to turn our backs on progress—to say "no, thank you" to a cure for AIDS, to return to the days when 50% of the population worked on farms to ensure an adequate food supply, to stop pursuing new and better ways to treat cancer—we must honestly admit to ourselves that we want technologies that make our life on Earth easier. No matter how antitechnology you are, would you be willing personally to pay the price for deterring advances in biotechnology? Probably not.

Given those facts, it is each individual's responsibility to see that biotechnology is developed and used wisely. We must do whatever we can to avoid the mistakes of the past and ensure that biotechnology benefits not only us but also the other organisms that share this planet with us.

Selected Readings

Antebi, Elizabeth, and David Fishlock. 1985. *Biotechnology: Strategies for Life*. MIT Press, Cambridge, Mass.

Bishop, J.A., and L.M. Cook (ed.). 1981. *Genetic Consequences of Man Made Change*. Academic Press, London.

Bishop, Jerry E., and Michael Waldholz. 1990. *Genome*. Simon & Schuster, New York.

Brock, Thomas, and Michael Madigan. 1988. *Biology of Microorganisms*. Prentice Hall, Englewood Cliffs, N.J.

Bronkowski, J. *Science and Human Values*. 1956. Harper & Row, New York.

Chilton, Mary-Dell. 1983. A vector for introducing new genes into plants. *Scientific American* 248:50–59.

*Congressional Office of Technology Assessment. 1981. *Impacts of Applied Genetics: Micro-organisms, Plants, and Animals*. U.S. Government Printing Office, Washington, D.C.

*Congressional Office of Technology Assessment. 1983. *The Role of Genetic Testing in the Prevention of Occupational Disease*. U.S. Government Printing Office, Washington, D.C.

*Congressional Office of Technology Assessment. 1984. *Human Gene Therapy: a Background Paper*. U.S. Government Printing Office, Washington, D.C.

*Congressional Office of Technology Assessment. 1987. *Field-Testing Engineered Organisms: Genetic and Ecological Issues*. U.S. Government Printing Office, Washington, D.C.

*Congressional Office of Technology Assessment. 1988. *Medical Testing and Health Insurance*. U.S. Government Printing Office, Washington, D.C.

*Congressional Office of Technology Assessment. 1990. *Genetic Monitoring and Screening in the Work Place*. U.S. Government Printing Office, Washington, D.C.

Davis, Bernard D. 1991. *The Genetic Revolution: Scientific Prospects and Public Perceptions*. Johns Hopkins University Press, Baltimore, Md.

Davison, John. 1988. Plant beneficial bacteria. *Bio/Technology* 6:282–286.

Florman, Samuel. 1981. *Blaming Technology: the Irrational Search for Scapegoats*. St. Martin's Press, New York.

Frankel, Mark, and Albert Teich (ed.). 1995. *The Genetic Frontier: Ethics, Law and Policy*. AAAS, Washington, D.C.

Gray, Paul E. 1989. The paradox of technological development. *In* Jesse Ausubel and Hedy Sladovich (ed.), *Technology and Environment*. National Academy Press, Washington, D.C.

Heiser, Charles B. 1981. *Seeds to Civilization: the Story of Food*. W. H. Freeman & Co., San Francisco.

Holden, Constance. Looking at genes in the workplace. *Science* 217:336–337.

Holtzman, Neil. 1989. *Proceed with Caution: Predicting Genetic Risks in the Recombinant DNA Era*. Johns Hopkins University Press, Baltimore, Md.

Kevles, Daniel, and Leroy Hood (ed.). 1993. *The Code of Codes. Scientific and Social Issues in the Human Genome Project*. Harvard University Press, Cambridge, Mass.

Kieffer, George. 1987. *Biotechnology, Genetic Engineering and Society*. National Association of Biology Teachers, Reston, Va.

Marx, Jean. 1989. *A Revolution in Biotechnology*. Cambridge University Press, Cambridge, England.

Nash, Roderick. 1989. *The Rights of Nature: a History of Environmental Ethics*. University of Wisconsin Press, Madison.

National Academy of Sciences. 1987. *Introduction of Recombinant DNA-Engineered Organisms into the Environment: Key Issues*. National Academy of Sciences Press, Washington, D.C.

National Research Council. 1989. *Field-Testing Genetically Modified Organisms: Framework for Decisions*. National Academy Press, Washington, D.C.

Nelkin, Dorothy, and Laurence Tancredi. 1989. *Dangerous Diagnostics: the Social Power of Biological Information*. Basic Books, Inc., New York.

Olson, Steve. 1989. *Shaping the Future: Biology and Human Values*. National Academy Press, Washington, D.C.

Singer, Charles, E.J. Holmyard, and Trevor Williams (ed.). 1954. *A History of Technology*, vol. I and II. Clarendon Press, Oxford, England.

Suzuki, David, and Peter Knudtson. 1989. *Genethics: the Clash Between the New Genetics and Human Values*. Harvard University Press, Cambridge, Mass.

Watson, James. 1990. The human genome project: past, present and future. *Science* 248:44–49.

Wenk, Edward. 1986. *Tradeoffs and Imperatives of Choice in a High-Tech World*. Johns Hopkins University Press, Baltimore, Md.

*To order publications from the U.S. Government Printing Office, write to the Superintendent of Documents, Mail Stop: SSOP, Washington, DC 20402-9328.

Genes, Genetics, and Geneticists

2

The Centrality of Genes

The remainder of this book is about genes: what they are, what they do, how they do it, and how we are using them. But before we delve into the workings of genes on a molecular level, let's step back and look at these extraordinary particles from a broader perspective.

Everyone recognizes that we must learn about genes to understand genetics and heredity, the transmission of inherited traits from one generation to the next. What you may not realize is that an understanding of genes is central to an understanding of all of biology. Each of the diverse branches of biology can be seen as a way of investigating questions about genetics.

- Developmental biology is, in large part, the study of gene regulation: what turns genes on and off.
- Physiology and its companion science, morphology, focus on the structural and functional manifestations of gene expression.
- Ecologists study the genetic adaptation of organisms to their environments.
- Taxonomy is the study of the genetic differences and similarities within and between species.
- Evolutionary biologists investigate changes in gene frequencies in populations over time.

You can take any biological phenomenon of interest and explore it from the angle of the gene.

First, using your mind's eye as a microscope, put the phenomenon under maximum magnification and ask about the molecular details of its genetic basis. Lower the magnification to the physiological level and ask how the function of this particular gene integrates with the functions of other genes to produce a living organism.

Now look at the whole organism. How much of what you see has a genetic basis? How much can be explained by environmental influences? How do the two interact?

Now take a giant step back and use your mind's eye like a telescope to explore the same gene over time and space. How has that gene's structure changed over time? Has the change in structure brought about a concomitant change in function? How much of that change has been meaningful in an evolutionary sense? How does the gene vary within the population? Between populations but within the species? Between species?

To an outsider, biology appears to be a fragmented field of study with many loosely related subdisciplines, and textbooks that dissect the branches of biology into separate, apparently unrelated chapters do little to correct this misconception. Yet all of the various threads of biological thought can be woven into a coherent whole with genetics as the unifying concept. Keep returning to the central landmark that genes provide, and you will be better able to integrate the field of biology into a coherent whole.

Genetics is central to biology, the study of life, because genes are central to life. The fundamental characteristic of living organisms is reproducibility. Living organisms produce more living organisms like themselves. For organisms other than those that reproduce by simple division, reproduction involves both the transmission of hereditary information from parent to offspring and a developmental process that transforms that information into a living organism. Both of these hallmarks of living organisms reflect the functioning of genes: faithful replication and information transmission. If life depends on genes, it only makes sense that the study of life, biology, has genetics at its core.

A Brief History of Genetics

In the beginning

The relationship between genes, reproduction, and development seems obvious to us now, but it was not always so. In fact, for thousands of years—99% of human history—we were totally ignorant of the biological basis of reproduction. Sperm and eggs are invisible to the naked eye, and we could understand only those phenomena we could see. The invisible details of reproduction were left to our imaginations, and imagine we did!

Some truly great scientists developed theories of reproduction that seem silly to us now. Hippocrates, Aristotle, and, much later, even Charles Darwin believed that offspring resulted from the blending of male and female genital secretions that contain "seeds" that had been shed by each body part: eye seeds, liver seeds, heart seeds. During reproduction, they thought, these seed-filled fluids coagulated and were transformed into an organism as each organ seed developed.

It seems to me that among the things we commonly see there are wonders so incomprehensible that they surpass all the perplexity of miracles. What a wonderful thing it is that this drop of seed from which we are produced bears in itself the impressions not only of the bodily form, but also the thoughts and inclinations of our fathers! Where can that drop of fluid contain that infinite variety of forms?

Michel de Montaigne, 1570

The invention of the microscope in the 1600s permitted observation of human sperm and eggs but did not quell the development of imaginative theories of reproduction. Many scientists believed that tiny humans resided in each sperm cell. They thought that these minuscule humans took root and grew in the female's uterus. In this scenario, the male provided all of the genetic material, and the female provided a nurturing womb for safe development. Some scientists even claimed that, using microscopes, they could see little chickens and horses in sperm taken from these animals. Other scientists were skeptical, however. In fact, one scientist ridiculed this theory of reproduction by saying he had seen a tiny human enclosed in a sperm take off its even tinier coat!

Even though our ancestors were ignorant of the precise details of reproduction at the microscopic level, they could easily see consistent hereditary patterns in the natural world. Life begets life, and offspring look like their parents.

As early as 10,000 years ago, the first agriculturists put these observations to work by systematically cross-breeding the plants and animals they had begun to domesticate. As soon as we domesticated living organisms, we began to genetically manipulate them through this selective breeding. Our agrarian ancestors accomplished this genetic manipulation by trial and error, relying only on minimal understanding acquired through observation. Their knowledge was incorrect and incomplete but nonetheless sufficient to enable them to domesticate and genetically improve virtually all of the crops we use today for food and fiber.

As successful as the early agriculturists were, their attempts at selective breeding were still hit-or-miss ventures. Sometimes the "magic" worked; sometimes it didn't. Such uncertainty is not very satisfying to humans. We want to explore and understand things from the inside out so we can exert more control over our lives.

The mysteries of heredity provided us with a fertile field for such exploration. Inheritance patterns often seem unpredictable. Traits disappear in one generation and reappear in another. Some observable traits have a genetic basis and therefore can be subjected to successful selection; others don't. Some traits are controlled by one gene, while others are controlled by hundreds of genes.

For centuries, most people invoked supernatural powers to understand the apparently inexplicable nature of biological inheritance. A few, however, persisted in trying to understand. Through their efforts we have come to understand so much about genes that now not only do we know what genes are made of and what they do but we can also manipulate them to our advantage. In other words, we know genes inside out.

How did we come to learn so much? Who were the pioneers in the field of genetics? What questions did they ask, and how did they answer them?

Placing our current understanding of genetics in historical perspective is a valuable exercise. Tracing the development of this field not only informs us about the workings of biological systems at a basic level but also provides insights into how science advances. Sometimes the next critical discovery in a field is obvious, and scientists race each other in an attempt to be the one who captures the prize. At other times, the most important discoveries are ignored because the timing isn't right. The field of genetics was born at such a time. No one could hear what the "Father of Genetics" was saying, because he was years ahead of his time.

Discovery: the discrete nature of genes

One evening in February 1865, a monk named Gregor Mendel presented a lecture to the local scientific society in Brno, Czechoslovakia. He described in great detail the results of 8 years of data on crossbreeding thousands of peas. No one was impressed, but the results he shared that night were some of the most important findings in the history of biology. We are fortunate that his paper, like all papers of that society, was published in a journal kept in libraries across Europe. Thirty-five years later, his work was rediscovered, and only then did scientists understand its importance.

Every biology textbook discusses Mendel's work in detail. All biology teachers and students know of Mendel's work and could probably describe the methods and results in their sleep. Yellow peas, green peas, round peas, wrinkled peas. Do the crosses; see the results; now do more crosses; count more offspring. But what was so significant about Mendel's work? Why do biologists speak of him in reverential tones?

Stated simply, Mendel changed the way we view the natural world. Although he never used the word "gene," his work revealed the inherent nature of genes and gave birth to a new branch of biology: genetics, the study of heredity. He demonstrated that the hereditary substance that passes from one generation to the next is organized as discrete packets of information. Heredity does not involve blending fluidlike contributions. Instead, heredity depends upon combining discrete particles from both parents (Fig. 2.1). When egg and sperm fuse during fertilization, the maternal and paternal hereditary particles do not become joined but retain their distinct identities.

What is so important about having hereditary information packaged as discrete particles? Bundling up genetic information into separable chunks provides a constant means of generating genetically variable offspring, even in the absence of **mutation.** Without genetic variation, life on Earth could not have evolved. Now, that's important.

Because genes are discrete particles, the maternal and paternal genes for a given trait (**alleles**) separate from each other during the production of gametes. To understand the importance of this separation in producing genetic variation, contrast the two models of heredity: fluid blending and discrete particle.

The Fluid Blending Model of Inheritance

Imagine pouring together milk (maternal genes) and chocolate syrup (paternal genes) to make chocolate milk (offspring). Now take the chocolate milk and divide it into four equal parts (gametes), and pour two of them together (fertilization). You still have chocolate milk. No matter how many "generations" you repeated this dividing and mixing, crossing offspring gametes to produce the next generation, you would still end up with chocolate milk offspring and only chocolate milk offspring. You would never again have a glass of pure chocolate or a glass of pure milk. In other words, the offspring would always be genetically identical.

Since genetic variation is the key to the evolution of life on Earth, such a system of genetic blending would be an evolutionary dead end. For evolution to proceed, genetically variable offspring must be generated constantly. When used in conjunction with sexual reproduction, the discrete nature of genes provides this variation.

The Discrete Particle Model of Inheritance

To demonstrate the discrete particle model, use discrete particles—jelly beans, M&Ms, marbles, index cards—as genes, and mimic sexual reproduction: produce gametes through meiosis, and join them in fertilization. Now mimic a second round of reproduction with the gametes of the offspring. For example, if you use two red (maternal) and two green (paternal) jelly bean genes, the first generation offspring will all have a single genotype: red/green. Cross those individuals, and you will have three genotypes: pure red, red/green, and pure green offspring, genetically variable offspring.

Mutation was not a factor in creating these genetically variable offspring. The variation results from reassorting existing alleles through meiosis and fertilization, the constituent elements of sexual reproduction.

Mendelian principles. This separation of the maternal and paternal genes for the same trait (male and female alleles) during gamete formation is known as Mendel's Principle of Segregation (Fig. 2.2).

Figure 2.1 Diagram of the different models of inheritance: fluiding blending and discrete particle.

Fluid blending   Discrete particle

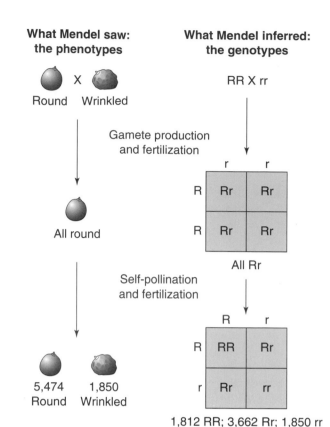

**What Mendel saw:
the phenotypes**

Round X Wrinkled

Gamete production
and fertilization

All round

Self-pollination
and fertilization

5,474 Round   1,850 Wrinkled

Phenotypic ratio = 3:1

**What Mendel inferred:
the genotypes**

RR X rr

	r	r
R	Rr	Rr
R	Rr	Rr

All Rr

	R	r
R	RR	Rr
r	Rr	rr

1,812 RR; 3,662 Rr; 1,850 rr

Genotypic ratio = 1:2:1

Figure 2.2 The Principle of Segregation. Mendel observed that experimental crosses between pea plants with round or wrinkled seeds yielded plants with round seeds in the first generation and a 3:1 mix of plants with round and wrinkled seeds in the second generation. From this result and observations of other characteristics, he inferred that maternal and paternal hereditary information is packaged as discrete particles and that maternal and paternal particles separate from each other during gamete formation.

A second important Mendelian principle also depends upon having hereditary information organized as discrete particles. This second principle, the Principle of Independent Assortment, involves separation of male and female genes for different traits (Fig. 2.3).

Its essence is this: on average, any gamete has an equal number of maternal and paternal alleles. On a purely mechanical level, this means that during meiosis, the paternal **chromosomes** do not line up on one side and the maternal chromosomes on the other. Maternal and paternal chromosomes do not stick together as a group during gamete production. In humans, each new gamete has 23 chromosomes. On average, 11 to 12 are of maternal origin, and 11 to 12 are paternal. (Of course, crossing over between homologous chromosomes makes the situation not as

tidy as just described when one looks at the specific maternal and paternal genes on chromosomes.)

Dominant and recessive alleles. The discrete nature of the gene permitted Mendel to observe and describe the concept of dominant and recessive characteristics. Think back to the chocolate milk versus jelly bean analogy. For the chocolate × milk cross, all offspring in all generations are chocolate milk. For jelly beans, assume that one color is dominant and the other is recessive. The first cross, red × green, tells us very little, because all of the offspring look the same and have the same genetic makeup. The revealing cross is the next one, in which red/green offspring are crossed with each other. This cross results in both red and green offspring as well as three genotypes. A trait that disappeared in the first generation reappears in pure form in the second. The traits do not blend into each other. One trait (**dominant allele**) simply overpowers the other (**recessive allele**) in the first-generation offspring.

Genotypes and phenotypes. Because Mendel chose traits that exhibited clear dominance relationships, he was able to elucidate another concept fundamental to our understanding of inheritance: the relationship between **genotype** and **phenotype**. The outward appearance of an organism (phenotype) may or may not directly reflect the genes that are present (genotype). When you want to understand the details of an organism's genetic makeup, sometimes the phenotype is helpful, and other times it can be very misleading.

Genetics, statistics, and probability. Another ancillary but very important feature of the discrete nature of genes involves not biological inheritance but the ease with which genetic traits can be scientifically investigated. Because genes are discrete entities (like coins, dice, or cards), the observed results of crosses can be scientifically analyzed by using the laws of statistics and probability (as with coins, dice, and cards). Using observable differences, scientists can quantify patterns of inheritance and infer the genetic basis of traits without knowing the details of the molecular biology of the gene of interest.

It would be difficult to overstate the importance of Mendel's findings to our understanding of the workings of biological systems from DNA to evolution. We could not learn all we know about the natural world until we could appreciate and understand his remarkable work on the particulate nature of inheritance. He could see patterns where everyone else saw disarray. From these patterns he made inferences and es-

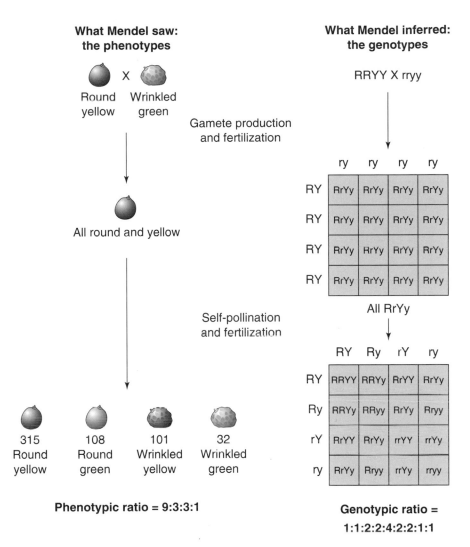

**What Mendel saw:
the phenotypes**

Round   Wrinkled
yellow   green

Gamete production
and fertilization

All round and yellow

Self-pollination
and fertilization

315   108   101   32
Round   Round   Wrinkled   Wrinkled
yellow   green   yellow   green

Phenotypic ratio = 9:3:3:1

**What Mendel inferred:
the genotypes**

RRYY X rryy

	ry	ry	ry	ry
RY	RrYy	RrYy	RrYy	RrYy
RY	RrYy	RrYy	RrYy	RrYy
RY	RrYy	RrYy	RrYy	RrYy
RY	RrYy	RrYy	RrYy	RrYy

All RrYy

	RY	Ry	rY	ry
RY	RRYY	RRYy	RrYY	RrYy
Ry	RRYy	RRyy	RrYy	Rryy
rY	RrYY	RrYy	rrYY	rrYy
ry	RrYy	Rryy	rrYy	rryy

**Genotypic ratio =
1:1:2:2:4:2:2:1:1**

Figure 2.3 The Principle of Independent Assortment. By observing the hereditary patterns of two separate traits, seed shape and color, Mendel inferred that during gamete formation, the alleles for one trait (seed shape) segregate independently of the alleles for a trait located on a different chromosome (seed color). What would Mendel have seen if the genes for seed color and shape were adjacent on a chromosome?

tablished the fundamental concepts of genetics before anyone knew genes existed.

And yet, no one was impressed the night he presented his results or for 35 years afterward. Mendel revolutionized biology 16 years after he died. Science works like that more often than you might guess.

Discovery: the chromosomal nature of inheritance

Once Mendel established that the hereditary material is organized into packets of information that separate from each other during gamete formation, scientists had to localize these particles within the cell.

Improvements in microscopy permitted the next set of discoveries, because we could actually see, on a cellular level, the phenomena Mendel had described. Using microscopy, we began to uncover the underlying mechanism of the principles Mendel had inferred from observing phenotypic variation. In other words, technological advances enabled the next generation of scientists to answer the next question: Where are these discrete particles located?

In 1902, 2 years after Mendel's work was rediscovered, Walter Sutton observed chromosomes behaving in ways that resembled the segregation of hereditary material Mendel had postulated. Studying meiosis in grasshoppers, Sutton, a graduate student at Columbia University, observed that chromosomes occur in morphologically similar pairs and that the two members of a chromosome pair separate from each other during gamete formation. He used this cytological evidence in conjunction with Mendel's findings to hypothesize that the hereditary material is associated with chromosomes. When he discovered this critical relationship, he ran excitedly to his major professor and mentor, E. B. Smith, who was one of the foremost biologists of his time. This is how Dr. Smith described what happened:

> *I well remember when in the early spring of 1902, Sutton first brought his main conclusion to my attention. I also clearly recall that at the time I did not fully comprehend his conception or realize its entire weight.*

Smith may not have understood what Sutton discovered, but Sutton was right on target. His theory was confirmed and elaborated by a remarkable group of biologists, also at Columbia University. In a 5-year period, they uncovered the major details of the chromosomal basis of heredity in Thomas Hunt Morgan's laboratory, an infamous room known to biologists as The Fly Room.

Thomas Hunt Morgan once said jokingly that God made the fruit fly just for him. With its relatively simple genetic makeup of only eight chromosomes (four pairs) and a very short generation time (12 to 14 days), the fruit fly, *Drosophila melanogaster,* is an excellent experimental animal for studying genetics. Much of our basic understanding of classical genetics comes from fruit fly studies; the great bulk of that early work came from Morgan's laboratory.

In the early 1900s, Morgan began to use *Drosophila* to address problems in evolution. He was not particularly interested in studying the cellular mechanics of inheritance. Instead, as an experimental scientist, he wanted to test existing theories about the role of mutation in evolution. Over and over he bombarded fruit flies with X rays or exposed them to chemicals in an attempt to mutate an easily observable trait.

One day, he finally got lucky. In the midst of thousands of red-eyed flies was a mutant at last: a white-eyed male. This single mutant mated with a red-eyed female, left its mutated gene, and in that act provided the raw material for a series of experiments that provided the foundation for classical genetics.

By all accounts, Morgan was quite an interesting character. On the one hand, he was quite informal, very humorous, and a stereotypical scatter-brained scientist. He was often mistaken for a beggar because his clothes were so ragged. His lab was just as messy as his clothes. Many desks and small tables were crammed into a tiny room, leaving little room to move. Hundreds of half-pint milk bottles filled with rotting bananas and thousands of flies were everywhere. An equal number of flies lived free in the room, hovering around ubiquitous stalks of bananas in various stages of decay. He recorded data on scraps of paper and the backs of envelopes, which promptly got buried under the piles of paper on his desk.

Despite his appearances, he was from one of the finest families in Kentucky and was an heir to the fortune of industrialist J. P. Morgan. He was known for giving money to graduate students who had a hard time making ends meet. He was also a brilliant investigator and an inspiring leader. He led an extraordinary group of young students to key findings that served as the footings on which to build our understanding of genetics. In a very few years, Morgan and his students, Alfred Sturtevant, Calvin Bridges, and Herman Muller,

- proved genes are chromosomally located
- introduced the concept of **sex-linked inheritance**
- introduced and elaborated the concept of genetic **linkage**
- originated the idea of **gene mapping**
- constructed the first genetic map of a chromosome
- demonstrated crossing over between homologous chromosomes

Discovery: the chemical nature of genes

Now that scientists knew where genes were located, they began to question the molecular makeup of genes.

In 1869, a German chemist, Frederick Miescher, had isolated a novel substance from the nuclei of white blood cells. He gave it the name "nuclein." Unlike proteins, nuclein had a high concentration of phosphorus. His colleagues tried to convince Miescher that nuclein was just another protein and phosphorus was a contaminant, but Miescher persisted in his belief that nuclein was another type of molecule. Eventually, chemical analysis revealed that chromosomes are made of both protein and Miescher's nuclein, which we now call deoxyribonucleic acid (DNA).

Of these two substances, which carries hereditary information? Almost everyone thought that proteins must be the hereditary material. Chemically, proteins are much more complicated than DNA, and everyone knew that the molecule of heredity had to be able to contain an extraordinary amount of information. It did not seem possible that a simple molecule like DNA could be the hereditary material.

Of course, now we know that they were wrong. The heredity material is not protein but DNA. But how did we discover this?

The Transforming Factor

The first chapter of the story occurred in 1928 in the laboratory of a man interested not in heredity or chromosomes but in vaccines. Frederick Griffith was trying to understand the differences between the strains of *Diplococcus pneumoniae* that cause pneumonia (virulent strains) and the strains that are harmless (nonvirulent). The virulent forms had polysaccharide capsules that gave them a smooth appearance. The nonvirulent bacteria had no capsules and were rough.

Griffith hoped that either heat-killed virulent strains or live nonvirulent strains could be used as a vaccine. He mixed heat-killed virulent and live nonvirulent strains together in hopes of making an effective vaccine. He injected the mixture into mice, and they died. How could dead bacteria be virulent?

He was able to retrieve virulent bacteria from blood samples taken from the dead mice. Something had been transferred from the dead virulent bacteria to the live nonvirulent bacteria. Griffith called this substance the "transforming factor." We now know the transforming factor was DNA, but Griffith didn't know that (Fig. 2.4).

The Griffith story also illustrates how science progresses. Often findings in one branch of science shed light on a different realm.

In 1943, O. T. Avery and his colleagues at the Rockefeller Institute purified the transforming factor and announced that it was DNA. Many scientists doubted this announcement and persisted in believing the hereditary material was made of protein. They doubted that DNA, with only four subunits, could carry enough information to turn a fertilized egg into a human being. Proteins, which have 20 subunits, can carry much more information.

In a way, it seems incredible now that scientists doubted that four subunits could provide sufficient information, when every day we transfer and process tremendous amounts of information via computer languages that have only two subunits!

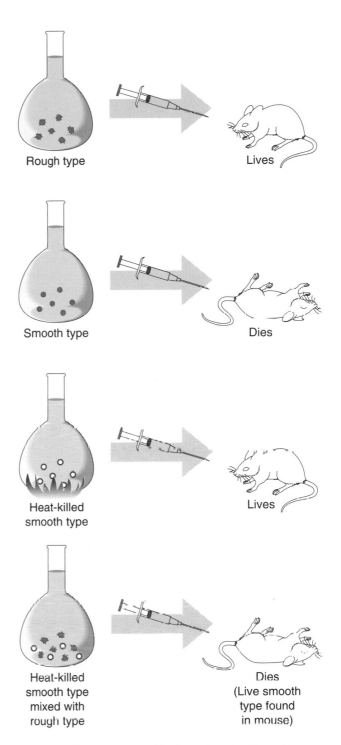

Figure 2.4 Discovery of the "transforming factor" by Griffith. The strain of pneumonia-causing bacteria (smooth) is virulent because of a gene that codes for a protective outer covering. A different strain (rough) does not cause pneumonia because it lacks the gene for the protective covering. When nonvirulent bacteria are mixed with heat-killed virulent bacteria, the nonvirulent bacteria are "transformed" into virulent bacteria. The gene for the protective outer layer is moved from the dead smooth bacteria to the living nonvirulent bacteria.

DNA or Protein?

The final answer to the DNA-or-protein argument was provided by definitive experiments conducted in 1952 by Alfred Hershey and Martha Chase. Because of work done on the atomic bomb in World War II, scientists had access to the radioactive substances that have proven invaluable in scientific investigation. These isotopes were crucial to the Hershey-Chase series of experiments on the molecular nature of the hereditary material.

By the 1950s, researchers had begun to use experimental organisms that were even simpler than bacteria, the bacteriophages. Bacteriophages are viruses that infect bacteria. Viruses consist of a coat protein with a small amount of genetic material, usually made of DNA, inside. They infect bacteria by injecting their genetic material into the bacterium while leaving the coat protein outside. Hershey and Chase exploited this bit of viral molecular biology to settle once and for all the question of the molecular nature of the genetic material.

Proteins and DNA differ chemically in many ways, but one simple chemical difference was critical to the Hershey-Chase experiment: DNA has phosphorus but not sulfur, and proteins have sulfur but not phosphorus.

Hershey and Chase grew viruses labeled with the radioactive isotopes of sulfur and phosphorus, ^{35}S and ^{32}P. These viruses were used to infect *Escherichia coli* cells. Then the solutions containing the infected bacteria and the viruses were agitated in a blender and centrifuged to separate the viral coats from the infected bacteria. The supernatant containing the viral coats was rich in ^{35}S, while the pellet of infected bacterial cells contained ^{32}P. These infected *E. coli* cells produced phage progeny that contained ^{32}P and no ^{35}S (Fig. 2.5).

Thus, the DNA versus protein debate was finally resolved. Because the phages injected their DNA into the bacteria while leaving their protein coats outside, Hershey and Chase concluded that DNA was the genetic material.

Again, note the recurring themes of scientific progress:

* Technological advances promote scientific discoveries.
* Findings in one branch of science shed light on others.

Discovery: the structure of DNA

As soon as the DNA versus protein debate was settled, another question was immediately born: What was the structure of the DNA molecule? Whatever the structure, scientists knew DNA's function depended on it. They also knew that whoever described the

Figure 2.5 The experiments of Alfred Hershey and Martha Chase. One group of viruses containing protein labeled with the radioactive isotope ^{35}S and a second group of viruses containing DNA labeled with the radioactive isotope ^{32}P infected bacterial cells by injecting their genetic material into the cell. Hershey and Chase separated the viral coats from the bacterial cells and found ^{35}S in the viral coats and ^{32}P within the bacterial cells. Viral progeny that resulted from the infection also contained radioactive ^{32}P. Hershey and Chase thus concluded that the genetic material of the virus was DNA and not protein.

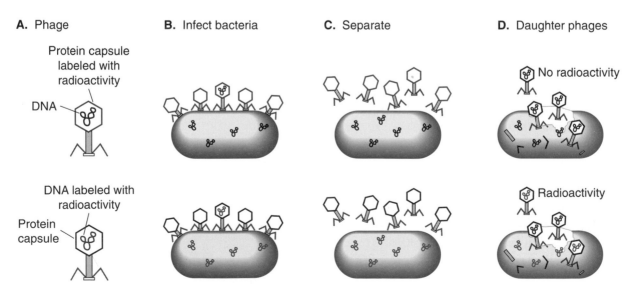

A. Phage   **B.** Infect bacteria   **C.** Separate   **D.** Daughter phages

structure would be guaranteed an important place in history. And so the race was on.

Before scientists proved conclusively that DNA is the genetic material, we actually knew a great deal about its biochemistry. We knew all of its components (phosphate, **deoxyribose,** and the four nitrogenous bases) and their molecular structures (see chapter 3). By 1930, we knew that each subunit (**nucleotide**) of DNA consists of one phosphate, one deoxyribose, and one nitrogenous base. Finally, in 1952, we discovered that the phosphate and deoxyribose moieties are linked together to form a chain and that the nitrogenous base is attached to this chain.

So what else was there to know? The three-dimensional structure. We could not deduce the three-dimensional structure from the information available, and we knew that the key to explaining DNA's function resided in its three-dimensional structure.

Findings from a number of branches of science coalesced and provided us with sufficient information to find this next piece of the puzzle.

- The physicists Maurice Wilkins and Rosalind Franklin provided data from X-ray diffraction analysis of DNA crystals.

- The chemist Erwin Chargaff discovered that DNA molecules always have equal amounts of **adenine** and **thymine** and equal amounts of **cytosine** and **guanine.**

- Another chemist, Linus Pauling, described the rules governing the formation of chemical bonds and published the details of a novel chemical structure certain proteins could assume, the alpha helix.

Watson, Crick, and the Double Helix

Frances Crick, a graduate student in physics at Cambridge, and James Watson, an American geneticist at Cambridge on a postdoctoral fellowship, both young and brash, were determined to be the ones who elucidated DNA's structure, and they did. Their success stemmed not only from their determination but also from a mutual distaste for sloppy thinking; their blunt, open, passionate style of communication; and a bit of luck.

They set out to discover DNA's three-dimensional structure by building a model. In 1953, when they happened upon the structure—a helix composed of two strands running in opposite directions with internal pairing between the bases, adenine, thymine, cytosine, and guanine—they knew they had it! They could

immediately see one DNA function, information replication, in its structure. Each strand could be used as a template for making another strand. In the brief, understated report in the journal *Nature* in which they described DNA's structure, Watson and Crick state:

> *This structure has novel features which are of biological interest. . . . It has not escaped our notice that the specific pairing we have postulated immediately suggests a possible copying mechanism for the genetic material.*

In his book *The Eighth Day of Creation,* Horace Freeland Judson eloquently describes their discovery in this way:

> *That morning, Watson and Crick knew, although still in mind only, the entire structure: It had emerged from the shadows of billions of years, absolute and simple, and was seen and understood for the first time.*

Ever since Watson and Crick discovered the structure of DNA, scientist after scientist has described DNA's structure as "beautiful." In reply, nonscientists usually ask why we see beauty in a molecule. DNA's beauty lies in its structural simplicity. It is a simple, straightforward molecule: simple in structure but astoundingly complex in its function.

The elegance of DNA has inspired awe in some nonscientists as well. A quote from Salvador Dali eloquently describes the feeling many scientists share with the painter regarding DNA:

> *And now the announcement of Watson and Crick about DNA. This, for me, is the real proof of the existence of God.*

In another report in *Nature* 2 months later, Watson and Crick provided insights into the second DNA function: the elaboration of the information it contains into a living organism:

> *It therefore seems likely that the precise sequence of the bases is the code which carries the genetic information.*

So Watson and Crick were able to deduce how DNA carried out its two functions, replication of the information it contains and elaboration of that information into an organism, by simply looking at its structure. A fundamental principle of architecture—"form follows function"—is also a fundamental truth in biology.

Theory: the central dogma

In 1957, Frances Crick proposed what was to become known as the central dogma describing the process of information translation:

DNA → RNA → protein

Information flows from DNA through RNA to proteins. The information in DNA is encoded in the linear sequences of nucleotides, just as the information on this page is encoded in linear sequences of letters. The information in DNA is translated into the language of proteins, which is composed of different "characters": amino acids in place of nucleotides. This translation is facilitated by a middleman, RNA, which, like DNA, is a nucleic acid composed of a sequence of nucleotides. Translating the information in DNA into a protein is somewhat like copying pages from a book and taking those sheets of paper to someone who could translate it into Japanese. (For more details, see chapter 3.)

Further studies supported the central dogma and led to the entrenched and very satisfying view of the gene as a stable, continuous segment of DNA whose linear nucleotide sequence corresponds to a linear sequence of amino acids. It was not long before this tidy picture of gene structure and function had to be revised, however. The more one learns about biological systems, the more one realizes how untidy they can be. If you look long enough, you can find exceptions to any rule or pattern in biology. Biological systems are too complex to always follow rules. This lack of predictability is a source of both joy and frustration for biologists.

Revising the Central Dogma

As we have learned more about gene structure and function, we have fine-tuned and elaborated our original concept of DNA translation. First, we found that the flow of information from DNA to RNA is not one way. Information can also flow from RNA to DNA. Then we discovered introns and exons. Often a portion of a gene's nucleotide sequence does not contain useful information, and this section is excised before DNA's information is translated into protein. The meaningful sequences of nucleotides in a stretch of DNA are known as an **exon;** the excised sequences are called **introns.**

We also found sequences of nucleotides that repeat themselves over and over and over, like a phonograph record that's stuck in a certain place (does anyone remember phonograph records?). Scientists labeled such sequences "junk DNA." Now, we're not so certain it's junk, but we still aren't sure why it's there.

Then we found that some nucleotide sequences in a gene don't stay put but hop around. They're known as transposable elements, **transposons,** or "jumping genes."

Perhaps most surprising and intriguing, we found that our perception of RNA as a passive middleman, relaying information between the "real" molecules, DNA and proteins, is not appropriate. RNA, like an enzyme, has the ability to catalyze chemical reactions. This discovery may help us explain the origin of life on Earth and, in particular, the evolution of nucleic acid and proteins.

Who knows what we will find next?

Understanding Genes

Because

- genes occupy a central place in the study of all biological sciences,
- the structure of the DNA molecule is so simple and yet so powerful,
- the mechanism of gene action is so awe inspiring,
- genetic variation is the key to the evolution of life on Earth,

biologists sometimes sound as if they are talking about a supernatural force when they discuss genes.

There's no doubt that genes are potent particles. If information is power, then genes are the powerhouses of the planet, for they contain the information of life. They are the repositories of the information that makes us who we are, that binds us to our species, and that separates us from other species. Genes forecast the future and reflect the past. Your genes contain information on your genealogical relationships and ethnic background and a chapter on the evolution of life on Earth.

Yes, they are powerful, but they are not omnipotent. We sometimes forget that fact, and it is important that we don't, particularly as we assess the applications and societal issues of gene-based biotechnologies.

Misunderstandings about genes

Biologists add to the misunderstanding surrounding genes by talking in shorthand and using insider's jargon that is easily misinterpreted by those not as familiar with the workings of genes.

Recall from chapter 1 that many of the fears and concerns surrounding biotechnology can be traced to public misunderstanding of the nature of genes. It is important that we do whatever we can to correct these misunderstandings and be more disciplined with the words and phrases we use to describe genes. Here are a few common misstatements made about genes and more accurate, albeit lengthier, explanations.

Genes make proteins.

Genes do not make proteins. Genes contain the information for making proteins. Each nucleotide sequence in a gene specifies the amino acid sequence in a protein. Proteins are made by other proteins (enzymes) that use the gene as a manufacturing guide. Proteins cannot manufacture proteins without this guide, but the guide is not sufficient for protein production.

Genes are on chromosomes.

Genes are not "on" chromosomes. A chromosome is composed of a very long molecule of DNA that is bound to proteins. A gene is a given length of that DNA molecule. The genes plus the bound proteins are the chromosome.

Genes replicate themselves.

Genes do not replicate themselves. Proteins (enzymes) replicate genes using the gene as a guide.

DNA is like a blueprint.

This is a very useful metaphor, up to a point. Then it, like all metaphors, becomes more constraining than useful.

DNA is like a blueprint because it contains information for building proteins just as architectural blueprints contain information for building a structure. But DNA is much more than that. DNA not only has information for building the structure of an organism but also contains information for making enzymes, hormones, receptor molecules, transport proteins, and antibodies.

In other words, to be like DNA, an architectural blueprint would have to contain instructions for

- *making* the tools and machines used to erect the building
- *fabricating* the electrical wires, polyvinylchloride pipes, dry wall, and insulation
- *securing* the money to fund the construction
- *designing* the precise details of heating and cooling units, security system and telecommunications networks, and even
- *creating* the workers, supervisors, attorneys, bankers, and real estate agents

and much, much more. That would be some blueprint!

Genes determine who we are.

Genes do not determine who we are. Genes *influence* who we are.

Envision all of the factors that could influence the process of constructing a building from the architectural blueprint to the appearance of the final product. A few immediately come to mind: choice of materials, worker competence, available funds. Because of these variables, nonidentical buildings could easily be constructed from the same blueprint. Now recall that much more information is included in a DNA "blueprint," providing many more opportunities for extraneous factors to shape the final product.

As we discussed in chapter 1, ignoring the complex realities of genetics—genes are only one factor that contribute to a trait (phenotype), many genes influence a single trait, and any one gene affects many traits—opens the door to simple-minded misuse of genetic information.

Understanding Evolution

In addition to being the repositories of information for synthesizing proteins, genes are the units of evolutionary change. Thus, a complete understanding of genes and genetics comes not only from learning about the molecular structure and function of genes but also from understanding and appreciating the role they play in evolution. After all, the study of evolution is the study of genetic variation over time and space.

Defining evolution

Evolution is simply the change in the frequencies of certain genes in a species' gene pool. Or, more specifically, it is the change in the relative proportion of alternative forms of a gene (alleles) in a population.

Evolution depends on two interrelated processes:

- the creation of genetic variation
- the selection of certain of these variants at the expense of others

The agents that create genetic variation are mutation and **recombination.** The primary agent that selects certain variants is called **natural selection.** Because the creation of genetic variation and the selection of certain variants are essential to evolution, we must look at each process in more detail if we are to understand evolution.

Creating genetic variation

As you might expect, because genetic variation drives evolution, nature has many mechanisms for generating genetic variation. All of these mechanisms are commonly grouped into two broad categories: mutation and recombination.

Types of Mutations

Biologists use the term "mutation" to refer to changes in the genetic information a cell carries. Two classes of mutations, based on the amount of genetic information (DNA) that is changed by the mutation, are usually recognized.

1. Changes in large segments of DNA molecules are termed **macromutations, macrolesions,** or **chromosomal aberrations.** Examples of macromutations include
    *   changes in the total *amount* of genetic information because of the loss (**deletion**) or addition (**duplication**) of a gene(s) or even entire chromosomes
    *   changes in the *positions* of the genes relative to one another without any change in the total amount of genetic information (**inversions** and **translocations**) (Fig. 2.6).

2. Changes in the sequence of nucleotides of single genes are termed **micromutations, microlesions,** or **point mutations,** because a relatively small amount of DNA is changed.

Effects of Mutations

What is the effect of a mutation on an organism? It depends on what has changed. Some mutations are beneficial; others are devastating; many are neutral. Whether it is harmless or devastating does not *necessarily* depend on whether or not it's a macro- or micromutation. Sometimes changing a single nucleotide has horrible effects on the phenotype. And yet, a complete doubling of genetic information has been very important in the evolution of many plants. In

Figure 2.6   Diagrams of the four types of chromosomal aberrations, or macromutations. The letters denote genetic loci.

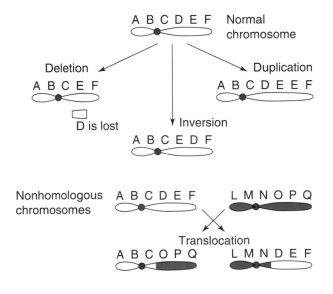

general, however, as you would expect, the larger the amount of genetic information that is changed, the more likely it is that the mutation will be harmful.

More information on the effects of mutations on protein structure and function and the resulting phenotypes is provided in chapter 3. Because the topic at hand is evolution, we focus here on mutations as sources of a genetic variation that is then fed into the evolutionary process.

From the point of view of the evolutionary biologist, the two types of mutation vary greatly because they affect the evolutionary process in different ways. Micromutations, or point mutations, create new forms of genes, or, as an evolutionary biologist would say, they can add new alleles to the existing gene pools if they are not harmful.

The effects of macromutations on the evolutionary process cannot be described in a single sentence because they vary in both type and significance according to the exact change that has occurred.

Deletions and duplications. When the total amount of genetic information changes, the evolutionary result may be profound, as in the immediate speciation that has sometimes followed chromosome number doubling, or it may be totally insignificant. For example, when a chromosome is lost or gained, the individual is often sterile, as is the case with XO and XXY humans. Although this condition is devastating to the individual, the mutation is irrelevant from an evolutionary perspective because the genetic change does not become incorporated into the gene pool.

Another type of macromutation, gene duplication, can be both directly and indirectly important in evolution. If the gene duplication results in increased production of a protein that contributes to the survival of the individual, it will be favored by natural selection. The indirect benefits of gene duplication result from a different scenario. If a single gene is represented by a number of copies, some of these duplicate genes are free to mutate into new allelic forms without the loss of the original gene's function.

Inversions and translocations. Macromutations that involve changing the positions of genetic loci relative to one another also vary in the effect they have on the evolutionary process. When a section of a chromosome breaks, rotates 180°, and then reinserts itself into the same chromosome (an inversion) or when nonhomologous chromosomes break and exchange pieces of chromosomes (translocation), the result is quite disruptive to normal meiotic chromosome pairing. Most gametes produced by individuals heterozygous for translocations do not contain a full complement of the genes involved in the translocation (Fig. 2.7). Consequently, macromutations that rearrange the position of genes on chromosomes are not a common source of genetic variation "usable" for driving evolutionary change. However, in certain cases, translocations and inversions have become established components of a species' gene pool. Because they deter appropriate pairing and crossing over in meiosis (see below), the genes in the inverted or translocated segment become tightly linked and are transmitted as a single unit. Sometimes these linked sets of genes are important evolutionarily because they represent specific gene combinations that are highly adaptive and confer a selective advantage.

Transposable Elements

A special form of mutation that occurs in both prokaryotes and eukaryotes involves transposable elements, or "jumping genes." These transposable elements, which are also called transposons, are short segments in the DNA that include a gene for a protein called a transposase. This protein causes the short piece of DNA to "jump" from its original location into a new, often random location somewhere else in the genetic material of the cell. Sometimes the jumping involves forming a new copy of the transposon; sometimes not (Fig. 2.8).

The insertion of a transposon at a new location can pose a problem: the transposon might land in the middle of an important gene, alter the nucleotide sequence within the gene, and prevent the synthesis of

Figure 2.7 Diagrams of the disruptive effect of a translocation on chromosome pairing in meiosis and of the gametes that result. The parental genotype possesses one normal version of each of the two chromosomes involved in the translocation event. The letters indicate genetic loci. Lowercase and uppercase letters denote different alleles for the same trait. Four of the six possible gametes derived from the parental genotype are inviable because they do not contain a complete complement of genes.

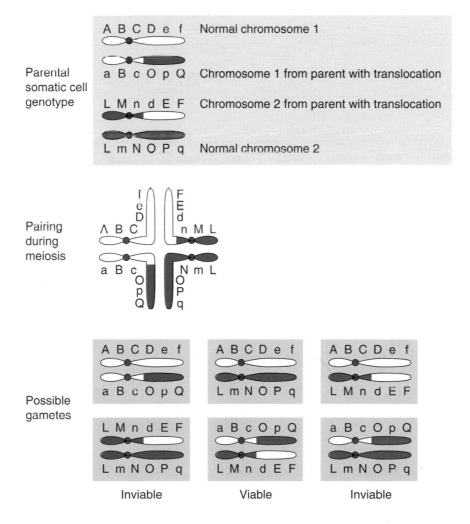

A. Transposons that move by **replicative transposition** are duplicated in the process of jumping, a copy-and-paste mechanism.

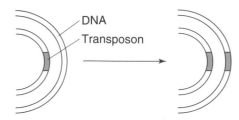

B. Nonreplicative transposons move in a cut-and-paste manner. No copy is made.

Figure 2.8   Replicative and nonreplicative transposons.

the protein. In 1992, scientists linked the occurrence of a genetic disease to a transposition event. Neither parent of the child with the genetic defect was a carrier of the gene for the disease. A detailed examination of the child's defective gene showed that a transposon had inserted into it, causing a mutation. To date, scientists have found transposons in plants, animals, yeasts, and bacteria.

Recombination

The second mechanism for generating genetic variation is recombination, the joining of genetic information from two sources to produce new genetic combinations. Recombination differs from mutation in that specific genes are not changed but simply reassorted. Exactly how the genetic material from two sources gets combined varies with the mode of reproduction of the organism.

Sexual reproduction and recombination. For almost all organisms that reproduce sexually, reproduction involves two processes:

- the production of gametes that differ genetically from the individual producing them
- the fusion of these gametes to create an individual that differs genetically from both parents

Recombination occurs during *both* processes. Let's start with the most straightforward example of recombination: fertilization.

Fertilization is the fusion of two gametes (eggs and sperm or pollen) into a single zygote. When the gametes, which contain genetic information, fuse, the resulting zygote's genetic information is a new combination of genetic information from the two parents. Thus, the offspring that develops from the zygote contains genetic information from two sources and is genetically novel: it is not genetically identical to either parent. All offspring of sexual reproduction, including you, are genetic recombinants.

Recombination also occurs during the production of these gametes through meiosis. Increased genetic variation in populations is the result of two associated meiotic events: the independent assortment and segregation of nonhomologous chromosomes and the crossing over that occurs between homologous chromosomes.

As a result of independent assortment and segregation of nonhomologous chromosomes in meiosis, a gamete differs genetically from the cell that gave rise to it by having one-half as much genetic material and a random assortment of maternal and paternal genetic material. So each gamete contains genetic mate-

Figure 2.9   Crossing over between homologous chromosomes occurs during gamete production. As a result, the gametes will differ from each other and the parental genotype.

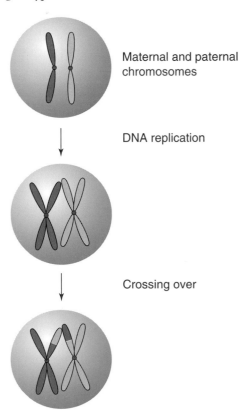

Maternal and paternal chromosomes

DNA replication

Crossing over

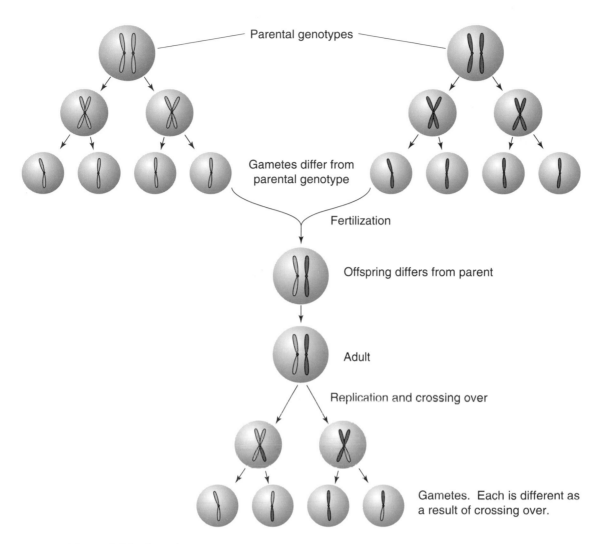

Parental genotypes

Gametes differ from
parental genotype

Fertilization

Offspring differs from parent

Adult

Replication and crossing over

Gametes. Each is different as
a result of crossing over.

Figure 2.10 Sexual reproduction, genetic variation, and recombination. Genetic variation is created during three stages of sexual reproduction. When gametes are produced, they differ genetically from the parental genotype because they have half the amount of DNA and a random assortment of paternal and maternal chromosomes. During fertilization, genetic material from two sources is combined, creating an offspring that differs genetically from both parents. Finally, during gamete production, crossing over between maternal and paternal (homologous) chromosomes occurs, creating "within-chromosome" genetic variation because the chromosome involved in the crossing over differs from its parent chromosome.

rial from two sources (maternal and paternal) in new combinations. The number of possible combinations of nonhomologous chromosomes that could be generated in the production of each human gamete is 2^{23}, or more than 8,000,000.

Independent assortment and segregation alone are a rich source of genetically variable gametes, but another recombination event occurs in meiosis. During the pairing of homologous chromosomes, genetic material from the paternal chromosome is exchanged with genetic material from the maternal chromosome through the process of crossing over (Fig. 2.9). The resulting chromosome is a recombinant: a hybrid containing genes of both maternal and paternal origins,

or a novel combination of genetic material from two sources. A graphical summary of the relationship between sexual reproduction and recombination is provided in Fig. 2.10.

Asexual reproduction and recombination. Many organisms do not reproduce sexually. In single-parent or asexual reproduction, a copy of the genome of the parent is passed along in its entirety to the offspring (Fig. 2.11). The offspring is therefore genetically identical to the parent organism and is often called a clone of the parent. The group of organisms resulting from repeated reproduction is called a clonal population, because every member of the population is genetically identical to every other member.

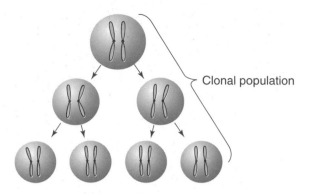

Figure 2.11 Asexual reproduction. Offspring are genetically identical to the parent. Genetic variation is created through mutation.

A typical example of asexual reproduction is the division of a bacterium to produce two bacterial cells. These two cells then divide, the four progeny divide, and so on until a large clonal population is generated. Bacteria are not the only organisms that reproduce asexually. Yeasts, protozoans, and algae can reproduce by simple fission or by sexual reproduction. A few species of animals reproduce asexually through parthenogenesis (in which females produce offspring without being fertilized by males). Many plants can propagate themselves asexually; humans take advantage of this to duplicate desirable plants by rooting leaves and cuttings.

You can see from comparing Fig. 2.10 and 2.11 that populations of asexually reproducing organisms would be far less genetically diverse than sexually reproducing populations if they had no other mechanism for generating genetically variable individuals. In fact, though, bacteria have three natural mechanisms by which genetic materials from two sources can be combined: conjugation, transformation, and transduction. Each of these processes is not only important in providing genetic variation for the evolution of bacteria but also plays an important role in human interac-

tions with microbes. Later in this book you will learn the importance of these three natural processes to research and technology and some of their medically important consequences.

Conjugation (Fig. 2.12) is a process by which one bacterium transmits a copy of some of its DNA directly to another bacterium. Conjugation is often called bacterial sex, which in a strict biological sense is the mixing of genetic information. In conjugation, the genetic mixing occurs in the recipient cell and does not, in and of itself, result in the production of offspring. So bacterial sex is not the same thing as sexual reproduction. After genetic exchange, a bacterium may undergo asexual reproduction as usual, and its new genetic information would be passed to its offspring. (See chapter 20, *Conjugative Transfer of Antibiotic Resistance in* Escherichia coli.)

In transformation (Fig. 2.13), a new combination of genetic material is created when cells take up DNA from the medium around them. In nature, some types of bacteria are easily transformed, while others are not. Scientists have learned to manipulate many kinds of cells to render them more susceptible to transformation in the laboratory. (See chapter 19, *Transformation of* Escherichia coli.)

In transduction (Fig. 2.14), a virus intermediary carries DNA from one bacterial cell to a second cell. During a typical virus infection, viral nucleic acid is replicated in the host cell, and viral coat proteins are made. Near the end of the infection cycle, the viral nucleic acid is packaged into the new virus coats, and new virus particles are released. For transduction to occur, some of the bacterial DNA must be packaged into a virus particle by mistake. This bacterial DNA is then injected into another bacterium and combines with the resident DNA. No infection occurs, since no virus genome was injected. (See chapter 21, *Transduction of an Antibiotic Resistance Gene.*)

Figure 2.12 Bacterial conjugation. Genetic material is exchanged between F⁺ and F⁻ cells.

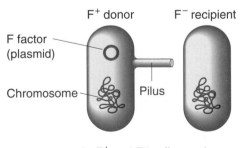

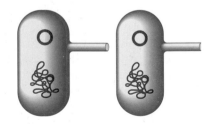

1. F⁺ and F⁻ cells are drawn together by the pilus of the fertile (F⁺) cell.

2. A copy of the fertility factor is transferred to the recipient cell.

3. Both cells are now F⁺.

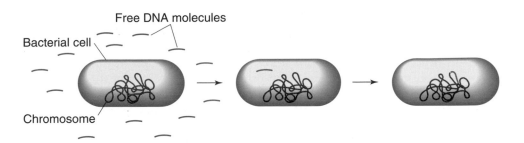

Figure 2.13 Transformation. A cell takes up free DNA from its environment, integrates it into its chromosome, and expresses the encoded products.

Selecting certain genetic variants

After genetically variable individuals are created by any of the methods just described, these individuals are exposed to the indifferent force of natural selection. Natural selection acts on individuals, but individuals do not evolve. Populations evolve. How does this happen?

What is "Survival of the Fittest"?

Natural selection is a measure of reproductive success but not necessarily of survival. Populations contain individuals that vary in their abilities to both survive and reproduce. This variation provides the raw material for evolutionary change. The fittest organisms are those that produce the most offspring that survive to reproductive maturity. If, however, an organism excels at skills required for survival but leaves relatively few offspring, it is not "fit."

On the other hand, traits that have questionable or even negative survival value persist in a population if they are beneficial in obtaining mates. Think of the peacock's tail. It is difficult to imagine how a 5-foot-long tail of brightly colored iridescent feathers could have positive survival value. But the peahens think those tails are gorgeous, so males with long, brightly colored tails are the "fittest."

Natural selection favors (selects for) those organisms that produce the most offspring and rejects (selects against) those that produce fewer offspring. The genes of the organisms that natural selection has favored are passed to the next generation in greater numbers than the genes of the organisms selected against. In this way, gene frequencies change over time, and this change is the process of biological evolution.

Observable Variation and Natural Selection

While populations are characterized by variation, only some of the individual variation that exists has a genetic basis. Some of the variation we observe in populations is exclusively phenotypic variation caused by environmental factors. Such variation may contribute to differences in survival and reproductive success and thus be subject to natural selection, but it is *not* the raw material for evolution. Traits that contribute to an increase in fitness will evolve only if they have a genetic basis.

Figure 2.14 Transduction. A virus serves as a conveyor of genetic material from one organism to another. In the example here, the organism is a bacterium, and the virus is a bacteriophage.

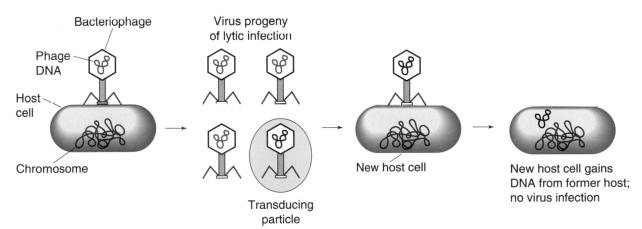

So, only some of the variation that exists in nature is significant in an evolutionary sense. Some isn't significant now and may or may not be significant in the future, depending on its degree of heritability and the selective pressures that act on that variation.

Natural Selection as a Honing Device

Some people have a somewhat sacred view of evolution. They have faith that through natural selection, evolution has honed all traits in all organisms so that there is a perfect fit between an organism and its environment. That opinion is appealing but naive. Most traits are present simply because they're not harmful—at that time in that environment. Given another environment or another time, they might be.

Even though evolution is a change in gene frequencies, natural selection acts on phenotypic traits. At any one time, only some of the traits of an organism are the focus of natural selection. Other traits are irrelevant to selection and just happen to get passed on with the favored genes. At that time, operating in that environment, natural selection is not "interested" in those traits. As a result, it is not molding a fit between those traits and the environment. Next year, however, those neutral traits might become the focus of natural selection, or a useful attribute one year may be a handicap, or linked to a handicap, the next.

For example, imagine that a mutation produces a color pattern that provides a lizard with better camouflage. If having that trait means the lizard lives longer and has more offspring, that gene will be passed to the next generation in greater numbers. It will drag with it other neutral genes that may have nothing to do with leaving more offspring.

On the other hand, the new gene and resulting color pattern might make the lizard more susceptible to predation. In that case, the gene would be selected against, as would the lizard's neutral genes and even other genes with positive survival value.

Now, imagine a situation in which the color pattern makes male lizards more susceptible to predation but is also the color pattern female lizards prefer when choosing mates. In nature, we often have selective pressures working in opposition to each other.

In addition, as described above, some of an organism's traits may not depend entirely, if at all, on its genetic makeup. Only traits that have a genetic basis are subject to natural selection. Therefore, if a trait is caused largely by environmental factors, it could not have been "honed" to perfection by natural selection.

Many people fail to realize that disorder, uncertainty, and, therefore, flexibility are common in biological systems, particularly at and above the level of the organism. That understanding may seem irrelevant, but in fact, in biotechnology it is critical for accurately analyzing the environmental issues of biotechnology.

Variation, chance, and change

The pivotal concepts of evolutionary biology are contained in the abbreviated description of evolution: variation, chance, and change. A true understanding of genes, biology, and even biotechnology is possible only if you comprehend and appreciate the role these factors play in driving evolution and shaping life on Earth. Therefore, they deserve a little more attention.

Variation

Genetic variation is the grist in the evolutionary mill. The great biologist Theodosius Dobzhansky, who was instrumental in elucidating the genetic basis of evolution, said that "nothing in biology makes sense except in the light of evolution." Therefore, understanding the nature of genetic variation is essential to understanding all of biology, not just evolution.

Evolutionary biologists have always studied genetic variation, even if they did not realize it. Charles Darwin, who did not know about genes and hereditary mechanisms, focused on phenotypic variation between species in an attempt to explain the diversity of species on Earth. He studied polymorphisms (different types) visible to the naked eye: birds' beaks, iguanas' feeding behavior, and the shapes of tortoises' shells. Later evolutionary biologists, understanding the relationship between genes and proteins, used electrophoresis to look at protein variation as a reflection of genetic variation. Today's evolutionary biologists are able to study genetic variation by looking directly at DNA variation through restriction fragment length polymorphism (RFLP) analysis.

Irrespective of the method used, when they ask questions about the amount and type of genetic variation, biologists are constantly changing focal lengths:

- How much variation is there within a population?
- How much variation is there between populations (within a species)?
- How much variation is there between species?

They also ask the flip sides of these questions: How much genetic similarity is there at each of these levels?

In studying biology, always keep in mind that both similarity and dissimilarity are informative and therefore important. All species are both genetically alike and genetically different from one another. Both genetic unity and genetic diversity characterize all of the Earth's species. Within a species, the same is true: all individuals of a species (or a population) are genetically alike and genetically diverse.

Because a continuum of genetic variation and similarities exists within a species and extends outside of a species to related species, determining where one species stops and another begins can be difficult and, at times, arbitrary. Where one chooses to demarcate a species boundary is flexible and, more often than you might expect, subject to lively debate. A large and dynamic branch of biology, taxonomy, is devoted to the question "Where should we draw the line?"

Because of that same genetic continuum, no one gene makes a fish a fish and an oak tree an oak tree any more than one gene that I have and you don't makes me more human than you.

Chance

Many people envision nature prior to human intervention as ideal and ordered: organisms perfectly adapted to unadulterated environments and living in balance with each other through time. These people operate on the tacit assumption that Mother Nature would keep a tidy house with a "place for everything and everything in its place" if humans would stop introducing mess and disorder. While this idyllic view of nature may be appealing, it bears no resemblance to reality. The planet and its species are constantly changing independently of any human presence.

Life on Earth is like a game of chance governed by constantly changing rules. An organism's opponent, natural selection, is also the game's unpredictable arbiter, rewriting the rule book on every play.

For biological organisms, every round of reproduction is like a new hand of cards. Chance determines

---

Scientific Models: Gateways or Barriers?

If you think that what happened to Mendel wouldn't happen today, think again. The story of another great geneticist, Barbara McClintock, parallels Mendel's so closely, it is eerie. Dr. McClintock is the scientist who discovered "jumping genes," or transposons, in the 1950s.

Like Mendel, she studied an organism so familiar it was almost dull—corn. Like Mendel, she was blessed with great powers of observation, an extraordinary intellect, and an uncommon amount of patience and persistence. For over 30 years she carried out the slow, laborious work of cross-pollinating Indian corn plants and, months later, observing the results of her experimental crosses. By simply using her uncanny ability to find patterns in the color of corn kernels, she developed a theory that called into question the fundamental nature of genes, and eventually revolutionized our understanding of genetics.

Unlike Mendel's audience, McClintock's was impressed—but not positively. In short, some thought a brilliant geneticist who was a member of the prestigious National Academy of Science had gotten a bit confused. What she proposed was heresy to the existing view of genes and chromosomes, and so they concluded that she was out of touch with reality, not that their model was wrong.

Often scientists forget that the models they develop are supposed to be useful tools for understanding

and interpreting the workings of the system under study. Instead they become wedded to the model and are unable to see and hear information that does not support it. Their model acts like a filter, letting expected information pass through to their brains but blocking the information they don't expect.

This surely was the case during the 30 years McClintock tried to explain to her colleagues what she saw in those corn kernels. Like Mendel, she saw patterns where everyone else saw disarray. She open mindedly asked what the patterns she saw revealed about the behavior of genes and chromosomes. She did not aim for a specific answer to support existing theory. Instead she sought true and deep understanding.

Other scientists were blind to the patterns and deaf to her reasoning. When they couldn't understand her message, they chose to ignore it because it did not fit with their preconceived notions of what was "supposed" to happen.

Eventually, after transposons were discovered in bacteria, yeast, and fruit flies, her fellow scientists could finally hear what she had to say. In 1981, Dr. McClintock received the MacArthur Laureate Award and the Lasker Award in Basic Medical Research; in 1982, Columbia University's Horwitz Prize; and in 1983, the Nobel Prize for her work on transposons.

---

which genes are combined. These new genotypes give rise to new phenotypes. The path from genotype to phenotype is susceptible to environmental influences, many of which are random occurrences.

These new phenotypes then enter the game uncertain of how they will fare, because some of the rules have changed since the last round. Certain phenotypes might be improvements—in that environment at that time. These phenotypes will be selected for, causing the proportion of certain cards in nature's genetic deck to shift before the next hand is drawn. That same phenotype in a different environment at another time might be the kiss of death.

Given such an uncertain scenario, what type of a team would you put together? You would probably hedge your bets and construct a team of diverse talents. Evolution has encouraged biological organisms to do just that. A species persists from one year to the next by producing genetically variable offspring, some of which manage to survive for another round.

And yet the greater the diversity within a species, the less perfectly adapted to its environment that species is. So organisms attempt to balance diversity—a measure of adaptability—and adaptedness. For any species, a constant and very dynamic tension exists between these two properties.

Change

The idea that environmental factors lead to evolutionary change by exerting selective pressure on organisms is familiar. These environmental forces acting on species are constantly changing, and a species' survival depends upon its ability to respond with evolutionary changes. When you think of environmental factors as selective agents, you probably envision aspects of the physical environment, like drought. And yet, other living organisms, the biological environment, make up the most important component of a species' environment. So if one species responds to physical environmental changes with evolutionary changes, its evolutionary change becomes an environmental change for a second species.

Also keep in mind that the relationship between organisms and their environment is not linear but circular. The activities of organisms continually alter the Earth's physical environment, often to the detriment of other species. Changes in the physical environment may then place new selective pressures on other species. Organism-driven environmental changes exert effects on individuals in the same species as well. Individuals in the same species compete for access to resources. Some win at others' expense.

Constant flux characterizes the natural world. Species are continually evolving—changing genetically—in response to the physical environment and to each other. It really is a jungle out there.

Understanding the details of evolutionary change may seem unrelated to biotechnology, but it is not. You will see that our underlying assumptions about evolution and the resulting state of the natural world drive some of the concerns people have about certain applications of biotechnology. These same assumptions encourage people to assign more power to biotechnology than it is due and to assume that the tasks in front of us are easier than they really are.

Selected Readings

Allen, Garland E. 1978. *Thomas Hunt Morgan: the Man and His Science.* Princeton University Press, Princeton, N.J.

Crick, Francis. 1988. *What Mad Pursuit: a Personal View of Scientific Discovery.* Basic Books, New York.

Dobzhansky, Theodosius. 1970. *Genetics of the Evolutionary Process.* Columbia University Press, New York.

Dobzhansky, Theodosius, Francisco Ayala, G. Ledyard Stebbins, and James W. Valentine. 1977. *Evolution.* W.H. Freeman & Co., San Francisco.

Futuyma, Douglas. 1979. *Evolutionary Biology.* Sinauer, Sunderland, Mass.

Iltis, Hugo. 1932. *The Life of Mendel.* Norton, New York.

Judson, Horace F. 1979. *The Eighth Day of Creation: Makers of the Revolution in Biology.* Simon & Schuster, New York.

Keller, Evelyn Fox. 1983. *A Feeling for the Organism: the Life and Work of Barbara McClintock.* W. H. Freeman & Co., New York.

Mayr, Ernest. 1969. *Populations, Species and Evolution.* Harvard University Press, Cambridge, Mass.

Stubbe, Hans. 1973. *History of Genetics.* MIT Press, Cambridge, Mass. (Translated by T. Waters.)

Watson, James. 1968. *The Double Helix.* Atheneum, New York.

Watson, James, and Francis Crick. 1953. Molecular structure of nucleic acids: a structure for deoxyribonucleic acid. *Nature* (London) 171:737.

Watson, James, and Francis Crick. 1953. Genetical implications of the structure of deoxyribonucleic acid. *Nature* (London) 171:964.

Watson, James, and John Tooze. 1981. *The DNA Story: a Documentary History of Gene Cloning.* W.H. Freeman & Co., San Francisco.

An Overview of Molecular Biology

Gene Structure and Function

A hallmark of living systems is that they reproduce themselves. For many years, one of the greatest mysteries of science was the puzzle of how the tiniest seed or fertilized egg could contain all the information needed for the development of an entire organism. Classical geneticists deduced that individual traits are determined by invisible information units they called genes. They presumed that each cell duplicated its genes before dividing so that each daughter cell could receive a complete set. They also presumed that the genes present in sperm and egg cells would carry the genetic information to the next generation. Then, somehow, those genes would direct the development of a new individual.

It was clear that the genetic material must be capable of two extremely important functions. First, it must be in a form that can be copied extremely accurately so that correct information is transmitted from cell to cell and generation to generation. Second, its information must somehow be translated into a living organism. What molecule could possibly fulfill these two complex and critical requirements? Scientists naturally assumed that the genetic material must be a very complicated molecule or molecules.

Today we know that DNA is the genetic material and that all life on Earth shares this wonderful molecule as its genetic material. However, at first, that discovery was so astonishing that many scientists refused to believe it. Why? Because DNA is such a simple molecule. How could such a simple molecule be responsible for the extraordinary diversity of life forms on this planet? The determination of the structure of DNA and the cracking of its genetic code are surely among the towering scientific achievements of our century. The elegant simplicity of DNA structure and the beautiful efficiency with which that structure provides for the two essential functions of the molecule can inspire both the poet and the engineer within each of us.

Structure of DNA

The genetic material of all life on this planet is made of only six components. These components are a sugar molecule (**deoxyribose**), a phosphate group, and four different nitrogen-containing bases: **adenine, guanine, cytosine,** and **thymine.** The essential building block of the DNA molecule is called a **nucleotide** or, more precisely, a **deoxynucleotide.** A deoxynucleotide consists of a deoxyribose molecule with a phosphate attached at one place and one of the four bases attached at another (Fig. 3.1). The carbon atoms of the deoxyribose sugar portion of a nucleotide are always numbered in the same way. The base is always attached to carbon 1, and the phosphate group is always attached to carbon 5.

In a DNA molecule, thousands or millions of these nucleotides are strung together in a chain by connecting the phosphate group on the number 5 carbon of one deoxyribose molecule to the number 3 carbon of a second deoxyribose molecule (a free water molecule is created in this process). Figure 3.2 shows an example in which three nucleotides are connected. The linkages formed between the deoxyribose molecules via the phosphate bridge are called **phosphodiester bonds.** Because the nucleotides are held together by bonds between their sugar and phosphate entities, DNA is often said to have a sugar-phosphate backbone. Notice that the ends of the molecule in Fig. 3.2 are labeled 5′ and 3′ for the carbon atoms that would form the next links in the chain at either end.

The sugar-phosphate backbone of DNA is an important structural element, but all of the information is contained in the four bases. The key to the transmission of genetic information lies in a characteristic of these bases: adenine and thymine together form a stable chemical pair, and cytosine and guanine form a second stable pair. The pairs are formed through weak chemical interactions called **hydrogen bonds.** These two pairs, adenine-thymine and cytosine-guanine, are called **complementary base pairs** (Fig.

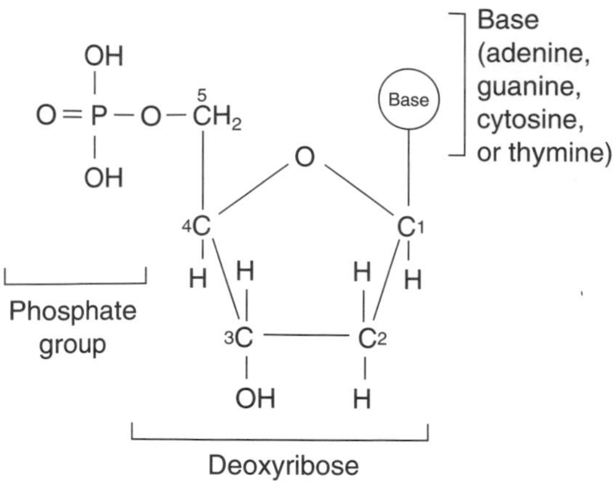

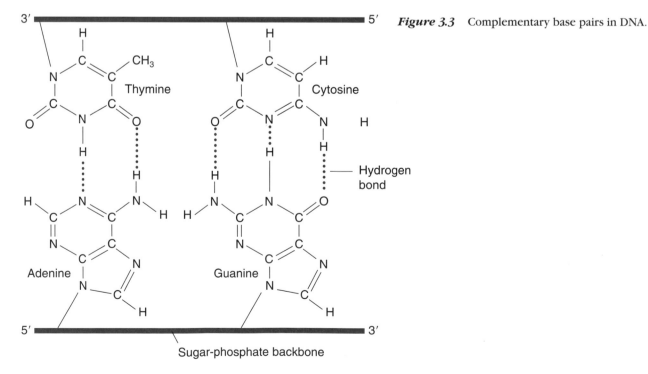

Figure 3.1 The nucleotide. Carbon atoms of the deoxyribose sugar portion are numbered according to chemical convention.

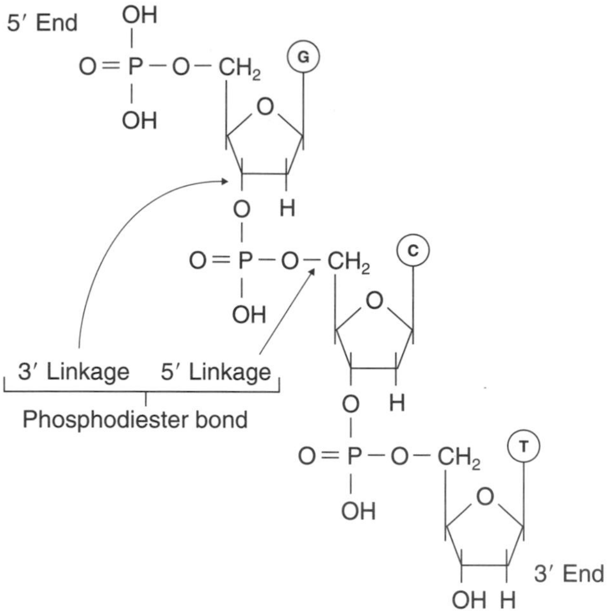

Figure 3.2 A trinucleotide.

3.3). In a DNA molecule, two sugar-phosphate backbones lie side by side, one arranged from the 5' end to the 3' end, and the opposite strand arranged from the 3' end to the 5' end. The bases attached to one strand are paired with their partner bases attached to the opposite strand (Fig. 3.3). Thus, the order of the specific bases on one strand is perfectly reflected in the order of the complementary bases on the other strand. Knowing the sequence of bases on one strand allows us to deduce the base sequence on the complementary strand. (See the activity *Constructing a Paper Helix.*)

DNA is often drawn as a flat molecule because that shape is easy both to draw and to look at. In reality,

each of the two sugar-phosphate backbones is wrapped around the other in a conformation called a double helix. The base pairs are on the inside of the helix, like the rungs of a ladder. For ease of representation and viewing, DNA can be presented by using a model in which each backbone is represented as a thin ribbon and the base pairs are represented schematically (Fig. 3.4).

Figure 3.3 Complementary base pairs in DNA.

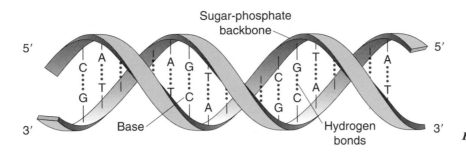

Figure 3.4   Ribbon model of DNA.

DNA function: faithful replication

The structure of DNA immediately suggests how DNA carries out the first critical function of genetic material: faithful replication. You can see that either of the two strands of DNA can be used as a template, or pattern, to reproduce the opposite strand by using the rules of complementary base pairing. When a cell is ready to replicate its genetic material, the two opposite strands are gradually "unzipped" to expose the individual bases. Each exposed strand is used as a template to form two new strands (Fig. 3.5A). The result is two daughter DNA molecules, each composed of one parental strand and one newly synthesized strand, and each identical to the parent DNA (Fig. 3.5B).

Figure 3.5   DNA replication. (A) Base pairing between an incoming nucleotide and the template strand of DNA guides the formation of a new daughter strand with a complementary base sequence. (B) In each round of DNA replication, each of the two DNA strands is used as a template for the synthesis of a new complementary strand, resulting in two daughter molecules, each with one "new" and one "old" strand.

A   Daughter strand   Template strand

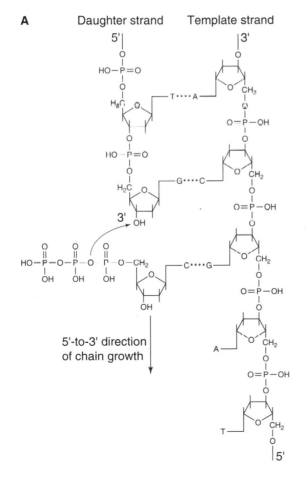

B   Parental DNA double helix

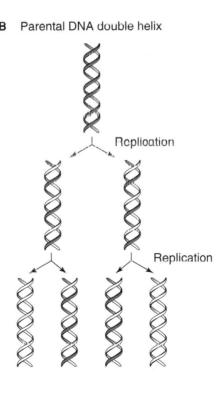

How does DNA replication occur in a cell? DNA is duplicated by enzymes, the protein workhorses of cells. Special cellular enzymes work together to unzip the DNA double helix, capture free nucleotides, pair the correct new nucleotide with the template base, and make the new bonds of the growing sugar-phosphate backbone. The central player of this protein team is called DNA polymerase. It is the enzyme that actually makes the correct base pairs and forms the new phosphodiester bonds. Finally, some of the DNA replication enzymes "proofread" the new DNA strand, checking for errors in base pairing and correcting any errors they find. These diligent enzymes ensure that very few errors occur during DNA replication so that genetic information is transmitted correctly. (See the activity *DNA Replication.*)

DNA function: information transmission

Although the structure of DNA immediately suggests how the molecule can fulfill the requirement for faithful transmission through duplication, it is not so obvious how such a simple molecule can determine the development of creatures as complex and varied as a blue whale or a rose. To understand how the structure of DNA elegantly fulfills this requirement, it is necessary to think about what makes a whale a whale or a rose a rose.

What does make a whale a whale? The answer is, its proteins. Just as proteins carry out the complicated task of duplicating the whale's DNA, other proteins carry out nearly every other function necessary to whale life. Structural proteins form the bricks and mortar of its skin, muscles, organs, and tissues. Transport proteins carry oxygen, nutrients, hormones, and other important molecules throughout its body and around its cells. Protein receptors embedded in cell surfaces bind with great specificity to the whale's hormones, enabling the whale to grow and develop properly. Proteins of the whale's immune system defend it from infection. Protein catalysts (enzymes) digest the whale's foods, synthesize fats for its blubber, replicate its DNA for transmission to baby whales, and carry out all the other metabolic tasks necessary for whale cell life. Thus, proteins provide the structure and carry out the functions of the whale. The same is true for a rose, a fruit fly, a bacterium, a human being, and every other life form on Earth.

So for DNA to dictate the development of organisms, the information in DNA must somehow be converted into proteins, the "stuff" of organisms. How does this conversion occur? To answer this question we must know a bit more about the makeup of proteins.

Proteins

So what exactly is a protein? A protein is a chain of amino acids. Amino acids are small organic molecules made mostly of carbon, hydrogen, oxygen, and nitrogen. Proteins are made from a pool of 20 different amino acids by connecting a few or thousands of these amino acids in various orders to form chains. For a simplistic analogy, think of making chains of

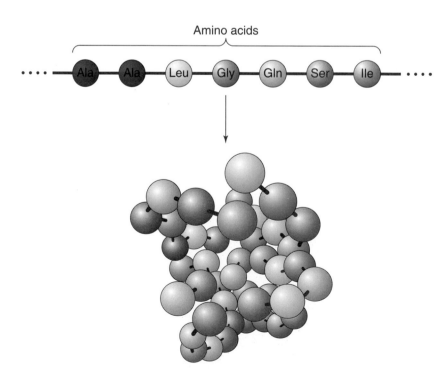

Figure 3.6 A protein is a chain of amino acids (represented by beads) that folds into a specific three-dimensional shape. The three-letter abbreviations on the beads are standard for specific amino acids (see Fig. 3.23).

pop-beads from an assortment of colored beads. But a protein is more than a straight chain of beads. You must also imagine that the chain is folded and coiled into a specific three-dimensional shape, because proteins' amino acid chains are folded and coiled into specific three-dimensional shapes (Fig. 3.6).

What enables a protein to perform its function? A protein's function is made possible by its unique three-dimensional structure. For an example, let's consider a single receptor protein embedded in the outer membrane of a cell. Our imaginary receptor protein looks like an irregularly shaped glob with interesting nooks and crannies. The shapes of the nooks and crannies are absolutely critical to the protein's function: one cranny on the outside of the cell membrane is the place where a growth hormone molecule must fit exactly to signal the cell to grow. Other nooks and crannies on the inside are sites that fit precisely to other molecules for communication with the rest of the cell. Through interaction at these sites, the receptor protein can tell the cell that the hormone signal has arrived.

Our imaginary receptor illustrates a crucial point about protein function. *A protein depends on its ability to fit to or bind to other molecules (sometimes other proteins) to carry out its functions. That ability is determined by its three-dimensional structure.*

What determines the three-dimensional structure of a protein? When amino acids are assembled into a protein chain, that chain immediately folds back upon itself to assume the most "comfortable" or energetically stable shape. The most energetically stable shape is determined by the interactions of the individual amino acids that make up the protein. Therefore, the identities of the component amino acids and the order in which they occur in the chain govern the final three-dimensional structure of the protein. The order of the amino acids in the chain is thus extremely important to the function of the protein. As you can imagine, the possibilities for constructing different and unique protein chains are almost limitless. (Imagine how many unique chains of pop-beads you could make by using 20 different colors of beads.) This variety is fortunate, considering the many and varied functions that proteins must perform. In fact, the forms and functions of proteins are so central to molecular biology that the second portion of this chapter is devoted to them.

But for now let us return to the question of how DNA controls the development of an organism. We have seen that an organism is the sum of its proteins. We have also seen that a protein's function depends upon its three-dimensional structure. Furthermore, its structure depends upon the sequence of amino acids in the protein chain. *DNA determines the characteristics of an organism because it determines the amino acid sequences of all the proteins in that organism.*

How does DNA determine an amino acid sequence? DNA contains a **genetic code** for amino acids in which each amino acid is represented by a sequence of three DNA bases (Table 3.1). These triplets of bases are called **codons.** The order of the codons in a DNA sequence is reflected in the order of the amino acids assembled in a protein chain (Fig. 3.7). The complete stretch of DNA needed to determine the amino acid sequence of a single protein is a **gene,** the

Table 3.1 The genetic code

First base in DNA triplet	A				G				C				T			
Second base in DNA triplet	A	G	C	T	A	G	C	T	A	G	C	T	A	G	C	T
Choices for third base in DNA triplet	A,G / C,T	A,G / C,T	A,C,G,T	A,C,G,T	A,G / C,T	A,C,G,T	A,C,G,T	A,C,G,T	A,G / C,T	A,C,G,T	A,C,G,T	A,C,G,T	A,G / C,T	A,G / C,T	A,C,G,T	A,C / T
Amino acid encoded	Lysine / Asparagine	Arginine / Serine	Threonine	Isoleucine / Methionine / Isoleucine	Glutamic acid / Aspartic acid	Glycine	Alanine	Valine	Glutamine / Histidine	Arginine	Proline	Leucine	Stop / Tyrosine / Stop	Tryptophan / Cysteine	Serine	Leucine / Phenylalanine

3. An Overview of Molecular Biology • 57

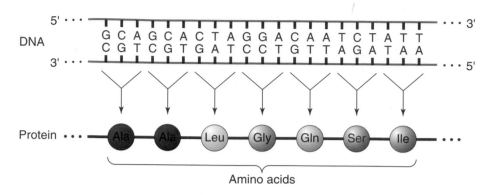

Figure 3.7 The base sequence of DNA determines the amino acid sequence of proteins.

unit of heredity defined by the classical geneticists. The complete set of genes in an organism is called its **genome.**

Protein synthesis

The process by which proteins are produced from the genetic code has several steps. DNA is essentially a passive repository of information, rather like a blueprint. The "action" of making a protein occurs at special sites in the cell called **ribosomes.** Therefore, the first step in protein synthesis is to relay the information from the DNA to the ribosomes. To accomplish this step, cellular enzymes synthesize a working copy of a gene to carry its genetic code to the ribosomes. This working copy is called **messenger RNA (mRNA)** (RNA is a close molecular relative of DNA). mRNA carries the genetic code for a protein to the ri-

bosomes. In the second step of protein synthesis, the codons in the mRNA must be matched to the correct amino acids. This step is carried out by a second type of RNA called **transfer RNA** (tRNA). Finally, the amino acids must be linked together to make a protein chain. The ribosome (which is made of proteins and RNA) performs this function. When the protein chain is complete, a genetic "stop sign" tells the ribosome to release the new protein into the cell.

RNA

Protein synthesis therefore requires a second type of nucleic acid molecule: RNA. Like DNA, RNA is made up of nucleotides composed of a sugar, a phosphate, and one of four different organic bases (Fig. 3.1). However, there are three important differences between DNA and RNA, two of them chemical and one of them structural. The chemical differences are that

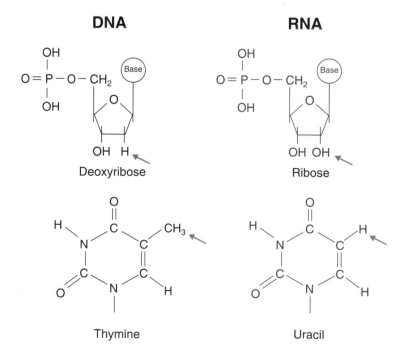

DNA   **RNA**

Deoxyribose   Ribose

Thymine   Uracil

Figure 3.8 Chemical differences between DNA and RNA.

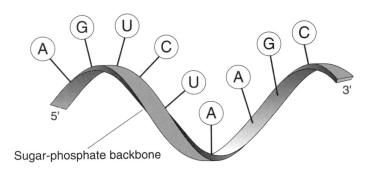

Sugar-phosphate backbone

Figure 3.9 A single-stranded RNA molecule.

(1) instead of the sugar deoxyribose, RNA contains the sugar **ribose** (hence the name *ribo*nucleic acid), and (2) instead of the base thymine, RNA contains the base **uracil** (Fig. 3.8). The sugar-phosphate backbone of RNA is linked together like the DNA backbone, and the bases are attached to the number 1 carbon, as in DNA. The important structural difference is that although RNA bases can also form complementary pairs, RNA is usually composed of only a single strand of sugar-phosphate backbone and bases. It does not have the base-paired double helix structure of DNA (Fig. 3.9), although it is capable of pairing with other single strands of DNA or RNA. As we shall see below, the single-stranded structure of RNA is ideally suited to its task of transferring information.

Synthesis of mRNA

The first step of protein synthesis is to make mRNA. This process resembles DNA replication in many ways. First, the DNA double helix must be unzipped to reveal the information-containing bases. Then, complementary nucleotides (that contain the sugar ribose, since we are making RNA) are paired with the exposed bases. During the synthesis of RNA, the base uracil substitutes for thymine and pairs with adenine. Phosphodiester bonds are made between the nucleotides, and the new mRNA contains a base sequence that is exactly complementary to the template DNA strand. The process of using a DNA template to create a complementary mRNA molecule is called **transcription** (Fig. 3.10).

Figure 3.10 Transcription. (A) Base pairing between an incoming ribonucleotide and the DNA template guides the formation of a complementary mRNA molecule. The DNA template closes behind the RNA synthesis site, releasing the new RNA molecule. (B) In transcription, a single DNA strand is used as a template. The RNA transcript is released, leaving the DNA molecule intact.

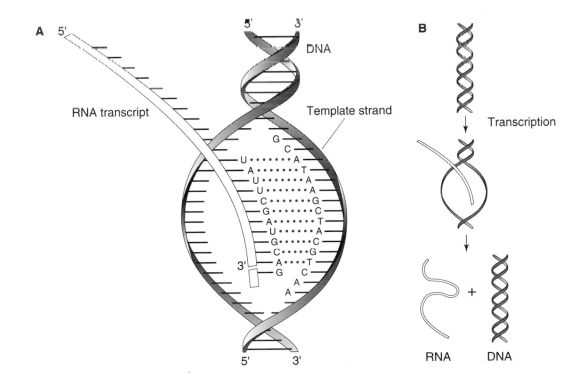

There are two major differences in transcription and DNA replication (compare Fig. 3.5 and Fig. 3.10). In DNA replication, both strands are used as templates to generate two new strands for two new helices. In RNA synthesis, only one DNA strand is used as a template, and only a single RNA strand is made. The second difference is that the new mRNA molecule is released from the DNA template as it is made. The DNA double helix "zips back up" as the mRNA is released. Newly synthesized DNA remains part of a new DNA helix, paired with its parent strand.

How is mRNA actually synthesized in the cell? By now you know at least part of the answer: enzymes. The RNA-synthesizing enzyme (**RNA polymerase**) has an interesting task. Not only must it select the correct complementary nucleotides and link them together (as does its counterpart, DNA polymerase), but it must also decide where a gene is. A DNA helix can contain thousands or millions of base pairs. The RNA polymerase must determine exactly where to start and stop synthesizing RNA so that it will transcribe a complete gene.

How does RNA polymerase know where to start making mRNA? The answer is that special genetic "traffic signals" are built into the DNA base sequence. One very important traffic signal is called a **promoter.** A promoter is a special sequence of DNA bases that tells the RNA polymerase to start synthesizing RNA. As you might imagine, other signals tell RNA polymerase to stop synthesizing RNA and leave the DNA template. These signals are called **terminators.** (See the lesson *From Genes to Proteins.*)

Using mRNA To Make a Protein

After transcription is complete, the mRNA moves to the ribosome, the site of protein synthesis. The ribosome recognizes the mRNA and holds it in proper alignment for its codons to be read correctly. Look back at Fig. 3.7 and think about this for a minute. The task at hand is to translate the DNA base code in the mRNA to amino acids, as shown in the figure. You can translate a DNA base code to amino acids by looking at the genetic code table (Table 3.1). How could a cell do this with molecules? What if you had a molecule whose one end fit exactly to one and only one codon and whose other end was connected to the correct amino acid? If you had one of these molecules for each of the codons in Table 3.1, you would have a molecular "key" for translating codons to amino acids. This is what cells do—your cells, bacterial cells, lizard cells, and the mold cells on the old sandwich in your refrigerator. The molecules that make up the molecular key are another type of RNA: tRNA molecules.

tRNA molecules are folded in on themselves to resemble cloverleafs. At the tip of one of the lobes is a sequence of three bases called an **anticodon** (Fig. 3.11). This anticodon pairs exactly with one of the codons on the single-stranded mRNA, using the rules of complementary base pairing. At the other end of the tRNA molecule is an amino acid. Each different tRNA is connected to the right amino acid for its anticodon. How does that happen? Enzymes, again. Every cell contains a host of exquisitely specific enzymes whose job it is to recognize individual tRNAs and attach the correct amino acid to them, so that each type of tRNA has a specific amino acid attached to it. The result is that when the anticodon on the tRNA pairs with its codon on the mRNA, the correct amino acid is brought to the ribosome.

The ribosome holds the mRNA molecule so the tRNAs pair with their complementary codons one at a time, in order. As the tRNAs bring in the correct amino acids, the ribosomes link the amino acids into a growing protein chain. Once an amino acid has been linked to the chain, the tRNA molecule is separated and released from the mRNA-ribosome complex (Fig. 3.12). This process, in which the DNA base sequence is translated to a protein amino acid sequence, is called (amazingly enough) **translation.**

Protein synthesis is a complicated process involving a variety of interactions between enzymes and RNA molecules. How does a ribosome distinguish an mRNA molecule from other RNA molecules in the cell? The answer is that every mRNA molecule con-

Figure 3.11 A tRNA molecule. Complementary base pairing between different portions of the tRNA molecule maintains its shape.

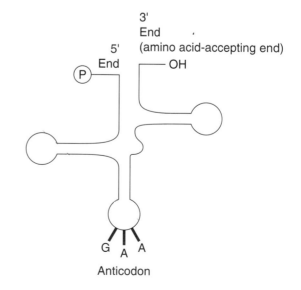

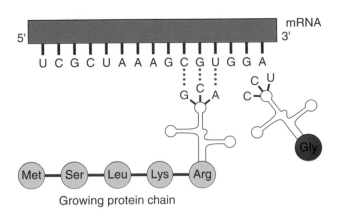

Figure 3.12 Translation. Complementary base pairing between the anticodons of incoming tRNA molecules and the codons of the mRNA guides the formation of the amino acid chain.

tains more than just the codons needed to make a protein. It also contains traffic signals for the ribosome.

Located on the mRNA before the coding sequence begins is a recognition sequence for the ribosome that is rather analogous to the promoter for RNA polymerase. It is followed by an **initiation codon** (usually AUG), where protein synthesis actually begins. At the end of the coding region, there is a stop signal (a **stop codon;** see Table 3.1) that tells the ribosome to release the mRNA. There are no tRNAs that pair with stop codons; that is what tells the ribosome to stop protein synthesis. Since the recognition sequence, initiation codon, and stop codon are present in the base sequence of the mRNA, they must also be encoded in the original DNA template that was transcribed to make the mRNA. (See the lesson *From Genes to Proteins.*)

Figure 3.13 Major genetic traffic signals in bacteria. These signals tell RNA polymerase where to begin and end transcription, enable the ribosome to recognize mRNA, and direct the ribosome to start and stop protein synthesis. RRE, ribosome recognition element.

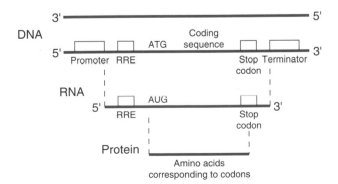

A summary of the major genetic traffic signals is presented in Fig. 3.13. The DNA molecule is a marvelous storehouse of complicated information. It contains not only the blueprints for the amino acid sequences of every protein an organism makes but also the traffic signals that direct the cell to interpret the information properly. As we shall see below, DNA even contains signals that allow a cell to regulate the synthesis of its proteins to meet the needs of its environment.

Although all cells on Earth use the same genetic code and synthesize proteins via mRNA and tRNA, there are important differences between broad categories of cells with respect to the flow of information from genes to proteins. These broad categories of cells are **eukaryotic** cells and **prokaryotic** cells.

Eukaryotic and prokaryotic cells

Eukaryotic cells are the cells present in most organisms with which we are familiar, such as fungi, plants, and animals. These organisms are also called **eukaryotes.** Their genetic material is located in an organized nucleus enclosed in a nuclear membrane. Eukaryotic cells also contain other membrane-bound organelles such as mitochondria, chloroplasts, lysosomes, and endoplasmic reticula. While most plants, animals, and fungi are multicellular organisms that may contain many specialized tissues and cell types, there are also single-celled eukaryotes. Examples of single-celled eukaryotes include yeasts, protozoans, and green algae.

The cells of **prokaryotes** have a much simpler structure. They lack membrane-bound organelles (including all of those listed above) and an organized nucleus. In prokaryotic cells, the genetic material is located in the cytoplasm. The two major groups of prokaryotes are eubacteria and archaea. Eubacteria are the familiar bacteria and blue-green algae (sometimes called cyanobacteria). Archaea (also called archaebacteria) are a group of prokaryotic organisms that inhabit extreme environments. They are genetically and biochemically similar to one another but are as different from eubacteria as the eukaryotes are, though they also use the same genetic code. Archaea are interesting in their own right but are generally not considered in this discussion. When we talk about prokaryotes or bacteria in this book, we are referring to eubacteria. Both eubacteria and archaea are single-celled organisms. There are no true multicellular prokaryotes.

Because bacteria grow rapidly, are easy to grow in large quantities, and have much smaller, more easily manipulated genomes than do eukaryotes, they are a favorite tool of biotechnologists. One of the major achievements of biotechnology has been to use bacteria to produce large quantities of proteins (such as human growth hormone) that are normally made in small quantities in eukaryotic organisms. This achievement is possible because the genetic code used in prokaryotes and eukaryotes is the same.

Differences in the Molecular Biology of Prokaryotes and Eukaryotes

Although their genetic codes are the same, the genetic traffic signals that direct the processing of the coded information are different in prokaryotes and eukaryotes. For example, promoters and ribosome recognition sequences in bacteria and eukaryotic cells differ greatly. In addition, prokaryotic genes usually exist as a continuous sequence on a DNA strand, and several are often transcribed from the same promoter. In contrast, eukaryotic genes are usually transcribed one at a time and are often encoded in many small pieces separated by stretches of noncoding DNA called **introns.**

When eukaryotic cells transcribe one of these split genes, a very long precursor RNA that includes all of the introns and the coding regions (**exons**) is synthesized first. Next, eukaryotic cells edit the RNA. In a process called **splicing,** the intron sequences are selectively cut out, and the exons are pieced together to make a functional mRNA (Fig. 3.14). As you might imagine, the mRNA contains coded signals at the beginning and end of each intron that direct the cell to remove the intron. After the introns have been removed and the exons have been spliced together, the functional mRNA moves to the ribosome for translation. Splicing does not occur in eubacteria, and bacteria lack the enzymes necessary for splicing eukaryotic RNA. (See the activity *From Genes to Proteins.*)

These differences in promoters, ribosome recognition sequences, and splicing not only have fascinated researchers but also have made life interesting for biotechnologists seeking to transfer information from eukaryotes to bacteria. As you might guess, simply transferring a gene from a mammal to a bacterium usually does not result in the production of a functional protein. In addition to the genetic information, biotechnologists must also provide the correct processing signals to the new host cell and "presplice" the gene if it contains introns. Many standard procedures have been developed to simplify these processes (see chapter 4).

Regulating gene expression

Cells must regulate the synthesis of their proteins in order to respond to environmental conditions. For example, most bacteria have genes encoding enzymes capable of breaking down quite a variety of sugars for energy. However, synthesizing these enzymes would be a waste of the cell's energy if the sugars were not available to the cell. So most bacteria synthesize an enzyme that breaks down a particular sugar only if that sugar is present in its environment.

Conversely, most bacteria encode enzymes capable of synthesizing all of the amino acids the bacteria need to make their proteins. Yet synthesizing these enzymes would be wasteful if the needed amino acids were already present in the environment, so most bacteria produce the amino acid-synthesizing enzymes only if amino acids are not available to them. Finally, many organisms, even bacteria, undergo some form of change during their life cycles. Plants and animals develop from a single fertilized egg. Some bacteria respond to adverse environments by forming spores. These structural changes also require controlled gene expression.

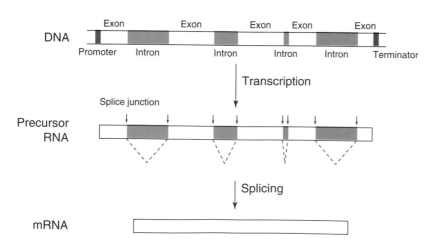

Figure 3.14 Splicing of precursor RNA to create mRNA.

How is gene regulation achieved? There are many mechanisms. The best-understood ones involve regulation of transcription, the synthesis of mRNA. The rate of degradation of specific mRNAs can also be controlled. The translation rate of mRNA molecules is often regulated. Any step between DNA and protein is a potential regulatory target. However, the most common model for gene regulation involves regulation of transcription.

Transcriptional Regulation through Repression

In addition to promoters, many genes contain sites to which regulatory proteins can bind. These regulatory sites are often very near the promoter. In the most typical scenario in prokaryotes, the binding of the regulatory protein prevents transcription of the gene either by blocking access to the promoter or by preventing progression of RNA polymerase along the gene. Regulatory proteins that bind to DNA and prevent transcription are called **repressors.**

The sugar lactose can be used by the bacterium *Escherichia coli* (and many other bacteria) as an energy source. *E. coli*'s lactose utilization genes are lined up in a row along its chromosome and are transcribed from a single promoter into one long mRNA. Several proteins are translated from this long message. The collection of lactose utilization genes is called the *lac* **operon.**

When lactose is present in the environment, *E. coli* synthesizes lactose utilization proteins from the *lac* operon. The lactose utilization proteins allow *E. coli* to derive energy from the sugar. When no lactose is present, these proteins are not synthesized. How does *E. coli* achieve this appropriate regulation of its *lac* genes? The following description is somewhat simplified but gives the basic idea.

E. coli synthesizes a *lac* repressor protein (Fig. 3.15). In the absence of the sugar lactose, this protein binds to the *E. coli* chromosome at a special site (the **operator**) near the promoter of the *lac* genes and prevents transcription of the genes. Consequently, the bacterium does not waste energy making lactose-utilizing enzymes when there is no lactose in the cell.

If lactose is present, however, the bacterium needs enzymes for using it. The *lac* regulation system allows the cell to respond beautifully to this new need. When lactose enters the cell, it interacts with a special site on the *lac* repressor protein. This interaction renders the repressor unable to bind to its site on the *E. coli* DNA, presumably by changing the shape of the protein. The repressor releases the DNA, leaving

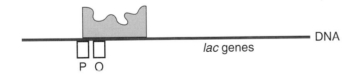

Active *lac*
roproooor protoin   Lactose   Inactive lactose-
                                       roproooor oomplox

Figure 3.15   Transcriptional regulation of the *lac* operon. P is the promoter; O is the operator.

A. No lactose in cells: active repressor prevents transcription.

DNA

lac genes

P O

B. Lactose in cells: inactive lactose-repressor complex allows transcription.

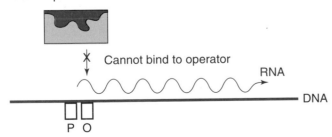

Cannot bind to operator

RNA

DNA

P O

Inactive *trp* repressor protein + Tryptophan → Active tryptophan-repressor complex

A. Low tryptophan concentration: inactive repressor allows transcription.

Cannot bind to operator

RNA

DNA
Tryptophan synthesis genes

P O

B. High tryptophan concentration: active repressor complex binds to operator and prevents transcription.

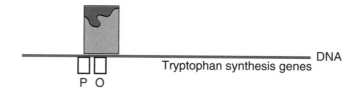

DNA
Tryptophan synthesis genes

P O

Figure 3.16 Transcriptional regulation of the *trp* operon. P is the promoter; O is the operator.

the gene free to be transcribed. The cell then can make the lactose-using enzymes and take advantage of the new energy source (Fig. 3.15).

The *lac* genes of *E. coli* are "turned on" when the appropriate sugar is present and are otherwise "turned off." Sometimes, however, it is better for a cell to have genes normally turned on and to turn them off only under special circumstances. An example of this type of regulation is found in the transcription of the tryptophan-synthesizing enzymes of *E. coli*.

Tryptophan is an amino acid that, like other amino acids, is essential for protein synthesis. *E. coli* has genes that encode enzymes for synthesizing tryptophan from scratch (the *trp* genes). Like the *lac* genes, the *trp* genes are also lined up on the chromosome and transcribed from one promoter. Since *E. coli* constantly needs tryptophan for making new proteins, the *trp* genes are normally turned on. Occasionally, however, a lucky *E. coli* might find itself in an environment where tryptophan is plentiful. In this case, the bacterium conserves energy by stopping the syn-

thesis of the *trp* proteins. Stopping expression of the *trp* genes in response to environmental changes is also achieved through a repressor protein (the following description is again somewhat simplified).

E. coli synthesizes a *trp* repressor protein, but the shape of the *trp* repressor does *not* allow it to bind to the chromosome (Fig. 3.16). In its native state, the *trp* repressor cannot attach to its site near the promoter of the *trp* genes and therefore does not interfere with their transcription.

If the concentration of tryptophan inside the cell rises, however, the excess tryptophan interacts directly with the inactive *trp* repressor. Tryptophan binds to a special site on the protein, and its binding changes the shape of the repressor. The altered shape of the protein allows it to bind to the regulatory site near the promoter of the *trp* genes. When the active repressor binds, transcription of the *trp* genes is turned off. Thus, the presence of excess tryptophan in the cell leads to a shutoff of the tryptophan-synthesizing enzymes.

Transcriptional Regulation through Activation

The *lac* and *trp* operons provide excellent examples of transcriptional regulation in prokaryotes. Bacterial cells are easy to culture and manipulate, and transcription and gene regulation were first studied in them. When techniques for studying these processes in eukaryotic cells became available, scientists expected to find similar mechanisms at work. However, many years of work have suggested that the typical regulation mechanism in eukaryotic cells involves activation rather than repression. In these cells, most promoters are apparently not recognized efficiently by RNA polymerase alone. Instead, eukaryotic RNA polymerase requires helper proteins called **transcriptional activators** or **transcription factors** to help it bind to a promoter and begin transcription. Some of these activators associate with the RNA polymerase enzyme and do not bind to DNA themselves. Other activators bind to special base sequences in DNA and then interact with RNA polymerase to help it bind to a promoter (Fig. 3.17).

The DNA-binding sites of these activators are often called **enhancers** because their presence enhances transcription from the associated promoter as long as the appropriate activator protein is present. The activity of genes can be regulated by the availability of the necessary transcriptional activators. For example, genes encoding the antibody proteins of the immune system have a specific enhancer, the immunoglobulin enhancer, associated with their promoters. This enhancer works only to enhance transcription in B lymphocytes (the antibody-producing cells of the immune system), because only B lymphocytes contain

Figure 3.17 Activator proteins are needed for transcription in eukaryotic cells.

A. No activators present: RNA polymerase cannot bind to promoter.

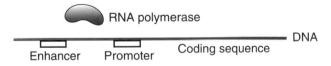

B. Activators present: RNA polymerase can bind to promoter and synthesize RNA.

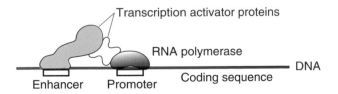

the transcriptional activator protein that binds the immunoglobulin enhancer. As a result, only your B lymphocytes make antibody proteins, even though all the cell nuclei in your body contain the antibody genes.

Another example of gene regulation through enhancers is the response of genes to steroid hormones such as estrogen (not all hormones are steroid hormones; other hormone types exert their effects on cells through different mechanisms). Estrogen and other steroid hormones pass through the cell membrane and bind to specific receptor proteins inside the cell. When the hormone binds, the receptor changes its shape in a way that allows it to move to the nucleus, where the receptor-hormone complex acts as a transcriptional activator by binding specific enhancers. Thus, specific genes are turned on in response to the presence of the hormone.

These examples of simple gene regulation mechanisms are typical. We want to point out that repression occurs in eukaryotic cells and activation occurs in prokaryotes, too. Many steps in protein synthesis other than transcription can also be regulated. However, a central theme of gene regulation is that it involves interactions between proteins and other molecules: additional proteins, small molecules, and DNA. These interactions are dependent (as are all protein functions) upon the three-dimensional structure of the proteins. At the end of the second part of this chapter (*Protein Structure and Function*), we will look closely at the structures of two DNA-binding regulatory proteins and how they interact with DNA.

Genomic organization

The genes encoded in DNA must be accessible to enzymes at the proper times for replication and transcription. Considering the extreme length of the DNA molecule in most organisms, this is a staggering requirement. Storage of DNA in cells is therefore not a haphazard affair, with the DNA "just lying there." DNA storage is a highly organized, complex phenomenon that is not well understood. Many scientists today are working to understand how cells manage their DNA information libraries.

Chromosomes

The physical packing of DNA inside cells presents a problem because of the extreme length of the DNA molecule relative to the size of the cell itself. If the DNA of *E. coli* were stretched out, it would be 1,000 times longer than the *E. coli* cell (Fig. 3.18). The problem faced by eukaryotic cells is even more amazing: the DNA of a single human cell would stretch 2 m, yet the cell itself has a diameter of only 1/50 of a millime-

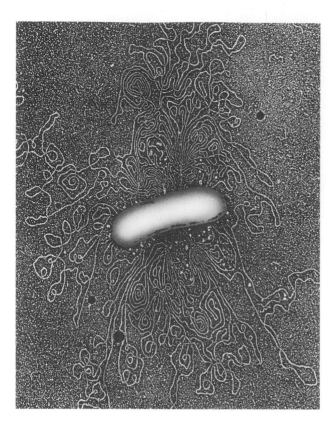

Figure 3.18 *Escherichia coli* osmotically shocked to release DNA (photograph courtesy of VU/©K.G. Murti).

ter. It is clear that cellular DNA must be very highly folded. In cells, DNA is folded and packed in association with proteins. The DNA-protein material is called **chromatin,** and the packed structure is called a **chromosome.**

In bacteria, DNA is present as one long circular molecule that is associated with special proteins in a single chromosome. These proteins are thought to hold the DNA in a condensed form so that it fits neatly into the cell. The circular chromosome is present in the cytoplasm, attached to the cell membrane.

In eukaryotes, the genome is usually divided among several different linear chromosomes located in an organized nucleus. Scientists think that the chromosomal DNA is packed in an extremely ordered manner with multiple layers of folding and coiling of the DNA molecules around the chromosomal proteins. Yet even though the DNA is tightly packed, all of the DNA encoding essential proteins is available for transcription at the proper time. How the cell organizes the complex problem of storage and accessibility is not understood. Some scientists think that the large amount of noncoding DNA present in higher organisms (see below) may play an important role in DNA organization.

Plasmids

Some cells, in particular bacteria, contain small rings of DNA outside of their chromosomes. These small circular pieces of DNA are called **plasmids.** Plasmids are replicated by the cell's enzymes and inherited by progeny bacteria. They typically contain a few thousand base pairs of DNA and encode a few proteins. None of these proteins is normally essential to the survival of the bacterium. Plasmids often contain genes for drug resistance or for poisons to kill rival bacteria. Because they are small and convenient to work with, plasmids are a favorite tool of biotechnologists. (See the activities *Recombinant Paper Plasmids* and *Transformation of* Escherichia coli.)

Virus Genomes

Viruses are not cells. They consist simply of genetic material enclosed in a capsule generally made of protein. Viral genetic material can be either DNA or RNA. Viruses require a host cell to make copies of themselves. When a virus infects a cell, it introduces its genetic material into that cell. The cell's enzymes and ribosomes transcribe and translate the viral genetic material and eventually reproduce virus particles. Outside of a host cell, viruses are completely dormant and are not even considered living.

It is easy to imagine a DNA virus genome substituting for the cell's own genes, but how does an RNA genome work? An RNA virus introduces its RNA genome into its host cell. When the viral RNA enters the cell, the host cell translates it as it would any other mRNA. Some RNA viruses encode in their RNA a message for a special enzyme that uses RNA as a template to synthesize new RNA, rather like DNA polymerase replicates DNA. While the viral enzyme is synthesizing more viral RNA, the host cell translates the RNAs to make viral proteins. Other RNA viruses encode an enzyme that uses RNA as a template to synthesize DNA. This enzyme is called **reverse transcriptase.** Once the viral RNA genome has been copied into DNA, the cell machinery of the hapless host transcribes it for the virus.

Viruses are extremely specific about the host cells they can infect. Not only do viruses usually infect only one type of organism (such as humans or cats), but they are limited to certain cell types within the organism. For example, a "cold virus" infects only the cells of the upper respiratory tract, while a "stomach virus" infects only the cells of the digestive tract. This specificity explains why a pet dog does not get its master's cold. However, a few viruses, such as rabies, normally infect several different host species. And once in a while, a virus may acquire the ability (possibly through mutation) to infect a new host.

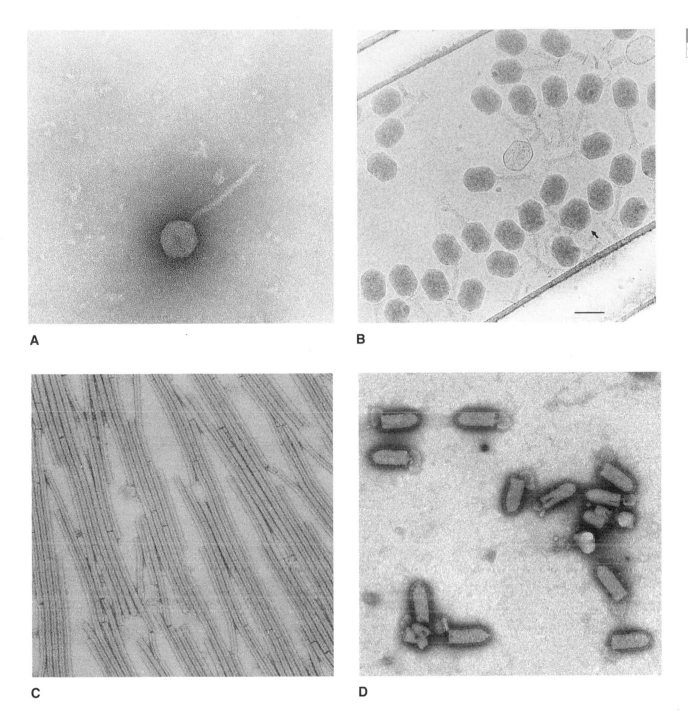

A

B

C

D

Figure 3.19 Electron micrographs of some different viruses. (A) Bacteriophage lambda (×275,000; photograph courtesy of VU/©K.G. Murti); (B) purified bacteriophage T4 (photograph courtesy of F. P. Booy; reprinted from J. D. Karam et al., ed., *Molecular Biology of Bacteriophage T4*, ASM Press, Washington, D.C., 1994); (C) tobacco mosaic virus (×144,000; photograph courtesy of VU/©K.G. Murti); (D) vesicular stomatitis virus (rabies group) (×100,000; photograph courtesy of VU/©K.G. Murti).

In addition to viruses that infect humans or other animals, there are plant viruses specific to certain plant tissues and bacterial viruses specific to certain bacterial strains. Bacterial viruses have played an important role in the development of molecular biology and have been given a special name: **bacteriophages.**

Electron micrographs of several viruses are shown in Fig. 3.19.

Whether a virus infects a plant cell, animal cell, or bacterial cell, the mechanism by which it recognizes its own host is similar. Proteins that make up the outer capsule of the virus recognize and bind to a specific molecule, often a protein, on the surface of the host cell. It is this specific molecular recognition that determines the extremely limited range of host cells a virus can infect.

Viruses as Tools in Biotechnology

Viruses have proven to be useful tools for biotechnologists because of some of the properties mentioned previously. Viruses are essentially containers of genetic material. They deliver that material into very specific cells. Scientists have developed ways of packaging new genetic material into several different types of viruses. The virus particles then inject that material into their host cells. Delivery of genetic material via viruses is an important method of gene transfer. (See the activity *Transduction of an Antibiotic Resistance Gene.*)

Noncoding DNA

How much DNA do organisms have? Naturally, it depends on the organism. The smallest viruses may have only a few thousand base pairs. Since the "average" protein requires 1,200 bases of coding sequence, these viruses encode only a few proteins. A bacterium such as *E. coli* contains about 4 million base pairs in its genome, while the human genome is composed of about 3 billion base pairs. Lest we grow smug about the complexity of the human genome, it bears noting that the genome of the mud puppy (an amphibian similar to newts and salamanders) is estimated to contain about 50 billion base pairs, and that of the lily contains 250 billion.

Does all of this DNA encode proteins? In bacteria and viruses, most of it appears to. Eukaryotes are a different story. In fact, only a small fraction of the DNA of multicellular eukaryotes (such as amphibians, mammals, and plants) is coding sequence.

What is all that extra DNA in eukaryotes? A large portion of it is repeated sequences that are present throughout the genome, sometimes in large clusters. Some of the repeated sequences appear to be transposons or parts of transposons. No one knows what the purpose of any of this DNA is or even if it has a purpose. This DNA has sometimes been called "junk DNA," though it seems presumptuous to label it junk before we understand it. A more neutral name for it is **noncoding DNA.** When you imagine the genomes of bacteria and viruses, you may imagine compact arrays of genes along the DNA. However, when you imagine your own genome or the genomes of plants or animals, you must imagine vast stretches of repeated DNA sequences occasionally broken up by genes. Puzzling, isn't it?

Mutations

Any change in a DNA sequence is called a **mutation.** It is a fact of life that mutations happen. They result from normal cellular processes and unavoidable environmental hazards. During DNA replication, the DNA polymerase enzyme occasionally makes errors that escape the proofreading that occurs during DNA replication. Environmental factors such as ultraviolet light or mutagenic chemicals damage DNA regularly. Most of this damage is corrected by **DNA repair enzymes,** which are proteins that recognize and repair abnormalities in DNA, but occasionally the repair enzymes miss something. When DNA polymerase attempts to use a damaged base as a template during DNA replication, it often cannot read the base properly and so inserts an incorrect base into the new strand, thus creating a mutation. Genetic events such as transposition (see *Transposons,* chapter 2) result in insertion or loss of segments of DNA, also changing the original sequence. Errors in cell division or recombination can lead to a rearrangement of the segments of chromosomes or even a change in chromosome number.

What is the effect of a mutation? It depends. It depends on where the mutation is in the DNA, exactly what it is, and, often, what environment the organism inhabits. For example, a sequence change in one of the many noncoding regions would probably not have any effect on the organism. Similarly, a mutation in an intron sequence in a eukaryote might not have any effect unless it involved a processing signal for the splicing enzymes. Even changes in coding sequences may not have any effect on a protein. Many amino acids are encoded by more than one codon. For example, the codons TTT and TTC each encode the amino acid phenylalanine (Table 3.1). A mutation that changed TTT to TTC would not have any effect on the protein. Mutations with no effect on a protein are often called "silent" mutations.

In addition, many amino acid changes may not alter the function of a given protein in a significant way. Recalling the example of the receptor protein in the cell membrane, imagine that a mutation occurred in a region of the protein apart from the specific hormone recognition area. As long as the change did not distort the overall shape of the protein or impair its interaction with the cell, it might not affect the function of the protein.

Although many mutations are harmless, others can be devastating. Mutations in genetic traffic signals such as promoters and ribosome recognition sequences can completely shut down the synthesis of a protein.

Sometimes a single base change in a coding region can result in an amino acid substitution that severely harms or destroys the protein's ability to perform its function. Recall again our example of the receptor protein embedded in the cell membrane. If anything distorted the shape of the receptor protein, the hormone might no longer fit and/or the receptor might no longer be able to communicate with the rest of the cell. The cell would lose its ability to respond to the body's signals for growth. Thus, a change in the protein's shape could have disastrous consequences for the organism. Similar examples could be given for proteins of the immune system that must recognize specific invaders, for transport proteins that carry specific nutrients, and so on.

Although most mutations are harmful or neutral, they can also be beneficial. Changes in the amino acid sequence of a protein might make it more resistant to heat, which could be an advantage if the organism's environment is becoming warmer. An alteration in the shape of another protein might allow it to bind to and break down a different type of sugar, which could be an advantage if the organism's environment contained that sugar. If a change in a protein's function is not immediately fatal, it might actually help the organism under the right environmental conditions.

One consequence of mutations and recombination is that the chromosomes of all sexually reproducing individuals differ in many ways. Although all humans have similar chromosomes and produce the same sets of essential proteins, the exact DNA sequences in those chromosomes varies. One obvious source of variation is the same variation that causes us to look different: one individual has genes encoding blue eyes, and another has genes encoding brown eyes. A less obvious source of variation is the accumulation of changes in noncoding regions of DNA. Great variety can be present in these regions, particularly in the number of repeated sequences present, with no outwardly observable effects. In fact, it is extremely unlikely that any two individuals (except identical twins) would share the same sequences in all of their noncoding DNA. It is this variety that makes possible the new procedures of DNA fingerprinting. (See chapter 4 and the lesson *Analyzing Genetic Variation: DNA Typing.*)

At the end of this chapter we will look at specific examples of a harmful and a helpful mutation: the nucleotide change, the amino acid changes, the effects on protein structure, the effects on the proteins' functions, and the effects of the changes in protein behavior on the phenotype of the organism. In order to put all of these things together, we first need to look at the link between genes and phenotype: proteins.

Protein Structure and Function

Genes are important because they supply information that directs the synthesis of proteins. It is the proteins that confer a phenotype on the cell: its biochemical capabilities, its shape, its communication channels, and so forth. The function of a protein is determined by its three-dimensional shape, which is determined by the nature and sequence of its amino acids. The relationship between amino acid sequence and three-dimensional structure has been called the "second half of the genetic code," because it is the three-dimensional structure that leads to function and phenotype. Unfortunately, the relationship between amino acid sequence and three-dimensional structure is not simple. It would be wonderful if we could predict the three-dimensional structure of a protein from its amino acid sequence, but we can't. At least not yet.

Over the past several years, the three-dimensional structures of hundreds of proteins have been painstakingly determined by using the techniques of X-ray crystallography and nuclear magnetic resonance (see chapter 4). These structures have greatly increased our understanding of how amino acid chains fold up to be energetically stable proteins. Here's a summary of the basic principles.

Amino acids and peptide bonds

There are 20 amino acids that normally make up proteins. Amino acids have the general chemical structure shown in Fig. 3.20. The R in the figure can be any of 20 different so-called side chains. These different side chains give the 20 amino acids their separate identities.

When amino acids are joined to make a protein chain, the OH group on one end of the amino acid reacts with the NH_2 group of another amino acid. A water molecule (H_2O) is lost, forming what is called a **peptide bond** (Fig. 3.21A). A protein consists of many amino acids joined together via peptide bonds. Some-

Figure 3.20 General structure of an amino acid. R signifies one of the 20 different side chains shown in Fig. 3.23.

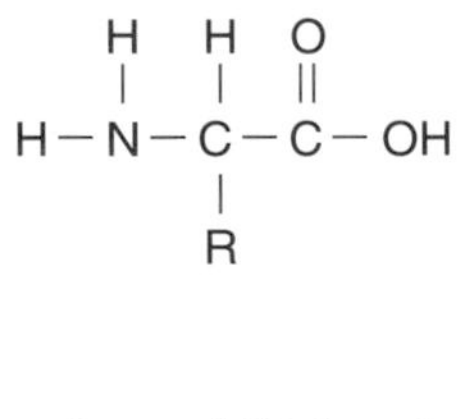

$$H - N - C - C - OH$$

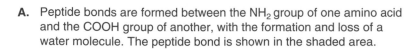

A. Peptide bonds are formed between the NH$_2$ group of one amino acid and the COOH group of another, with the formation and loss of a water molecule. The peptide bond is shown in the shaded area.

$$
\begin{array}{ccccccccc}
 & H & & H & & O & & & & & & & H & & H & & O \\
 & | & & | & & || & & & & & & & | & & | & & || \\
H- & N- & & C- & & C- & OH & + & H- & N- & & C- & & C- & OH \\
 & & & | & & & & & & & & | \\
 & & & R_1 & & & & & & & & R_2
\end{array}
$$

$$
\begin{array}{ccccccccc}
 & H & & H & & O & H & & H & & O \\
 & | & & | & & || & | & & | & & || \\
H- & N- & & C- & & C- & N- & & C- & & C- & OH & + H_2O \\
 & & & | & & & & & | \\
 & & & R_1 & & & & & R_2
\end{array}
$$

B. A protein is a polypeptide backbone with various amino acid side chains (R$_n$).

N terminus   C terminus

$$
\begin{array}{ccccccccccccccccccc}
H & & H & & O & H & & H & & O & H & & H & & O & H & & H & & O & H & & H & & O \\
| & & | & & || & | & & | & & || & | & & | & & || & | & & | & & || & | & & | & & || \\
H-N- & & C- & & C- & N- & & C- & & C- & N- & & C- & & C- & N- & & C- & & C- & N- & & C- & & C- & OH \\
 & & | & & & & & | & & & & & | & & & & & | & & & & & | \\
 & & R_1 & & & & & R_2 & & & & & R_3 & & & & & R_4 & & & & & R_5
\end{array}
$$

Figure 3.21   To form proteins, amino acids are joined by peptide bonds.

times a single chain of amino acids is even called a polypeptide. Like a DNA strand, a protein backbone has a direction, too. One end has a free NH$_2$ group, and the other has a free COOH group. These ends are called the *N terminus* and the *C terminus,* respectively. The overall effect is that a protein has a uniform peptide backbone with various amino acid side chains (Fig. 3.21B). The identity and order of the side chains in a protein are called the **primary structure** of the protein. Primary structure is a direct consequence of the DNA base sequence in the gene encoding that protein.

You have probably already figured out that if all proteins have a uniform peptide backbone, then the nature of the amino acid side chains must be the factor that governs how an individual protein folds into a three-dimensional structure. That is true, and fortunately, one particular property of side chains is the most important for influencing three-dimensional structure. This property has to do with how well the individual side chains interact with water molecules. To understand this property, let's start with a quick review of covalent bonds.

Polar and nonpolar covalent bonds

The atoms in a protein molecule are held together by **covalent** chemical bonds, which consist of electrons shared by two atomic nuclei. If the electrons are shared equally by the nuclei, then their negative charge is distributed evenly over the area of the bond and balanced by the positive charges in the nuclei. However, nuclei of certain elements attract electrons more strongly than other nuclei do (the ability to attract electrons is called **electronegativity**). If a strongly electronegative element forms a covalent bond with a less electronegative element, the electrons tend to be found near the strongly electronegative nucleus. You can think of the electronegative nucleus as pulling the bond electrons away from the less electronegative nucleus. This uneven distribution of electrons creates a partial negative charge around the electronegative nucleus and a partial positive charge at the other nucleus. Chemical bonds with this type of uneven charge distribution are called **polar** bonds because they have positive and negative poles (Fig. 3.22).

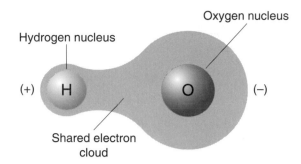

Figure 3.22 A polar covalent bond. Although the oxygen and hydrogen nuclei are sharing two electrons, the highly electronegative oxygen nucleus tends to draw them away from the weakly electronegative hydrogen nucleus. As a result, the oxygen end of the bond acquires a partial negative charge, while the hydrogen end is partially positive.

Amino acids (and proteins) are made primarily of carbon, hydrogen, oxygen, and nitrogen. Of these four elements, oxygen and nitrogen are strongly electronegative, and carbon and hydrogen are not. Thus, bonds between oxygen and hydrogen, oxygen and carbon, nitrogen and hydrogen, and nitrogen and carbon are polar. However, many of the bonds in a protein molecule are between carbon and hydrogen and are not polar. Whether a larger chemical group like the side chain of an amino acid is polar or not depends on its constituent chemical bonds.

The chemical structures of the side chains of all 20 normal amino acids are shown in Fig. 3.23. The structures are grouped according to whether the side chains are fully charged (ionized) at physiological pH, are polar (partially charged), or are nonpolar. A fourth group, the aromatic amino acids, are called that because, like other aromatic compounds, they have a ring structure in their side chains. Of this group, phenylalanine is nonpolar, and the other two are somewhat polar.

Polarity and Stability

The polarity of the amino acid side chains is fundamentally important to protein structure, because polarity determines how stably a side chain interacts with other elements in the protein and in the cellular environment. A compound that has an overall polar character or is actually electrostatically charged is energetically stable when it associates with other compounds with complementary charges or partial charges. The complementary charges neutralize each other. Nonpolar molecules do not associate with charged or polar molecules in an energetically favorable manner. Instead, they associate comfortably with other nonpolar molecules.

We stated above that the single property of amino acids that most determines three-dimensional protein structure is the ability to interact stably with water. Water is so important because the intracellular environment is water based, as are other body fluids. How stably an amino acid side chain interacts with water therefore determines how "comfortable" it is when exposed to the intracellular fluid.

Here is the bottom line about water. Water consists of two polar oxygen-hydrogen bonds and is a very polar molecule (Fig. 3.24). It thus associates comfortably with other polar or charged molecules. For this reason, molecules that are electrostatically charged or that are polar are called **hydrophilic** (water loving). Since nonpolar molecules do not associate comfortably with water, they are called **hydrophobic.** Hydrophobic amino acid side chains (the nonpolar ones in Fig. 3.23) do *not* associate stably with the intracellular fluid. Hydrophilic amino acid side chains (the charged and polar ones in Fig. 3.23) *do* associate stably because their charges or partial charges can be neutralized by complementary partial charges of polar water molecules.

Hydrogen Bonds

One type of neutralization that is particularly important in considering protein structure is the **hydrogen bond,** the same kind of bond found between base pairs in DNA (see Fig. 3.3). Hydrogen bonds are not covalent bonds. They are much weaker and form when two highly electronegative nuclei "share" a hydrogen atom that is formally bonded to only one of them. The partial positive charge on the bonded hydrogen neutralizes the partial negative charge on the second electronegative nucleus (Fig. 3.25). Hydrogen bonds form only between the three most electronegative elements: oxygen, nitrogen, and fluorine. Of these three elements, only oxygen and nitrogen are common in biological systems, so you can forget about fluorine when thinking about protein structure. In proteins, the most important groups involved in hydrogen bonding are N, NH, O, OH, and CO groups. Water can form hydrogen bonds with all of them (Fig. 3.26). Look at the hydrogen bonds in DNA in Fig. 3.3, and you will see the same groups.

The fundamental consideration of protein structure

Now that we have looked at the features of covalent bonds that are important for understanding protein structure, let's see what it all boils down to. A protein molecule is a long peptide backbone with a mixture of charged, polar, and nonpolar amino acid side chains. Cytoplasm is a watery environment, so the

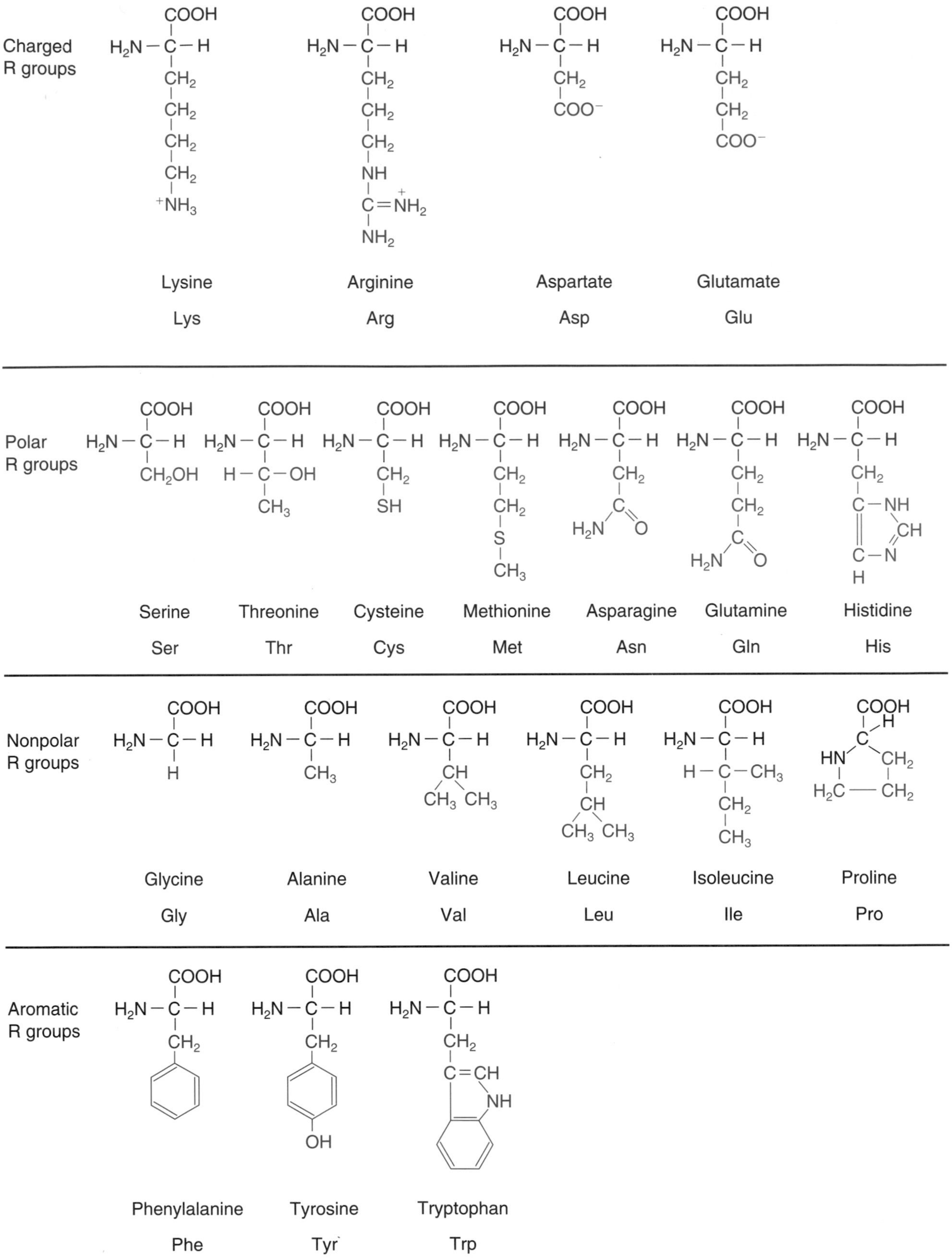

Charged R groups

Lysine
Lys

Arginine
Arg

Aspartate
Asp

Glutamate
Glu

Polar R groups

Serine
Ser

Threonine
Thr

Cysteine
Cys

Methionine
Met

Asparagine
Asn

Glutamine
Gln

Histidine
His

Nonpolar R groups

Glycine
Gly

Alanine
Ala

Valine
Val

Leucine
Leu

Isoleucine
Ile

Proline
Pro

Aromatic R groups

Phenylalanine
Phe

Tyrosine
Tyr

Tryptophan
Trp

Figure 3.23 The amino acids commonly found in proteins. The three-letter abbreviation for each is shown beneath its full name.

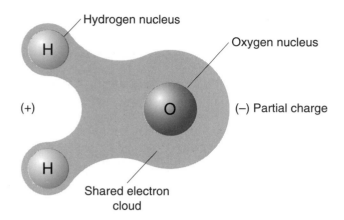

Hydrogen nucleus

Oxygen nucleus

H

(+)

O

(−) Partial charge

H

Shared electron
cloud

Figure 3.24 Water is a very polar molecule. The strongly electronegative oxygen nucleus hogs the electrons it shares with the hydrogen nuclei.

charged and polar side chains will be stabilized through interactions with water molecules there. However, the hydrophobic amino acid side chains do not associate stably with water; they are more stable when clustered together away from water.

It appears that the basic rule underlying protein structure is, as much as possible, to fold up hydrophobic amino acid side chains in the interior of the protein, creating a water-free hydrophobic environment. Hydrophilic side chains, meanwhile, are stable when exposed to the cytoplasm on the surface of the protein molecule. This is not to say that you would never find a hydrophilic amino acid on the interior of a protein or a hydrophobic one on the surface, but in general, the rule holds good. A protein is therefore said to have a hydrophobic core. The three-dimensional structure of each individual protein can be thought of as a solution to the problem of creating a stable hydrophobic core, given that protein's primary structure. However, every protein structure must solve one common problem. As you might guess, solving that problem forms another theme in protein structure.

Figure 3.25 A hydrogen bond (dotted line) is a weak electrostatic attraction between opposite partial charges.

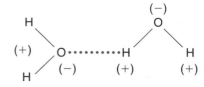

H
(−)
O

(+) O ·········· H H
H (−) (+) (+)

Taking Care of the Hydrophilic Backbone

There is one major problem in folding a protein to create a hydrophobic core: the backbone. Look at Fig. 3.21. The peptide backbone is full of NH and CO bonds, and both kinds of bonds are highly polar. On the surface of a protein, these partially charged bonds can be readily neutralized through hydrogen bonding with water. However, for a protein structure to be stable, the partial charges of the peptide backbone must also be neutralized inside the protein core, where there is no water. The solution to this problem is a major factor in determining protein structure.

The fundamental solution to the problem of the peptide backbone in the hydrophobic interior is for the backbone to neutralize its own partial charges. The NH groups can form hydrogen bonds with the CO groups (Fig. 3.26), neutralizing both. Since every amino acid contributes one NH group and one CO group to the backbone, this solution is very convenient. However, because of geometric constraints, the NH and CO groups from the same amino acid are not in position to form a hydrogen bond with one another. Instead, the peptide backbone must be carefully arranged so that the NH and CO groups along it are in position to form hydrogen bonds with complementary groups elsewhere along the backbone. Two basic arrangements work well, and these two arrangements form major components of protein structure.

The first self-neutralization arrangement for the peptide backbone is for the backbone to form a helical coil, as if it were winding around a pole. The amino acid side chains point outward, away from the imaginary pole. The NH and CO groups along the back-

Figure 3.26 Common hydrogen bonds in biological systems.

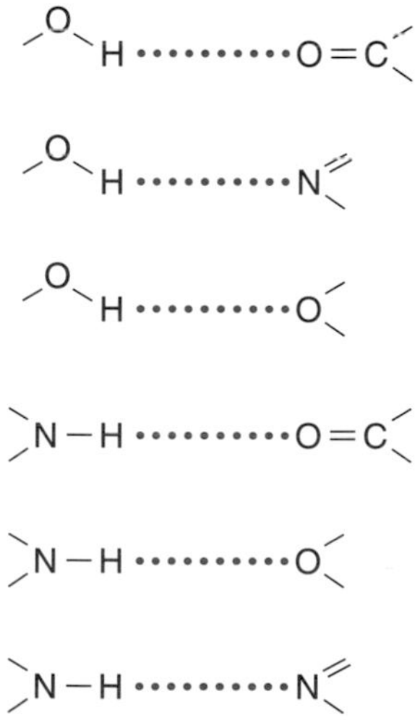

O—H ·········· O=C

O—H ·········· N

O—H ·········· O

N—H ·········· O=C

N—H ·········· O

N—H ·········· N

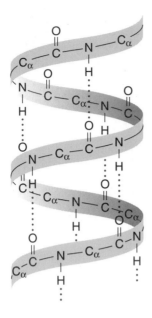

Figure 3.27 The alpha helix. C$_\alpha$ indicates the carbon atoms with side chains, which are not shown.

bone form hydrogen bonds with complementary groups above or below them on the pole, as shown in Fig. 3.27. This arrangement is called an **alpha helix.**

In the second self-neutralization arrangement, stretches of the peptide backbone lie side by side, so that a CO group on one backbone can form a hydrogen bond with an NH group on the adjacent backbone (Fig. 3.28). The amino acid side chains point alternately above and below the plane of the backbones. This arrangement is called a **beta sheet,** and the individual stretches of backbone involved in the sheet are called *beta strands*. Beta sheets are usually not flat, but twisted.

Secondary structure

Within a protein molecule, particular stretches of the amino acid chain may assume an alpha-helix or beta-sheet conformation. Certain amino acid sequences favor the formation of each one, although we cannot predict these structures from the primary structure with perfect accuracy. The regions of alpha helix and beta sheets within a protein are referred to as **secondary structure.**

Figure 3.28 A beta sheet. C$_\alpha$ indicates the carbon atoms with side chains, which are not shown.

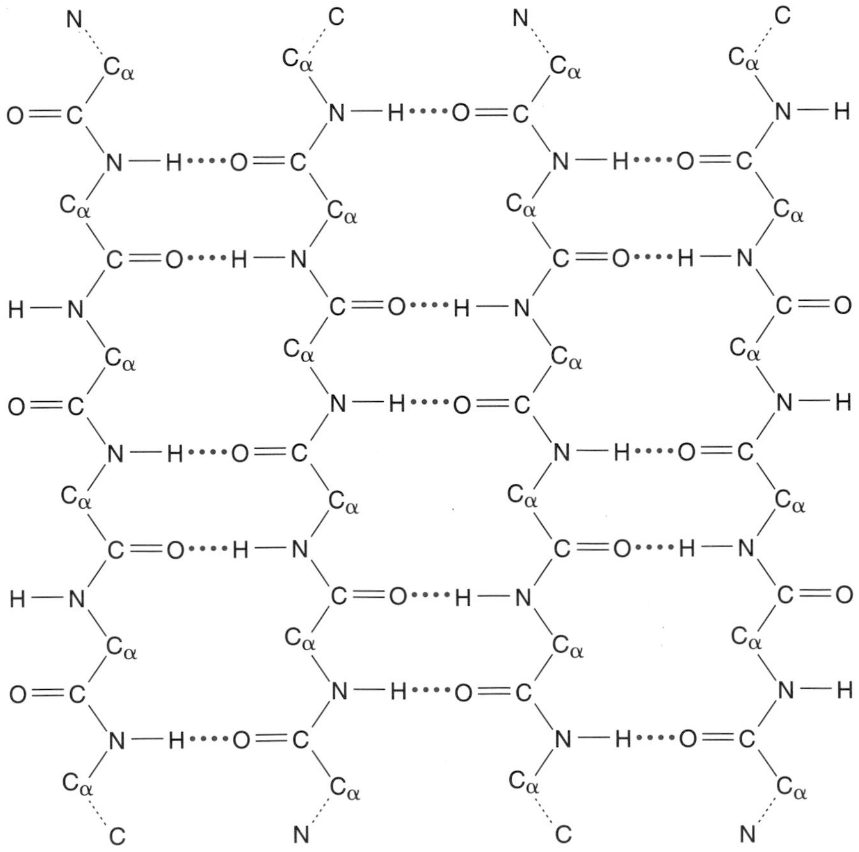

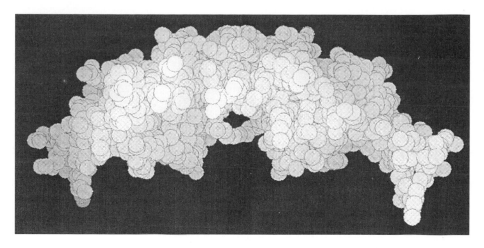

Figure 3.29 In this drawing of the replication termination protein of *E. coli,* each sphere represents an atom. Even though this protein is small, its structure is complex. (Drawing courtesy of Stephen White, in whose laboratory the structure was determined.)

So within the hydrophobic core of a protein, some segments of the backbone may be found in the alpha-helix conformation while other segments are arranged as beta sheets. (Some proteins are formed entirely from alpha helices; others, entirely from beta strands.) These secondary structures are often connected to one another via stretches of amino acids on the surface of the protein, where the partially charged backbone does not need to assume a particular secondary structure because it is neutralized by water in the cellular environment. The active sites of enzymes often involve these unorganized loops of amino acids, probably because the loops are freer to change conformation to bind a substrate.

Drawing Proteins

If you look at a picture of a protein in which all the atoms are shown, you can't tell much (Fig. 3.29). The picture contains too much information, too many atoms to tell what the big picture is. For purposes of understanding the overall structural plan, it is most helpful to look at just the configuration of the backbone. One popular way of showing this configuration is to draw the backbone as a ribbon, with alpha helices coiled and beta strands indicated with arrowheads. The arrows point toward the C terminus of the amino acid chain. Free loops are uncoiled regions of ribbon without arrowheads. Examples of some protein structures drawn in this manner are shown in Fig. 3.30.

Look at the drawings in Fig. 3.30 closely. Plastocyanin (panel A) is composed of beta strands connected by loops. The central part of flavodoxin (panel B) is a twisted beta sheet. The strands of the beta sheet are connected by regions of alpha helix. The center of triose phosphate isomerase (panel C) also consists of beta strands, but they are arranged somewhat differently. Try using your finger to follow the peptide backbones of all three structures from N to C. You can see that adjacent strands in a beta sheet do not have to come from contiguous stretches of the backbone but may be segments that are widely removed in the primary structure.

You can also see from the flavodoxin and triose phosphate isomerase structures that alpha helices are apparently at the surfaces of these proteins. This arrangement is fairly common because of a handy property of alpha helices. Since the amino acid side chains stick out from the center of the imaginary barber pole, one side of the pole can have mostly hydrophilic side chains while the other has mostly hydrophobic ones. With such an arrangement, the helix can sit comfortably at the protein surface, its hydrophobic side buried in the protein core and its hydrophilic side exposed to the cellular fluid. Neat, isn't it?

Structural Motifs

Simple combinations of a few secondary-structure elements occur frequently in proteins. Protein structure scientists call them *motifs*. You can think of a motif as a little module of protein structure; many proteins are put together by assembling combinations of these modular motifs. Of course, that statement makes it sound as if the structure is independent of the primary amino acid sequence, which it is not. But if you simply compare lots of protein structures, it does seem as though various structural motifs form a sort of "tool kit" for assembling larger proteins. A couple of examples of structural motifs are the helix-loop-helix, the beta turn, the beta-alpha-beta, and the beta barrel. These motifs are shown in Fig. 3.31. There are many others.

A. Plastocyanin

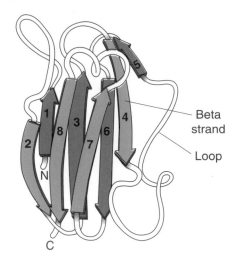

Beta
strand

Loop

B. Flavodoxin

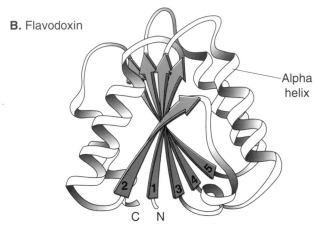

Alpha
helix

C. Triose phosphate isomerase

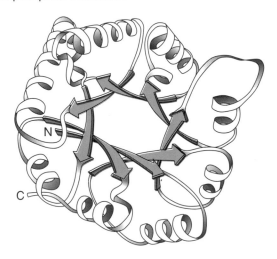

Figure 3.30 Ribbon drawings of protein structures. The beta strands in panels A and B are numbered in order from the N terminus to the C terminus of the amino acid chain. (A) Plastocyanin; (B) flavodoxin; (C) triose phosphate isomerase. (Drawings courtesy of Jane Richardson.)

Look at Fig. 3.30. You can find three of the structural motifs shown in Fig. 3.31 within the proteins of Fig. 3.30. Plastocyanin has many beta-turn motifs. Both flavodoxin and triose phosphate isomerase contain many beta-alpha-beta motifs (trace the backbones; you will discover there is a helix between each pair of beta strands). The center of triose phosphate isomerase is a beta barrel. Some structural motifs are associated with functional activities, though most are not. For example, one variety of helix-loop-helix (also called helix-turn-helix) is a DNA-binding motif that occurs in a variety of DNA-binding proteins. When these proteins bind their target DNA sequences, the helices of this motif sit next to the DNA, and the side chains on one of the helices reach into the major groove and make hydrogen bonds with the edges of specific bases there. (For an example, see *Structure and function* later in this chapter.)

Domains

Several secondary-structure motifs usually combine to form a stable, compact three-dimensional structure called a domain. Small proteins may consist of a single domain (such as those pictured in Fig. 3.30); larger proteins may fold into several separate domains. The structures of individual domains within a protein and the way multiple domains fit together are called the **tertiary structure** of the protein.

Domains appear to be fundamental units of protein structure and function. Domains are usually formed from continuous stretches of amino acids and therefore are translated from continuous regions of mRNA. In multifunctional proteins, it is not uncommon to find that the protein folds into several domains and that each domain is associated with one function. Sometimes it is possible to separate the domains of a protein (through brief enzymatic digestion), and we sometimes find that the separate domains retain their individual functions.

An example of a bifunctional protein with two domains is the repressor protein of the bacterial virus lambda. The lambda repressor folds into two domains; the first 92 amino acids fold into the N-terminal domain, and amino acids 132 through 236 fold into the C-terminal domain. The other 40 amino acids connect the two domains (Fig. 3.32A). Each domain of the lambda repressor has a separate function.

The two functions of lambda repressor are to bind to the correct operator DNA and to form **dimers** by binding to a second molecule of itself. A dimeric protein is a stable association of two copies of the same polypeptide chain. (See *Quaternary structure*

A. Helix-loop-helix

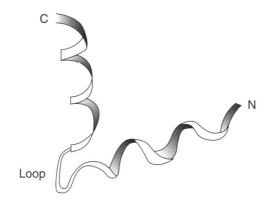

B. Beta turn

C. Beta-alpha-beta

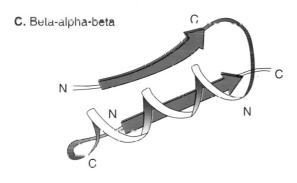

D. Beta barrel

Figure 3.31 Some protein structure motifs. (A) Helix-loop-helix; (B) beta turn; (C) beta-alpha-beta; (D) beta barrel. (Panels A and C are from C. Branden and J. Tooze, *Introduction to Protein Structure,* Garland Publishing, Inc., New York, 1991; panels B and D are from A. Lehninger, D. Nelson, and M. Cox, *Principles of Biochemistry,* 2nd ed., Worth Publishers, Inc., New York, 1993.)

below.) Each of the chains is called a **monomer.** In lambda repressor, the folded C-terminal domain of one monomer binds to the C-terminal domain of a

second repressor monomer to create the dimer. The N-terminal domains bind to DNA via a helix-turn-helix motif (Fig. 3.32B). The N- and C-terminal domains can be separated by digesting the 40-amino-acid linker region, which is more vulnerable to digestion because it is not folded tightly into a tertiary structure. After separation, the N-terminal domain still binds its DNA recognition sequence but cannot dimerize, and the C-terminal domain can dimerize but cannot bind DNA. Not all domains retain a function when they are separated, but many do.

Quaternary structure

Many functional proteins consist of a single amino acid chain, but many contain more than one polypeptide chain. These chains can be multiple copies of the same chain, as in the dimeric lambda repressor described previously. They can also be assemblies of different polypeptides; *E. coli* RNA polymerase contains five different chains encoded by five different genes. The identity and number of the polypeptide chains and how they fit together in the final protein are called the protein's **quaternary structure.**

Figure 3.32 Domain structure of the bacteriophage lambda repressor protein. (A) The N-terminal domain consists of amino acids 1 through 92, and the C-terminal domain consists of residues 132 through 236. (B) The repressor forms dimers through the interaction of the C-terminal domains. The N-terminal domains bind to a specific DNA sequence. (From M. Ptashne, *A Genetic Switch,* 2nd ed., Blackwell Scientific Publications and Cell Press, Cambridge, Mass., 1992.)

A

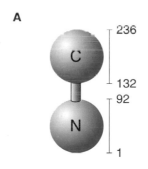

B

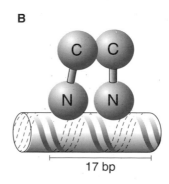

Do polypeptide chains automatically fold into the correct secondary and tertiary structures as they are synthesized and then associate with the correct additional polypeptides into quaternary structures? Many of them do. However, the folding of some proteins is assisted by other proteins that have been named *chaperone proteins*. Some molecular chaperones seem to act by increasing the rate of final folding; others actually guide the folding itself as well as the assembly of multiple polypeptide chains into complex quaternary structures.

Stability of protein structure

Three-dimensional protein structures are held together largely through the relatively weak chemical interactions of hydrogen bonds and the favorable interactions of hydrophobic side chains in the interior. Anything that disrupts these weak interactions—heat, extremes of pH, organic solvents, detergents—can alter the folding of the protein. The most extreme form of alteration is the complete unfolding of the amino acid chain, a process called **denaturation.** Denaturation of a protein is often irreversible. You can observe denaturation by frying an egg. The egg white protein albumin is soluble in its native state. As you heat it, the albumin denatures and coagulates, forming a white solid. Cooling the cooked egg does not reverse the process.

Proteins vary in their stability. One way to quantify stability is to measure the temperature at which a given protein denatures. This is sometimes called the melting temperature (T_m) of the protein. Some proteins require much higher temperatures to unfold than others do. For example, the enzymes of organisms that inhabit hot springs and ocean thermal vents are stable at very high temperatures. No one single thing makes these enzymes more thermostable. It appears that many different aspects of their primary and tertiary structures contribute to their heat resistance.

One feature of protein structure, however, makes a significant contribution to stability. This feature exploits the special properties of one amino acid, cysteine. In an oxidative environment, two properly positioned cysteine residues can react with each other to make a disulfide bridge (Fig. 3.33A). The intracellular environment is not oxidative, so disulfide bridges are rare inside cells, but many extracellular proteins contain them. Disulfide bridges anchor regions of the protein in a specific configuration, stabilizing the structure (Fig. 3.33B). Disulfide bridges can form between distant portions of the same domain or between different polypeptide chains within a quaternary structure.

Similar domains are found in different proteins

We said previously that domains appear to be fundamental units of protein structure and function. Scientists believe this statement because similar domains appear in different proteins, sometimes many different proteins. Some of the domains appear to be mostly structural; others are connected with specific functions. For example, a DNA-binding domain called the homeodomain (rhymes with Romeo-domain) is found in a large number of transcriptional activator proteins that interact with a specific type of enhancer sequence. (See *Transcriptional Regulation through Activation* above.) These proteins activate different sets of genes and are found in a diverse set of organisms, including worms, fruit flies, and humans. Even so, the proteins all bind to DNA via their homeodomains. The amino acid sequences of the homeodomains are very similar in these proteins.

Another example of a functional domain found in many different proteins is the domain that binds the enzyme cofactor nicotinamide adenine dinucleotide (NAD). A number of enzymes use NAD as a cofactor in oxidation-reduction reactions. Each of these enzymes has a domain that binds NAD, and the structure of that domain is practically identical from enzyme to enzyme. In fact, many of these enzymes have two domains, the common NAD-binding domain and a second unique domain containing the active site at which the substrate binds. Surprisingly, the amino acid sequences of NAD-binding domains vary, but the structure is almost perfectly conserved from enzyme to enzyme.

A repeated domain with no known biochemical activity is the kringle. It consists of about 85 amino acid residues folded into a shape that reminded some scientists of a certain Danish pastry, the kringle; thus, it was named. Kringle domains are found in a variety of proteins.

Modular Proteins

You may be wondering now if it is possible to put together proteins by connecting domains like modules or tinker toys. The answer is yes. Although not all proteins are made this way, many are, with perhaps a few unique regions thrown in. Figure 3.34 shows some modular protein structures. These modular structures suggest that many genes did not evolve from scratch. Rather, it looks as if many genes were patched together from pieces or copies of preexisting genes, resulting in proteins that contain domains common to many other proteins. It would be esthetically pleasing if protein domains were encoded by exons, with in-

A. Two cysteine side chains can form a disulfide bridge.

Peptide
backbone

$$-CH_2-SH + HS-CH_2- \longrightarrow -CH_2-S-S-CH_2-$$

B. A disulfide bridge stabilizes the structure of this domain of an immunoglobulin protein (antibody).

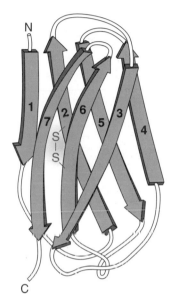

Figure 3.33 Disulfide bridges stabilize protein structure. (Panel B is from C. Branden and J. Tooze, *Introduction to Protein Structure,* Garland Publishing Inc., New York, 1991.)

trons providing the bridges between them. For some domains, such as the epidermal growth factor domain in Fig. 3.34, this possibility seems to be true. The domain is encoded by a single exon. Unfortunately, the gene segments encoding more domains are distributed between several exons, with no apparent pattern to the exon-intron structure. So the modular structure of proteins is not always related in a logically obvious (to us at this time) way to the modules of coding sequence in eukaryotic genes.

The modular structure of some proteins has implications for biotechnology. Nature has produced proteins with many different functions by joining similar domains in different ways over the course of evolution. Using recombinant DNA technology (described in chapter 4), we can now swap domains, too, by swapping portions of genes. For example, the portion of the lambda repressor gene encoding the first 92 amino acids (the DNA-binding domain) can be replaced with a similar domain from a different repressor protein. The hybrid protein works as a repressor, and it recognizes the DNA-binding site of the second protein.

Structure and function

Protein structure is an interesting topic in and of itself, but it is so important because protein function depends on it. Let's look at a few specific examples: a structural protein, two DNA-binding proteins, and an enzyme.

Keratin: a Structural Protein

The **keratins** are a family of similar proteins that make up hair, wool, feathers, nails, claws, scales, hooves, and horns and are part of the skin. Keratin fibers also form part of the cytoskeleton. To fulfill their function, these proteins must be very strong. In addition, they must not be soluble in water (it would be quite unhandy if your hooves dissolved), even though most proteins are. Let's look at how the structure of keratin makes strong, water-insoluble fibers possible.

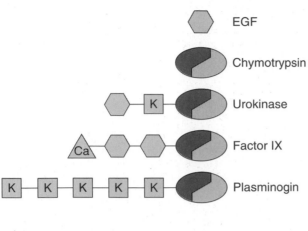

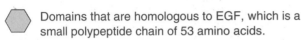

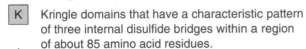

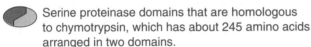

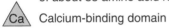

EGF

Chymotrypsin

Urokinase

Factor IX

Plasminogin

Domains that are homologous to EGF, which is a small polypeptide chain of 53 amino acids.

Serine proteinase domains that are homologous to chymotrypsin, which has about 245 amino acids arranged in two domains.

K Kringle domains that have a characteristic pattern of three internal disulfide bridges within a region of about 85 amino acid residues.

Ca Calcium-binding domain

Figure 3.34 Domain structures of some modular proteins. Epidermal growth factor is a protein that signals several cell types to divide. The other four proteins are protein-cleaving enzymes with a variety of physiological roles. (From C. Branden and J. Tooze, *Introduction to Protein Structure*, Garland Publishing Inc., New York, 1991.)

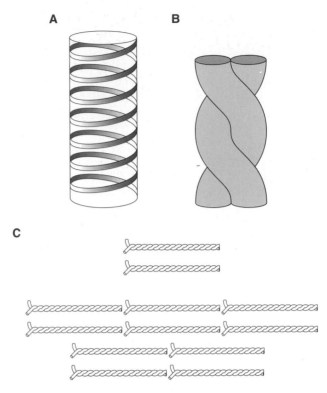

Figure 3.35 Keratin, a structural protein. (A) A single keratin molecule forms a long alpha helix. (B) Two keratin alpha helices then wrap around each other. (C) Two-chain coils lie end to end and side by side, forming fibers. (Panel C is from A. Lehninger, D. Nelson, and M. Cox, *Principles of Biochemistry*, 2nd ed., Worth Publishers, Inc., New York, 1993.)

The amino acid chain of keratin folds into one long alpha helix. Almost all of its amino acids—alanine, isoleucine, valine, methionine, and phenylalanine— are hydrophobic, and they extend outward from the helical backbone. The presence of all these hydrophobic side chains everywhere violates the general rule that hydrophobic side chains must be buried in the protein's core. As you probably expect, this rule violation is important: it means that keratin is not energetically comfortable surrounded by water molecules and therefore is not soluble in water. Imagine, the reason your fingernails don't dissolve when you wash them is all those hydrophobic side chains.

Instead of associating with the watery environment inside cells, keratin molecules associate with each other in large groups. Their structure, a single long helix, lends itself to forming fibers. First, two alpha helices of keratin wind around each other. Their hydrophobic surface side chains interact favorably in this conformation. These two-chain coils lie end to end and side by side with other coils, forming fibers (Fig. 3.35).

These fibers are not only held together by favorable hydrophobic interactions between side chains but are also stabilized by disulfide bridges between the coils. The bridges make the fibers strong and rigid. Different keratin proteins have different amounts of cysteine to make the bridges: the harder the final structure (hooves versus hair, for example), the more disulfide bridges are present. In the toughest keratins, such as tortoise shells, up to 18% of the amino acids are cysteines involved in disulfide bridges.

If you have ever had a permanent wave in your hair, you have manipulated the disulfide bridges of your keratin hair fibers. Recall that the disulfide bridges form only in the right kind of chemical environment (an oxidizing environment; see above). With the right chemicals, an oxidizing environment can be changed to its opposite (a reducing environment), causing the disulfide bridges to break apart.

When you get a permanent, the hair stylist first wraps your hair around small rods. Next, a smelly solution is applied to your hair. The smelly solution contains a *reducing agent,* a chemical that changes the oxidizing environment in your hair and breaks the disulfide bridges holding the keratin fibers side by side (the reducing agent is the component of the solution with the strong odor, too). While this is going on, the hair stylist has also arranged for your hair to be warm, either by placing you under a hair dryer or by putting a plastic bag over your hair. The moist heat breaks some hydrogen bonds that keep the keratin alpha helices stiff, allowing them to relax a little. The net effect is that the keratin helices move a little with respect to one another while they are wound around the rods.

After your hair has incubated sufficiently in the warm reducing environment, the stylist rinses out the reducing agent and applies a neutralizing solution. This solution restores the oxidizing environment, allowing disulfide bridges to re-form. But here's the catch. Your hair has been relaxed around the curling rods, and many of the cysteine SH groups now form disulfide bridges with new cysteine SH groups. These brand new disulfide bridges hold the hair fibers in the conformation they were in around the curling rods—a permanent wave (Fig. 3.36). Rinsing and cooling your hair allows the keratin helix hydrogen bonds to

reestablish themselves, returning your hair to normal except that new disulfide bridges now hold it in a wavy shape.

Lambda and *trp* Repressors: DNA-Binding Proteins

The bacteriophage lambda repressor protein binds to a specific operator DNA sequence in its bacteriophage genome and prevents RNA polymerase from transcribing certain genes, as do the bacterial *lac* and *trp* repressor proteins. As described previously, a single lambda repressor polypeptide chain folds into two domains (Fig. 3.32). The C-terminal domains of two polypeptides bind to one another, creating a dimeric protein.

The N-terminal domains of the lambda repressor bind to operator DNA sequence via a helix-turn-helix motif. The lambda operator DNA sequence is symmetrical; each N-terminal domain of repressor interacts with identical bases. Figure 3.37A shows how the helix-turn-helix motifs of the two N-terminal domains sit on the operator DNA. The amino acid side chains of helix 3 (part of the helix-turn-helix motif) contact specific bases within the DNA-binding sequence (Fig. 3.37B). When the protein binds to DNA, helix 3 sits alongside the DNA molecule so that these specific contacts can occur. Figure 3.37C shows one of these amino acid-base contacts in detail.

Figure 3.36 The biochemistry of a permanent hair wave.

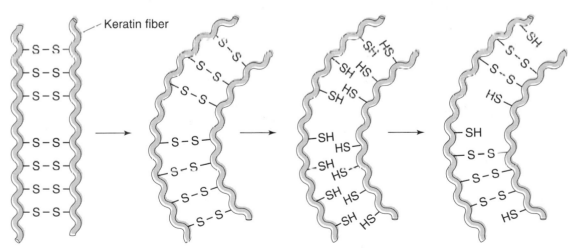

1. Normal hair
2. Hair is wrapped around a curling rod.
3. Reducing agent breaks disulfide bonds.
4. Neutralization allows disulfide bridges to re-form; new disulfide bonds hold hair in a wavy conformation.

A. The orientation of helix-turn-helix DNA-binding regions on operator DNA

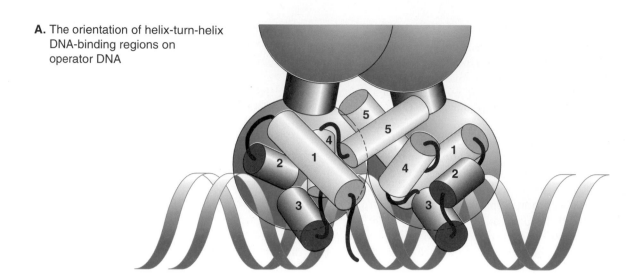

B. The amino acid sequence of helix 3. The specific bases in the operator sequence that are contacted by amino acid side chains are indicated by arrows.

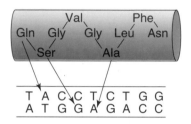

C. The contact between the first glutamine side chain in helix 3 and its target base

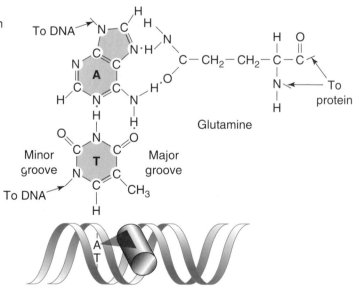

Figure 3.37 Binding of lambda repressor protein to DNA. (From M. Ptashne, *A Genetic Switch,* 2nd ed., Blackwell Scientific Publications and Cell Press, Cambridge, Mass., 1992.)

From this example you can see that both the overall structure of the lambda repressor and its specific amino acid sequence are important to its function. The helix-turn-helix motif is oriented within the protein so that helix 3 can sit alongside the DNA molecule. The specific amino acids within helix 3 must contact specific bases for the binding to work, so the protein binds only to its recognition sequence. (For further information, see the works by Ptashne listed in *Selected Readings.*)

Now let's revisit a protein we met earlier in this chapter, the repressor of the tryptophan operon, to see how a DNA-binding protein itself can be regulated by a second molecule. Recall that the *trp* repressor protein can bind both to the amino acid tryptophan and to a specific DNA base sequence, the *trp* operator sequence, but that the repressor binds *only* to the operator DNA when it is also binding to tryptophan (Fig. 3.16).

Like the lambda repressor, the *trp* repressor is a dimeric protein that binds to DNA via a helix-turn-helix motif. The amino acid sequence of the DNA-binding helix is different, so it binds a different sequence of bases. Unlike the lambda repressor, the *trp* repressor's structure doesn't position the DNA-binding helix correctly for interacting with DNA.

Each of the *trp* repressor monomers has the helix-turn-helix motif; however, the two monomer chains interact in such a way that the DNA-binding helices are folded in toward the main body of the protein and aren't positioned to fit alongside the DNA molecule (rather like the folded-in claws of a crab). Here's where tryptophan comes in. The repressor protein interacts with two molecules of tryptophan (one per monomer). The tryptophan molecules fit into the structure like wedges between the DNA-binding helices and the body of the protein, forcing the DNA-binding helices to swing out (imagine the crab's claws unfolding outward from its body; Fig. 3.38). Now the helices are positioned to bind their DNA recognition sequences. The tryptophan-repressor complex binds the *trp* operator and blocks further transcription of the *trp* operon.

Chymotrypsin: an Enzyme

Since enzymes are proteins that cause a chemical reaction, any discussion of their structure and function has to include the chemical reaction they promote. Chymotrypsin is a digestive enzyme that breaks down other proteins; it is a **proteinase** (or protease, pronounced PRO-tee-ace). Proteinases cleave the peptide bonds in protein backbones.

Chymotrypsin belongs to the family of proteinases called serine proteinases. They are called serine proteinases because a serine side chain participates in the actual cleavage of the substrate molecule (serine is pictured in Fig. 3.23). The digestive enzyme trypsin is also a serine proteinase; the domain structures of some other serine proteinases are shown in Fig. 3.34.

Chymotrypsin cleaves one peptide bond at a time, cutting a single polypeptide chain into two shorter

Figure 3.38 Binding of the amino acid tryptophan to the *trp* repressor protein changes the conformation of the repressor so that it can bind to DNA. (From B. Alberts, D. Bray, J. Lewis, M. Raff, K. Roberts, and J. Watson, *Molecular Biology of the Cell*, 3rd ed., Garland Publishing, Inc., New York, 1994.)

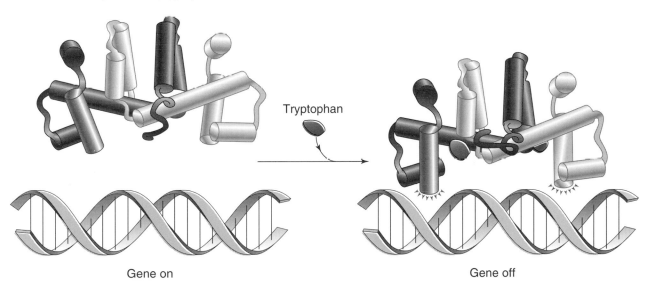

Gene on

Tryptophan

Gene off

A. Overall reaction catalyzed by the enzyme. A and B represent the rest of the protein molecule on either side of the peptide bond to be cleaved (see Fig. 3.21).

$$A-\overset{\overset{\displaystyle O}{\|}}{C}-\overset{\overset{\displaystyle H}{|}}{N}-B + H_2O \longrightarrow A-\overset{\overset{\displaystyle O}{\|}}{C}-OH + H-\overset{\overset{\displaystyle H}{|}}{N}-B$$

B. The overall reaction proceeds in two steps. E—OH represents the chymotrypsin enzyme with—OH on the catalytic serine side chain (see Fig. 3.23).

Step 1. $E-OH + A-\overset{\overset{\displaystyle O}{\|}}{C}-\overset{\overset{\displaystyle H}{|}}{N}-B \longrightarrow E-O-\overset{\overset{\displaystyle O}{\|}}{C}-A + H-\overset{\overset{\displaystyle H}{|}}{N}-B$

Step 2. $E-O-\overset{\overset{\displaystyle O}{\|}}{C}-A + H_2O \longrightarrow E-OH + HO-\overset{\overset{\displaystyle O}{\|}}{C}-A$

C. Close-up of step 1, showing the role of the catalytic serine side chain and its histidine helper.

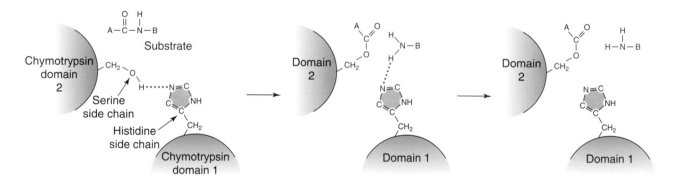

D. The domain structure of chymotrypsin, showing the positions of the serine (S195) and histidine (H57) side chains.

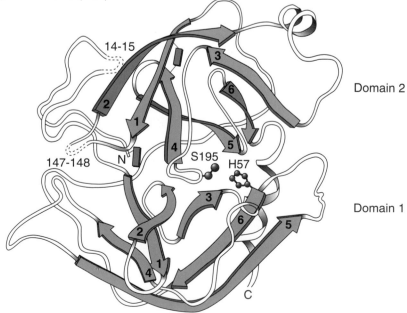

chains (Fig. 3.39A; see Fig. 3.21 to review peptide bonds and polypeptides). The cleavage reaction occurs in two steps (Fig. 3.39B). First, the peptide bond is cut, freeing one half of the original molecule but leaving the other half covalently bonded to the catalytic serine side chain. In the second step of the reaction, the bound substrate chain is released, restoring the enzyme to its original form. Both of these steps require a basic amino acid side chain in the proper position to hold onto a hydrogen. In chymotrypsin, the basic side chain is a histidine (Fig. 3.23). A close-up of the first step of the reaction is shown in Fig. 3.39C.

The structure of the serine proteinases positions the catalytic serine and the basic side chain so that the cleavage reaction can occur. A schematic of the structure of chymotrypsin is shown in Fig. 3.39D. The protein has two domains, and the serine and histidine are each in a different one. They are located within loop regions where the domains fit together. Other amino acids in the loop help bind the polypeptide substrate in the proper position for cleavage, and some specific amino acids, which vary from proteinase to proteinase, confer substrate specificity. Chymotrypsin, for example, prefers to cleave peptide bonds adjacent to aromatic amino acid side chains, while trypsin prefers positively charged side chains. All these features of serine proteinase activity are understood in detail. We decided it was beyond the scope of this book to describe them all and instead refer you to the excellent book by Branden and Tooze that is listed in *Selected Readings*, where you will find a more complete discussion.

Effects of mutations on protein structure and function

Earlier in this chapter, we defined a mutation as any change in a DNA sequence and stated that the effects of a mutation depend on what, if any, effect it has on the production or function of proteins. In this discussion, we focus on some specific effects of amino acid changes on proteins.

From the examples given previously, you have probably realized that the function of a protein really depends both on its three-dimensional structure and, often, on specific amino acids within that structure.

For example, there are many helix-turn-helix DNA-binding proteins. For them to bind DNA, the helix-turn-helix structure has to be maintained. However, the base sequence these proteins bind to depends on the identities of the amino acids within the DNA-binding helix. In addition, many proteins have multiple jobs to do, and these jobs are usually carried out by different regions of the protein, often different domains (recall the lambda repressor protein described previously: one domain binds DNA, and the other forms dimers). So the effect of any amino acid change depends on what (if anything) that change does to critical protein structures, whether or not that particular amino acid is specifically involved in a function, and how that function or structure relates to the rest of what the protein does.

The accumulating data about DNA and protein sequences indicate that protein structures and functions can often be maintained through many amino acid changes. The keratin proteins are a good illustration. Their specific amino acid sequences vary; their cysteine content varies; yet they are all recognizable, though different, versions of keratin. The genetic mutations that resulted in different versions of the protein did not destroy structure or function. Another example of this kind is hemoglobin, which has more than 300 known genetic variants in the human population alone. Most of these variants are single amino acid changes with only minor structural and functional effects. Protein structure, it seems, is reasonably robust in the face of many amino acid changes.

A Harmful Mutation
A specific example of a harmful mutation is found in the disease sickle cell anemia. This disease is the result of 1 of the more than 300 variations in hemoglobin. In sickle cell anemia, a single A-to-T mutation changes the sixth codon from GAG to GTG. This mutation changes the sixth amino acid in the 146-amino-acid protein from glutamic acid to valine. In the three-dimensional structure of hemoglobin, the sixth amino acid sits on the surface of the protein. Glutamic acid (the normal amino acid) is hydrophilic, so its side chain is stable when exposed to the intracellular fluid. Valine, however, is hydrophobic (Fig. 3.23). Its side chain is more energetically stable when interacting with other hydrophobic molecules.

Figure 3.39 Mechanism of action of the proteinase chymotrypsin, an example of a serine protease. (Panel D is from C. Branden and J. Tooze, *Introduction to Protein Structure,* Garland Publishing, Inc., New York, 1991.)

The problem with the valine side chain is not simply that it is hydrophobic. It doesn't really change the three-dimensional structure of hemoglobin for it to be on the surface, and the mutant hemoglobin can still bind oxygen. The problem is how the hydrophobic surface valine affects the way hemoglobin molecules interact with each other.

As it happens, the hydrophobic valine side chain on the surface just fits into a hydrophobic pocket that is exposed on the hemoglobin molecule when it is not bound to oxygen. The surface valine is not positioned to fit into its own pocket, but it can fit into the pocket on a second molecule. When the mutant hemoglobin molecules give up oxygen in the capillaries, the deoxygenated molecules fit together in a lock-and-key fashion. The surface valine fits into the pocket of another molecule, which itself has a surface valine to fit into another molecule, and so on (Fig. 3.40). The mutant hemoglobin molecules aggregate into long fibers, changing the red blood cells' shapes to a sickle form and interfering with circulation of the cells through the capillaries. The impaired circulation gives rise to a number of deadly problems in the afflicted individual.

Thus, the fatal disease sickle cell anemia is a consequence of hydrophobic interactions and altered protein structure. If the hydrophobic valine did not happen to fit into the hydrophobic pocket on the deoxygenated hemoglobin molecule, you would not get polymerization of the mutant hemoglobin, sickling of cells, and impaired circulation.

A Beneficial Mutation

An example of a possibly beneficial mutation is the inherited condition benign erythrocytosis. Individuals with this condition have highly elevated levels of red blood cells. Far from being ill, these individuals have greatly enhanced stamina. One such person, the Finnish athlete Eero Maentyranta, won three gold medals for cross-country skiing in the 1964 Winter Olympics. Scientists recently determined the molecular basis for the athlete's success.

Red blood cells arise from progenitor cells called stem cells that are found in bone marrow. Stem cells are stimulated to mature into red blood cells by a hormone. The hormone communicates to the stem cell through a 550-amino-acid receptor protein embedded in the stem cell's outer membrane. The N-terminal portion of the protein lies outside the cell, forming a docking site for the hormone. When the hormone binds, the C-terminal portion of the receptor, positioned inside the cell, transmits the maturation signal. The C-terminal end of the receptor also contains a docking site for a cellular protein that prevents transmission of the maturation signal. Docking of this cellular protein thus acts as a molecular brake on red blood cell production (Fig. 3.41A). The Finnish athlete and other members of his family carry a mutant version of the receptor gene. A G-to-A mutation changes codon 481 from TGG, for tryptophan, to TAG, a stop codon. This single base change causes the athlete's ribosomes to stop synthesis of the receptor protein 70 amino acids early.

Losing the 70 C-terminal amino acids does not disrupt the extracellular hormone-binding domain of the protein, its cell membrane-spanning domain, or the intracellular region responsible for transmitting the maturation signal. However, loss of those 70 amino acids removes the docking site for the braking protein. Thus, Eero Maentyranta's red blood cell production has no molecular brakes, and he and other mutation-bearing family members have higher-than-normal levels of red blood cells (Fig. 3.41B). Their blood can carry more oxygen than normal, so they have enhanced stamina. However, it is possible that their elevated red cell levels would be detrimental in some circumstances. Some reports state that "afflicted" family members have normal life spans, while other reports claim that their average life spans are reduced.

Figure 3.40 Representation of sickle cell hemoglobin aggregation. (A) Normal hemoglobin molecules do not stick together. (B) The hydrophobic patch on the surface of sickle cell hemoglobin caused by the glutamate-to-valine substitution at position 6 (Val-6) fits neatly into a hydrophobic pocket on a second molecule. Thus, sickle cell hemoglobin molecules can polymerize in a head-to-tail fashion.

A. Normal hemoglobin

B. Sickle cell hemoglobin

Val-6

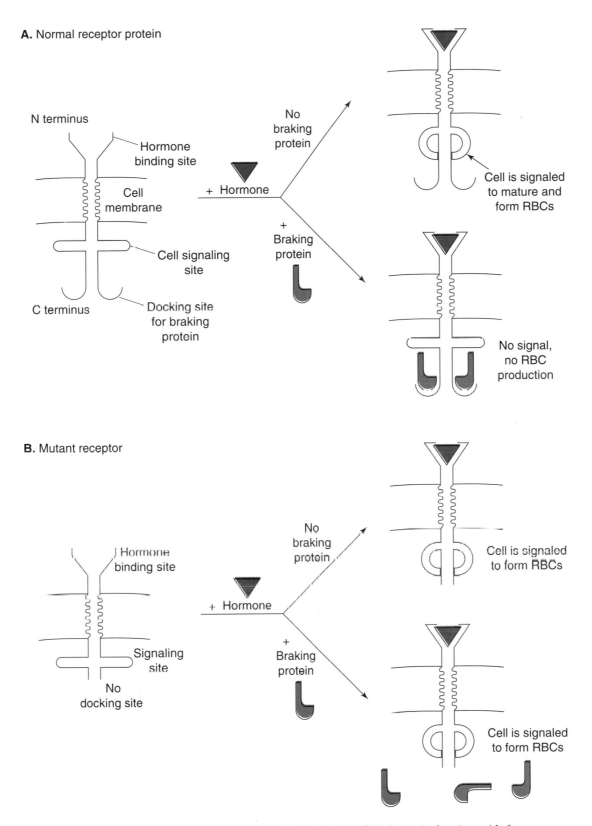

A. Normal receptor protein

N terminus

Hormone binding site

Cell membrane

Cell signaling site

C terminus

Docking site for braking protein

+ Hormone

No braking protein

+ Braking protein

Cell is signaled to mature and form RBCs

No signal, no RBC production

B. Mutant receptor

Hormone binding site

Signaling site

No docking site

+ Hormone

No braking protein

+ Braking protein

Cell is signaled to form RBCs

Cell is signaled to form RBCs

Figure 3.41 Schematic representation of how the loss of 70 C-terminal amino acids from a receptor protein results in increased red blood cell (RBC) production.

Predicting three-dimensional protein structure

Predicting the three-dimensional structure of a protein from its amino acid sequence is a major unsolved problem in structural biology. Structural biologists have searched and are searching for clues in solved protein structures, looking for amino acid patterns that correlate with specific structures. Structure prediction computer programs based upon these statistical studies have been developed, and they are useful but not perfect. If you are trying to predict a three-dimensional structure for a protein whose gene you have recently discovered, you are most likely to be successful if someone else has already discovered a similar protein and determined its structure.

DNA and protein sequence information is stored in large public databases, and computer software is available to compare these sequences. Proteins with homologous amino acid sequences (sequences that match fairly well) have similar three-dimensional structures and generally have similar functions. So the first step in determining the three-dimensional structure of a protein is to compare its amino acid sequence to those of all proteins in the public databases to see if the structure of a homologous protein has already been solved. If you are lucky and find one whose structure has been determined, you have a good model for your protein. This model can then be used as a basis for identifying where the active sites of your protein are and can guide further experiments. If you don't find a homologous protein with a solved structure, the best you can do is to use the imperfect computer algorithms for structure prediction.

Sometimes you can get clues about what a protein does from amino acid sequence comparisons. For example, scientists identified the gene that is defective in cystic fibrosis (CF) patients in 1989. At the time, they did not know exactly what the protein product of the gene did, although they knew from the symptoms of CF that the protein's function was related to the movement of salt across cell membranes. (This movement is defective in CF patients, causing some cellular secretions to contain abnormal levels of salt and water.) When scientists compared the predicted amino acid sequence of the CF gene product to the amino acid sequences of known proteins, they found that part of the CF protein looked exactly like helical protein domains that span cell membranes. So their amino acid sequence comparison strongly suggested that the CF protein sits in the cell membrane, a reasonable position for a protein involved in transporting things in and out of cells.

It is important to note that we are discussing *predicting* protein structure from the amino acid sequence, not *determining* that structure. The *determination* of a protein's structure is a scientific specialty all its own, requiring expensive intrumentation and highly specialized training. Most molecular biologists who study the function of genes and proteins do not have the training or the instruments to determine a protein's structure. Even if they did, the methods available at this time do not work on all proteins (see chapter 4). So predicting protein structures from DNA (and therefore amino acid) sequences is part of trying to figure out what a protein does and how it does it or what part of a protein is doing a specific thing.

Testing Structure-Function Predictions

Molecular biologists have come up with ways to test predictions about what parts of a protein are involved in specific functions even when they don't know what the structure is. How can they do this? In a way, they imitate nature. They introduce mutations into proteins (via manipulating the DNA sequence of the gene) and determine what the changes do to the protein's function.

For example, if you are studying a newly discovered DNA-binding protein and find that part of its amino acid sequence is consistent with a helix-turn-helix motif, you might suspect that the putative helix-turn-helix is the portion of the protein that binds DNA. Being a scientist, you would like to test your hypothesis.

First, you would have to clone the gene for your protein so you could work with it (see chapter 4). Then you could introduce specific base changes into the gene to cause specific amino acid changes in the protein. To test your hypothesis, you could change specific amino acids in the region you think might be a DNA-binding helix. If you found that your mutations caused the protein to bind to *different* specific DNA sequences, you would have good evidence that the region of the protein you altered was the DNA-binding region. Would you have proved that the structure was actually a helix-turn-helix? No, but your evidence that that particular area of the protein is directly involved in DNA binding would support the structure prediction.

Molecular biologists use approaches like the mutation experiment just described, often guided by predictions about a protein's structure, to try to learn which regions of proteins are involved in performing specific functions. In the end, though, the only way to be completely certain of a protein's structure is to use one of the instrumental methods described in chapter 4 for determining it, a long and exacting process.

Molecular biology, recombinant DNA technology, and scientific knowledge

Here at the end of this chapter is one final thought. You have just waded through a lot of information about basic molecular processes involving DNA and proteins. We started with the structure of DNA and genes and moved to protein synthesis, then to protein structure and function, and finally back to connections between changes in DNA sequence and changes in protein function. Along the way, we tried to emphasize that all the cellular operations involving DNA are performed by enzymes. The next chapter is a description of how scientists have learned to use some of these enzymes to manipulate DNA and, thereby, proteins. The whole business of using cellular enzymes to manipulate DNA is often called recombinant DNA technology.

Recombinant DNA technology as such seems to make the news in connection with controversial biotechnology products, gene therapy, and scary science fiction, but thousands of scientists around the world are using recombinant DNA technology every day simply to learn more about how life works. Think about the preceding discussion on predicting protein structure and probing protein function. These efforts involve computerized comparisons of DNA sequences from hundreds of genes. They involve identifying and cloning the gene for the hypothetical DNA-binding protein. They involve altering the sequence of the cloned gene and producing protein from the altered DNA sequence to see what the effects of the amino acid changes are. In real life, accomplishing these steps would all involve recombinant DNA technology.

One of the wonderful things (at least for a scientist) about the field of molecular biology is that the technologies it has spawned have made it possible to learn even more basic biology. It seems that the more we learn, the more neat technologies we develop. The new technologies let us design new kinds of experiments for learning even more—almost a chain reaction of knowledge. As you read chapter 4 and beyond, notice how the molecular technologies are applied to many scientific questions as well as to medical and other applications in day-to-day human life.

Summary

All of Earth's living creatures use DNA as their genetic material. The elegant structure of this wonderful molecule enables it to perform its two functions: replication and transmission of information. DNA controls the form of an organism by specifying the amino acid sequences of all the proteins in that organism. The proteins, in turn, form many of the organism's building blocks and carry out nearly all of its metabolic processes.

Protein synthesis in all known organisms uses the same genetic code and the same general process. The universality of this most basic life process demonstrates the interrelatedness of all life on the planet. The cells of all creatures synthesize mRNA. The codons of all their mRNAs carry the same meaning. The ribosomes of the simplest bacterium are capable of reading the genetic code and synthesizing the protein from a human brain cell.

All forms of earthly life arose as a result of genetic variation followed by natural selection. Mutation and recombination can result in the formation or addition of new genes, the destruction or deletion of old ones, major and minor changes in protein structures, and changes in gene regulation.

Now that scientists understand the inner workings of heredity and genetic variation, they have begun to experiment with those processes. People are beginning to harness the natural mechanisms of genetic variation and protein synthesis to manipulate the genetic content of organisms and direct their gene expression. Are these experiments fundamentally different from natural evolution? No, except in the sense that humans are at the controls. Are they occurring on a time scale that is immeasurably faster than nature's time scale? Yes.

The realization that humans now have the power to manipulate the genes of living organisms is exhilarating, awe inspiring, and potentially disturbing. It offers the hope of treatment for previously incurable genetic diseases. It holds out promise for an increased and better food supply. At the same time, our newfound power raises a multitude of hard questions. It has now become the responsibility of all of us to understand this power of genetic manipulation. Only through understanding can we hope to use and regulate this power wisely.

Selected Readings

Beardsley, T. 1991. Smart genes. *Scientific American* 265:86–95. Gene regulation and development.

Branden, C., and J. Tooze. 1991. *Introduction to Protein Structure.* Garland Publishing, Inc., New York. Well-illustrated, readable book about protein structure.

Cuerces-Amabile, C., and M. Chicurel. 1993. Horizontal gene transfer. *American Scientist* 81:332–341. Transfer of genetic information outside of parent to offspring.

Doolittle, R., and P. Bork. 1993. Evolutionarily mobile modules in proteins. *Scientific American* 269:50–56. Protein domain structure.

Hall, S. 1995. Protein images update natural history. *Science* 267:620–624. How protein structure determinations and new computer models of protein structure are changing the science of biology.

Holtzman, D. 1991. A "jumping gene" caught in the act. *Science* 254:1728–1729. Two cases of human hemophilia A apparently caused by the movement of a transposon.

McGinnis, W., and M. Kuziora. 1994. The molecular architects of body design. *Scientific American* 270:58–66. Genes, proteins, and development.

Ptashne, M. 1989. How gene activators work. *Scientific American* 260:40–47.

Ptashne, M. 1992. *A Genetic Switch,* 2nd ed., Blackwell Scientific Publications and Cell Press, Cambridge, Mass. Excellent short book describing in molecular detail how bacteriophage lambda switches from lysogenic to lytic growth. Protein structure, protein-protein interactions, protein-DNA interactions and gene regulation are combined in one well-understood biological system.

Rennie, J. 1993. DNA's new twists. *Scientific American* 266:122–132. Current thinking about DNA structure.

Rhodes, D., and A. Klug. 1993. Zinc fingers. *Scientific American* 268:56–65. Protein structure and gene regulation; zinc fingers are one structural motif used for DNA binding.

Richards, F. 1991. The protein folding problem. *Scientific American* 264:54–63. An overview of the attempt to understand what determines three-dimensional protein structure.

Roush, W. 1995. An "off switch" for red blood cells. *Science* 268:27–28. Description of the molecular biology of the stamina-enhancing mutation causing benign erythrocytosis.

Tjian, R. 1995. Molecular machines that control genes. *Scientific American* 272:54–61. Description of the complex assembly of proteins needed for transcription of eukaryotic genes.

Watson, J., M. Gilman, J. Witkowski, and M. Zoller. 1992. *Recombinant DNA.* Scientific American Books, W. H. Freeman & Co., New York. Excellent general reference.

Applying Molecular Biology: Recombinant DNA Technology

The Tools of DNA-Based Technologies

The 20th century has witnessed a veritable explosion of knowledge about molecular biology. This new knowledge has demonstrated more clearly than ever that all life on Earth is related: every living creature uses DNA and RNA to store and transfer genetic information, and every living creature uses the same genetic code for making its proteins. As their understanding of the inner workings of the hereditary machinery increased, scientists began to wonder if the genetic material and the expression process could be manipulated. They asked questions such as these: Is it possible to transfer a gene from one organism to another and have that gene expressed in the new host? Is it possible to combine two genes into one? Is it possible to regulate gene expression? Can we use the uniqueness of every person's DNA for identification? Can DNA changes tell us anything about evolution? Can we look at DNA and detect mutations that will lead to disease? Can we identify microorganisms on the basis of their DNA sequences? Is it possible to change the genetic makeup of an organism in a way that is advantageous for human society?

Before these questions could be answered, ways to manipulate DNA had to be found. This task turned out to be relatively simple, because DNA is constantly manipulated in nature. It is copied, cut, and rejoined over and over again in living cells. Nature's agents of DNA manipulation are enzymes. DNA-based technology employs these enzymes, which scientists have identified and purified for use in the laboratory.

Cellular enzymes

What are the cellular enzymes used by biotechnologists to manipulate DNA and protein expression? We discuss some of the more important ones here.

Restriction Endonucleases

Restriction endonucleases recognize specific base sequences in a DNA molecule and cut the DNA at or near the recognition sequence in a consistent way (Fig. 4.1). The most commonly used restriction enzymes recognize palindromic sequences (sequences in which both strands read the same in the 5′-to-3′ direction). Restriction endonucleases are made by bacteria and are thought to defend their bacterial producers against invading DNA, such as bacteriophage genomes. Their names indicate the organism from which they were purified (*Eco*RI from *Escherichia coli*, *Hin*dIII from *Haemophilus influenzae*, and so forth.) (See the activity *DNA Scissors: Introduction to Restriction Enzymes* for more information.)

DNA Polymerases

DNA polymerases are the enzymes that copy DNA. They synthesize a single new strand that is complementary to the template strand of the parent molecule. In order to synthesize new DNA, DNA polymerases absolutely require a template strand and a primer. A primer is any piece of DNA that is base paired to the template strand in such a way that the 3′ end of the primer is available to serve as the starting point for the new DNA. The first base of the new DNA is attached via a phosphodiester bond to the 3′ end of the primer and is complementary to the base on the template strand. Synthesis proceeds as more bases are attached to the primer (Fig. 4.2). DNA polymerases have been purified from a wide variety of organisms, which is not surprising, since all organisms must copy their DNA. (See the activity *DNA Replication* for more information.)

One group of organisms whose DNA polymerase enzymes have become very important in the laboratory is bacteria that live at very high temperatures (such as in hot springs and thermal vents in the ocean floor). Since their natural environments are very hot, their DNA polymerases function well at high temperatures. Scientists now use these heat-tolerant polymerases to

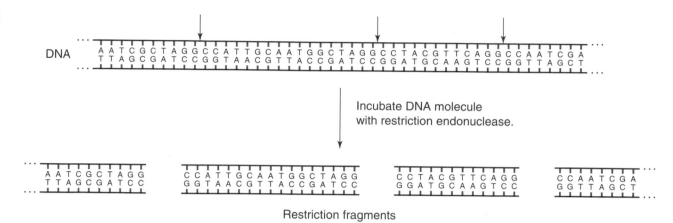

DNA

```
... ┌─┬─┬─┬─┬─┬─┬─┬─┬─┬─┬─┬─┬─┬─┬─┬─┬─┬─┬─┬─┬─┬─┬─┬─┬─┬─┬─┬─┬─┬─┬─┬─┬─┬─┬─┬─┬─┬─┬─┬─┬─┬─┬─┐ ...
    A A T C G C T A G G C C A T T G C A A T G G C T A G G C C T A C G T T C A G G C C A A T C G A
    T T A G C G A T C C G G T A A C G T T A C C G A T C C G G A T G C A A G T C C G G T T A G C T
... └─┴─┴─┴─┴─┴─┴─┴─┴─┴─┴─┴─┴─┴─┴─┴─┴─┴─┴─┴─┴─┴─┴─┴─┴─┴─┴─┴─┴─┴─┴─┴─┴─┴─┴─┴─┴─┴─┴─┴─┴─┴─┴─┘ ...
```

Incubate DNA molecule
with restriction endonuclease.

```
... ┌─┬─┬─┬─┬─┬─┬─┬─┬─┐        ┌─┬─┬─┬─┬─┬─┬─┬─┬─┬─┬─┬─┬─┬─┬─┐        ┌─┬─┬─┬─┬─┬─┬─┬─┬─┬─┬─┐        ┌─┬─┬─┬─┬─┬─┬─┬─┐ ...
    A A T C G C T A G G        C C A T T G C A A T G G C T A G G        C C T A C G T T C A G G        C C A A T C G A
    T T A G C G A T C C        G G T A A C G T T A C C G A T C C        G G A T G C A A G T C C        G G T T A G C T
... └─┴─┴─┴─┴─┴─┴─┴─┴─┘        └─┴─┴─┴─┴─┴─┴─┴─┴─┴─┴─┴─┴─┴─┴─┘        └─┴─┴─┴─┴─┴─┴─┴─┴─┴─┴─┘        └─┴─┴─┴─┴─┴─┴─┴─┘ ...
```

Restriction fragments

Figure 4.1   Restriction endonucleases recognize and cut specific sites in a DNA molecule. The arrows indicate the cleavage sites of one such endonuclease.

"Xerox" DNA fragments via the polymerase chain reaction (PCR), which requires repeated heating of the enzyme-DNA mixture. (See *Amplifying DNA: PCR* later in this chapter and the activity *The Polymerase Chain Reaction: Paper PCR* for more information.)

RNA Polymerases

RNA polymerases are the enzymes that read a DNA sequence and synthesize a complementary RNA molecule. RNA polymerases require a special sequence of

bases on the DNA template, called a promoter, to signal to them where to begin transcription, but they do not require a primer (Fig. 4.3). Like DNA polymerases, RNA polymerases have been purified from many organisms, since all organisms must transcribe their genes.

DNA Ligases

Ligases join pieces of DNA (or RNA) together by forming new phosphodiester bonds between the pieces (Fig. 4.4).

Figure 4.2   Activity of DNA polymerase.

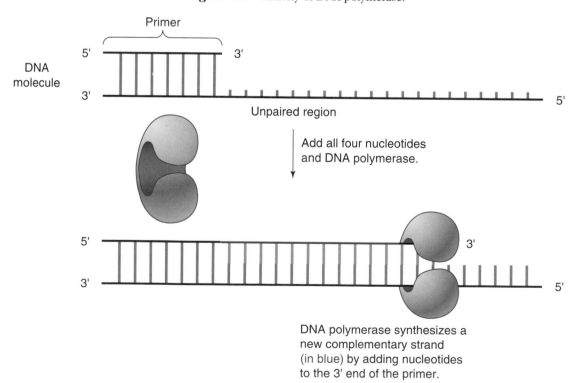

Primer

DNA molecule

5' ——————— 3'
3' ———————————————————————— 5'

Unpaired region

Add all four nucleotides
and DNA polymerase.

5' ——————————————————— 3'
3' ———————————————————————— 5'

DNA polymerase synthesizes a new complementary strand (in blue) by adding nucleotides to the 3' end of the primer.

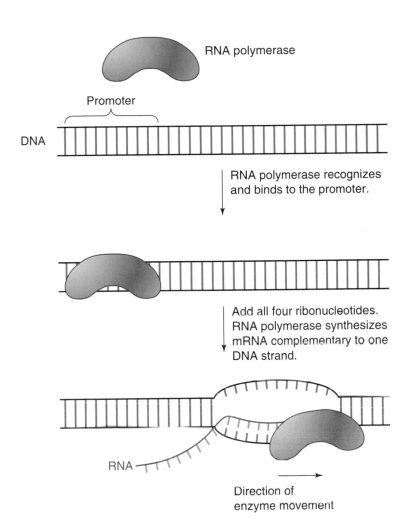

RNA polymerase

Promoter

DNA

RNA polymerase recognizes
and binds to the promoter.

Add all four ribonucleotides.
RNA polymerase synthesizes
mRNA complementary to one
DNA strand.

RNA

Direction of
enzyme movement

Figure 4.3 Activity of RNA polymerase.

Reverse Transcriptases

As their name suggests, reverse transcriptases read an RNA sequence and synthesize a complementary DNA (often abbreviated as cDNA) sequence (Fig. 4 5). These enzymes are made by RNA viruses that convert their RNA genomes into DNA when they infect a host (see *Virus Genomes* in chapter 3). Reverse transcriptases allow scientists to synthesize a DNA gene from an RNA message. This ability is useful for dealing with eukaryotic genes, since the original genes are often split into many small pieces that are separated by introns in the chromosome. The messenger RNA (mRNA) from these genes has undergone splicing in the eukaryotic cell, and the introns are gone, leaving only the coding sequences (see *Differences in the Molecular Biology of Prokaryotes and Eukaryotes*

Figure 4.4 Activity of DNA ligase.

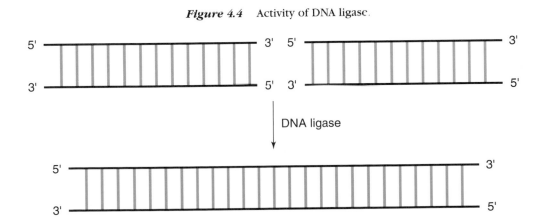

DNA ligase

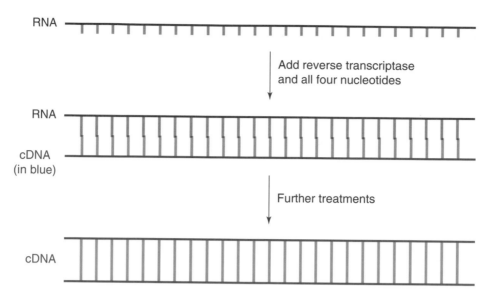

RNA

Add reverse transcriptase
and all four nucleotides

RNA

cDNA
(in blue)

Further treatments

cDNA

Figure 4.5 Activity of reverse transcriptase.

in chapter 3). Reverse transcriptase can convert this mRNA into a "prespliced" gene consisting only of protein-coding sequences. Why would someone want a prespliced gene? Because bacteria possess no equipment for splicing, they must be given a prespliced version if they are to express the correct protein product from a eukaryotic gene. Thus, the process of making cDNA from mRNA is important for expressing eukaryotic genes in prokaryotes.

Translation Extracts

Although not enzymes, translation extracts have also been useful to biotechnologists. These extracts are made from eukaryotic cells or bacteria, and they contain ribosomes, transfer RNAs, and other necessary translation components. When mRNA is added to such an extract, translation occurs, and the protein encoded in the RNA message is synthesized.

One of the more time-consuming tasks for the pioneers of biotechnology was purifying enzymes such as those described previously for laboratory use. Not surprisingly, genes for most of these enzymes have now themselves been cloned and engineered for large-scale protein production. Many companies produce and sell these valuable proteins for commercial and research uses. Essentially all of the enzymes used in DNA-based technologies can now be easily purchased.

Natural vectors

In addition to manipulating DNA outside the cell, biotechnologists must be able to introduce the manipulated DNA into the host cell of their choice and have it stably maintained there. Once again, nature provides

the tools. The term **vector** is used to describe any vehicle that carries DNA into a host cell. Plasmids and viruses are useful natural vectors both for delivering DNA to a new host cell and for providing for the maintenance of that DNA after the delivery.

Naturally occurring plasmids are used for transfer of DNA by transformation (the uptake of naked DNA by a cell; see *Asexual reproduction and recombination* in chapter 2). In addition to being small (and thus easy to work with), plasmids are replicated by the host cell and distributed to daughter cells during cell division. Therefore, a new gene introduced on a plasmid vector will be stably maintained in progeny cells. The most common form of gene cloning is the insertion of a gene into a plasmid and the subsequent introduction of the plasmid into the desired host. Scientists now have many plasmids tailor-made for cloning procedures. (See the activity *Transformation of* Escherichia coli for more information.)

One particularly useful natural plasmid is the Ti plasmid from the plant pathogen *Agrobacterium tumefaciens*. After the bacterium infects a plant, the Ti plasmid transfers itself from the bacterial cell into the invaded plant and inserts certain of its genes into the plant DNA. Biotechnologists take advantage of the Ti plasmid by adding genes of interest to the plasmid. The altered Ti plasmid is reintroduced into *A. tumefaciens*, the plasmid-containing bacterium is inoculated into a plant, and the altered Ti plasmid then transfers the new genes to the plant's DNA. Unfortunately, only certain plants are susceptible to infection by *A. tumefaciens*. (See the activity Agrobacterium tumefaciens: *Nature's Plant Genetic Engineer* for more information.)

Biotechnologists also employ viruses as agents of DNA transfer. Many viruses normally operate by combining their DNA with that of their host (see the activity *Transduction of an Antibiotic Resistance Gene* for more information). Biotechnologists can insert genes of interest into the genomes of these viruses and then let the virus carry the new DNA into the host cell and recombine it with the host DNA. One advantage of viruses as vectors is their extreme host specificity. Foreign genes can be targeted to specific cell types by the appropriate virus. This characteristic of viruses may make them important players in gene therapy procedures. For example, a respiratory virus has already been used experimentally to introduce a healthy cystic fibrosis gene into the lungs of rats and the nasal passages of people.

Other tools for transferring DNA

Although transformation and transduction are useful means of introducing foreign DNA into cells, they do not work in every case. Especially with eukaryotic cells (both plant and animal), it can be difficult to introduce a foreign gene. Two recent technological methods of getting DNA inside cells are electroporation and the so-called DNA gun. In electroporation, an electric current is used to force DNA across the cell membrane. Although this procedure is also used with bacteria, it is particularly helpful for introducing plasmids into eukaryotic cells.

The other means of DNA introduction is the DNA gun. To use the gun, the desired DNA must first be coated onto microscopic pellets. These pellets are then fired into the target cells. At some frequency, the DNA on the pellets may be maintained and expressed in the target cells. The DNA gun has proven useful for introducing DNA into plant cells that are not susceptible to infection by *A. tumefaciens* and was recently used successfully to introduce DNA into animals.

A different approach to the introduction of new genetic material is cell fusion, the actual fusing of two cells into one. Cell fusion is usually achieved by treating the two different cell types with a chemical that affects the cell membranes and promotes the fusion of cells in close contact. For plant cell fusion, the thick outer walls must first be digested away with enzymes, leaving the membrane-bound **protoplasts.** The fused cells do not maintain both sets of chromosomes. Instead, they lose chromosomes until they reestablish the correct number. Chromosomes from either parent can be lost, apparently randomly, resulting in a cell containing some chromosomes from each parent cell. Cell fusion is the basis of monoclonal antibody technology.

Important chemical methods

Biotechnologists and molecular biologists are not the only scientists who study nucleic acids and proteins. Chemists and biochemists have investigated the structures of these molecules for many years, and their efforts have resulted in some extraordinarily useful technologies.

Automated DNA Synthesis

Scientists studying the structure of DNA worked out reaction conditions for chemically synthesizing single-stranded DNA molecules. These methods have been standardized to the point of automation. DNA synthesizers can now make single-stranded DNA molecules over 100 bases long from scratch. Such relatively short single-stranded DNA segments are called **oligonucleotides.** The desired oligonucleotide sequence is typed into a computer that controls the synthesizer, and the synthesizer connects the nucleotides in the proper order. How can we get double-stranded DNA? Synthesize two complementary single strands and allow them to hybridize (form base pairs). Chemically synthesized oligonucleotides are often referred to as synthetic oligonucleotides.

Automated Protein Sequencing

A vital piece of information about any protein is the sequence of its amino acids. Initially laborious chemical methods have now been automated, and the latest protein-sequencing equipment can determine the amino acid sequences of tiny amounts of purified protein. Limitations of protein sequencing are that the protein must be pure and that the accuracy of the sequencing deteriorates as the number of sequential amino acids determined increases. To avoid the problem of inaccuracy, scientists often determine only the first several amino acids on one end of a protein, or they cleave the protein molecule into several smaller fragments that can each be sequenced separately and accurately.

Three-Dimensional Protein Structure Analysis

The function of a protein depends on its three-dimensional structure. Chemists and biochemists now use the sophisticated techniques of X-ray crystallography and nuclear magnetic resonance (NMR) spectroscopy to determine protein structure. These methods were originally used to determine the structures of relatively small molecules. They have been refined and augmented for application to complicated molecules such as proteins. Both methods are difficult and have significant limitations.

X-ray crystallography, the more widely used technique, requires the use of extremely pure protein crystals. Crystals are very regular, packed arrays of molecules, and since proteins can have quite irregular shapes, most are not easy to crystallize, and many cannot be crystallized at all. Even for those that can be crystallized, growing protein crystals large enough for crystallography studies can take months or years. If a suitable crystal is obtained, a beam of X rays is directed into it. The regular array of protein molecules within the crystal diffracts the X rays in a pattern that is recorded on film and used to deduce the arrangement of atoms within the molecules. X-ray diffraction data obtained by Rosalind Franklin from DNA crystals were used by Watson and Crick to determine the three-dimensional structure of DNA.

Protein NMR uses highly concentrated, pure solutions of protein, avoiding the need for crystals. However, many proteins do not remain soluble at the required concentrations but instead aggregate into insoluble clumps of molecules. In addition, the technique works only with very small proteins.

NMR exploits the magnetic properties of certain atomic nuclei (usually ^1H in protein NMR) to determine molecular structure. These nuclei have a magnetic moment or spin. When protein molecules are placed in a strong magnetic field, the spins of the hydrogen nuclei align along the field. The nuclei are then excited by applying electromagnetic pulses to the sample. When the nuclei return to their ground state, they emit electromagnetic energy, which can be measured. The frequency of the emitted electromagnetic radiation depends on the precise molecular environment of the individual nucleus. By varying the nature of the applied electromagnetic pulses and measuring the subsequent emissions, we can deduce various molecular properties of the sample.

The structures of the proteins analyzed to date dramatically illustrate the relationship of form and function. Proteins have grooves, pockets, and even pincerlike structures for binding to DNA or other molecules. Some proteins assume a different shape when bound to their targets. Knowing the structure of a protein often allows us to see how the protein performs its function. It also helps molecular biologists think about how other proteins might work. Knowing the structure of a protein helps biotechnologists design ways to improve the protein.

Fundamental procedures

The tools and technologies described previously are employed in several basic procedures that form a core of standard techniques for molecular biologists and biotechnologists. These standard techniques are rather like the basic procedures that a chef learns: making a white sauce, sautéing, beating egg whites, poaching, cooking a sugar syrup. The chef then applies selected techniques to different ingredients to produce a wonderful variety of dishes. Biotechnology is rather like that. Using these basic procedures with all kinds of organisms and experimental systems, biotechnologists and molecular biologists working around the world are producing an amazing array of knowledge, products, and even new basic techniques. What are some of these basic procedures?

Separating DNA Fragments: Gel Electrophoresis

In the course of almost any manipulation of DNA, it is necessary (or at least desirable) to have a look at the various pieces of DNA you are working with. The standard method used to separate DNA fragments is electrophoresis through agarose gels.

Agarose is a polysaccharide (as are agar and pectin) that dissolves in boiling water and then gels as it cools, like Jello-O. To perform agarose gel electrophoresis, a slab of gelled agarose is prepared, DNA is introduced into small pits in the slab, and then an electric current is applied across the gel. Since DNA is highly negatively charged (because of the phosphate groups), it is attracted to the positive electrode. To get to the positive electrode, however, the DNA must migrate through the agarose gel.

Smaller DNA fragments can migrate through an agarose gel faster than large fragments. In fact, the rate of migration of linear DNA fragments through agarose is inversely proportional to the $\log_{10}$ of their molecular weights. What it boils down to is that if you apply a mixture of DNA fragments to an agarose gel, start current flowing, wait a little, and then look at the fragments, you will find that the fragments are spread out like runners in a race, with the smallest one closest to the positive electrode, the next smallest following it, and so on (Fig. 4.6). Because of the mathematical relationship given previously, it is possible to calculate the exact size of a given fragment on the basis of its migration rate. (For more information, see the activity *DNA Goes to the Races*.)

After gel electrophoresis, the DNA fragments in the gel are usually stained to render them visible. DNA fragments can also be isolated and purified from agarose gels.

Finding Complementary Base Sequences: Hybridization

Hybridization is a natural phenomenon that provides another essential tool for biotechnologists. It

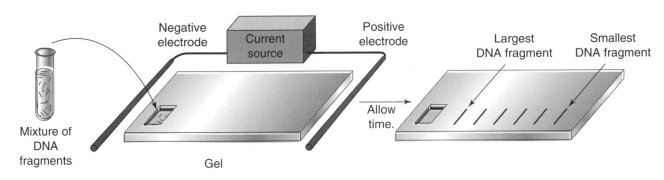

Figure 4.6   Gel electrophoresis of DNA fragments.

occurs as a consequence of the structure of DNA. Hybridization (also called **annealing** or **renaturation**) is the term used to describe the process in which two single DNA strands with complementary base sequences stick together to form a correctly base-paired double-stranded molecule. Hybridization occurs spontaneously: if two complementary single DNA strands are mixed together and left alone, they will hybridize. The time it takes for hybridization to occur is directly related to the length of the DNA sequences involved; as one might expect, short complementary sequences can line up correctly and base pair much faster than long sequences can. (For more information, see the activity *Detection of Specific DNA Sequences: Hybridization Analysis*.)

Why is hybridization an essential biotechnological tool? It provides a way to look for a specific DNA sequence in a mixture or to operate selectively on specific sequences. For example, suppose we want to know whether a specific DNA sequence has been successfully inserted into a bacterium's genome. To find out, we first make a **probe** for that sequence. A probe is a piece of single-stranded DNA that will hybridize only to the DNA of interest (because of its unique base sequence). Next, we denature the bacterium's DNA (often by heating to 95°C, which causes base pairs to come apart), thus exposing the bases on single strands. Then we add the probe. If the new DNA sequence is present, the probe will hybridize, or "stick," to the sample DNA. A hybridized probe can be detected by a number of methods (Fig. 4.7).

Another important use of hybridization is to provide a starting place for additional techniques. A segment of DNA with a hybridized probe is the starting material for a variety of other procedures (see below for examples). Hybridization also works with RNA molecules; complementary RNAs can hybridize to form a double-stranded RNA molecule, and a complementary RNA can hybridize to DNA. As you will see, hybridization is an important step in many of the other basic

methods. The ability to synthesize oligonucleotides with any desired base sequence to use as probes has greatly increased the options for applications of all of them.

Transfer of DNA Fragments: Blotting

Often it is important to know which DNA fragment in a mixture contains a sequence of interest. Gel electrophoresis and staining alone cannot answer this question, because all DNA fragments look alike when stained. By now you know that if we had a specific probe for the sequence of interest, we could perform hybridization on the DNA fragments and determine which one contained our sequence. Unfortunately, hybridization doesn't work well on DNA fragments embedded in an agarose gel. To get around this problem, scientists use a procedure called blotting.

Figure 4.7   Hybridization analysis.

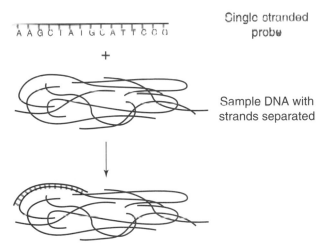

If sample DNA contains the base sequence complementary to the probe sequence, the probe will form base pairs with the sample DNA, or hybridize to the sample, and physically stick to it.

Blotting DNA is analogous to blotting ink writing. If you write something with a fountain pen and cover the writing carefully with a sheet of blotting paper, the pattern of your writing will be exactly transferred to the blotter. DNA blotting works in the same manner. DNA fragments are separated by agarose gel electrophoresis, and then the gel itself is covered with a membrane and blotting paper. The DNA fragments in the gel transfer to the membrane in exactly the same arrangement as they were in the gel. Once the DNA fragments are on the membrane, they can be hybridized to probes to test for the presence of specific sequences (Fig. 4.8).

This DNA transfer and hybridization technique was first described by a scientist named Southern in 1975. It acquired the name "Southern blotting" for its originator. Later on, other scientists modified the procedure to use with RNA fragments. They named their procedure for transferring RNA from a gel to a membrane "Northern blotting." Later still, other researchers figured out how to transfer proteins from gels to

membranes to do tests on them. The name of this procedure? "Western blotting." No Eastern blot has yet been invented. (For more information, see the activity *Detection of Specific DNA Sequences: Hybridization Analysis.*)

Amplifying DNA: PCR

PCR is a clever procedure that takes advantage of DNA polymerase enzymes and synthetic oligonucleotides to make many copies of a specific segment of DNA. In brief, we synthesize two primers that hybridize to opposite strands of the DNA molecule at the boundaries of the segment to be copied. DNA polymerase then copies both of the strands, starting at the two primers (Fig. 4.9). Because the reaction mixture contains primers complementary to both strands of the DNA, the products of the DNA synthesis can themselves be copied with the opposite primer. PCR is actually a chain reaction that results in thousands or millions of copies of a given DNA segment. It is used to amplify DNA for cloning, to detect

Figure 4.8 Blotting and hybridization analysis. DNA fragments separated by gel electrophoresis are transferred from the gel to a membrane. The membrane is then exposed to a probe to test for the presence of specific DNA sequences.

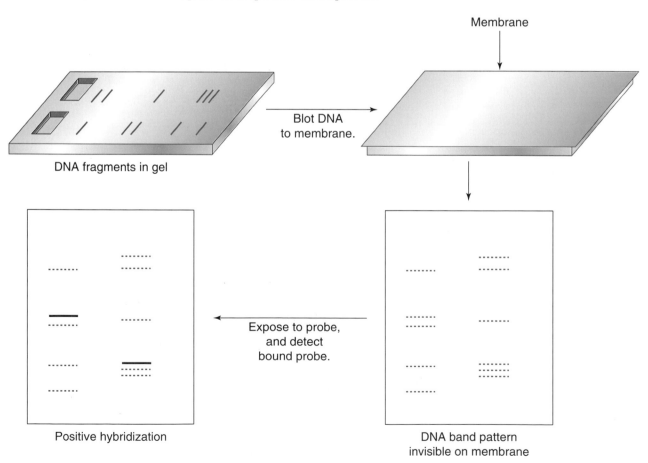

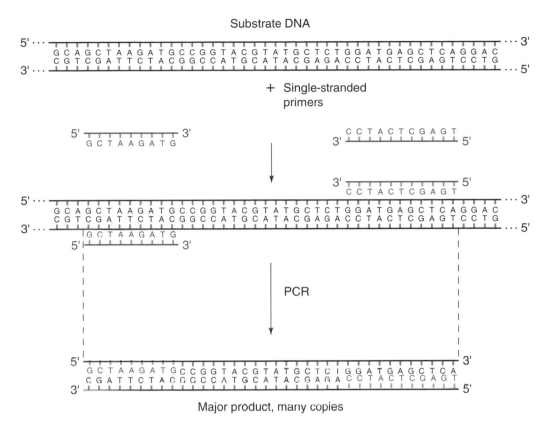

Substrate DNA

+ Single-stranded primers

PCR

Major product, many copies

Figure 4.9 PCR makes many copies of a DNA segment lying between and including the se quences at which two single-stranded primers hybridize to the substrate DNA molecule. The primers are usually synthetic oligonucleotides.

scarce DNA, and to distinguish different DNA samples (for example, from related viruses). (See the activity *The Polymerase Chain Reaction: Paper PCR* for more information.)

Determining the Base Sequence of a DNA Molecule

The sequence of bases in a gene determines the sequence of amino acids in a protein. If we can read the base sequence of a gene, we can learn about the protein it encodes. Furthermore, if we want to manipulate the gene, it helps to know what we are starting with. The ability to determine the sequences of bases in a DNA molecule is an important tool for research as well as for applied science.

The first attempts to determine the sequence of bases in a given piece of DNA were completely chemical approaches. These chemical methods were successful but slow and labor-intensive. The advent of biotechnology has greatly simplified the process. Even so, determining a DNA sequence is still a multistep process that requires precision and care. Sequencing relatively short stretches of DNA (a few hundred bases long) can usually be accomplished in a

few hours or days, but sequencing long segments takes much longer. DNA sequencing was recently automated. Most large laboratories that focus heavily on sequencing DNA have acquired the expensive sequencing instruments, but most research laboratories that sequence DNA only as relatively minor parts of other projects still rely on manual procedures.

Current DNA sequencing methods employ DNA polymerase enzymes, synthetic oligonucleotides, and hybridization techniques. The DNA polymerase synthesizes a new DNA strand on a single-stranded template using special nucleotide derivatives. These nucleotide derivatives allow scientists to determine the order in which new bases are added to the growing chain. The order in which bases are added to the new strand reveals the DNA sequence of the template strand through the rules of complementarity. (See the activity *DNA Sequencing* for more information.)

Remember that DNA polymerases must have a primer from which to begin synthesis (see *DNA Polymerases* above). For DNA sequencing, we need a primer that is hybridized to the single-stranded template adjacent to the area that we wish to sequence. If any of the adja-

cent sequence is known, a complementary oligonucleotide primer can be chemically synthesized. Another option is to clone the region of interest into a
vector of known sequence. A primer complementary
to adjacent vector sequences can then be used.

RNA can also be sequenced. RNA sequencing is similar to DNA sequencing but uses the enzyme reverse
transcriptase. Reverse transcriptase reads the base sequence of single-stranded RNA and synthesizes a
complementary DNA molecule (see above). The same
strategy of using altered nucleotides to give a detectable sequence is used with reverse transcriptase
and an RNA template. Reverse transcriptase also requires a primer. As with DNA sequencing, chemically
synthesized oligonucleotides are usually hybridized to
the template RNA to provide a starting point for the
enzyme.

Cloning DNA
The term **cloning** means the production of identical
copies of something through asexual reproduction.
When applied to DNA, cloning is usually understood
to mean the insertion of a piece of DNA into a cell in
such a way that the DNA will be replicated (copied)
and maintained. Simply forcing a segment of DNA
into a cell does not usually result in the copying and
maintenance of that DNA. Cells, whether prokaryotic
or eukaryotic, contain a variety of DNA-degrading enzymes that destroy unprotected DNA segments. To
clone a segment of DNA (a gene, perhaps), the DNA
must be placed into a carrier DNA molecule (a vector) whose structure is immune to the degrading enzymes. The vector protects the cloned segment and
provides for the replication and maintenance of the
DNA in its new host cell. Natural virus and plasmid
vectors were described previously.

The most basic method of cloning DNA begins with
cutting the DNA into segments with a restriction enzyme (or a combination of enzymes). The selected
vector is cut with the same enzyme(s). The restriction fragments and the cut vector are mixed with the
enzyme DNA ligase (see above), which connects the
desired fragment with the vector. Plasmids that contain a segment of DNA originally from another source

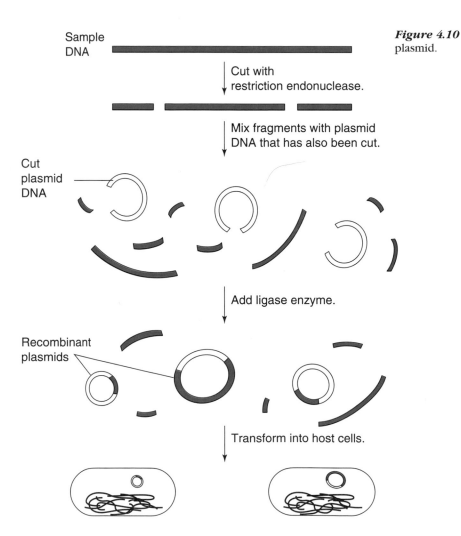

Figure 4.10   Cloning a DNA segment into a plasmid.

are often called *recombinant plasmids* (Fig. 4.10). (See the lesson *Recombinant Paper Plasmids* for more information.)

You are probably wondering what happens to all the other fragments when you ligate vector and insert molecules and whether the vector couldn't simply be reconnected to itself or other fragments. After all, the ligase enzyme is "blind"—it merely connects one DNA end to another. The answer to both questions is yes. Part of the process of cloning is identifying the recombinant molecule you want out of the variety that can be produced. (See the activity *Recombinant Paper Plasmids*.)

Making DNA Libraries

A common application of cloning is to produce what is called a **DNA library** or **genetic library.** To make

a library, the entire genome of an organism is digested with restriction enzymes, and the entire batch of products is mixed with vector DNA and ligated. The resulting mixture of recombinant plasmids is transformed into a host organism. The pool of transformed cells containing the collection of plasmids is called a DNA library of that organism (Fig. 4.11A). The idea is that the resulting recombinant plasmids or viruses will contain fragments covering the whole genome of the organism. A DNA library can be used like a resource. If a scientist knows what gene he wishes to study, he has only to look for it in the library and pull out a cloned version. (See below and the activity *Detection of Specific DNA Sequences: Hybridization Analysis.*)

A special kind of DNA library is often made from eukaryotic organisms. Much of the DNA in eukaryotes is

Figure 4.11 DNA libraries.

A. Genomic DNA library. The insertions in the recombinant plasmids represent the entire DNA content of the organism.

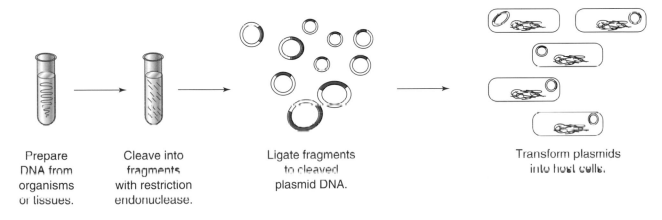

| Prepare DNA from organisms or tissues. | Cleave into fragments with restriction endonuclease. | Ligate fragments to cleaved plasmid DNA. | | Transform plasmids into host cells. |

B. cDNA library. The insertions in the recombinant plasmids represent genes that were being expressed in the sample.

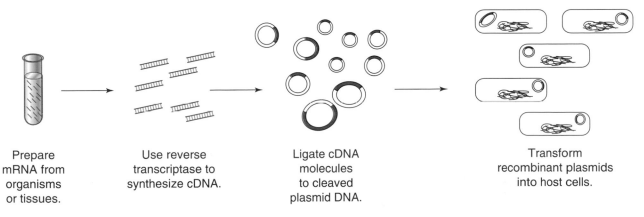

| Prepare mRNA from organisms or tissues. | Use reverse transcriptase to synthesize cDNA. | Ligate cDNA molecules to cleaved plasmid DNA. | | Transform recombinant plasmids into host cells. |

noncoding and would not be of interest to scientists studying the proteins encoded by genes. To avoid cloning vast regions of noncoding DNA, scientists make what is called a **cDNA library.** The first step in making one of these libraries is to isolate mRNA from the organism. Then, reverse transcriptase is used to generate cDNA copies of all the mRNA molecules. Finally, these cDNA copies are inserted into vector DNA molecules en masse, resulting in a cDNA library. The cDNA library contains DNA with the coding sequences of all the proteins the organism was producing when the mRNA was isolated (Fig. 4.11B).

Putting the Knowledge and Tools to Work

Given the tools and procedures described previously, we are now able to use our understanding of the structure and function of DNA to solve problems and make useful products. What kinds of things can we do? The following discussion gives a sample of some of the possibilities.

Analyzing genetic variation

As discussed in chapter 2, genetic variation, both between and within species, is pervasive in the natural world. It provides the raw material on which natural selection acts to create new species and change existing ones. While its importance to evolutionary change is obvious, the ways in which the existence of genetic variation is used in human endeavors is probably less so. We diagnose certain diseases, determine the compatibility of tissue types in organ transplants, produce new and better agricultural crops, find the perpetrators of crimes, and probe the origins of species by investigating genetic variability.

Until recently, determinations such as these were based on phenotypic characteristics, that is, some tangible manifestation of the information contained in the DNA molecule. Phenotypic manifestations, while important, are limited by our ability to observe and measure them. Furthermore, our ability to assess genetic variation by measuring phenotypic characteristics depends on the relationship between genotype and phenotype. Sometimes genetic differences are not detectable at the phenotype level (for example, people with blood group genotypes AA and AO both express the phenotype of blood group A, despite the genetic difference).

About 30 years ago, more detailed methods of looking at variations in proteins came into widespread use. Investigators could use electrophoresis (see above;

electrophoresis of proteins is fairly similar to electrophoresis of DNA) to compare sizes and electrophoretic mobilities. Amino acid sequence determinations gave scientists an indirect way of looking at variation in individual genes. Protein sequence comparisons are still used to compare species, particularly in evolutionary science.

Now we are able to assess genetic variation by looking at the DNA itself. There is much more variation in the DNA of different individuals than we can observe and measure phenotypically, even through amino acid sequence determinations. By looking directly at the DNA, we can make our measures of genetic variation more accurate and more precise.

How is genetic variation analyzed with DNA-based technologies? One method is DNA sequence comparisons. This method gives detailed and accurate information but cannot be applied to whole genomes because of their great size. DNA sequence comparisons focus on small, specific regions. To give meaningful data, these target regions have to be areas of significant variation; comparisons of most human genes would show very few if any differences between individuals. For this reason, sequencing studies of individuals of the same or closely related species often focus on mitochondrial DNA. Mitochondria have small circular genomes that undergo mutation at a much higher rate than the nuclear genome; thus, mitochondrial DNA sequences display much more variation between individuals than do most nuclear sequences.

One way to compare whole genomes is through hybridization studies. In this approach, genomic DNAs from two species are prepared and then hybridized to each other. The more similar the nucleotide sequences of the two species, the more extensive the regions of base-paired duplex they will form. To measure the extent of the duplex DNA, the hybrid molecules are heated until the strands separate again. The temperature at which the strands separate (the **melting temperature**) is a function of the extent of base pairing. The more perfectly matched the two genomes, the higher the temperature required to separate the strands of the hybrid molecules. The melting temperature thus gives a gross measurement of DNA sequence similarity.

A shortcut to DNA sequence comparisons is to focus only on changes to the short base sequences at which restriction enzymes cut the DNA molecule. Changes in the DNA sequence can create or eliminate these sites. By digesting two DNA molecules with the same restriction enzyme and comparing the lengths of the fragments generated, we can estimate how similar the two molecules are. Two identical molecules will give identical patterns. Two fairly similar molecules will give similar patterns, with perhaps a few differences in fragment size (caused by addition or removal of cut sites through mutations). Scientists estimate the similarity of DNA molecules from the similarity of their restriction fragment lengths. This approach permits sampling of variation across the entire genome.

Just as an observable phenotypic variation is called a polymorphism (from the Greek for "many forms"), these differences in restriction fragment lengths are also called polymorphisms. Restriction fragment length polymorphisms (RFLPs) are used in DNA fingerprinting, paternity determinations, genetic disease diagnosis, plant breeding, and evolution and conservation research. (See the activity *Analyzing Genetic Variation: DNA Typing.*)

Evolutionary Studies

Evolution is a unifying theme in biology, and molecular biology has provided powerful new tools for studying it. Evolutionary biologists seek to discover how individual species arose from earlier forms and to understand the mechanisms of the process of evolution itself. Biotechnology has literally revolutionized both endeavors.

Biologists who seek to describe how modern species arose had to rely formerly on morphological, ecological, and behavioral comparisons between various modern species and between modern and fossil forms to deduce degrees of kinship. DNA and protein sequence comparisons have given these scientists an entirely new set of data to consider. To use molecular data in evolutionary studies, biologists first assemble protein or DNA sequence data from a specific protein or region of the genome in the group of organisms under study. They then measure the degree of difference in the sequences. On the assumption that changes accumulate slowly and relatively steadily, they construct various "trees" that show how the different sequences could have been generated from a common ancestor. These processes are assisted by computer programs. An evolutionary tree based on the amino acid sequence of the protein cytochrome *c* is shown in Fig. 4.12.

Another molecular evolutionary analysis found that the kiwi, a flightless New Zealand bird, is more closely related to the flightless birds of Australia than to the moas, the other major group of New Zealand flightless birds. DNA and protein sequence comparisons as well as hybridization studies found that humans and chimpanzees are more similar to one another than to any other species. These studies suggest that ancestors of modern humans and chimps probably diverged a mere 5 million years ago. Using molecular techniques to probe evolutionary relationships between species is a large and active area of current research.

An interesting twist from the field of molecular evolution is the discovery that we can extract DNA fragments from appropriate ancient samples. DNA is rapidly degraded to small fragments after an organism dies, and only special combinations of circumstances will preserve soft-tissue cells with DNA. However, samples preserved in bogs, including 17-million-year-old magnolia leaves, have yielded enough DNA for comparisons with modern species, as have amber-encased insects. It is also possible to recover DNA fragments from bones and teeth, which are more commonly preserved than is soft tissue. PCR is particularly useful in studying ancient DNA because it can amplify the small amounts typically present in specimens. We stress that the DNA recovered in these processes is extremely fragmentary and is useful only for comparisons with modern sequences to measure the amount of change. So far, experience with ancient DNA suggests that using it to resurrect extinct species is impossible.

For scientists who seek to understand the mechanisms behind evolution, molecular biology and biotechnology have provided a gold mine of new information. First, studies of genes and genomes from many organisms have underscored how fundamentally similar we all are. For example, studies using hybridization, RFLP analysis, and sequence comparisons suggest only about a 1.6 to 3% difference in the genomes of humans and chimpanzees. From earlier protein and enzyme studies, we already knew that all animals produce about the same complement of enzymes and proteins. Now that we are mapping genes to specific locations on chromosomes, we find that animals' genes are arranged similarly from species to species. Differences in chromosome number and structure seem to have been generated in a process in which chromosomes were cut or broken into large pieces and then put back together in many different ways into different final numbers of chromosomes.

Looking at the DNA sequences of individual genes, we see that nature appears to be quite economical: once a protein function that works has evolved, it is often used over and over. We find similar domains in many different proteins. We find proteins that appear to have diverged from two original copies of the same gene. We find similar proteins filling similar roles in widely disparate species. For example, the proteins that govern body plan in worms, flies, mice, and hu-

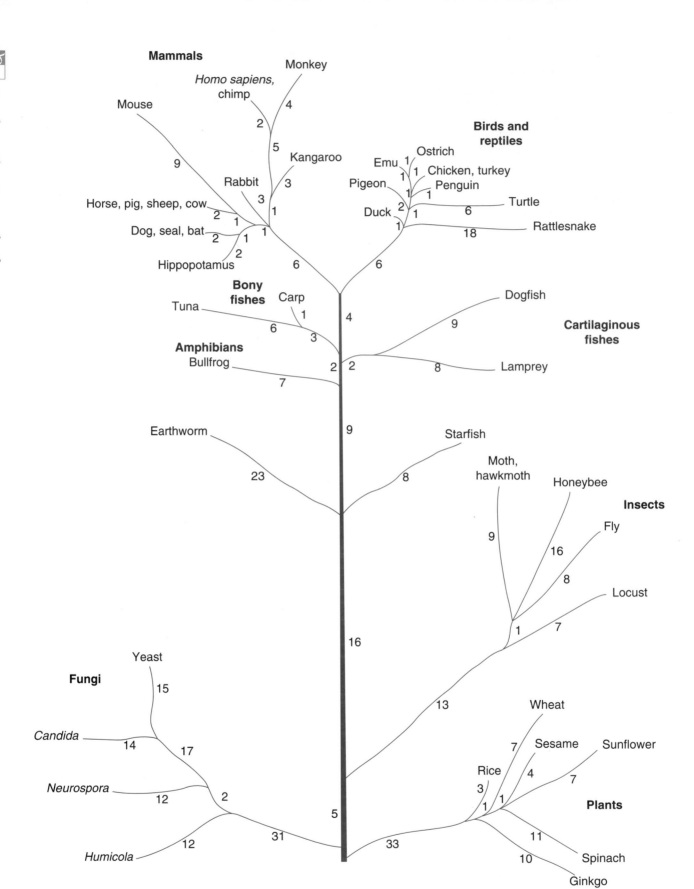

Figure 4.12 An evolutionary tree based on the amino acid sequence of the protein cytochrome *c*. Numbers represent the numbers of amino acid changes between two nodes of the tree. For example, when the common ancestor of cartilaginous fish diverged from the line leading to bony fish, mammals, etc., it evolved and accumulated two amino acid changes in its cytochrome *c* protein before the lines leading to dogfish and lamprey diverged. The dogfish line accumulated nine more amino acid changes in becoming the modern organism; the lamprey accumulated eight changes. Thus, the amino acid sequences of cytochrome *c* in lamprey and dogfish have 17 differences. (From A. Lehninger, D. Nelson, and M. Cox, *Principles of Biochemistry,* 2nd ed., Worth Publishers, Inc., New York, 1993.)

mans are very similar. They are even arranged in the same way in the chromosomes of flies, mice, and humans. This flood of new information about genome structure is giving evolutionary scientists a wealth of fuel for theories about genome evolution.

DNA Typing

Identifying people who will not or cannot identify themselves has long been a challenge in the courtroom and other arenas. Identification has traditionally been made through phenotype: appearance, voice patterns, blood types, fingerprints, and such. DNA-based technologies now make it possible to "go to the source" of individual identity: a person's DNA.

No two individuals (except identical twins) have identical DNA sequences. It is theoretically possible to identify nearly every individual on Earth from his or her DNA sequence, though doing so is not realistic. It *is* possible, however, to make very good predictions about individual identity on the basis of a limited and practical examination of DNA with RFLPs.

DNA-based identification (DNA typing) is based on regions of DNA that vary greatly from individual to individual. When these regions are cut by a restriction enzyme, many different patterns are generated, depending on the sequence of the individual's DNA. The restriction fragment patterns generated from these highly variable regions thus constitute a sort of genetic fingerprint of the individual.

To examine an individual's DNA fingerprint, a sample of his or her DNA is cut with a restriction enzyme, and the resulting fragments are separated by gel electrophoresis. Because the human genome is so large, literally thousands of fragments are generated. To look selectively at the fingerprint regions, the restriction fragments are blotted to a membrane and hybridized to labeled probes for the fingerprint regions (a Southern blot; see *Transfer of DNA Fragments: Blotting* above). The labeled probes reveal the fragments generated from those regions, and the sizes of those fragments constitute the fingerprint (Fig. 4.13). Alternatively, highly variable regions of the genome are selectively amplified by PCR, and the amplified products are examined by hybridization to probes or by simple staining.

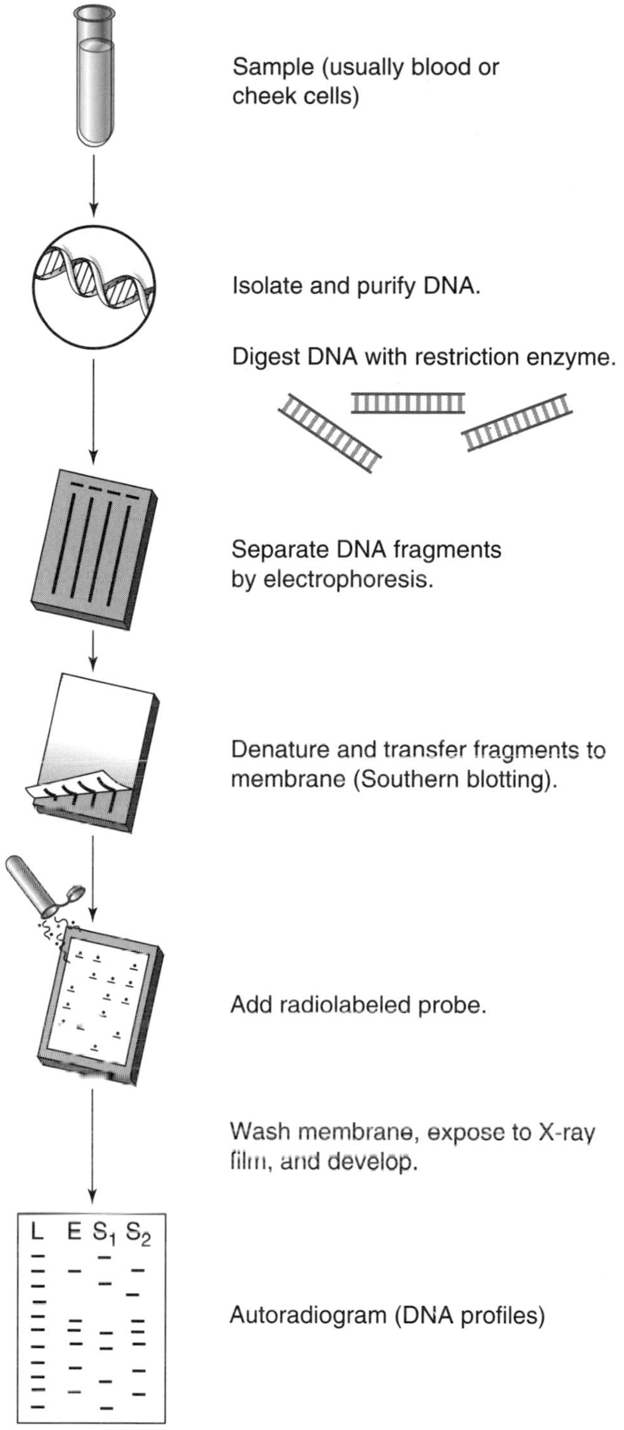

DNA fingerprinting is used for comparisons and to rule out suspects. For example, if blood is found at a crime scene, a DNA fingerprint can be generated from the blood and compared to DNA fingerprints from any suspects. If the crime scene fingerprint does not match a suspect's, that suspect cannot have left the sample and is cleared. In fact, about 30% of the FBI's

Sample (usually blood or cheek cells)

Isolate and purify DNA.

Digest DNA with restriction enzyme.

Separate DNA fragments by electrophoresis.

Denature and transfer fragments to membrane (Southern blotting).

Add radiolabeled probe.

Wash membrane, expose to X-ray film, and develop.

L E S₁ S₂

Autoradiogram (DNA profiles)

Figure 4.13 DNA fingerprinting by RFLP analysis. Radioactive probes are detected by exposing the membrane to X-ray film after hybridization is complete.

DNA typing cases have cleared the prime suspect, and DNA typing has cleared people who were in prison serving time for violent crimes.

If the crime scene sample fingerprint does match a suspect's DNA fingerprint, the relevant question becomes how likely it is that a person other than the suspect could have left it. Since DNA fingerprinting looks at only a portion of an individual's genome, two people could have very similar fingerprints. Scientists look at databases of DNA fingerprints and calculate the frequency of the patterns in the fingerprint in question. They then use those numbers to calculate the odds that another individual will have the same DNA fingerprint. Forensic laboratories try to conduct their testing so thoroughly that random matches are extremely unlikely. (For more information, see the activities *Analyzing Genetic Variation: DNA Typing, A Mix-Up at the Hospital, A Paternity Case,* and *The Case of the Bloody Knife.*)

Although DNA typing gets media coverage for its use in forensic cases, the technique is also widely used in conservation biology and ecological research. DNA typing can reveal the degree of kinship of individual animals. This knowledge can be critical to the success of captive breeding programs for endangered species, for which biologists need to select genetically different individuals as breeding pairs. Knowing degrees of kinship between animals in a living group is also essential to behavioral ecology studies. In many species, it is impossible to determine paternity simply by watching the living group; DNA typing provides a way to solve this problem (see the activity *Analyzing Genetic Variation: DNA Typing*). Analysis of genetic variability can also provide insight into the status of wild populations. Biologists have assumed that vigorous wild populations contain genetically variable individuals rather than genetically identical ones. DNA typing, together with protein comparisons, provides the means for testing this hypothesis.

In a recent application of DNA typing, scientists were able to solve the mystery of the Mexican loggerhead turtles. Pacific loggerheads nest in Japan and Australia, not Mexico, yet young loggerheads could always be found off the coast of Baja California. Many biologists did not believe the young turtles could have come the 10,000-mile distance from Japan (Australia is even farther), and so the origin of the Baja turtles was a mystery. Now DNA comparisons have established that the Baja population is made up of turtles from both the Japanese and the Australian groups. Apparently the young turtles are carried to Mexico by ocean currents, and they then swim back to Japan or Australia to breed. Some swim!

DNA-Based Detection of Pathogens and Disease Diagnosis

The diagnosis and treatment of a particular disease often requires identifying a particular **pathogenic** (disease-causing) microorganism. Traditional methods of identification involve culturing these organisms from clinical specimens and performing metabolic and other tests to identify them. DNA technology is making its presence felt in the clinical laboratory, speeding up and simplifying many identification procedures.

The idea behind DNA-based diagnosis of infectious disease is simple: if the pathogen is present in a clinical specimen, its DNA will be present. Its DNA has unique sequences that can be detected. DNA-based diagnosis involves the detection of pathogen DNA, often in the clinical specimen itself.

As you probably expect, detection of pathogen DNA relies on hybridization with probes specific to the pathogen. There are two main approaches to hybridization tests. One is a direct hybridization, in which labeled probe is added to the sample, and hybridization is detected directly. This method may not be useful if the pathogen is present in very low numbers, so that very few molecules of probe can hybridize.

To get around this problem, scientists and clinicians are turning to PCR (see *Amplifying DNA: PCR* above). PCR also starts with the hybridization of specific DNA sequences. In PCR, however, the oligonucleotide probes are not detected but instead are used as primers for a chain reaction of DNA synthesis. If the primer-probes can hybridize, large amounts of a specific DNA fragment will be synthesized. If the primer-probes do not find complementary DNA sequences, no DNA synthesis occurs. Whether or not a new DNA fragment is synthesized thus provides a means of detecting hybridization. The advantage of the PCR method is that hybridization of probes to even a tiny number of DNA molecules (far too few to be detected by direct methods) yields enough new DNA to be detected easily. PCR is therefore a much more sensitive method of DNA-based diagnosis. (See the activity *The Polymerase Chain Reaction: Paper PCR.*)

Finding genes

Much of biotechnology concerns the manipulation of genes. How are genes identified in the first place? The strategy for finding a particular gene depends on several factors, one of which is whether the protein product for the gene is known. If you know what the protein product of the gene is, you have a handle on finding the gene. It's a different ballgame if you don't know what the product of an unknown gene does. Let's first consider how to find a gene for a known protein.

Finding Genes for Known Products
Using genetics. If you are looking for a gene in a microorganism, the easiest way to find it may be by

using genetics. Let us imagine that we would like to find the gene for a specific enzyme involved in the biosynthesis of the amino acid histidine in *Escherichia coli*. What we need is a mutant *E. coli* that lacks this enzyme. Since literally thousands of *E. coli* mutants have been characterized, it is probably easy to get one, for example, from the *E. coli* Genetic Stock Center at Yale University. Our mutant cannot make histidine and therefore cannot live unless its medium is supplemented with that amino acid. Now we go to our *E. coli* DNA library (see above). The plasmids in this library contain inserted DNA fragments from the entire chromosome of normal *E. coli*. We transform a batch of the mutant, histidine-requiring *E. coli* with this mixture of plasmids. Since transformation occurs at a low frequency, each *E. coli* cell will receive at most one plasmid. Now we look for a transformed *E. coli* that has acquired the ability to make histidine. We can find it easily because it can now grow on histidine-free medium (Fig. 4.14). The assumption is that any bacterium now able to make histidine must have received the missing gene on its new plasmid. We isolate the plasmid from that bacterium, and, with luck, have found the histidine biosynthesis gene we

Figure 4.14 Using genetics to find an *E. coli* gene for histidine biosynthesis.

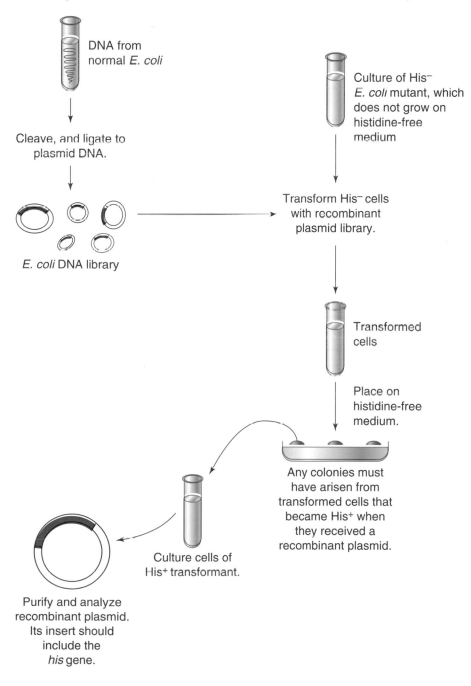

DNA from normal *E. coli*

Cleave, and ligate to plasmid DNA.

E. coli DNA library

Culture of His⁻ *E. coli* mutant, which does not grow on histidine-free medium

Transform His⁻ cells with recombinant plasmid library.

Transformed cells

Place on histidine-free medium.

Any colonies must have arisen from transformed cells that became His⁺ when they received a recombinant plasmid.

Culture cells of His⁺ transformant.

Purify and analyze recombinant plasmid. Its insert should include the *his* gene.

were looking for within the bacterial DNA insert in the plasmid.

Using the product. The genetic method outlined previously is simple and efficient, but it is applicable only under limited circumstances. Obviously, if you are looking for an animal gene, you cannot collect a batch of mutant animals, transform them with plasmids, and look for restoration of a normal phenotype. Instead, we must pull some different techniques from our biotechnology toolbox, as in the example below.

Suppose you wish to locate the gene for polar bear hemoglobin, a known gene product. If you cannot find another scientist who has already made a library of polar bear DNA, you first make one of these. Now you need specific information about the amino acid sequence of the protein, so you obtain a specimen of polar bear blood and isolate the hemoglobin from the red cells. You give the purified hemoglobin to your friend the protein chemist, who analyzes a portion of its amino acid sequence. Then, working backward from the amino acid sequence, you determine the sequence of nucleotides that encodes that portion of polar bear hemoglobin. Next, you go to another friend, a nucleic acid chemist, and have her synthesize an oligonucleotide with that sequence. You now have a probe for the hemoglobin gene.

Returning to your laboratory, you use one of a number of methods to label the probe for detection (such as with a radioactive isotope). Then you hybridize the probe to your polar bear DNA library to identify molecules that have the hemoglobin DNA sequence. Barring complications, you will eventually find the gene (Fig. 4.15).

This thumbnail sketch of one approach to finding the gene for polar bear hemoglobin makes it sound like a fairly easy proposition. It is not. Even when you know what you are looking for (polar bear hemoglobin), the procedures are technically demanding, and unexpected difficulties are almost certain to arise. Finding a gene can be one of the most difficult and time-consuming aspects of molecular biology projects.

Finding Genes with Unknown Products

The task of finding a gene when the protein product is not known is much harder than finding a gene for a known protein. Why would anyone want to do it at all? One reason is to find the genes involved in inherited diseases. In many cases, the identity of the defective protein involved in a genetic disease is not known. Identifying the defective gene would be a starting place for understanding the cause of the disorder.

To find a disease gene with an unknown product, scientists must have DNA samples from a large number of related individuals who carry the disease in their family. From the family's medical history, the scientists know which members have the disease and which members are carriers (if the disease is a recessive trait). The DNA samples are digested with many different restriction enzymes, and the digestion products are examined for variations between individuals (RFLPs; see *Analyzing genetic variation* above). The goal is to find a restriction fragment or fragments seen only when the disease gene is present. Such a distinctive fragment or pattern is called a **marker.** If a marker is found, it is assumed that the region of the chromosome containing the marker is very close to the disease gene.

After a marker for a disease gene has been identified, a new phase of work begins. Long stretches of DNA (often hundreds of thousands of base pairs) around the marker must be sequenced and searched for sequence patterns that look like genes (with promoters and coding regions). When potential genes are found, their sequences from the healthy and sick family members must be compared. If the sequences of a putative gene from sick individuals are always different from the sequences of the same putative gene from healthy individuals, then the defective gene is probably the culprit in the disease.

This entire process can take years and is not foolproof. Recently, two well-regarded scientific groups reported that they had found the gene for a particular genetic disease. The only problem was that the two groups had identified completely different genes on different chromosomes. Once a candidate for a disease gene has been identified, further study of more affected people is usually required to confirm that the identification is correct.

Screening for Genetic Diseases

After a genetic marker for a disease has been found and confirmed, that marker can be used as a tool to predict the presence of the disease gene. If an individual wishes to know whether she carries the disease gene, scientists can prepare a sample of her DNA and look for the RFLP that constitutes the genetic marker. If the marker is present, she is likely to carry the disease gene. The laboratory procedure for screening for genetic disease is essentially the same as that for DNA fingerprinting (see above), with the restriction enzymes and the probe selected to reveal the marker region.

Why do we say *likely* to carry the genetic disease? Because genetic markers are usually not the disease-

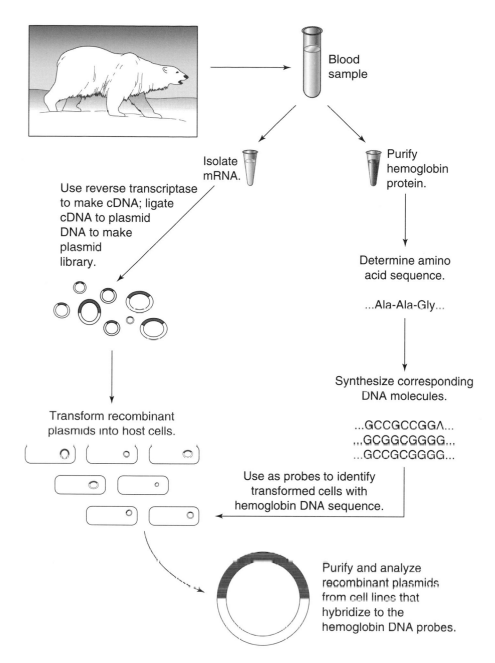

Figure 4.15 Finding the gene(s) for a known product: polar bear hemoglobin.

Labels within figure:

Blood sample

Isolate mRNA.

Use reverse transcriptase to make cDNA; ligate cDNA to plasmid DNA to make plasmid library.

Transform recombinant plasmids into host cells.

Purify hemoglobin protein.

Determine amino acid sequence.

...Ala-Ala-Gly...

Synthesize corresponding DNA molecules.

...GCCGCCGGA...
...GCGGCGGGG...
...GCCGCGGGG...

Use as probes to identify transformed cells with hemoglobin DNA sequence.

Purify and analyze recombinant plasmids from cell lines that hybridize to the hemoglobin DNA probes.

causing mutations themselves but only RFLP patterns that are close to those mutations and tend to be inherited with them, so the presence of a marker is not proof that the mutation is present. Genetic recombination events could separate the marker RFLP from the disease mutation. Likewise, the absence of the genetic marker may not be absolute proof that a disease gene is not present. In addition, there may be many different mutations that can cause a given genetic disease. The vast majority of an affected population may carry one particular mutation and its associated genetic marker. A rare, alternative disease-causing mutation may not be associated with the marker RFLP; in these cases, the marker is not a useful diagnostic tool. Or a marker might be present in some families but not others. However, if the limitations are understood, genetic testing for a disease gene can provide valuable information for physicians and potentially affected individuals. In addition, once mutant genes are positively identified, more-specific probes for diagnosis can be made. Since more and more disease genes are being identified, direct examination of those genes is becoming increasingly common as a means of diagnosing genetic disease. (See the activity *The Molecular Basis of Genetic Diseases.*)

Genetic engineering

The term **genetic engineering** refers broadly to the process of directed manipulation of the genome of an organism. Because genetic engineering usually involves the combining of genes from two or more sources, it is also commonly referred to as **recombinant DNA (rDNA) technology.** Genetic engineering usually involves the manipulation of a specific gene. The goal of genetic engineering, of course, is not to manipulate an organism's DNA per se but to change something about the proteins produced in that organism: to cause it to produce a new protein, to stop producing an old protein, to produce more or less of a protein, and so on. Manipulating the genome is merely the way to influence protein production.

In general, the gene in question must first be identified and cloned into a plasmid or other vector so that it is easy to work with. The cloned gene is manipulated in the laboratory by using the tools and techniques described previously. The altered gene is then inserted into the target organism. The target organism has now been genetically engineered. What are some genetic manipulations that we can make and insert into a host cell?

Regulating the Expression of Existing Genes

Increasing gene expression. Inserting several copies of a useful gene into an organism is one way to increase production of the gene product. This technique has been used by the company Novo Nordisk to increase production of bacterial amylase, an enzyme used in the baking and soft drink industries. For years, the company used the bacterium *Bacillus licheniformis* to produce amylase. Scientists at the company have now given the bacterium extra copies of its own amylase gene, resulting in greatly increased yields of the protein.

Turning genes off. In many cases it is advantageous to decrease the expression of a particular gene. An example can be found in the case of the familiar flavorless grocery store tomato. Grocery store tomatoes lack flavor because they are picked green. Green tomatoes are hard and can be harvested by machine, packaged, and shipped without damage. Ripe tomatoes taste much better but are soft and would very likely be damaged during handling. Scientists discovered that the softening of tomatoes during ripening is caused by the enzyme polygalacturonase. Blocking production of this enzyme would allow a tomato to ripen and develop flavor without becoming soft.

Although there are many approaches to blocking production of an enzyme, scientists at Calgene, Inc., decided to use an exciting new approach to gene control. They introduced into the tomato plant cells a gene encoding an **antisense RNA** for polygalacturonase. Antisense RNA is RNA that is exactly complementary in sequence and opposite in polarity to the normal mRNA of a gene. Its structure allows the antisense RNA to hybridize to the mRNA, creating a double-stranded RNA molecule. Double-stranded RNA cannot be translated by ribosomes, so expression of the normal gene is reduced (Fig. 4.16). After constructing this genetically engineered tomato, Calgene asked the Food and Drug Administration to evaluate its safety. The tomato was approved for sale as food and at the time of this writing is being test-marketed. Soon everyone may be able to test whether these tomatoes are better tasting than their standard grocery store relatives.

Many scientists view antisense RNA as an exciting new way to fight disease. The idea is simple: turn off genes specific to the disease-causing organism. A short complementary RNA or DNA molecule should hybridize specifically to the disease organism's RNA and could inactivate expression of proteins essential to the organism without having any undesirable effect on the host. The challenges are to identify the appropriate target RNA sequences and then to deliver the antisense oligonucleotide to the appropriate cells. Biotechnology companies are exploring ways to modify DNA and RNA chemically so that these molecules will withstand the cellular enzymes that normally digest free nucleic acids and will easily enter cells from the circulation. In 1992, the Food and Drug Administration approved of the first antisense RNA drug for clinical trials in humans, a drug to fight the human papillomavirus that causes genital warts. (See *Advanced Activity: Antisense Regulation* for more information.)

Adding New Genes

People have long used selective breeding to introduce desirable traits into plants and animals. Selective breeding is, of course, limited by the ability of organisms to mate and by the characteristics of natural variants of the species. Genetic engineering is not limited in either of these ways. By using genetic engineering, we can transfer genes for desirable characteristics from one organism to any other. For example, a bacterial gene can be transferred into a plant.

Scientists have introduced the gene for the insect resistance protein of *Bacillus thuringiensis* (the Bt used by home gardeners in many "natural" pest control preparations) into corn and other plants. The Bt gene was inserted into the corn genome by using the

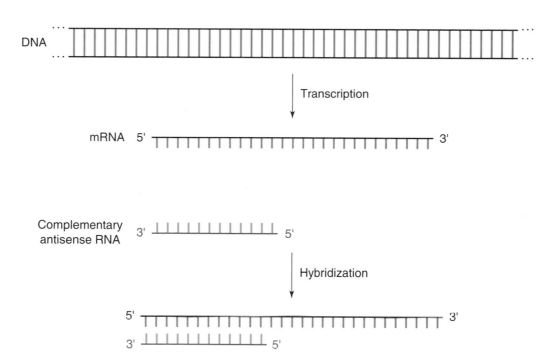

DNA

Transcription

mRNA 5' ——————— 3'

Complementary antisense RNA 3' ——————— 5'

Hybridization

5' ——————— 3'
3' ——————— 5'

Figure 4.16 Antisense RNA turns off gene expression by hybridizing to the 5' (front) end of a complementary mRNA. Double-stranded RNA is not recognized by ribosomes and cannot be translated.

gene gun, since corn is not susceptible to infection by *A. tumefaciens* (see *Natural vectors* above). Field tests of the genetically engineered corn plants showed that the plants were able to resist caterpillar pests while neighboring unaltered plants were devastated. This particular feat of genetic engineering could reduce the use of pesticides on farms as well as save farmers expense and worry.

Another example of adding new genes by genetic engineering is the introduction of a human gene into a microorganism. This kind of genetic engineering is done to obtain large amounts of the gene product. For example, insulin to treat diabetics used to be obtained from animal cadavers. In the 1980s, scientists isolated the gene for insulin from human cells and introduced it into bacteria. To get expression of the human gene in the bacteria, it was necessary to substitute a bacterial promoter and other control sequences in the human gene. Now large quantities of insulin can be harvested from bacterial cultures. Other human proteins produced in bacteria include human growth hormone and hemoglobin.

Gene Therapy

A medical application of biotechnology called gene therapy has received a great deal of media attention. And no wonder. Gene therapy holds out the hope that previously incurable, fatal inherited diseases may be treatable. The goal of gene therapy is to treat genetic

diseases by giving the patient healthy copies of the defective gene. Why is gene therapy listed under the heading *Genetic engineering* in this chapter? Because gene therapy is actually a form of genetic engineering. Patients' genomes are manipulated so that they can produce a protein they were not previously producing (or were not producing in adequate quantities). The first trials of gene therapy are now under way.

Gene therapy sounds logical and straightforward, but it is not a simple undertaking. First, the defective gene must be identified, and a healthy copy must be cloned. Then the healthy gene must be appropriately delivered to the patient. The delivery is the tricky part: the new gene needs to be delivered to the correct tissue. For example, it does no good to deliver an essential liver receptor gene to blood cells. If you can deliver the new DNA to the appropriate tissue, the gene must still be expressed in those cells. As yet, we have no reliable method for actually replacing a defective gene on a chromosome with a healthy one. Genes can be inserted randomly into the chromosome, with the risk of causing some damage. In addition, genes inserted at random are sometimes not expressed well; the reasons for this failure are not understood.

The first test of human gene therapy used blood cells. Blood cells were a natural first case because they are easy to obtain and easy to reintroduce into the body.

The patient was a little girl with severe combined immunodeficiency disease. This child had an extremely defective immune system and could not defend herself against infection. Her condition was the same as that of the "bubble boy," who lived his short life in a sterile environment. The cause of her condition was a defect in the gene coding for the enzyme adenosine deaminase (ADA). Scientists at the National Institutes of Health removed blood from her, separated the lymphocytes (white blood cells), and used a retrovirus to introduce a healthy copy of the *ada* gene into them. They then reintroduced the lymphocytes into their patient. The altered cells produced the missing enzyme, and at this time, the little girl is healthier than she has ever been.

Scientists are experimenting with ways to deliver genes to other body cells. Some of the methods under study involve different viruses. Specifically, altered cold viruses have been used to deliver a healthy cystic fibrosis gene to the respiratory systems of rats and people. An altered liver virus has been used to deliver a gene involved in cholesterol metabolism to liver cells. Nonviral systems are also being tested.

Although genetic therapy was originally envisioned as a remedy for inherited diseases, the scope of genetic medicine has already moved far beyond that concept. For example, researchers are looking to antisense RNA as a means of fighting cancer, viral infections, and even baldness. Other scientists are experimenting with genetic engineering of tumors to make them more vulnerable to drug treatments. Still others are using genetic approaches in hopes of stimulating the immune system to fight cancer. In the next few years, gene therapy trials will undoubtedly multiply in number and scope.

Protein engineering

As is probably evident from the foregoing examples, almost any manipulation of DNA sequence is possible. The goal of most of these manipulations is the same: to affect a protein. At present, our efforts are largely centered on protein *production:* blocking it, increasing it, or causing it to happen in a new cell type. A future direction of biotechnology will be to improve the *functioning* of proteins by altering their structures or even to design new proteins to fulfill a specific need. Increasing the use of enzymes and biological processes in manufacturing offers significant advantages in reducing industrial pollution and utilizing renewable rather than nonrenewable resources (see chapter 1), so protein engineering offers important potential benefits to society.

One desirable way in which industrial enzymes could be modified would be to increase their stability so that they would be less vulnerable to changes in temperature, pH, or other reaction conditions. Hardier enzymes could be used under a greater variety of industrial conditions, and higher temperatures could mean faster reaction rates. Increasing the stability of a protein means increasing the stability of the protein's tertiary and, possibly, quaternary structures. So before a scientist can undertake to engineer a protein for increased stability, he or she must know its three-dimensional structure. Once that structure has been determined, the scientist looks for ways to increase stability without altering the protein's function.

The most obvious way to increase a protein's stability is to introduce disulfide bridges (see *Stability of protein structure* in chapter 3) to hold the tertiary structure together. These bridges have to be geometrically compatible with the protein's structure and must not alter crucial amino acids. The bacteriophage enzyme lysozyme has been engineered in this way. This enzyme has two domains. Using recombinant DNA technology, scientists introduced codons for cysteines in positions within the gene that were selected after careful study of the protein's structure. The altered gene was reintroduced into cells, which duly produced the protein with the new cysteines. The cysteines combined to form three disulfide bridges that tied the domains together in their appropriate three-dimensional configuration (Fig. 4.17). These alterations significantly increased the thermal stability of the enzyme. The engineered protein withstood heating to a temperature 23°C higher than the native protein could endure.

Scientists are already looking beyond simply modifying existing proteins to creating new ones from scratch. Significant advances in protein design will not be achieved until we gain a greater understanding of how amino acids determine protein structure and exactly how that structure determines activity. Gaining this understanding is a difficult undertaking, considering the enormous variety in both protein structure and functions. However, progress is being made. With the improving technologies of NMR spectroscopy and X-ray crystallography, determining the actual three-dimensional structure of a protein molecule is becoming more common (although not easy). Knowing the structure of a protein provides valuable clues about how the protein actually works. Some day we hope to be able to predict the three-dimensional structure of a protein from its amino acid sequence and understand exactly how that structure interacts with other compounds to produce the protein's activity. This knowledge could then be used

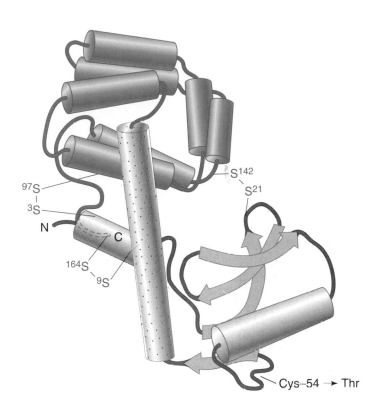

97S

3S

N

C

164S

9S

S142

S21

Cys—54 → Thr

Figure 4.17 Protein stability engineering. Three engineered disulfide bridges (blue) tie the two domains of bacteriophage T4 lysozyme in the proper configuration. The two domains are blue and gray and are connected by the stippled helix. The numbers are the positions of the cysteines in the 164-amino-acid protein. The cysteine at position 54 was changed to a threonine to keep it from interfering with proper formation of the engineered bridges. (Adapted from M. Matsumura, G. Signor, and B.W. Matthews, *Nature* 342:291, 1989, with permission.)

to design proteins for specific tasks, such as breaking down toxic chemicals or manufacturing products from renewable resources that are currently manufactured from petroleum. The possibilities are limitless.

Has anyone already produced an "artificial" enzyme? Yes. A group of researchers in Denver, Colorado, designed a hypothetical protein-digesting enzyme. The design was assisted by computer programs that make predictions about protein structure. When the protein was chemically synthesized, it had the predicted enzyme activity. This achievement, published in 1990, is the first clear example of synthesis of an artificial protein with catalytic activity.

The Denver group's protein was made by chemical synthesis. Protein engineers of the future will probably use recombinant DNA technology to produce their designer enzymes. Once the desired amino acid sequence of the new enzyme has been determined, they could synthesize DNA encoding that amino acid sequence, clone the DNA into a plasmid, and transform an appropriate host. The host cell would then synthesize the protein. This process of starting with a protein sequence and making a gene for it turns the linear "central dogma" of biology (DNA to RNA to protein) into a circle. It is amazing that only 4 decades after the discovery of the structure of DNA, we have learned so much.

Summary

As our understanding of the genetic machinery and protein biochemistry of cells grows, so does the number of technologies based on that understanding. We learned the chemistry of the DNA molecule; now we can synthesize it ourselves. We learned how genetic variation is determined by DNA sequence variation; now we look directly at variation in DNA for identification. We learned how nature copies, recombines, and transfers genes; now we copy, recombine, and transfer genes ourselves. When we understand the forces that determine protein structure and function, we will create our own custom enzymes.

Where is this new technology taking us? In many cases, to clear-cut benefits. More precise identification techniques will minimize mistakes in paternity and criminal cases. Faster diagnostics will result in more effective disease treatments. Genetic engineering will give us crops that are more nutritious and disease resistant. We will use biological systems to diagnose and then clean up environmental pollution. Gene therapy will alleviate much suffering.

There are also perplexing questions about appropriate use of these technologies. Decisions about appropriate use will inevitably (and should) involve values and moral judgments (see Part III of this book). An understanding of the technologies themselves will provide some factual basis for making these important judgments.

Apart from the social implications of DNA-based technologies, their development offers us an exciting, unfolding story of science. The more we learn, the more we find that nature is vastly more complicated than we originally imagined. The intricacies of heredity and gene expression dwarf our first conceptions of the processes. New surprises are constantly announced. Regardless of the practical applications that may come from our increasing understanding of life's basic processes, the wonder engendered by the incredible complexity of the processes must surely inspire and enrich all of us.

Selected Readings

Cohen, J.S., and M.E. Hogan. 1994. The new genetic medicines. *Scientific American* 271:76–82.

Hall, S. 1995. Protein images update natural history. *Science* 267:620–624.

Mullis, K. 1990. The unusual origin of the polymerase chain reaction. *Scientific American* 262:56–65.

Paabo, S. 1993. Ancient DNA. *Scientific American* 269: 86–92. Isolating and studying DNA from ancient materials. Anyone with questions about the feasibility of Jurassic Park should read this article.

Saiki, R.K., S. Scharf, F. Faloona, K. Mullis, G. Horn, H. Erlich, and N. Arnheim. 1985. Enzymatic amplification of beta-globin genomic sequences and restriction site analysis for diagnosis of sickle cell anemia. *Science* 230:1350–1354.

Watson, J., M. Gilman, J. Witkowski, and M. Zoller. 1992. *Recombinant DNA,* 2nd ed. Scientific American Books, W. H. Freeman & Co. New York.

White, R., and J. Lalouel. 1988. Chromosome mapping with DNA markers. *Scientific American* 258:40–48.

PART II

Classroom Activities

Classroom activities for wet and dry laboratories are provided. Presented in lesson plan format, these activities are grouped into four sections that stress fundamental concepts. The first section teaches concepts basic to molecular biology: the structure and function of DNA. The next two sections focus on manipulating DNA and transferring it between organisms. These sections illustrate how a deep understanding of biological systems allows us to use such systems to our advantage. The last section emphasizes the importance of genetic variation in biological systems and teaches techniques that allow us to look at genetic variation directly.

The lesson plans include lists of resource materials available from commercial suppliers. These lists are not exhaustive, nor are they intended as endorsements, but we hope they will be helpful. The addresses of the companies, along with additional sources of information and materials, are given in Appendix I.

Classroom Activities

DNA Structure and Function

A basic principle of biology is that structure must serve function. We can see examples of this principle when we consider the structure of a bird's wing in relation to flight or a human eye in relation to sight. The same relationship between structure and function holds true at the molecular level. Nowhere is this relationship more apparent than in the DNA molecule. Its structure is wonderfully suited to its two functions: replication and information translation. It also permits easy regulation through a number of mechanisms.

These lessons focus on the DNA molecule and how its information is replicated and translated into proteins. Models, simulations, and discussions illustrate the structure, packaging, storage, and replication of DNA; the processes of transcription and translation; and aspects of gene regulation, including the exciting field of antisense technology. Simple wet laboratory procedures for extracting DNA from *Escherichia coli* and store-bought baker's yeast conclude the section.

In teaching about the structure and functions of DNA, there is no substitute for models, whether you are teaching middle school students or adults. Many wonderful models for illustrating DNA structure and replication, as well as transcription and translation, are commercially available. If you already use some of these materials to teach DNA structure and protein synthesis, you may find some ideas in the activities on DNA replication, gene regulation, and antisense technology that could easily be incorporated into your lessons. If you do not have a commercial DNA model set, this section begins with an inexpensive paper model that all of your students can help make.

Constructing a Paper Helix

5

About This Activity

Many students, whether children or adults, often have a difficult time visualizing the structure of DNA. DNA model kits are excellent teaching aids but are often too expensive for schools to purchase. If you do not already have a kit, this lesson provides an inexpensive model. In addition, this model allows each student to participate in building a DNA helix. Having students actually assemble a model of DNA is probably the most useful way of communicating information about DNA structure to the large population of students who learn by doing.

The following activity uses paper to make a colorful, easy-to-assemble DNA helix. The activity can be used with students of many ages and different abilities. We have used it successfully in classroom settings ranging from middle school to college. Even though it may seem too simple for advanced or college students, our experience suggests otherwise. Though advanced students are usually familiar with the base pairs, they often do not understand the antiparallel orientation of the two DNA strands and usually have not thought about the fact that 5′ and 3′ ends of strands are different. Advanced procedures such as DNA sequencing and the polymerase chain reaction depend on these differences, and students need to understand these aspects of DNA structure before they can understand the procedures. Examining the model seems to get the information across to many students better than looking at pictures in a book does.

Hanging the DNA model from the ceiling or attaching it to a wall is useful in teaching the rest of the activities in this book, nearly all of which call upon students to visualize a DNA molecule. The model gives students a concrete structure on which to base more abstract, difficult concepts. You will find the paper helix to be especially useful in teaching replication, transcription, and restriction of DNA.

Class periods required: 1–2

Introduction

The following discussion is intended purely for information to the teacher. It is not necessary to share all of it (or even most of it!) with your students. Use your judgment as to how much you are comfortable with and how much you think your classes can absorb.

DNA is the blueprint of life. It controls body form, functions, and appearance by coding for the proteins that form the bricks and mortar of tissue, carry out metabolic activity, fight infection, regulate growth, and synthesize fats and pigments. Chemically, DNA is made up of small repeating units. The repeating unit of DNA is quite simple; it is composed of three parts: the sugar deoxyribose, a phosphate group, and one of four organic bases, adenine, cytosine, thymine, or guanine. This unit is called a nucleotide or, more properly, a **deoxynucleotide** (Fig. 5.1). (RNA is also composed of nucleotides but uses the sugar ribose; hence RNA is said to be composed of **ribonucleotides**.) To make a DNA polymer, many deoxynucleotides are joined together by phosphodiester bonds between the phosphate and sugar groups.

The phosphodiester bonds form a bridge between the number 5 carbon of the deoxyribose portion of one nucleotide and the number 3 carbon of the deoxyribose portion of the adjacent nucleotide (see Fig. 5.1 for numbering). By using the carbon numbers, we can talk about direction with respect to DNA polymers. In Fig. 5.2, the trinucleotide is shown 5′ to 3′ from top to bottom, because the 5′ carbon of the first nucleotide comes first. Imagine flipping the trinucleotide upside down. Now the 3′ carbon with the OH group of the T nucleotide comes first; this is the 3′-to-5′ orientation.

The four bases of DNA are of two types: purines and pyrimidines. Adenine and guanine are purine bases, and cytosine and thymine are pyrimidine bases. These bases are capable of forming two specific pairs: adenine with thymine and cytosine with gua-

Classroom Activities

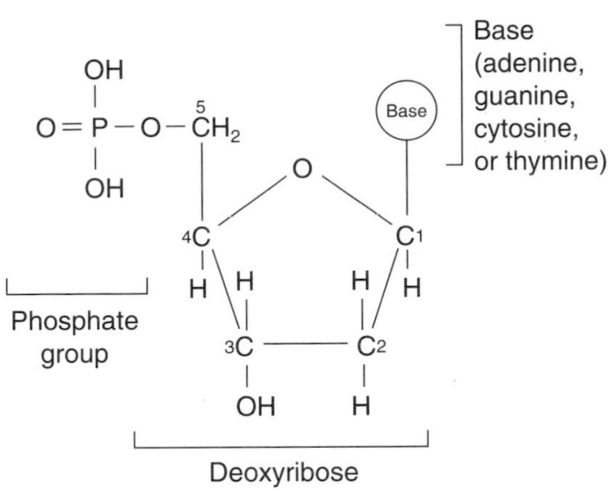

Figure 5.1 The deoxynucleotide.

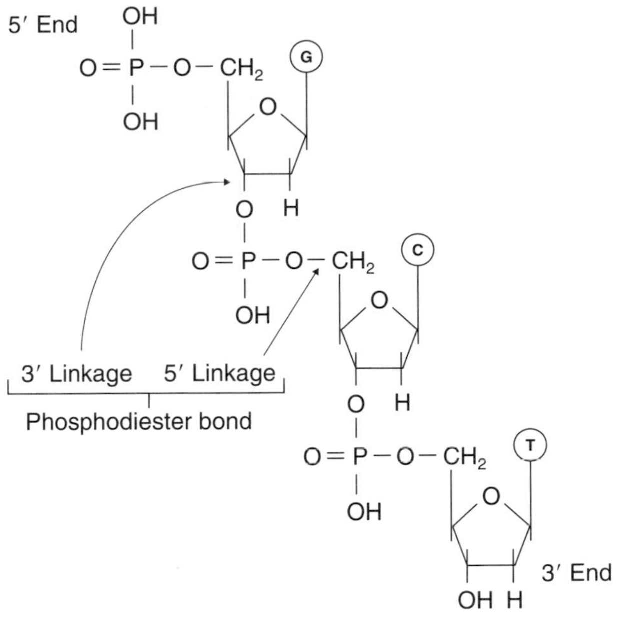

Figure 5.2 A trinucleotide.

nine. Notice that a purine always pairs with a pyrimidine. Purine bases are larger than pyrimidines. A purine-purine pair would be much larger than a pyrimidine-pyrimidine pair. However, the two purine-pyrimidine base pairs are the same size (Fig. 5.3). Since they are the same size, they can fit into the interior of the DNA helix neatly in any random order. The base pairs are connected by weak chemical bonds called hydrogen bonds (see chapter 3 for more information). Because of the chemical structures of the bases, adenine and thymine are connected by two hydrogen bonds, and cytosine and guanine are connected by three.

In a DNA molecule, two complementary DNA polymers are connected by the hydrogen bonds between these base pairs. The sugar-phosphate backbones of the polymers are oriented in opposite directions: one runs 5′ to 3′, and the other runs 3′ to 5′. Finally, the two polymers (usually called strands) are twisted around each other to form the famous double helix (Fig. 5.4).

Figure 5.3 Complementary base pairs in DNA.

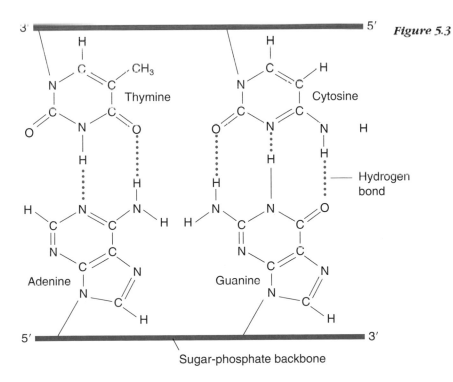

Sugar-phosphate backbone

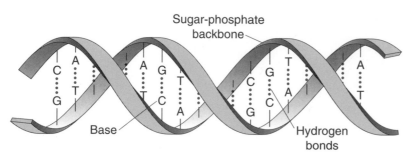

Sugar-phosphate backbone

Base

Hydrogen bonds

Figure 5.4 Ribbon model of the DNA helix.

The double helix structure of DNA is ideally suited to the cellular environment of the molecule. The cell is a water-based environment, and molecules that are charged or electrically polar (like the charged phosphate groups and the sugar molecules) interact readily with it. The organic bases of DNA, however, are nonpolar and therefore hydrophobic (water fearing). They are more stable when interacting with other hydrophobic molecules than when interacting with water. In the double helix, the bases are turned toward each other, away from the watery cytosol. Between the sugar-phosphate backbones, the flat hydrophobic base pairs stack together, and these so-called "stacking interactions" further stabilize the double helical structure.

Even though the base pairs are on the inside of the helix, cellular proteins can sense the identities and sequences of those base pairs. The "edges" of the base pairs are "visible" to cellular proteins between the spiraling strands of the sugar-phosphate backbone. The edge of each base presents a different chemical configuration to the outside, so a protein that interacts directly with DNA can "read" the base sequence. In this way, RNA polymerase can recognize a promoter, repressor proteins can find their binding sequences, and DNA replication proteins can identify their start sites, the replication origins (see the activity *DNA Replication*).

If you have a three-dimensional model of DNA, you can see the two grooves spiraling up the outside of the helix between the sugar-phosphate backbones. Because of the geometry of the helix, one of these grooves is wider than the other. The wider one is called the *major groove* of DNA; the smaller one is the *minor groove*. (Yes, DNA *is* groovy.) The important thing about the grooves of DNA is that they are the place where proteins interact with specific DNA sequences. Most of the sequence-specific DNA-protein interactions that are currently understood involve a protein binding to DNA in the major groove. (For an example of a specific protein-DNA interaction, see *Lambda and* trp *Repressors: DNA-Binding Proteins* in chapter 3.)

When the structure of DNA was first proposed by Watson and Crick in 1953, the helical structure they described was thought to be fixed and unchangeable. Since then, we have learned that the DNA helix can assume slightly different forms, can bend and kink, and can even unwind. All of these different forms appear to be important in the functioning of the cell. The "standard" form of DNA is called the B-form helix. It has 10.5 base pairs per complete turn, and its center of symmetry is down the middle of the base pairs. The B-form helix is the kind that your class will build. However, two additional helix forms, called A and Z, have also been shown to exist.

A-form DNA has about 11 base pairs per complete turn, and the center of symmetry is not down the middle of the base pairs but along the outside of them between the "wrappings" of the sugar-phosphate backbone (in the major groove). Z-form DNA has even more differences: A- and B-DNA are right-handed helices, but Z-DNA is left-handed. It has about 12 base pairs per helical turn, and the geometry of the sugar-base bonds is altered. If you would like details on the structures of A and Z helices, please refer to a text such as Watson's *Molecular Biology of the Gene* (see Appendix H).

The paper DNA model described below allows students to construct a large DNA helix. The model can be used in subsequent lessons on DNA replication, transcription, restriction enzymes, DNA sequencing, and so forth. If you reduce the template patterns on a photocopier, students can make miniature paper DNA models to use in small groups. (Depending on the size of the models and the stiffness of the paper you use, the smaller models may not twist well, but they would work for DNA replication activities.) The *Tips* section on the next page provides suggestions about adapting the model for younger or more advanced classes.

Objectives

The objective of this lesson is to construct a paper model of the DNA helix. Students will do this by making individual nucleotides. Class members will then join their nucleotides together to form a double helix.

At the end of this activity, students should be able to

1. Describe the structure of DNA;
2. State which bases form the complementary pairs in DNA;
3. Identify the purine and pyrimidine bases and state which are larger;
4. Describe what is meant by 5' and 3' strand ends (advanced students only).

Materials

- Photocopies of template patterns in Appendix K
- Five colors of construction paper
- Scissors, ideally for each student
- Glue
- Stapler and staples

Preparation

- Write on the template patterns the color each base, the sugar, and the phosphate should be (for example: cytosine, yellow; guanine, green; adenine, pink; thymine, blue; phosphate, black; deoxyribose, red).

- Photocopy the templates.

- Photocopy the *Student Activity* pages if you are going to use them and your students do not have manuals (see *Tips* below).

Procedure

To assemble nucleotides (students do this)

1. Cut out the pattern for the assigned nucleotide(s).
2. Place the pattern on construction paper of the appropriate color, and cut it out.
3. Label the construction paper piece the way the pattern is labeled (omit color name).
4. Glue the nitrogenous base to the sugar molecule by matching up the dots.
5. Glue the phosphate group onto the model by matching up the stars.
6. The teacher will join the nucleotides together to form a helix.

To assemble the helix (teacher does this)

For the helix to come out even, you will need to keep track of how many of what kind of nucleotides are available. An easy way to do this is to use exactly half of all the A's, T's, C's, and G's to make the first strand

(put the bases in any order); then you will be sure to have the right number of complementary bases to assemble the second strand.

To make the first strand, staple the phosphate group to the 3' carbon position on the deoxyribose molecule (see Fig. 5.1 to 5.4 for carbon numbers; the positions are denoted by squares on the templates). Assemble this one with the sugars "right side up" (the phosphate groups will be pointing upward; this orientation will be 5' to 3').

For the second strand, have one of your students hold the first strand for you. Ask your students what nucleotide is needed to base pair with the first nucleotide. When they tell you, select that nucleotide. Turn that nucleotide upside down to simulate the 3'-to-5' direction (the phosphate group should be pointing down). Remind the class that the two strands of a helix run in opposite directions. Make the base pair by stapling the two bases together; overlap them to make a strong connection. Ask students what the next nucleotide should be. When they tell you, select that nucleotide. Connect it to the previous nucleotide by stapling the phosphate group to the 3' sugar position and then connect the base pair. Allow overlap for strength. Continue until you have completed the second strand. A segment of the assembled model is shown in Fig. 5.5.

To make the helix, twist the paper ladder. It will work best if you keep the ladder vertical while you twist; have a student hold one end for you, or hang the helix from the ceiling. Some teachers festoon their classroom with the model, letting it run up the walls and across the ceiling. Keep the helix in your room as a twisted model or in the flat form. Although the helix is a better representation of DNA, the flat form illustrates the most important features, that is, the linear sequence of nucleotides and the complementarity of the strands.

Tips

Pass out the *Student Activity* pages only to younger students, low-achieving classes, or average 9th- or 10th-grade biology classes. The questions can be given to any class, although some of them are quite easy for advanced students; use your judgment about whether to assign the more advanced questions to your students.

In low-achieving classes, it may be beneficial to cut out models during one class period and paste them together the second day. Divide the class into groups.

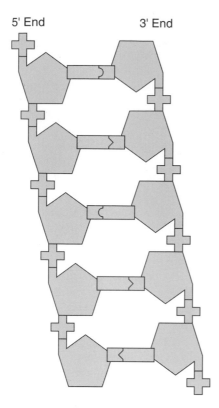

5' End   3' End

Figure 5.5 A segment of the assembled model
untwisted.

Assign each group a nucleotide, the phosphate, or the
sugar to cut out. Make sure they label each piece.

In an average class, have each student cut out all four
nucleotides.

In an above-average class, do not tell students how to
put the model together (don't give them the activity
sheet). Let them discover for themselves. The students can be divided into small groups and asked to
make miniature DNA molecules with reduced templates. Two 64% reductions of the templates make a
nice size, but be aware that a DNA molecule on this
scale made of heavy construction paper does not
twist well. You may want to experiment with flimsier
paper if you want students to make twisted models to
use at their desks. (Flat desk-scale models would be
nice for the DNA replication activity.)

Advanced-placement classes can be asked to modify
this very simple model to be more accurate. They

should label all functional groups on the sugar and
number the carbons to determine the 5′ and 3′ ter-
mini (see introductory material for numbering). They
can be asked to create better templates for the nitrogenous bases in order to correctly simulate the sin-
gle- and double-ring structures of the purines and
pyrimidines. They can also make small models at
their desks.

DNA replication can be taught with this model by
having students make extra nucleotides, simulate un-
winding of the helix, and build new strands (see lesson 6, *DNA Replication*). The large helix model is also
useful in introducing restriction enzymes (see *DNA
Scissors*); you can build in a palindrome and "cleave" it
to create blunt or sticky ends. You can also refer to
the large helix during the activity *Sizes of the* Escherichia coli *and Human Genomes* to remind students of the structure of the DNA molecule. Use your
imagination to find additional ways to employ this inexpensive model.

Answers to Student Questions

1. Thymine
2. Cytosine
3. The nucleotide (deoxynucleotide)
4. A double helix
5. Adenine and guanine
6. Cytosine and thymine
7. Because of the chemistry of the bases; because of
   the space requirements of the helix
8. Deoxyribose

Answers to Advanced Questions

9. TCGAGTC

10. C, because the opposite strand is antiparallel to
    the given strand. The sequence in the answer to
    question 9 would be in the 3′- to - 5′ direction.

11. 55. In 100 base pairs, there are 200 bases. There
    are 45 cytosines, so there are also 45 guanines
    (90 bases accounted for). There are 110 more
    bases (A's and T's) in the molecule. There must
    be equal numbers of A's and T's, since these two
    bases pair. Therefore, there are 55 of each.

Constructing a Paper Helix 5

Introduction

DNA is called the blueprint of life. It got this name because it contains the instructions for making every protein in your body. Why are proteins important? Because they are what your muscles and tissues are made of; they synthesize the pigments that color your skin, hair, and eyes; they digest your food; they make (and sometimes are) the hormones that regulate your growth; they defend you from infection. In short, proteins determine your body's form and carry out its functions. DNA determines what all of these proteins will be.

The DNA molecule is a double helix. Think of it as a ladder that has been twisted into a spiral. The outside of the ladder is made up of alternating sugar and phosphate groups. The sugar is called **deoxyribose.** The rungs of the ladder are made up of nitrogen-containing bases. There are four different nitrogen-containing bases in DNA: **adenine, guanine, cytosine,** and **thymine.** These four bases are of two types: purines and pyrimidines. Purines are large double ring structures. Adenine and guanine are purines. Pyrimidines are smaller single-ring structures. Cytosine and thymine are pyrimidines. For more information on DNA and its component parts, see chapter 3.

Inside the DNA ladder, two bases pair up to make a "rung." One base sticks out from each sugar-phosphate chain toward the inside of the ladder. It forms a pair with a base sticking out from the opposite sugar-phosphate chain. Only three rings can fit between the two sugar-phosphate chains, so a pyrimidine (one ring) and a purine (two rings) form a pair. Because of the chemical structures of the bases, adenine always pairs with thymine, and cytosine always pairs with guanine.

Activity

The goal of this lesson is to construct a paper model of a DNA helix. You will do so by making the fundamental unit of DNA. This unit, called a **nucleotide,** consists of one sugar molecule, one phosphate group, and one nitrogenous base. Each member of the class will make nucleotides and then join them to form the ladder-like helix.

Procedure

1. Cut out the pattern for the nucleotide(s) assigned to you.
2. Place the pattern on construction paper of the appropriate color.
3. Label your pieces the way the pattern is labeled.
4. Glue the nitrogen base to the sugar molecule by matching up the dots.
5. Glue the phosphate group onto your model by matching up the stars.
6. Your teacher will join the nucleotides together to form a helix.

Questions

1. What base does adenine pair with?

2. What base does guanine pair with?

3. What is the smallest unit of DNA called?

4. What is the shape of the DNA molecule?

(continued on next page)

5. Which bases are purines?

6. Which bases are pyrimidines?

7. Why must a purine pair with a pyrimidine?

8. What is the name of the sugar in the DNA backbone?

Advanced Questions

9. Suppose you know that the sequence of bases on one DNA strand is AGCTCAG. What is the sequence of bases on the opposite strand?

10. Referring to question 9, suppose that the 5′-most base on the given strand is the first A from left to right. What would be the 5′-most base on the opposite strand?

11. Assume that a 100-base-pair DNA double helix contains 45 cytosines. How many adenines are there?

DNA Replication

6

About This Activity

DNA replication is a topic usually presented in 9th-grade biology. The essential fact of DNA replication is that the base-pairing rules make it very easy to generate two identical new helices from one helix. This basic piece of information is all that is really necessary for young students to know about replication. More advanced students who will be learning about DNA sequencing and/or the polymerase chain reaction (see the activities *DNA Sequencing* and *The Polymerase Chain Reaction: Paper PCR*) need to know a little more about DNA replication so that they can understand these interesting applications. The approach taken in this lesson is to provide detailed information to the teacher in the introductory material and then give a two-part activity. The first part is appropriate for young students; more advanced students will perform both parts of the lesson.

The first activity described below is a simple (and necessarily inaccurate) paper simulation of DNA replication. It is quite sufficient for most 9th graders, since it makes the point that two strands of the parent double helix are used as templates to synthesize two daughter strands. The paper DNA models described in this section can be used in the exercise. If you are teaching more advanced students, use the basic activity as a starting point and then introduce a more accurate picture of DNA replication through the second activity.

The second activity is a student reading about DNA polymerase, the central DNA replication enzyme. This reading provides the detail students will need to understand subsequent lessons. It contains questions, one of which involves using the information in the reading to identify inaccuracies in the simple model they have just used. You may supply them with additional information as you feel necessary. *The background information in the introduction that follows contains far more detail about DNA replication than you need to share with students. It is there for your information and enjoyment.* The two as-pects of DNA synthesis that your advanced students need to know (if they will be doing the activities on DNA sequencing and/or the polymerase chain reaction) are that synthesis is unidirectional and that it *absolutely requires a primer.*

Class periods required: 1

Introduction

All organisms must copy their genetic information, both during cell division and for transmission to their offspring. This critical task is carried out by groups of proteins working together, but the central player is the enzyme DNA polymerase. DNA polymerase selects the correct new nucleotide by checking that the nucleotide pairs correctly with the template base, and then the enzyme forms the new phosphodiester bond linking the new nucleotide to the growing DNA chain. Not surprisingly, the characteristics of DNA polymerase determine the overall features of DNA replication inside the cell and in the test tube.

In the DNA synthesis reaction itself, an incoming deoxynucleotide triphosphate binds to the polymerase. This nucleotide must be in the triphosphate form; the enzyme will not bind mono- or diphosphonucleotides for incorporation into DNA. (A bit of terminology: a nucleotide-like molecule consisting of only the sugar and the base is called a **nucleoside;** nucleotides are sometimes referred to as nucleoside monophosphates, diphosphates, or triphosphates to specify how many phosphate groups are attached.) Next, the polymerase checks to see whether the incoming nucleotide pairs properly with the template base. If it does, the enzyme forms a bond between the first of the three phosphates on the 5′ carbon of the new nucleotide and the 3′ hydroxyl (OH) group on the last nucleotide of the growing chain (Fig. 6.1). The formation of this bond leaves one of the phosphate groups in the growing DNA chain and liberates an inorganic phosphate molecule, a "pyrophosphate," with the other two phosphorous atoms.

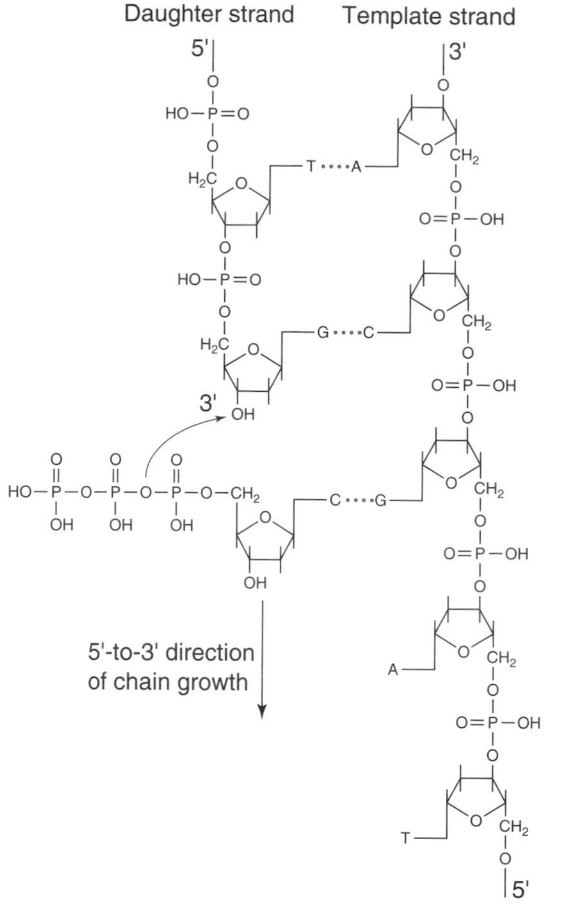

Daughter strand   Template strand

5'-to-3' direction
of chain growth

Figure 6.1 DNA polymerase (not shown) checks the pairing of an incoming deoxynucleoside triphosphate with the template base and then forms a new phosphodiester bond between the 5' phosphate group of the new nucleotide and the 3' hydroxyl group of the previous nucleotide.

Finally, many, but not all, polymerases also possess a proofreading function that checks the new base pair to see whether it is accurate. If the new base does not pair correctly with the template base, then the new nucleotide is removed from the chain, and the enzyme tries again. Polymerases that lack proofreading ability, such as reverse transcriptases, accumulate more mistakes during the synthesis of DNA than polymerases that can proofread.

How frequently do DNA polymerases make mistakes in replicating DNA? According to in vitro studies, the *Escherichia coli* enzyme adds an incorrect base once every 10^5 or 10^6 base pairs, but its ability to proofread and correct its mistakes brings its final error rate down to one error per 10^8 base pairs. Inside the cell, additional DNA mismatch correction enzymes lower the in vivo error rate to one in 10^{10} base pairs. *E. coli*'s genome has about 3×10^6 base pairs, so the bacterium makes less than one error in DNA replication per cell division. In contrast, the reverse transcriptase

of retroviruses (such as the human immunodeficiency virus that causes AIDS) is a less careful polymerase than *E. coli*'s enzyme and also lacks proofreading ability. These enzymes make an error in replication nearly every 10^4 base pairs. This error-prone replication of genetic material is thought to be the cause of the rapid mutation rate of the human immunodeficiency virus.

Two important features of the DNA polymerase reaction are the direction of DNA synthesis and the requirement for a primer. Notice (Fig. 6.1) that the enzyme always connects the 5' phosphate of the incoming nucleotide to the 3' OH group of the growing chain. This feature of DNA replication means that DNA synthesis is unidirectional, from 5' to 3'. Unidirectionality presents a problem for chromosome replication that is discussed below.

The other feature of the reaction is that DNA polymerase must have a growing new strand to connect with the incoming nucleotide. No known DNA polymerase can begin synthesizing a complementary strand on a naked single-stranded DNA molecule. However, if the single strand has a short, complementary oligonucleotide annealed (base paired) to it somewhere, DNA synthesis can begin at the 3' end of that oligonucleotide and continue in the 5'-to-3' direction down to the end of the single strand. In this example, the long single strand is the template strand, and the short complementary oligonucleotide annealed to it is called the primer. In general, a primer is an oligonucleotide (either DNA or RNA; many DNA polymerases use RNA primers) annealed to the template with a 3' OH group available at the end as a place to begin synthesis. The template molecule usually extends far beyond the end of the primer. Nature has devised many ways of providing primers for DNA synthesis; some are mentioned below.

When working with DNA polymerases outside the cell (as in biotechnology applications), it is necessary to provide a template with an annealed primer, deoxynucleoside triphosphates, and an appropriate buffer to achieve DNA synthesis. The direction of synthesis will always be 5' to 3' beginning at the 3' end of the primer. Scientists use the selection of the primer to determine where or whether DNA synthesis will occur.

Very often, the primer is a synthetic oligonucleotide (see chapter 4) added to the reaction mixture separately. The template DNA and the primer can be annealed by heating and cooling (more about this in chapter 16, *Detection of Specific DNA Sequences: Hybridization Analysis*). Thus, the scientist can decide

126 • Recombinant DNA and Biotechnology

where she wants DNA synthesis to begin and can synthesize a complementary primer that has its 3′ end at that site (provided she knows the DNA sequence of the region).

Primer specificity can also be used to determine whether a particular DNA molecule is present in a mixture, for example, whether a disease-causing microorganism is present in a tissue sample. In tests based on the polymerase chain reaction (see *The Polymerase Chain Reaction: Paper PCR*), primers that hybridize only to the DNA of the organism of interest are synthesized. If the organism is present in the sample, the primers can anneal, and DNA synthesis will occur. If the organism is not present, the primers will not anneal, and no synthesis can occur. The presence or absence of DNA synthesis is used to determine whether the organism of interest is in the sample. The requirement for a hybridized primer is a powerful tool that gives scientists a lot of control over when and where DNA synthesis occurs in vitro.

Cellular DNA does not normally exist as single strands with annealed primers. The chromosomes of organisms are generally completely double-stranded DNA molecules. In vitro, DNA polymerases will not begin DNA synthesis on a perfectly double-stranded DNA molecule. Scientists use manipulations such as denaturing double-stranded molecules and hybridizing short primers to the single strands to enable DNA synthesis to occur in the test tube. How do cells get around the problem?

The question of how cells manage to initiate DNA replication (and time and control it) is the subject of ongoing research in many laboratories around the world. For chromosomal DNA replication to begin, three major events must take place: (1) the replication proteins must assemble on the DNA, (2) the DNA helix must be opened to expose unpaired bases for use as the template, and (3) a primer must be provided. The following description of DNA replication initiation is a generalization from findings from several systems and does not apply in every single case.

Chromosomal DNA replication is usually initiated at specific sites along the DNA called replication origins. In general, these sites contain specific base sequences at which special replication initiator proteins bind. Origins also often contain a region of mostly A-T base pairs, presumably because it is easier to open a region of DNA helix that is rich in A-T pairs (only two hydrogen bonds pair A and T, but three bonds pair G and C; see chapter 16). The initiator protein or proteins bind to the recognition site in the replication origin, and this binding triggers the assembly of the replication proteins. The double helix is opened at or near the origin, and a primer anneals or is made. The primer is usually RNA. It can be a transcript of the origin region that was synthesized previously, or it can be a special short RNA synthesized on the spot. As you can imagine, the use of RNA for DNA replication primers means that DNA replication initiation is often entangled with transcription, and some of the best-understood initiation mechanisms are fairly complicated. Many scientists are entertaining themselves trying to sort out different methods of DNA replication initiation in different organisms.

After the helix is opened and the primer is available, DNA polymerase can begin to replicate DNA. The polymerase is assisted in its task by helper proteins; typical functions of these proteins are to "unzip" the double-stranded template ahead of the polymerase, to protect any exposed single-stranded regions of DNA (these regions are very vulnerable to degradation by cellular enzymes), and to help solve the problem of replication of the opposite strand.

What is the problem of the opposite strand? The problem is the direction of DNA synthesis and the lack of primers. The primer laid down at the origin enables DNA polymerase to synthesize DNA in the 5′-to-3′ direction away from the primer and the replication origin, using the strand annealed to the primer as the template. What happens to the opposite strand? We know that the opposite strand is replicated along with the first strand as the replication complex moves away from the origin. Yet since DNA synthesis can occur only 5′ to 3′, replication on the opposite strand should head back toward the replication origin, in the opposite direction. How does the cell coordinate DNA synthesis in opposite directions?

We have known since the 1960s that replication of the opposite strand generates short Okazaki fragments. While a complementary strand for the first strand (the *leading* strand) is synthesized in one long piece starting at the origin primer, the partner strand for the second strand (the *lagging* strand) is synthesized in pieces (see Fig. 6.2). How is this piecewise synthesis of the lagging strand accomplished, and where do the primers come from?

Research on DNA replication in bacteriophage T4 has shown that when the phage's replication proteins assemble, *two* molecules of DNA polymerase join the complex. One is responsible for synthesis of the leading strand, and the other is responsible for the lagging strand. The way in which these two molecules are believed to work together to replicate both strands of the DNA template is shown in Fig. 6.3. This figure is com-

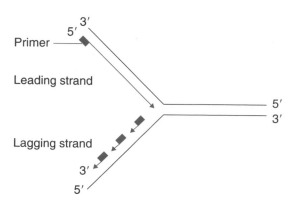

Figure 6.2 The lagging strand is synthesized as Okazaki fragments during DNA replication.

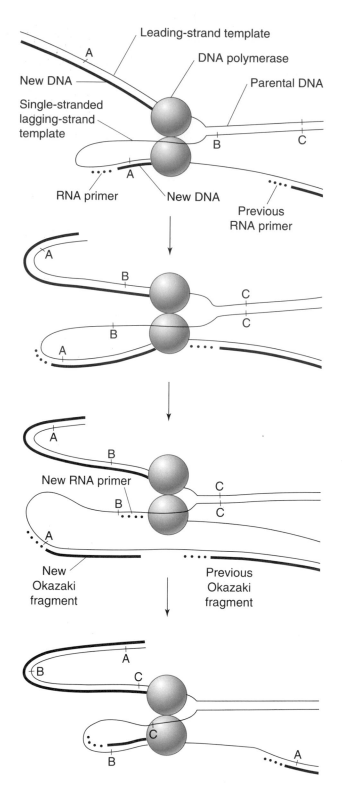

Figure 6.3 This current model for the simultaneous replication of both DNA strands (described in the text) has been christened the "trombone model."

plicated; look at it as you read the explanation in the text. A, B, and C on the DNA strands in the figure are simply position markers to help you follow the movement of the polymerase molecules along the DNA.

The polymerase complex (with its helper proteins) travels down the parental DNA for a few hundred nucleotides, synthesizing the leading strand as it goes. During this time, the other template strand is left single stranded; it is coated with special proteins to protect it. At a certain point, one of the helper proteins in the complex synthesizes a short RNA primer on the second template strand. The second DNA polymerase molecule (the first one is tending to business on the leading strand) begins synthesis of an Okazaki fragment at the 3′ end of this primer.

Look at the first two images in the figure. The top strand of DNA in each image is the template for the leading strand. Imagine the two linked polymerase molecules moving from left to right down the DNA (from A to B along the molecule), synthesizing the leading strand in one smooth piece. Meanwhile, the template for the lagging strand is folded into a hairpin. This orients the template correctly for 5′-to-3′ synthesis. Notice that the lagging-strand DNA is being synthesized from A away from B—in the opposite direction from the leading strand. The polymerase proteins do not move back toward the replication origin during this time; rather, the template for the lagging strand is pulled through the replication complex. Notice that the loop protruding to the left of the polymerase gets larger from the first to the second image. This movement reminds some scientists of the way the slide of a trombone moves out from the instrument.

When the newly synthesized lagging strand segment reaches the previous primer, the new duplex DNA is released, and another primer is synthesized to start the next Okazaki fragment (look at the second and third images in Fig. 6.3). This new Okazaki fragment will extend from B back to the primer near A (fourth image). The RNA primers will later be removed and replaced with DNA, and then the Okazaki fragments will be joined together by the enzyme DNA ligase.

This seemingly complicated method of DNA replication is a consequence of the directionality of DNA synthesis and the requirement for primers. Those same constraints govern the use of DNA polymerase enzymes outside the cell. However, these very limitations can give scientists ways of controlling DNA synthesis in vitro (as discussed above).

A final replication problem caused by unidirectionality and the requirement for primers is the "end" problem. Look at Fig. 6.2 and imagine what happens when the replication complex reaches the end of the DNA molecule (the right end in the picture). It is clear that the leading strand could be replicated up to the end of a double-stranded helix, but the lagging strand again has a problem. In general, an RNA primer is not synthesized at the extreme end of the chromosome, so how is the end of the lagging strand replicated? Once again, nature has found many ways to circumvent the problem created by the specificity of her enzymes. Some organisms have circular chromosomes. Higher organisms have developed special structures called telomeres at the ends of their chromosomes that solve the problem for them. Still other organisms (viruses in particular) use interesting combinations of DNA synthesis and genetic recombination to achieve complete replication of their DNA.

If you have become intrigued by the process of DNA replication and want more information and details, a good discussion of the process can be found in *Molecular Biology of the Gene,* vol. I, by Watson et al. (See Appendix H for specific information.)

Objectives

After the simulation activity, all students should be able to

1. Describe how the complementary base-paired structure of DNA allows two identical new helices to be made from a single template helix;

2. State that DNA replication is carried out by proteins inside the cell.

After the student reading activity, advanced students should be able to

3. Explain what is meant by the statement "DNA replication is unidirectional";

4. Explain what is meant by the statement "DNA replication requires a primer";

5. Name the central enzyme involved in DNA replication (DNA polymerase);

6. State that purified replication proteins can synthesize DNA in the laboratory under the right conditions.

Materials

1. Simulation of DNA replication
   DNA models showing the sugar, phosphate, and bases separately. These can be colored paper cutouts from the model presented in an earlier lesson, commercially available pop-bead models, or other materials. You will need enough for the students to work with in small groups or as individuals.

2. Student reading activity
   Sufficient photocopies of the *Student Reading* pages if students do not have manuals

Resource Materials

- Demonstration kit "DNA Made Easy," sold by Connecticut Valley Biological (catalog no. AP-425) and Carolina Biological Supply Co. (catalog no. 17-1040)

- DNA simulation bead kits; several available, including bulk beads; Carolina Biological Supply

Preparation

- Assemble kit materials or paper cutouts. If you are using cutouts, photocopy the templates onto different colors of paper and have each student cut out several "sugars," "phosphates," and "bases" (all four) before beginning the activity.

- If you made paper models of DNA and saved them, use them for replication templates with the paper cutouts.

- Photocopy the *Student Reading* pages if you plan to use that part of the lesson and students do not have manuals.

Procedure

1. Simulation of DNA replication

- Ask your students why a cell might want to make a copy of its DNA. Ask them if they think it would matter whether the copy were an exact copy. (The point is to have them realize that if an organism is to have offspring that are like itself, it must pass on an accurate copy of its genetic material.)

- Ask your students what copies DNA inside the cell. They may not know, but the correct answer is enzymes, and enzymes are proteins. We want them to realize that all body and cellular processes involve proteins and that DNA's central role is to carry the information needed to make all these important proteins.

- Make sure that each student (or student group) has its starting DNA molecule (whether it is a paper model they made or one from a kit; the base sequence is totally unimportant). You may enjoy asking younger students how the starting molecule might be used as a pattern to make new DNA.

- Tell students that the DNA building blocks the cell uses are nucleotides (a sugar, a phosphate, and a base). Have them assemble some of each kind of nucleotide from their model components. It is not necessary to bring up the point about nucleoside triphosphates with younger students.

- Explain to your students (if they haven't arrived at the idea already) that the cell uses each strand of the parental DNA molecule as a template (pattern) to make a second strand. If necessary, explain that the parental molecule is "unzipped" (the correct term is denatured or melted) to expose unpaired bases and that new nucleotides are brought in one at a time, in order, and formed into a complementary chain that pairs with the template.

- Have students "unzip" their DNA molecules so that the unpaired bases are exposed. Then have them build a complementary strand on each parental strand, using the nucleotide precursors. Check your students' work to make sure that they have correct base pairs in their new molecules.

- Talk to the students about their new molecules. Are the new ones exactly like the parent molecule? (They should be.) How many "old" strands are in each new molecule? (One.) How many "new" strands? (One.)

This is a sufficient treatment of DNA replication for younger students.

2. Student reading activity

For a more detailed treatment of DNA replication, tell your advanced students that although the exercise they have just completed gets a very basic point across, it is inaccurate in nearly every detail. Have students read *DNA Polymerase: the Replicator* and answer the questions. Make sure they understand that the stick figure portion of the nucleoside triphosphate in the figure represents the deoxyribose sugar and the base with 5' phosphate groups and the 3' OH group shown (since they are the functional groups that participate in new bond formation). You may want to show them Fig. 6.1 from the teachers' introduction.

Ask your students what features of the simple replication exercise are inaccurate (e.g., DNA polymerase binds only nucleoside triphosphates for incorporation, yet the above model uses nucleoside monophosphates; the model uses no primers; the model does not take into account the directionality of DNA synthesis). Be sure that they have noted the requirement for a primer and the directionality of DNA synthesis. You may want to use one of their models to show that DNA polymerase would have to synthesize DNA from left to right on one strand but from right to left on the other to maintain the 5'-to-3' direction. Use your judgment as to how much information to present about mechanisms of chromosomal replication. It is not necessary (and would be very difficult) to try to execute the detailed replication model in Fig. 6.3 with paper cutouts, but if you like, you could draw or reproduce the model for students to see.

Answers to Student Questions

1. DNA synthesis will occur on templates b and d.

   Templates a and c do not have primers. In both template b and template d, synthesis will begin at the 3' end of the primer and continue rightward with respect to the picture (only the short strand on the left in template d can serve as a primer, since there is no protruding template beyond the right-hand short strand).

2. Discussed above.

Classroom Activities

DNA Polymerase: the Replicator

The accurate copying of DNA is one of the most important jobs an organism must do during its life. Why do you think this statement is true? What would happen to you if your ancestors' cells had not taken the trouble to make accurate copies of their DNA?

For such an important task, cells employ not one but a whole team of enzymes. However, the star of the team is the enzyme DNA polymerase. This protein builds the new daughter strand from nucleotides in the cell and checks the new base pairs for accuracy. The other protein team members help DNA polymerase do its job.

All bacteria and higher organisms and many viruses have their own DNA polymerase proteins, which are encoded by their DNA. All of these DNA polymerases work in a similar way and even resemble one another. Scientists therefore often study the complicated process of DNA replication in simple systems, such as bacterial and animal viruses, and apply what they have learned as they look at higher organisms. From these simple systems, many general facts about DNA polymerases have emerged. Let's look at some of these facts and then consider how they were determined.

1. DNA polymerases use deoxynucleoside triphosphates as precursors for the synthesis of DNA. These molecules are held at a special binding site on the polymerase before they are incorporated into a new DNA molecule. Nucleotides with only one or two phosphate groups will not bind there and are not incorporated into new DNA.

2. DNA polymerases cannot begin synthesizing DNA without a starting point called a **primer.** A primer is a piece of DNA or RNA base paired to the template strand so that the template strand sticks out past the 3′ end of the primer (see the figure). DNA synthesis begins at the 3′ end of the primer, where DNA polymerase attaches the first new nucleotide.

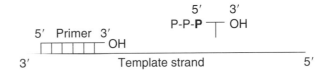

DNA replication. The 5′ phosphate group shown in boldface will react with the 3′ OH group of the primer to form a new link in the growing DNA strand.

3. DNA polymerases synthesize DNA in one direction only. They start by attaching the 5′ phosphate group of the new nucleotide to the 3′ OH group of the last nucleotide in the primer, and they continue in this manner. They cannot connect the 3′ OH group of a new nucleotide to the 5′ phosphate group of a primer. DNA synthesis is therefore said to occur in the 5′-to-3′ direction (relative to the new strand).

How were these facts deduced? By looking at reactions with purified DNA polymerase proteins in vitro ("In vitro" means "in glass"; it refers to the use of glass test tubes in the laboratory and basically means "in the test tube," even though almost everyone uses disposable plastic tubes for this kind of work now). A scientist could take purified DNA polymerase protein, add a DNA template, and then experiment by adding deoxynucleoside monophosphates, diphosphates, or triphosphates and looking for DNA synthesis. By using radioactive nucleotides, a scientist can detect DNA synthesis by the appearance of radioactive DNA strands. In tests like this, it is clear that only nucleoside triphosphates are incorporated into DNA. It is also possible to mix DNA polymerase with radioactive nucleotides and ask whether the radioactivity becomes associated with the protein. In this manner, it is possible to detect the binding of nucleotides to DNA polymerase.

Besides looking at the effects of different forms of nucleotides, scientists looked at different forms of DNA templates. They constructed different types of DNA molecules and asked whether DNA synthesis occurred when polymerase and deoxynucleoside triphosphates were added. Through these experiments they learned that DNA polymerase must have a primer on the template strand before it can synthesize DNA. Finally, by examining the DNA that was synthesized, they learned that DNA polymerase can add nucleotides only in the 5′-to-3′ direction. The enzyme could never synthesize a complementary strand to a portion of the template that was "upstream" (on the 5′ side) of the primer.

DNA polymerases are now important tools in molecular biology research and in biotechnology applications. They are used in cloning, copying, and sequencing DNA. In later lessons, you may have the opportunity to learn how DNA polymerases can help reveal the base sequence of a piece of DNA or how they are used to make a "DNA Xerox machine" that is extremely useful in detecting organisms that cause disease, among other applications. When you do these activities, notice that in each case, the DNA polymerase is given a primer and that DNA synthesis occurs in the 5′-to-3′ direction.

Questions

1. Given what you have learned about DNA polymerase, on which of the DNA template molecules shown below could the enzyme synthesize a new strand if given nucleoside triphosphates? Show where DNA synthesis would begin on each molecule and in what direction it would proceed.

2. Given what you have learned about DNA polymerase, what is wrong with the simple model of DNA replication that you used earlier?

A
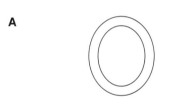

B

C

D
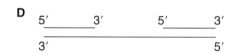

From Genes to Proteins

About This Activity

In this lesson, students act out transcription and translation either at their desks with paper models or by assuming roles and moving around the room. You may simply model the process in prokaryotes or eukaryotes, or you may expand the activity as a means of exploring the differences in genetic traffic signals and gene structure in prokaryotes and eukaryotes. This activity also makes a wonderful lead-in to a discussion of the various means of gene regulation (in as much detail as you wish).

Class periods required: *1–3*

Introduction

DNA determines the characteristics of an organism by specifying the amino acid sequences (and therefore the structure and function) of its proteins (see chapter 3 for review). In order to direct the synthesis of a protein, the DNA must contain not only codons for each of the amino acids in that protein but also regulatory sequences that tell the cell's protein-synthesizing machinery where to start and stop. Although prokaryotes and eukaryotes use the same genetic code, they have evolved different regulatory signals.

The first important regulatory region for protein synthesis is the **promoter,** the sequence of DNA bases that RNA polymerase recognizes and binds to before beginning transcription. Without a promoter, transcription does not occur. Prokaryotic and eukaryotic promoters are different, as are prokaryotic and eukaryotic RNA polymerases. A typical bacterial promoter consists of the sequence 5'-TTGACA separated by 17 bases from the sequence TATATT-3' (some variation is permitted).

Eukaryotic promoters are more variable. One component of a eukaryotic promoter is the sequence TATA, which is about 30 bases upstream of where transcription begins in yeasts and 60 bases upstream in mam-

mals. This component is often called the TATA box. Usually, two more promoter components are found somewhat further upstream of the TATA box. These are the sequence CCAAT and a GC-rich sequence. However, remember from chapter 3 that eukaryotic RNA polymerase alone is not efficient at starting transcription from a promoter. Eukaryotic promoters are associated with a variety of different **enhancer** sequences, sites on the DNA where various transactivating transcription factors bind and stimulate transcription (see chapter 3).

Transcription begins downstream of the promoter and continues until a transcription terminator is reached. The DNA sequences of terminators are complicated; different classes of terminators have different structures, and prokaryotic and eukaryotic terminators also differ. The function of a terminator is to cause RNA polymerase to stop transcribing DNA and release the DNA template.

The nucleotide sequence of messenger RNA (mRNA) is translated into protein at the ribosome. Translation does not simply begin at the 5' end of a message and end at the 3' end. Additional regulatory elements that direct the ribosomes to begin and end translation are contained within the RNA itself. The RNA bases at the extreme 5' end and the extreme 3' end of the message are not translated into protein.

Foremost among these regulatory elements in bacteria is the signal for ribosomal recognition. In bacteria, this element commonly has the sequence 5'-GAGG-3' or 5'-AGGA-3' located 8 to 13 nucleotides upstream of the initiation codon. The ribosome recognizes the element and binds to the mRNA there. How does binding occur? The bacterial ribosomal RNA (rRNA) contains a base sequence complementary to the ribosomal recognition element (RRE) on the mRNA, and the rRNA and mRNA hybridize there. Because the RRE is close to the initiation codon, the initiation codon is brought into proper position for translation to begin.

The picture is not as clear in eukaryotes. In the simplest model, eukaryotic ribosomes are thought to bind at the 5′ end of the mRNA and to search down the mRNA for the first initiation codon in order to begin translation. Experiments in some laboratories suggest that this simple model does not explain all situations.

In both prokaryotes and eukaryotes, translation begins at the initiation codon AUG. This codon specifies the amino acid methionine. Prokaryotic and eukaryotic ribosomes use a special modified form of methionine (formylated methionine in *Escherichia coli*) to begin protein synthesis. In *E. coli,* initiation of protein synthesis also requires three proteins called **initiation factors.** In eukaryotes, many more protein factors are required. Once initiated, protein synthesis continues down the mRNA until a stop codon is reached, at which point the ribosome releases the mRNA.

The genetic traffic signals involved in converting DNA sequences into proteins (in prokaryotes) are summarized in Fig. 7.1.

In prokaryotes, the product of transcription is mRNA. In eukaryotes, however, RNA polymerase synthesizes a primary transcript that must be processed further before it can function as a message. One of these processing steps is **splicing.**

Many, but not all, eukaryotic genes contain **introns** (see chapter 3). An intron is a section of DNA that is embedded in the protein-coding region of a gene but is not represented in the mRNA or protein sequence. The intron DNA is transcribed into RNA along with the coding regions to form a long precursor RNA. Introns are removed from the precursor RNA by splicing, which occurs in the nucleus.

The mechanisms of splicing are the subject of current research. There appear to be at least three mechanisms: one for transfer RNA (tRNA), one for mRNA, and one for rRNA. Each mechanism uses a different set of signals to determine precisely which sequences should be removed from the precursor RNA. A generic model of splicing is shown in Fig. 7.2. Accurate splicing is very important; several types of thalassemia (a blood disorder) in humans appear to be caused by mutations in the splicing signals of the hemoglobin gene that lead to splicing errors in the hemoglobin RNA and thus incorrect protein synthesis.

For this lesson

The description presented here is superficial; further reading in a molecular biology text will provide more details, should you wish to delve deeper. It is of course not necessary to convey all of this information to students. You may teach about as many of these regulatory elements as you deem appropriate; the most important is undoubtedly the promoter.

The cutouts provided allow you to simulate protein synthesis in prokaryotes and eukaryotes. The most complete model of bacterial protein synthesis would incorporate the promoter, the terminator, and the RRE. The most complete eukaryotic model would incorporate the promoter, the terminator, and the splicing start and stop signals (the splice junctions).

Objectives

After this lesson students should be able to

1. Describe DNA's function as the basic hereditary material controlling cellular activity via control of the cell's enzyme systems;
2. Describe transcription and translation;
3. Describe the function of the promoter and the terminator;
4. (Optional) Contrast prokaryotic and eukaryotic genetic "traffic signals."

Materials

- Scissors
- DNA-RNA genetic code triplet sheets; any desired regulatory element sheets

Figure 7.1 Major genetic traffic signals in bacteria. These signals tell RNA polymerase where to begin and end transcription, enable the ribosome to recognize mRNA, and direct the ribosome to start and stop protein synthesis.

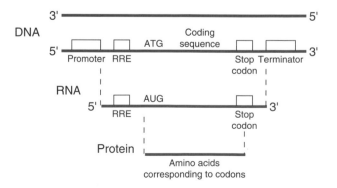

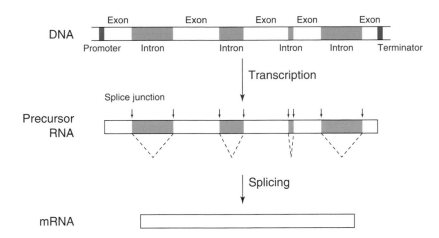

Figure 7.2 Splicing of precursor RNA to produce mRNA. The DNA exons contain the coding sequence for the protein.

Resource Materials

Demonstration kit "DNA Made Easy," sold by Connecticut Valley Biological (catalog no. AP 425) and Carolina Biological Supply Co. (catalog no. 17-1040)

Variations on This Activity

Most of the discussion below is related to a form of the activity in which students act out transcription and translation by moving around the classroom. Some teachers use this form of the activity regularly and enjoy it. The activity also works as a demonstration; you can tape the DNA template cards to the blackboard and show how RNA is formed.

Another option is to have students work in pairs at their desks. Reduce the cutout patterns on a photo copier, and group them onto a few sheets. Give each team a set of the sheets. Students can cut out the patterns as simple squares of paper to save time and then use the cutouts to simulate transcription and translation at their desks.

Preparation

Modify the preparation according to which lesson variation you plan to use.

Before class, cut out the genetic code cards in Appendix K (or have your students cut them out in class). The DNA cutouts come in complementary pairs: one cutout has filled-in letters, and its complement has outlined letters. The outlined-letter strand will represent the coding strand of the DNA; the filled-letter strand is the noncoding strand. Paperclip the complementary DNA cards together in sets of two. Arrange the cutouts into four stacks: DNA, mRNA, tRNA, and amino acids. Put the start sequence (TAC) at the second position in the DNA stack, and put the stop sequence (ATC) on the bottom. Then place the promoter card on top of all the DNA cutouts, and place the terminator on the bottom.

Decide which model (prokaryote or eukaryote) you will use and which traffic signals you want to incorporate. Cut out the traffic signal cards. If you use the RRE card, it should go between the first DNA cutout and the start (TAC) card. If you use splicing start and stop cards, they can go anywhere between the start and stop cards as long as you do not splice out the entire gene! Sample lineups for prokaryote and eukaryote models are shown below (Fig. 7.3).

Figure 7.3 Sample DNA sequence lineups for transcription. The splice junctions and splicing step are optional. DSC, DNA sequence card with any three bases; RRE, ribosome recognition element; TAC, noncoding DNA card with TAC base sequence (the corresponding coding card is ATG); . . . , any number of DNA sequence cards; stop, the stop codon.

Prokaryotic model

Promoter	DSC	RRE	TAC	DSC	DSC...DSC	DSC	stop	DSC	terminator

Eukaryotic model

Promoter	DSC	TAC	DSC...DSC	splice	DSC...DSC	splice	DSC...DSC	stop	DSC	terminator

Tips

1. If possible, photocopy the triplet codes onto colored paper, using different colors for DNA, mRNA, tRNA, and amino acids.

2. If possible, laminate the cards so they can be used over and over.

3. Before the activity begins, make sure that students understand the importance of proteins in determining the characteristics of an organism (see introductory material).

4. Some teachers glue the cutouts onto larger squares of paper and then thread yarn through reinforced holes in the squares to make placards. Students can hang them over their shoulders as labels during the simulation. These placards can be used from year to year.

Procedure for Acting Out Transcription and Translation

These instructions are relevant even if you plan to do the activity as a demonstration or have the students use paper models at their desks.

1. Setting the scene. If you are using the eukaryotic model, it can be helpful to set the scene for students as follows.
   • The classroom's floor, walls, and ceiling are analogous to a cell membrane.
   • Designate one area as the nucleus, where transcription occurs.
   • Designate another area as the ribosome, where translation occurs.
   • All other nonnucleus areas in the room are cytoplasm.

Important: If you want to use the *E. coli* model, remember that *bacteria have no nuclei*. The chromosome and the ribosomes are associated with the cell membrane. Designate a ribosome area accordingly, and let the DNA be next to the membrane. Also remember that *E. coli* uses the RRE but does *not* use splicing. (You don't have to talk about ribosome recognition, but *do not* show splicing in bacteria.) The eukaryotic model does *not* use the RRE (let the ribosome recognize the 5′ end of the message) and may use splicing. If you are very ambitious and have time, you can have students go through both models to see the similarities and differences involved.

2. Distribute the DNA sequence card pairs and desired regulatory elements and the matching mRNA codes. The large letters on the cards refer

to the first letters of the nucleotide bases (A = adenine, C = cytosine, G = guanine, T = thymine, and U = uracil). Do not distribute the tRNA and amino acids cards until step 10.

3. Review DNA structure as needed. The informational part of DNA is within the paired bases, so transcription does not begin until RNA polymerase unzips the two DNA strands. RNA polymerase catalyzes the pairing of DNA's exposed bases with complementary RNA nucleotides that happen by.

Students may get confused about which strand of DNA is doing what. One strand (in this case represented by cards with outlined letters) is the *coding strand*. It contains the three-base codons that will be found in mRNA. To move the coding information from a gene to the ribosome for translation, the cell makes a working copy of the gene's coding strand. This copy is mRNA. How can the cell make a copy of the coding strand? By synthesizing mRNA complementary to the noncoding strand. The process is similar to DNA replication except that only one strand of the DNA molecule is used as a template. The mRNA generated is a copy of the coding strand. How does the RNA polymerase "know" which strand to use as a template for synthesis? The promoter sequence orients the enzyme properly.

Transcription

4. If you are simulating gene expression in prokaryotes, line the students up from left to right as follows. Each student holds a pair of cards, one with filled letters and the other with complementary outlined letters.
   • The first student holds the promoter cards.
   • The second student holds a DNA sequence card pair (anything but ATG/TAC or the stop codon TAG/ATC).
   • The third student holds the ATG/TAC DNA card pair. The TAC DNA code will produce the mRNA initiation codon AUG.
   • The fourth student holds a DNA sequence card pair.
   • Optional: The fifth student holds the RRE cards.
   • The next several students hold DNA sequence card pairs. The order is not important.
   • The third from last student holds the TAG/ATC card pair. This is the stop codon.
   • The next to last student holds any DNA sequence card.
   • The last student holds the terminator.

Please refer to Fig. 7.3 for pictures of lineups for both prokaryotic and eukaryotic gene expression. If you are simulating gene expression in eukaryotes, be sure to use the eukaryotic model.

Have the DNA students hold their card pairs in front of their bodies with the letters facing the class. The cards with filled letters should be held above the cards with the outlined letters. You should see a row of filled-letter cards across the top and a complementary row of outline-letter cards across the bottom of two rows of DNA sequence. These two rows represent the base-paired DNA molecule. The student with the promoter should stand to the left as described previously.

5. Designate one student to represent RNA polymerase. This student should walk to the promoter student and shake hands, signifying that the RNA polymerase has recognized the beginning of a gene. After the handshake, the DNA students should lower the outline-letter cards (the lower row) to their sides, exposing a single-stranded template for RNA polymerase. They should hang on to the outline-letter cards for later comparison with the mRNA.

6. Review the following rules with the students who have the mRNA cards.
   - RNA cytosine always pairs with DNA guanine.
   - RNA uracil always pairs with DNA adenine.
   - RNA adenine always pairs with DNA thymine.
   - RNA guanine always pairs with DNA cytosine.

7. Using these rules, students with mRNA cards must find their DNA match. Have them pair in order, starting with the first DNA sequence card and progressing to the terminator, where pairing stops. When the RNA and DNA have correctly paired and the terminator has been reached, compare the sequence of the RNA to the sequence of the outline-letter DNA cards the DNA students are holding (the coding strand of the DNA). The students should see that the base sequences of the outline-letter cards and the RNA are identical except for the substitution of U for T in the RNA. The sequence of the RNA is complementary to that of the filled-letter DNA cards.

8. The DNA students can sit down, leaving a chain of RNA. (You may want to remind students that in real life, the DNA chain extends beyond the terminator to the next gene, so it is important to have a "stop sign" for RNA polymerase.) Tell the class that they have performed transcription. They have made a very short (shorter than in real life) complementary section of RNA that perfectly reflects the sequence of bases in the DNA molecule.

Note: After students match up their cards, they will not necessarily be in the order shown in Fig. 7.4, but the DNA-mRNA pairs should match vertically as in the figure.

9. If you are simulating splicing, do it now. To simulate splicing, have the RNA stay in the nucleus. The two splice junctions and everything between them leave and sit down, and the RNA sequence immediately before the first and after the second splice junctions are left adjacent to each other.

Translation

Students will notice that each tRNA card contains a group of three letters. The three-base sequence of tRNA represents its anticodon. In addition, each tRNA card has a three-letter abbreviation (in the "arrow" part of the card) for one of the 20 amino acids (Table 7.1). This lesson includes only 7 of the 20 amino acids used for making proteins.

10. Separately distribute the tRNA and amino acid cards (one per student). Students with either type of card should be randomly scattered in the "cytoplasm." If you are going to include ribosome recognition in your simulation, designate one student to be the ribosome and perform the recognition function. That student should stand at the designated ribosome area.

11. Have the students with the tRNA cards find the students with their specific amino acids and stand together. For example, the tRNA card GGC with the letters PRO should find the amino acid PROLINE. In the cell, a specific enzyme for each type of tRNA molecule joins the tRNA to its correct amino acid.

12. Students with mRNA cards should walk to the area designated "ribosome" and stop just outside

Figure 7.4 DNA-mRNA sequence matches.

DNA (coding)	CCG	ATG	AAT	GTC	GAG	CTA	TCC	TAC	GGC
DNA (noncoding)	GGC	TAC	TTA	CAG	CTC	GAT	AGG	ATG	CCG
mRNA	CCG	AUG	AAU	GUC	GAG	CUA	UCC	UAC	GGC

Table 7.1 Amino acid abbreviations[a]

Abbreviation	Amino acid
Ala	Alanine
Arg	Arginine
Asn	Asparagine
Asp	Aspartic acid
Cys	Cysteine
Gln	Glutamine
Glu	Glutamic acid
Gly	Glycine
His	Histidine
Ile	Isoleucine
Leu	Leucine
Lys	Lysine
Met	Methionine
Phe	Phenylalanine
Pro	Proline
Ser	Serine
Thr	Threonine
Trp	Tryptophan
Tyr	Tyrosine
Val	Valine

[a]These standard three-letter abbreviations are used frequently in scientific literature to denote amino acids.

it. In the eukaryotic cell, the student with the very first (the 5′) mRNA card should shake hands with the ribosome recognizer to signify that the ribosome has found the 5′ end of the message. In the *E. coli* model, the ribosome shakes hands with the RRE.

13. Review the following rules.
    - tRNA cytosine always pairs with mRNA guanine.
    - tRNA uracil always pairs with mRNA adenine.
    - tRNA adenine always pairs with mRNA uracil.
    - tRNA guanine always pairs with mRNA cytosine.

14. The linked tRNA and amino acid students should walk together to the ribosome area. Starting with the AUG codon, have the correct tRNA pair with its codon. When the second tRNA finds its place on the mRNA strand, the first and second amino

acids link hands. Then the first tRNA detaches from its amino acid and returns to the cytoplasm. The amino acids remain at the ribosome. Continue down the mRNA molecule, using the tRNAs to link each new amino acid to the growing chain. As each new amino acid is added, the previous tRNA leaves. Translation stops when the stop codon is reached. At that point, the amino acid chain is released, and the mRNA and the new amino acid chain leave the ribosome.

Amino acids linked in this way are called **peptides.** Long chains of linked amino acids are often called **polypeptides.** Translation is complete when a sequence of mRNA information translates into a polypeptide. A protein can be one polypeptide or an association of more than one polypeptide (chapter 3).

The students' mRNA-tRNA pairings should vertically match those shown in Fig. 7.5.

15. Use the *Student Question* section as a basis for class discussion, or assign the questions as homework.

Follow-up (Optional)

- This activity is a good lead-in to a discussion of regulation of transcription (see chapter 3 for information). If you plan to talk about transcriptional regulation, be sure to include the promoter in your activity.

- For advanced classes, it is fun to examine all of the steps in the pathway leading from gene to protein and to imagine ways of regulating protein synthesis at every step. It is very likely that natural examples of every imagined regulatory mechanism actually exist. If you discuss gene regulation here, be sure to bring up the idea of antisense RNA (see the advanced activity *Antisense Regulation*).

- If you and your class like this sort of thing, think about making a videotape of the activity and showing it to a lower-grade-level class to explain transcription and translation.

Figure 7.5 mRNA, tRNA, and amino acid matches. The two codons with no tRNA are unpaired because they come before the start codon and after the stop codon. If these codons were between the start and stop translation signals, they would be paired with GGC/Pro and CCG/Gly, respectively.

mRNA	CCG	AUG	AAU	GUC	GAG	CUA	UCC	UAG	GGC
tRNA	none	UAC	UUA	CAG	CUC	GAU	AGG	AUC	none
Amino acids		Met	Asn	Val	Glu	Leu	Ser	Stop	

DNA	C	ATG	TCC	CCG	GAG	AAT	GTC	GAG	CTA	TCC	GGC	TAG
	NC	TAC	AGG	GGC	CTC	TTA	CAG	CTC	GAT	AGG	CCG	ATC
mRNA		AUG	UCC	CCG	GAG	AAU	GUC	GAG	CUA	UCC	GGC	UAG
tRNA		UAC	AGG	GGC	CUC	UUA	CAG	CUC	GAU	AGG	CCG	none
Amino acid		Met start	Ser	Pro	Glu	Asn	Val	Glu	Leu	Ser	Gly	none stop

Figure 7.6 Diagram answer to question 7. The noncoding DNA and tRNA sequences are identical except that tRNA contains uracil instead of thymine. For amino acid abbreviations, see Table 7.1. C, coding strand; NC, noncoding strand.

Answers to Student Questions

1. The sequences are complementary.

2. The sequences are the same except that mRNA uses uracil instead of thymine.

3. The noncoding strand is complementary to the mRNA.

4. Certainly the noncoding strand cannot be used to make the same protein as the one encoded by the coding strand; the DNA sequence is different. It is unlikely that the noncoding strand of a gene would contain a promoter and an ATG in proper alignment for transcription and translation. In addition, chances are that even if you could transcribe and then initiate translation, the resulting peptide would be short because of random stop sequences.

(Depending on the sophistication of your students, you may want to discuss the following aspect of transcription with them. The RNA polymerase cannot transcribe the wrong strand after recognizing the promoter. RNA polymerase, like DNA polymerase, synthesizes nucleic acid in the 5′-to-3′ direction. To do so, it must read the template in the 3′-to-5′ direction. Only one strand of the DNA is in the proper orientation to the promoter for use; the other strand essentially points backward.)

5. An extra base inserted near the beginning of a gene would throw off the entire coding sequence. Completely different amino acids would be coded for, and you would probably encounter a stop codon long before the correct end of the protein. Examples will vary.

6. The obvious answer is that translation would be affected because the deficient amino acid would not be available to bind to its tRNA and could not be incorporated into the peptide chain. A less obvious answer is that ultimately, transcription would also be affected because RNA polymerase could no longer be synthesized properly on account of the missing amino acid.

7. See Fig. 7.6 and 7.7. C indicates the coding strand of DNA, represented by sequence cards with outlined letters; NC indicates the noncoding strand, represented by filled-letter cards.

Figure 7.7 Answer to question 8. For amino acid abbreviations, see Table 7.1. C, coding strand; NC, noncoding strand.

DNA	C	ATG	CTA	GTC	CCG	GGC	AAT	TCC	GAG	CCG	GTC	TAG
DNA	NC	TAC	GAT	CAG	GGC	CCG	TTA	AGG	CTC	GGC	CAG	ATC
mRNA		AUG	CUA	GUC	CCG	GGC	AAU	UCC	GAG	CCG	GUC	UAG
tRNA		UAC	GAU	CAG	GGC	CCG	UUA	AGG	CUC	GGC	CAG	none
Amino acid		Met	Leu	Val	Pro	Gly	Asn	Ser	Glu	Pro	Val	stop

Gene Regulation

All cells regulate the expression of their genes. Essentially every bacterial gene that has been studied is regulated at some step during its expression. A stunning variety of regulation mechanisms is used, including regulation of transcription initiation (as described for the *lac* and *trp* operons in *Regulating gene expression* in chapter 3), regulation of the termination of transcription, degradation of mRNA, control of translation initiation (through antisense oligonucleotides or proteins), and alteration of promoter recognition. It may be safe to assume that any gene regulation mechanism we can imagine is in use somewhere in the natural world.

Specific examples of various gene regulation mechanisms include the following.

Activation and repression of transcription

Regulation of the frequency of transcription is the most important level of gene control exercised in bacterial systems. The most common form of transcriptional control is carried out by specific regulatory proteins that bind within or near the promoter sequence and either prevent or enhance transcription. Two examples of ways that repressors of transcription are used to respond to environmental conditions are found in the *lac* and *trp* operons, as described in chapter 3. Transcriptional activators generally bind near the promoter and interact directly with RNA polymerase to increase the frequency of transcription. This type of interaction appears to be of particular importance in eukaryotes.

Alteration of promoter recognition

Transcriptional regulation by alteration of promoter recognition occurs when a drastic change in gene expression is needed: a new set of genes must be transcribed and/or a currently transcribed set must be turned off. For example, many viruses encode proteins that bind to the host RNA polymerase. The modified RNA polymerase can subsequently recognize only the virus's promoters, which have base sequences different from those of the normal host promoters. In this way, the virus switches transcription from the host genes to its genes. Another example of this type of regulation is found in the bacterium *Bacillus subtilis*. This organism forms durable, dormant spores in response to adverse environmental conditions. Spore formation requires the expression of a number of genes that are not active during the normal life cycle of the bacterium. In addition, normal gene activity all but ceases during the spore stage. To achieve this gross change in gene expression, *B. subtilis* synthesizes a special protein that binds to its RNA polymerase and causes it to recognize only the special promoters controlling sporulation genes.

Repression of translation

Regulation of gene expression can be exerted through control of the rate of translation of an mRNA. A good example of translational repression can be found in the synthesis of the ribosomal proteins of *E. coli*. The ribosomes of *E. coli* are made up of large rRNA molecules and several proteins that bind to specific regions of the RNA. The genes for the ribosomal proteins are lined up in operons. A single mRNA is transcribed from each operon and translated into several proteins. As the proteins are translated, they find free rRNAs in the cytoplasm and bind to their recognition sites. When all available rRNA is complexed with protein, one of the proteins from that operon binds to the translation initiation region of the operon mRNA. The binding of that protein to the mRNA prevents further translation. For each one of the ribosomal protein operons, one of the encoded proteins acts as a repressor of translation. Through this mechanism, a balance between the amount of available rRNA and ribosomal proteins is achieved.

Regulation by antisense RNA

Antisense gene regulation was discovered in the 1980s. Several natural examples of this form of regulation have now been described. In antisense gene regulation, a short RNA exactly complementary to the 5' end of an mRNA is synthesized. This antisense RNA hybridizes to the translation initiation region of the mRNA and prevents translation. Two examples of antisense regulation in nature are found in the synthesis of two outer membrane proteins in *E. coli* and in the synthesis of the transposase protein of transposon *10*.

Antisense gene regulation has engendered a great deal of excitement because of the potential for artificial gene control. Several biotechnology companies are focusing on antisense regulation as a means of controlling gene expression in crop plants, controlling viral disease, and treating cancer and other human diseases. This technology is potentially so important that it receives separate treatment in the following lesson plan.

After your class has completed *From Genes to Proteins* and understands how the genetic information in DNA is manifested in the structure of proteins, introduce basic concepts of gene regulation. Some of these gene regulation concepts can be illustrated through role playing in an extension of *From Genes to Proteins* (as described below) or simply by classroom discussion.

Ideas for Gene Regulation Activities

- Assign roles and act out the *lac* repressor preventing transcription of the *lac* genes when lactose is not present and allowing transcription when lactose is present (refer to chapter 3). This process could be illustrated simply with the DNA codons, promoter, RNA polymerase, a repressor protein, and lactose; there is no need to actually make the RNA and go through translation.

- Assign roles and act out the *trp* repressor controlling expression of the tryptophan synthesis genes (refer to chapter 3).

- Have students brainstorm about ways in which a cell might regulate the synthesis of a particular protein. You could use their ideas as a jumping-off point for introducing some of the gene regulation mechanisms mentioned previously.

- Have students act out translational repression. For this, you could start with the mRNA, the ribosome area (with a "recognizer" in it), and someone assigned to be the translational repressor protein.

- Illustrate the regulatory power of changing the promoter recognition by representing a series of genes as lines on the blackboard. Indicate the promoters at the left-hand ends of the lines with P_N for normal promoters and P_S for sporulation promoters. Show the students that when RNA polymerase is modified to recognize only P_S, all the P_N genes are shut off.

Antisense Regulation

Antisense oligonucleotides are segments of DNA or RNA that are exactly complementary in base sequence and antiparallel in orientation to a gene of interest. In the natural examples of antisense gene regulation, an antisense RNA hybridizes to the 5′ end of an mRNA molecule, preventing the ribosome from recognizing the translation initiation site and thus blocking translation of the mRNA. Antisense gene regulation has caused great excitement among biotechnologists because of its potential for allowing artificial control of gene expression. Theoretically, the expression of any gene could be suppressed through introduction of a sufficient amount of an appropriate antisense oligonucleotide, allowing the prevention or modulation of any process driven by the expression of specific genes.

The first application of antisense technology we will encounter in daily life will be the Flavr Savr tomato. We are all familiar with tasteless grocery store tomatoes. These tomatoes lack flavor because they are picked when they are hard and green. In that state, they are mechanically packaged and shipped throughout the country. They are then force "ripened" by exposure to ethylene oxide gas, which turns them red but does not cause the development of ripe-tomato flavor.

Commercial tomatoes are not allowed to ripen before harvest because as tomatoes ripen, they soften. Soft ripe tomatoes do not have the durability and shelf life of hard green tomatoes and are therefore deemed impractical by large commercial growers.

If the softening process could be separated from the development of ripe flavor, tomatoes could be allowed to ripen on the vine. One of the enzymes responsible for tomato softening is polygalacturonase. This enzyme breaks down the structural polysaccharides of the tomato. Scientists at Calgene, Inc., designed an antisense RNA specific to this gene. They then constructed a "gene" for this anti-RNA: it had a promoter, a coding region for the anti-RNA, and other normal plant genetic traffic signals. They introduced their artificial gene into tomato plants, and the anti-

sense regulation worked. The genetically engineered plants produce tomatoes that can ripen on the vine without becoming soft.

Scientists at the University of Massachusetts have reported success in using antisense oligonucleotides to stop the thickening of rat artery walls after balloon angioplasty surgery. This surgery is used to open fat-clogged arteries in humans. In about one-third of cases, the technique backfires, triggering cell growth in the artery walls and reclosing of the blood vessels. These scientists performed angioplasty on rats. Afterward, they smeared the artery walls with a gel containing antisense molecules directed at a gene that promotes cell growth. Rats that received this treatment showed no expression of that gene at the treatment site, and their artery walls did not thicken. Untreated artery walls expressed the gene and, as a result, thickened after the surgery.

Potential applications of antisense technology in health care hold great promise for improving treatments for many of our most dreaded illnesses. Cancer and viral infections are among those that result from inappropriate expression of the body's own genes or from expression of foreign genes inside our cells. Current chemotherapy for cancer and viral infections, such as AIDS, interferes with the enzymes necessary for cell growth and affects both diseased and healthy cells. Patients undergoing these treatments frequently become seriously ill from side effects. In contrast, treating these diseases through specific prevention or reduction of expression of disease-causing genes represents an ideal therapy.

To develop an antisense drug or gene regulator, precise understanding of the genetic system to be regulated is essential. In the case of the Flavr Savr tomato, an understanding of the reasons for tomato softening was required. For therapeutics against viral diseases, the function of viral genes in infection must be known. To treat a specific cancer, the genetics of that particular cancer must be worked out. The bulk of molecular biology research over the last decade or so has been directed at answering these genetic and mechanistic questions. Antisense technology now offers the prospect of a direct payoff in a variety of applications.

There are technical difficulties to be overcome before oligonucleotides can be widely used as drugs. Cells contain enzymes (nucleases) that normally degrade free DNA and RNA, so biotechnology companies have been experimenting with modified oligonucleotides that will evade these nucleases but still hybridize with their targets. Several different changes in the phosphodiester backbone are being explored, including replacing it with a peptide backbone like those found in proteins.

A final obstacle to be overcome in antisense drug development is technological: how to manufacture and purify oligonucleotides in large enough quantities and at low enough cost to make therapeutic use feasible. Until now, oligonucleotides were needed only in minute quantities for research purposes, and the available synthesis and purification technologies are suitable only to such small-scale needs. Advances in large-scale oligonucleotide synthesis and purification must occur before these drugs can be widely available.

A variety of therapeutic oligonucleotides are in the clinical trial stage of development. In 1992, an antisense therapeutic against human papillomavirus (the causative agent of genital warts) went into clinical trial. A Houston researcher received permission to give 14 lung cancer patients a gene for an antisense molecule directed at an oncogene (see the reading *The Molecular Genetics of Cancer*). Other scientists have used an antisense gene against human immunodeficiency virus genes to make cultured human cells resistant to the AIDS virus.

Answers to Student Questions

There are no correct answers to these questions. The questions are designed to get students to think about the way antisense regulation works, the enormous potential of the method, its inherent limitations, and the kinds of things scientists would have to consider in developing a particular antisense gene regulation strategy.

1. Here are some examples of things the Calgene scientists would have had to know and to test. The scientists had to know which tomato genes were involved in softening and that knocking out one of these genes would not cause the tomatoes to develop an undesirable flavor. They had to know the sequences of the mRNAs of their potential target genes. They had to make antisense genes, put

them into test plants, and see whether the production of the softening enzyme was reduced. They had to determine whether reducing the production of the softening enzyme actually helped the tomatoes stay firm as they ripened. They also had to make sure that the tomatoes produced by their test plants tasted good.

2. In order to begin designing an antisense drug, it would be critical to have already identified the viral gene whose expression is essential for infection and to know the sequence of that gene. To identify such a gene would require some work. The virus would have to be isolated and analyzed genetically to identify its genes. Studies of virus mutants would show which genes were essential to its life cycle. The nucleotide sequences of these genes would have to be determined. Once this information was available, you could begin to design an antisense molecule. (The idea is to get students to think about how much you must know about a system in order to apply antisense technology, or any technology, to it.)

3. Antisense technology reduces gene expression. It would probably not be effective against diseases that are caused by the lack of a gene product (or the lack of a correct gene product) rather than by inappropriate expression of a gene. An example of such a disease is cystic fibrosis, which is caused by two defective copies of a gene for a cell membrane protein. Turning off expression of these defective genes would leave the patient with no membrane protein at all, which would be worse than a defective protein. Genetic diseases caused by dominant mutations (such as Huntington's disease) may be amenable to antisense drugs. In these diseases, the patient usually has one good copy of the gene and one mutant copy that dominates and causes the illness. If an antisense molecule that blocked expression of only the mutant form of the gene could be designed, such a molecule could be an excellent therapy. You may want to discuss this possibility with your class. Ask students what information would have to be available to design such an antisense molecule (examples: the disease gene must be found; the disease gene and the normal gene must be sequenced, and the differences must be understood; it must be determined whether an antisense molecule is effective in blocking only the mutant form.)

Selected Reading

Ezzell, C. 1992. Blood-vessel growth genes stop making sense. *Science News* 14(10):151.

From Genes to Proteins

You Are Your Proteins

You have probably heard many times that "your DNA determines your characteristics." Did you ever wonder how DNA does that? DNA determines all your characteristics (and the characteristics of every plant, animal, fungus, bacterium, etc., in the world) by determining what **proteins** your cells will synthesize.

Why are proteins so important? Because of the many things they do. There are many types of proteins, and each type performs an important kind of job in your body. For example, *structural proteins* form the "bricks and mortar" of your tissues. Two of them, actin and myosin, enable your muscles to contract. Another structural protein, keratin, is the basic component of hair. *Carrier proteins* transport important nutrients, hormones, and other critical substances around your body. One of these proteins is hemoglobin, which carries oxygen through your blood to your tissues.

Another large class of proteins is the **enzymes.** Enzymes are the body's workhorses. They carry out chemical reactions in your body. Enzymes digest your food, synthesize fats so your body can store energy, and carry out the work of making new cells. They make molecules and perform activities necessary for life.

In fact, if you want to sum up the importance of proteins in your body, you could say: "*Nearly every biological molecule in my body either is a protein or is made by proteins.*" So by telling your cells what proteins to synthesize, DNA controls your characteristics.

What Proteins Are

What exactly is a protein? A protein is a biological molecule made of many small units linked together in a chain (rather like beads on a string). The units are **amino acids,** small molecules composed of carbon, hydrogen, oxygen, and nitrogen. Twenty different amino acids can be used in making your proteins.

Imagine sitting down with 20 containers, each filled with a different type of bead, to make a string of beads. You may use as many or as few of each kind of bead as you wish, you may string them in any order you wish, and you may use any number of beads you wish. This situation is a reasonable representation of the possibilities for protein synthesis within a cell. The beads represent amino acids, and any string of them represents a protein. You can get an idea of the infinite variety of proteins that could be made!

However, in the cell, one thing is different from our example. The cell is handed an instruction sheet that tells it which beads (amino acids) to string to make a particular chain. The instructions come from the cell's DNA.

How DNA Directs Protein Synthesis

DNA contains a *genetic code* for amino acids, with each amino acid represented by a sequence of three DNA bases. These triplets of bases are called **codons.** The order of the codons in a DNA sequence is reflected in the order of the amino acids assembled in a protein chain. The complete stretch of DNA needed to determine the amino acid sequence of a single protein is called a **gene.**

In your cells, the DNA is located within the chromosomes in the nucleus. DNA contains all the instructions for making every protein in your body. However, your cells make proteins at the ribosomes, located in the cytoplasm. An individual ribosome makes only one protein at a time. So when your cell needs to make a protein, a "working copy" of the instructions for that one protein is copied from the DNA and sent to the ribosome for use. This working copy is messenger RNA (mRNA).

After the base sequence of a gene (DNA) is copied into mRNA, the mRNA travels to the ribosome, where its code is translated into protein. The translation step is carried out by a second type of RNA called transfer RNA (tRNA). The tRNA matches the correct amino acids to the codons in the mRNA. The amino acids are linked together to make the new protein.

Wait a Minute—Isn't Protein a Kind of Food?

Most of us have heard that protein is part of a healthy diet. In fact, most people think of food when they hear the word "protein." So what's going on? How does protein in the diet fit into a discussion of genes?

The protein you eat is composed of individual protein molecules made by plant and animal cells. Animal muscle tissue (meat) is particularly rich in protein (re-member actin and myosin, mentioned above?). When you eat protein, regardless of its source, your digestive enzymes break the individual protein molecules down into amino acids (like taking the beads off the string). The individual amino acids are used by your cells to make your proteins—a form of biological recycling—or can be further broken down for energy.

Classroom Activity

Today, you and your classmates will act out the steps involved in translating the DNA code into proteins. You will see the roles played by mRNA and tRNA and may also be introduced to some of the "traffic signals" that direct their action.

Questions

1. DNA is double-stranded. One strand is the coding strand, and the other is the noncoding strand. The noncoding strand is used as the template to make the mRNA. What is the relationship between the base sequences of the coding and noncoding strands?

2. What is the relationship between the base sequence of the coding strand and the base sequence of mRNA?

3. What is the relationship between the base sequence of the noncoding strand and the base sequence of mRNA?

4. Would there be a problem if the RNA polymerase transcribed the wrong strand of DNA and the cell tried to make a protein?

5. A frameshift mutation is caused by the insertion or deletion of one or two DNA bases. What would be the effect on the amino acid sequence of a protein if one extra base were inserted into the gene near the beginning? Use an example to show what you mean.

6. Suppose an individual has a nutrient deficiency due to poor diet and is missing a particular amino acid. How would transcription and translation be affected?

(continued on next page)

7. Given below are some tRNA anticodon-amino acid relationships and a stretch of imaginary DNA. Fill in the empty boxes in the chart by writing the DNA sequence of the coding strand and the correct mRNA codons, tRNA anticodons, and amino acids. Use the following tRNA-amino acid relationships:

GGC	UUA	CAG	CUC	GAU	AGG	CCG
Pro	Asn	Val	Glu	Leu	Ser	Gly

DNA											
Coding											
Noncoding	TAC	AGG	GGC	CTC	TTA	CAG	CTC	GAT	AGG	CCG	ATC
mRNA											
tRNA											
Amino acid	start										stop

What are the similarities between the noncoding DNA sequence and the tRNA sequence?

8. A new and exciting branch of biotechnology is called protein engineering. To engineer proteins, molecular biologists work backward through the protein synthesis process. They first determine the exact sequence of the polypeptide they want and then create a DNA sequence to produce it. Use the rules of transcription and translation to "engineer" the peptide sequence below. Fill in the rows for tRNA anticodons, mRNA codons, and the two DNA strands. Use the tRNA-amino acid relationships given in question 7.

DNA											
Coding											
Noncoding											
mRNA											
tRNA											
Amino acid	Met start	Leu	Val	Pro	Gly	Asn	Ser	Glu	Glu	Pro	Val

Classroom Activities

Does Antisense Make Sense?

Imagine what you might be able to do if you could prevent or decrease the expression of any gene. Prevent cancer from developing? Prevent viral diseases? Control certain genetic diseases?

Biotechnologists are pondering these and other questions largely because of a new technology for gene regulation: antisense. What is antisense technology, and how does it work? Antisense technology, like the rest of biotechnology, is based on a natural phenomenon. In antisense gene regulation, a cell synthesizes a very short piece of RNA that is exactly complementary to the ribosome recognition region (usually the 5' or "front" end) of the messenger RNA (mRNA) of the gene to be regulated. Because of its complementary base sequence, the antisense RNA can base pair with the mRNA. When it does this, the 5' end of the mRNA becomes double stranded. The ribosome can no longer recognize and translate it. By preventing translation of mRNA, antisense RNA can decrease gene expression (see figure).

How antisense regulation works. The gene to be regulated is transcribed normally, producing an mRNA molecule. The antisense oligonucleotide is exactly complementary in sequence to the area of the message where the ribosome normally binds. When the antisense molecule base pairs to the mRNA, the ribosome cannot bind to its recognition site, and translation does not occur. (The bacterial gene sequence shown is taken from the *Escherichia coli lacZ* gene.) RRS, ribosome recognition site; ic, initiation codon; aRNA, antisense RNA.

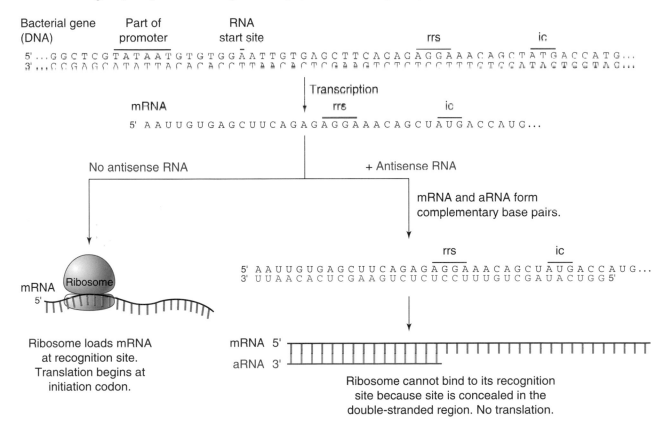

Antisense technology copies this natural approach except that the antisense molecule can be RNA, DNA, or even a chemically modified version of either one, just as long as it will base pair with the target molecule. A general term used to refer to these short molecules is "antisense oligonucleotides" (the prefix "oligo" means "several"; "nucleotide" can refer to the building blocks of DNA or RNA). It is theoretically possible to reduce the expression of any gene by introducing an appropriate antisense oligonucleotide into the cell.

Regulation by antisense is very precise because of the specificity of base pairing. For any oligonucleotide, the chance that its complementary sequence will occur randomly is 1 in 4^n, where n is the number of bases in the oligonucleotide. So the chance of a 15-nucleotide antisense molecule accidentally pairing with an unintended target is 1 in 4^{15}, which is less than 1 in 1 billion (multiply it out for yourself). For longer antisense molecules, the odds are even more remote.

How is it possible to introduce antisense oligonucleotides into a cell or an organism? There are basically two choices: give the molecule as if it were a drug, or genetically engineer a gene encoding the antisense molecule into the organism. To genetically engineer an antisense molecule, a biotechnologist must synthesize a piece of DNA that encodes the proper sequence, then attach appropriate genetic traffic signals such as a promoter and a terminator (to tell RNA polymerase to start and stop making RNA), and then put the new gene into the organism. Once the artificial gene is introduced, the new host cells will synthesize the antisense RNA.

Is antisense technology being used for any practical applications? Yes. One genetically engineered food that will soon be available is the Flavr Savr tomato. The Flavr Savr is an "antisense tomato." It will be tastier than regular grocery store tomatoes because scientists have designed it so that large-scale growers can let the Flavr Savr ripen on the vine before harvesting it. Normal grocery store tomatoes are picked when they are green and hard. They are then shipped all over the country and turned red by exposure to ethylene gas. The tomatoes never develop ripe tomato flavor. The reason growers do not let the tomatoes ripen before picking them is that as they ripen, tomatoes become too soft for mechanical handling and a long shelf life.

Scientists at the biotechnology company Calgene, Inc., reasoned that if a tomato could ripen without softening, growers could let them ripen and develop flavor before picking them. They developed an antisense molecule for a gene encoding a softening enzyme. An artificial gene for the antisense RNA was constructed and introduced into tomato plants, where it prevents softening during ripening. Calgene claims the result is a much tastier grocery store tomato. You will soon be able to test this claim for yourself.

Questions

1. What kinds of things did the Calgene scientists have to know in order to make their antisense tomato? What kinds of things did they probably test during the development of the Flavr Savr?

2. Imagine that you want to control a viral disease with an antisense drug. What kinds of things would you need to know to begin to design your antisense molecule?

3. Can you think of a disease or a type of disease that would probably not be treatable by antisense drugs?

Classroom Activities

Sizes of the *Escherichia coli* and Human Genomes

About This Activity

This activity uses models to show the relative sizes of an *Escherichia coli* cell, the *E. coli* chromosome, a typical plasmid, and a gene. The models demonstrate graphically how much longer the *E. coli* chromosome is than the cell. Calculations included in the student questions use the analogies of letters in a book and miles of railroad track to suggest the size of the human genome. This lesson pairs nicely with the next activity (chapter 9) *Extraction of Bacterial DNA*. It can be performed as a student activity or as a teacher demonstration.

Class periods required: 1/2-1

Introduction

DNA is stored in cells in the form of chromosomes and plasmids. The amount of DNA required to store the information necessary for making even a simple organism such as a bacterial cell is very large. One of the wonders of biology is that cells are able to store and access the great lengths of DNA needed to encode their hereditary information.

The bacterium *E. coli* is estimated to have about 2,000 or more genes in its genome. The average bacterial gene is considered to be 1,200 base pairs long; allowing for noncoding regions, the DNA of *E. coli* is estimated to be about 4 million base pairs long. A typical plasmid might be 3,000 base pairs long (some, in particular those coding for multiple drug resistances, are quite a bit larger) and encodes just a few genes.

What are the physical sizes of these DNA molecules? The *E. coli* chromosome consists of one large circular DNA molecule that if stretched out would be approximately 10^{-3} m (1 mm) long. By comparison, the *E. coli* cell is only 1×10^{-6} to 2×10^{-6} m long. The *E. coli* DNA molecule is thus 1,000 times longer than the cell! Even a lowly plasmid of 3,000 base pairs would be 10^{-6} m long, or approximately the length of the cell, if it were linear. Yet the chromosome of *E.*

coli constitutes only 2 to 3% of the cell's weight and occupies only 10% of its volume.

DNA occupies such a small fraction of the cell's volume (considering the enormous length of the DNA molecule) because it is an extremely slender molecule. It is capable of a high degree of folding and coiling, an essential feature for packing it into the cell. Although the degree of folding required to fit the DNA of *E. coli* into the bacterium is impressive, the folding necessary for packaging DNA into a human cell is even more remarkable.

The human cell is approximately 2×10^{-5} m in diameter. The human genome is estimated to consist of about 3 billion base pairs. If the DNA of a single human cell were stretched out, it would be about 2 m long, or 100,000 times longer than the cell! Yet all this DNA not only fits into the cell but is also accessible to the cell's enzymes for information transfer and replication.

This activity uses models that are 10,000 times life size. It is instructive to note that ×10,000 magnification is analogous to inflating an ant to the size of a tractor-trailer rig. (It is fun to ask students to guess how big a ×10,000 ant would be.) A convenient model for a ×10,000 *E. coli* cell is a 2-cm (20-mm) gelatin capsule (these can be purchased at pharmacies or health food stores). The capsule is a good model for reasons besides its length: it is rod shaped, like *E. coli*, and it has an outer wall that is analogous to the outer membrane of the bacterium. If you do this activity in conjunction with DNA extraction, you can use the capsule to illustrate that the cell membrane must be dissolved to release the cell contents, including the DNA.

The DNA models consist of appropriate lengths of thread, yarn, or string. Thread is most accurate because it is the most slender, but yarn or string is easier to see if you are doing a demonstration. The ×10,000 *E. coli* chromosome is represented by a 10-m length of thread. Use the thinnest thread you can find, and remind students that it represents a double helix. The size of a ×10,000 plasmid is represented by 10 mm of thread, yarn, or string. A single ×10,000 gene is only 4 mm long: the length of a piece of fuzz if you use yarn.

Objectives

At the end of this activity, students should be able to

1. Describe the relative sizes of the *E. coli* cell, chromosome, genes, and plasmids;
2. Explain that strands of DNA are extremely thin and must be very tightly coiled, since an immense length of DNA fits into a relatively tiny cell volume;
3. List some differences between the *E. coli* and human genomes;
4. (Optional) Use scientific notation to make relevant calculations about the dimensions of DNA.

Materials

- Felt-tip pen
- Scissors
- Glue
- Meter stick (for measuring thread)

For each laboratory team of four students, you need to gather a letter-size envelope and the following materials.

- A 2-cm gelatin capsule (no. 0; purchase at pharmacy or health food store)
- 10 m of white thread; 10 m of colored thread
- Two 20-mm-long strands of white and colored thread, yarn, or twine
- A 4-mm-long piece of colored thread, yarn, or twine
- Two 3-by-5 index cards with yarn glued to each one as described below

Preparation

1. Place one gelatin capsule into each envelope.

2. Use a meter stick and scissors to cut the following for each student team:
   - One 10-m-long segment of thread
   - One 4-mm-long segment of colored thread, yarn, or twine
   - One 10-mm-long segment of colored thread, yarn, or twine

3. Glue the 4-mm piece of colored thread, yarn, or twine to a 3-by-5 card, and label it "Average length of a single bacterial gene."

4. Glue the 10-mm segment to the second 3-by-5 card, and label it "Length of typical bacterial plasmid." If you use thread or string, glue it into a circle to simulate the circular plasmid. You may make a "double-stranded" thread model of a plasmid if you wish.

5. Put a 10-m segment of thread and one each of the two different cards into each envelope with the gelatin capsules.

Tips

1. If time allows, have your students prepare the "gene" and "plasmid" cards and measure out the 10-m lengths of thread. This exercise allows them to practice metric measurement.

2. Student questions involving scientific notation should be assigned according to the mathematical abilities of your students. The questions can be answered by the class, thus serving as a review of scientific notation.

Procedure

1. Review or present any background material.

2. For a class activity, split students into groups of four, and have them follow the procedure on the activity sheet. (In step 7, the best way to fit the thread into the capsule is to fold the lengths of thread in half several times until they make a "wad." Insert one end of the wad into the longer section of the capsule, and push the remainder in with a twisting motion.)

3. Students answer questions in class or at home.

Alternative Procedure

This activity works well as a demonstration coupled with the next activity, *Extraction of Bacterial DNA*. A good time to do it is during the 65°C incubation in that activity. It will not hurt the DNA preparation if you incubate the cells for more than 15 min.

Answers to Student Questions

1. Each thread could represent one strand of the DNA double helix.

2. If each gene is 4 mm long, there would be 250 genes in a meter, or 2,500 in 10 m. Some *E. coli* DNA is noncoding; the bacterium is estimated to have about 2,000 genes.

3. Real *E. coli* DNA occupies only 10% of the cell volume. The thread occupies a greater portion of the capsule volume; therefore, the thread is too thick for an accurate representation of DNA.

4. 10 m for the *E. coli* genome
10,000 m for the human genome
10,000 m ÷ (1,609 m/mi) = 6.3 mi of thread to represent the human genome

5. $(3.0 \times 10^6) \times (3.4 \times 10^{-10}$ m$) = 10.2 \times 10^{-4}$ m $= 1.02 \times 10^{-3}$ m

6. $(1.2 \times 10^3) \times (3.4 \times 10^{-10}$ m$) = 4.08 \times 10^{-7}$ m

7. $(3.0 \times 10^3) \times (3.4 \times 10^{-10}$ m$) = 10.2 \times 10^{-7}$ m $= 1.02 \times 10^{-6}$ m

8. $(1.02 \times 10^{-3}$ m$) \times (1.0 \times 10^4) = 1.02 \times 10^1$ m $= 10.2$ m

9. $(4.08 \times 10^{-7}$ m$) \times (1.0 \times 10^4) = 4.08 \times 10^{-3}$ m $= 4$ mm

10. $(1.02 \times 10^{-6}$ m$) \times (1.0 \times 10^4) = 1.02 \times 10^{-2}$ m $= 10.2$ mm

11. $(3 \times 10^9$ ties$) \times (2$ ft/tie$) = 6 \times 10^9$ ft
6×10^9 ft $÷ (5,280$ ft/mi$) = 1,136,364$ mi
$1,136,364$ mi $\times (1.61$ km/mi$) = 1,829,546$ km

12. $1,136,364$ mi $÷ (24,000$ mi per circumference$) = 47.3$ trips around the Earth

13. Will vary depending on the textbook. A book with 100 characters per line (a high estimate), 50 lines per page, and 600 pages will have 3×10^6 characters. It would take 1,000 of these books to contain enough characters (3×10^9) to represent the human genome.

Note: Discrepancies occasionally appear in discussions of the size of the human genome versus the amount of DNA in the human cell. These discrepancies arise because human cells contain two copies of each chromosome. An estimate of the amount of DNA in one copy of each chromosome is 3×10^9 base pairs; if you sequence all of these bases, you have sequenced the human genome. Each cell contains a second copy of the genome on the homologous chromosomes. Therefore, the DNA content of a human cell is $2 \times 3 \times 10^9$ base pairs, and the length of that DNA would be $2 \times 3 \times 10^9 \times 3.34 \times 10^{-10}$ m $= 20.04 \times 10^{-1}$ m, or 2 m.

Sizes of the *Escherichia coli* and Human Genomes

Classroom Activities

Introduction

DNA is stored in cells in the form of chromosomes and plasmids. The amount of DNA required to store the information necessary for making even a simple organism such as a bacterial cell is very large. One of the wonders of biology is that cells are able to store and access the great lengths of DNA needed to encode their hereditary information.

The bacterium *Escherichia coli* is estimated to have about 2,000 genes in its genome. The average bacterial gene is considered to contain 1,200 base pairs; allowing for noncoding regions, the DNA of *E. coli* is estimated to be about 4 million base pairs long. A typical plasmid is about 3,000 base pairs long and encodes just a few genes.

What are the physical sizes of these DNA molecules? The *E. coli* chromosome consists of one large circular DNA molecule that if stretched out would be approximately 10^{-3} m (1 mm) long. By comparison, the *E. coli* cell is only 1×10^{-6} to 2×10^{-6} m long. The *E. coli* DNA molecule is thus 1,000 times longer than the cell! Even a lowly plasmid of 3,000 base pairs would be 10^{-6} m long, or approximately the length of the cell, if it were linear. Yet the chromosome of *E. coli* constitutes only 2 to 3% of the cell's weight and occupies only 10% of its volume.

DNA occupies such a small fraction of the cell's volume (considering the enormous length of the DNA molecule) because it is an extremely slender molecule. It is capable of a high degree of folding and coiling, an essential feature for packing it into the cell. Although the degree of folding required to fit the DNA of *E. coli* into the bacterium is impressive, the folding necessary for packaging DNA into a human cell is even more remarkable.

The human cell is approximately 2×10^{-5} m in diameter. The human genome is estimated to consist of about 3 billion base pairs. If the DNA of a single human cell were stretched out, it would be about 2 m

long, or 100,000 times longer than the cell! Yet all this DNA not only fits into the cell but is also accessible to the cell's enzymes for information transfer and replication.

This activity uses models that are 10,000 times life size to demonstrate the relationship between the sizes of an *E. coli* cell, its chromosome, its plasmid, and a single gene. The $\times 10,000$ *E. coli* is represented by a 2-cm gelatin capsule, the $\times 10,000$ *E. coli* chromosome is represented by a 10-m length of thread, the plasmid is represented by 10 mm of thread, and a gene is represented by 4 mm of thread.

Materials

Obtain from your teacher a letter-size envelope containing the following items.

- A 2-cm (20-mm) gelatin capsule
- 10 m of thread
- An index card with 4 mm of thread labeled "Average length of bacterial gene"
- An index card with 10 mm of thread labeled "Length of a typical bacterial plasmid"

At the direction of your teacher, form groups of four for the next steps.

1. Remove the gelatin capsule from the envelope. It represents a single *E. coli* bacterium that has been enlarged 10,000 times.

2. Remove the two index cards from the envelope. The lengths of thread or string on the cards represent the length (but *not* the diameter) of an *E. coli* gene and plasmid magnified 10,000 times.

3. Remove the thread from the envelope, and stretch it out. The thread represents the bacterial chromosome magnified 10,000 times.

4. Let two people make a circle with the thread. The *E. coli* chromosome is circular and is attached to the cell membrane.

5. A third person can now hold up the index card with the ×10,000 bacterial gene next to the DNA loop. The average bacterial gene contains about 1,200 base pairs. Remember that a gene is composed of all the segments of DNA that instruct the cell to make a single protein, whether those segments are continuous or not.

6. With the bacterial chromosome and bacterial gene models still in view, a fourth person can hold up the index card with the × 10,000 bacterial plasmid. Plasmids carry one or a few genes neces-

sary for their own replication and stability and often carry genes that give the bacterium important characteristics such as antibiotic resistance. You can see that the plasmid is tiny in comparison to the chromosome.

7. Now that you have compared the sizes of the chromosomes, plasmids, and genes, try to reconstruct the bacterium by inserting the "chromosome" into the capsule. It isn't easy! The real *E. coli* chromosome occupies about 10% of the cell volume.

Questions

1. How could two 10-m lengths of thread represent the *E. coli* chromosome more accurately?

2. How many bacterial genes would fit on your DNA circle (formed in step 4)?

3. Is the thread that you tried to stuff in the capsule too thick to represent the DNA's actual thickness? What is the reason for your answer? (Hint: What percentage of "bacterial cell" volume does the thread occupy in your model, and what is the actual volume that DNA occupies in *E. coli*?)

4. In this activity, how many meters of thread did it take to represent the *E. coli* genome? If the human genome is 1,000 times longer than the *E. coli* genome, how many meters would it take to represent the human genome? How many miles of thread would that be?

Mathematical Calculations

- The distance between DNA base pairs is 3.4×10^{-10} m.
- The *E. coli* chromosome contains about 3×10^6 base pairs.
- The average *E. coli* gene contains 1,200 base pairs.
- A typical plasmid contains about 3,000 base pairs.

Using the information given above, calculate the following.

5. How long (in meters) is the *E. coli* chromosome?

6. How long (in meters) is the average *E. coli* gene?

7. What is the circumference (in meters) of a typical *E. coli* plasmid?

8. If *E. coli* were magnified 10,000 times, how long (in meters) would its chromosome be?

9. If *E. coli* were magnified 10,000 times, how long would its average gene be?

10. If *E. coli* were magnified 10,000 times, how long would a typical plasmid be?

The human genome can be related to a length of railroad track. The railroad ties represent the base pairs, and the rails represent the sugar-phosphate backbone of the DNA molecule. The railroad ties are 2 ft apart.

11. The human genome contains 3×10^9 base pairs. How many miles of track will it take to represent the human genome?

12. The circumference of the Earth is 24,000 miles. How many times would the railroad track representing the human genome wrap around the Earth at the equator?

(continued on next page)

13. Another way to represent the size of the human genome is to relate the base pairs to characters on a page in a book. Calculate how many of these books it would take to represent the human genome in the following manner.
    - Choose a page in your text that is mostly print.
    - Count the number of characters on five randomly selected lines. Find the average number of characters per line (C). Record C.
    - Count the number of lines on the page (L). Record L. Calculate the average number of characters (N) on a page by multiplying C times L. Record N.
    - Calculate the number of characters in your text (T) by multiplying N by the number of pages in your text. Record T.
    - To determine how many books like your text it would take to represent the human genome if every character represented a base pair, divide the number of base pairs in the human genome (3×10^9) by the number of characters in your text (T). How many books would be required?

Extraction of Bacterial DNA

About This Activity

In this activity, students can actually see a mass of stringy DNA fibers precipitate from bacterial cells. The laboratory work is very easy to perform. You can grow your own bacteria and use materials from a drugstore, or you can order pregrown cells and solutions from scientific supply houses. Two alternative but very similar procedures are given. One is for use with pregrown cells (procedure 1). The other is for home-grown cells (procedure 2). An additional activity for extraction of DNA from yeast cells is also provided.

Class periods required: 1

Introduction

The preparation of DNA from any cell type involves the same general steps: (1) breaking open the cell (and nuclear membrane, if applicable), (2) removing proteins and other cell debris from the nucleic acid, and (3) doing a final purification. Each step can be accomplished in several different ways, and the method chosen generally depends on the purity needed in the final DNA sample and the relative convenience of the available options.

If a cell is enclosed by a membrane only (as *Escherichia coli* or a human cell is), the cell contents can be released by dissolving the membrane with detergent. Cell membranes are made of proteins and fats. Just as detergent dissolves fats in a frying pan, a little detergent dissolves cell membranes. (The process of breaking open a cell is called cell **lysis**.) As the cell membranes dissolve, the cell contents flow out, forming a soup of nucleic acid, dissolved membranes, cell proteins, and other cell contents that is referred to as a cell **lysate**. Additional treatment is required for cells with walls, such as plant cells and many bacterial cells. These treatments can include enzymatic digestion of the cell wall material or physical disruption by means such as blending or grinding.

Detergent treatment provides an additional benefit in DNA preparation: denaturation of proteins. When a protein is denatured, the amino acid chain unfolds, and its three-dimensional structure is altered or lost. Denaturation blocks enzyme activity, which is important because all cells contain DNA-digesting enzymes called DNases (deoxyribonucleases). If the DNases in a cell are not denatured after cell lysis, they will digest the cellular DNA into small pieces. Heat also denatures most proteins.

After cell lysis, the next step in a DNA preparation usually involves removing proteins from the nucleic acid. Treatment with protein-digesting enzymes (proteinases) and/or extractions with the organic solvent phenol are two common methods of protein removal. Proteins are soluble in phenol, but DNA is not. Extracting an aqueous DNA-protein mixture (such as a cell lysate) with phenol separates the protein into the phenol and leaves the DNA in water. Following removal of the protein, DNA is usually subjected to additional purification. Final purification methods include precipitation, dialysis, and high-speed centrifugation. The level of purification required depends on what the DNA will be used for.

In the activity described here, no attempt is made to purify the DNA, since all that is required is to see it. Students will lyse *E. coli* with detergent and layer a small amount of alcohol on top of the cell lysate. Either ethanol or isopropanol (rubbing alcohol) can be used. DNA is insoluble in either alcohol and will form a white, weblike mass (precipitate) at the interface of the alcohol and water layers. By moving a glass rod up and down through the layers, students can collect the precipitated DNA on the rod. This DNA is very impure; the mass contains cellular proteins and other debris, but the stringy fibers are DNA. This easy procedure is fun, it lets students see DNA with their own eyes, and it shows the fibrous nature of DNA.

This inexpensive laboratory can reinforce discussions in the areas of enzymes, cell structure, lipid membranes, denaturation, solubility, detergent-lipid interaction, density, and the nature of the bacterial genome.

Objectives

At the end of this laboratory, students should

1. Understand how to extract a visible mass of DNA;

2. Be familiar with certain physical and chemical properties of DNA, such as solubility and high molecular mass.

Materials

Equipment
- Centigrade thermometers
- Hot plate and large pot

Supplies
- *E. coli* cells
- Eyedroppers
- Ethanol (95%) or rubbing alcohol (isopropanol)
- 15-ml culture tubes
- 50% solution of dish detergent (Palmolive works well) in water, or 10% sodium dodecyl sulfate (SDS), or shampoo
- glass stirring rods

If you grow your own cells (see *Preparation* below), you will also need

- Medium such as tryptic soy broth, Luria broth, or nutrient broth
- *E. coli* culture
- Inoculating loop(s)
- Large flask (if you grow one big culture)
- Incubator (optional)

Note: Medium recipes and procedures for growing *E. coli* are given in the appendices.

Resource Materials

- Edvotek kit 203, "Isolation of *E. coli* Chromosomal DNA"
- DNA Extraction Kit, Carolina Biological Supply Co. (catalog no. 17-1091)

Preparation

About the two procedures

The two commercial kits supply freeze-dried cells and a buffer to suspend them in along with other materials. The recommended procedure (which minimizes the amount of cells and detergent you use) is

given in the Student Activity pages as procedure 1. It calls for you to add 250 μl of detergent. Use graduated disposable droppers or large-volume micropipettes for this procedure.

You may grow your own cells if you like. The DNA extraction works fine with cells suspended in broth. The advantage of home-grown cells is that there will be plenty for every student to extract a sample of DNA.

If you grow your own, you must start a few days ahead. Streak out a few plates with *E. coli*. Have the students use sterile technique to inoculate 4 ml of tryptic soy or Luria broth in a 15-ml tube. Also have them set up 4 ml of uninoculated broth as a control (one tube of each per laboratory team). Allow the cells to grow until they are fairly dense (at least overnight). Proceed with the extraction.

Alternatively, inoculate one large culture yourself, let it grow up, and dispense 4 ml of cells to the students when you are ready.

Home-grown cells may be kept in the refrigerator for a day before use.

For the activity

1. Photocopy the *Student Activity* pages for your class if students do not have manuals.

2. If you are using dishwashing liquid or shampoo, make a 50% solution in water. (Approximately 6 ml per student team). The dishwashing liquid is diluted because otherwise it is too viscous to pour easily. The 10% SDS (another detergent) is ready to use as is.

3. When the students are ready to extract the DNA, they will need a 60 to 70°C hot-water bath. You can use a large kitchen pot, a thermometer, and a hot plate. The only real requirement is to get a volume of water large enough to prevent the temperature from dropping below 60°C during the denaturing process.

Tips

1. Demonstrate the proper procedure for layering alcohol and spooling the DNA while the students' samples incubate in the hot-water bath. Layering is easiest when you slant the tube and let the alcohol run in slowly.

2. The 15-min incubation is also a good time to remind students of the length of *E. coli*'s DNA (from

the activity *Sizes of the* Escherichia coli *and Human Genomes*) and to make sure they understand that what they will see is the DNA from millions of *E. coli* cells sticking together in a fibrous mass.

3. The precipitation works better if the alcohol is cold. Store it in a freezer if you have access to one. (The procedure works with room temperature alcohol, too.)

4. Question 2 provides a good lead-in to a discussion of how density affects the layering of liquids, if you wish to get into that. Fun demonstrations using different concentrations of sucrose (for example, 0, 20, and 50%) tinted with different food colorings can further illustrate the point.

Procedure

1. Review or present any necessary background material. This activity makes a nice follow-up to *Sizes of the* Escherichia coli *and Human Genomes*. Explain the basic steps in DNA extraction (lysing the cell, destroying DNases, precipitation).

2. Tell students which procedure they will use. If materials permit, let each student prepare his own DNA by following the instructions on the activity sheet. Demonstrate the layering-spooling technique.

3. Answer questions.

Answers to Student Questions

1. The double-stranded complementary structure makes it easy to replicate the molecule. It also makes it easy to synthesize an RNA copy of one strand by using the other strand as a template. If one strand is damaged, the sequence information is retained on the other strand. The undamaged strand can be used as a template to repair the damaged strand.

2. Alcohol is less dense than water, so it can float on top of the water. If the alcohol were denser than the broth or the cell lysate, it would sink to the bottom of the tube.

Additional Activity: Extraction of Yeast DNA

This procedure is described fully in an article by Larry Wegmann in *The Science Teacher,* December

1989. It uses household materials that students can bring from home. The procedure is simple and reliable and can be used with 9th-grade or younger students.

Materials

- Fleischmann's All Natural Yeast
- Athletic Shoe Cleaner-Deodorizer
- Adolph's 100% Natural Tenderizer, unseasoned
- Ethanol or isopropanol (rubbing alcohol)
- Glass rods
- 250-ml (or similar size) beakers
- Hot plate or other means of heating water

Procedure

1. Heat 100 ml of tap water in a beaker to 50 to 60°C.

2. Mix in one-half package of baker's yeast until it is thoroughly dissolved.

3. Add 5 ml of athletic shoe cleaner.

4. Maintain solution at 50 to 60°C for 5 min, stirring occasionally.

5. Dissolve 3 g of meat tenderizer in the solution. (The meat tenderizer contains papain, a proteinase.)

6. Maintain the temperature at 50 to 60°C for 20 min. At this point, the solutions can be stored in the refrigerator overnight.

7. Allow the solution to come to room temperature.

8. Tip the beaker at an angle, and slowly add 100 ml of ethanol or isopropanol so that two layers are formed.

9. Slowly insert the glass rod through the alcohol into the yeast lysate, and stir the layers gently just at the interface. Do not mix the layers. A precipitate of DNA will form at the interface.

10. After stirring for a few minutes, allow the layered solution to stand for several additional minutes.

The yeast DNA will not spool onto the glass rod. Presumably, nuclease activity in the preparation (either from the yeast cells themselves or from the meat tenderizer) cuts the DNA into pieces too small to spool.

However, the procedure reliably yields a mass of visible DNA.

Extraction of DNA from Animal Tissue

This procedure was shared with us by an experienced high school teacher. We have not tried it, but she says it works well.

Materials

- Frozen cat or dog testes
- Mortar and pestle
- 10% SDS or 50% detergent solution, as described previously
- Cold ethanol or isopropanol
- Glass stirring rod

Contact a local veterinarian, and have him or her freeze cat or dog testes from neutering operations. After you collect them from the vet, keep them in the freezer.

To extract DNA, remove one organ or a piece of one organ from the freezer, and grind it in the mortar and pestle. Add some detergent solution, stir, and transfer the entire contents of the mortar to a test tube. Incubate in a 65°C water bath for about 15 min. Layer cold alcohol, and spool as for the bacterial DNA.

Be sure to ask your students why there is so much DNA in testis tissue.

Extraction of Bacterial DNA

9

Introduction

In this activity, you will extract a visible mass of DNA from bacterial cells.

The preparation of DNA from any cell type involves the same general steps: (1) breaking open the cell (and nuclear membrane, if applicable), (2) removing proteins and other cell debris from the nucleic acid, and (3) doing a final purification. These steps can be accomplished in several different ways, and the method chosen generally depends on the purity needed in the final DNA sample and the relative convenience of the available options.

If a cell is enclosed by a membrane only, the cell contents can be released by dissolving the membrane with detergent. Cell membranes are made of proteins and fats. Just as detergent dissolves fats in a frying pan, a little detergent dissolves cell membranes. (The process of breaking open a cell is called cell **lysis.**) As the cell membranes dissolve, the cell contents flow out, forming a soup of nucleic acid, dissolved membranes, cell proteins, and other cell contents that is referred to as a cell **lysate.** Additional treatment is required for cells with walls, such as plant cells and many bacterial cells. These treatments can include enzymatic digestion of the cell wall material or physical disruption by means such as blending or grinding.

After cell lysis, the next step in a DNA preparation usually involves purification by removing proteins from the nucleic acid. Treatment with protein-digesting enzymes (proteinases) and/or extractions with the organic solvent phenol are two common methods of protein removal. Proteins dissolve in phenol, but DNA does not. Furthermore, phenol and water, like oil and water, do not mix but instead form separate layers. If you add phenol to an aqueous (water-based) DNA-protein mixture like a cell lysate and mix well, the protein dissolves in the phenol. After you stop mixing, the phenol separates from the aqueous portion, carrying the protein with it. The DNA remains in the aqueous layer. To remove the

protein, simply remove the phenol layer. Following removal of the protein, DNA is usually subjected to additional purification.

In this activity you will not attempt any DNA purification: your goal is simply to see the DNA. You will lyse *E. coli* with detergent and layer a small amount of alcohol on top of the cell lysate. Because DNA is insoluble in alcohol, it will form a white, weblike mass (precipitate) at the interface of the alcohol and water layers. By moving a glass rod up and down through the layers, you can collect the precipitated DNA. This DNA is very impure; the mass contains cellular proteins and other debris along with the stringy fibers of DNA.

Before you begin the DNA isolation, make sure you know whether to follow procedure 1 or procedure 2. They are essentially the same but differ in the volume of cells and the volumes and nature of the reagents you will use.

Procedure 1

1. Obtain from your teacher 2.5 ml of *Escherichia coli* cells in a salt solution. Add 250 µl (1/4 ml) of 10% sodium dodecyl sulfate (SDS), and mix well. SDS is a detergent and an ingredient of many detergents we buy at the store, such as Woolite.

2. Your teacher will provide a 60 to 70°C water bath. Place each tube into the water bath for 15 min. *Note:* Maintain the water bath temperature above 60°C but below 70°C. A temperature higher than 60°C is needed to destroy the enzymes that degrade DNA.

3. Cool the tube (on ice if you have it) until it reaches room temperature.

4. For the DNA to be visible, it must be taken out of solution, or precipitated. Watch your teacher demonstrate the following technique. Use a pipette to carefully layer 6 ml of 95% ethanol (or

isopropanol) on top of the suspension in each tube. The alcohol should float on top and not mix. (It *will* mix if you stir it or squirt it in too fast, so be careful.) Water-soluble DNA is insoluble in alcohol and precipitates when it comes in contact with it.

5. A weblike mass (precipitate) of DNA will float at the junction of the two layers (the interface). Push a rod through the alcohol into the soup and turn the rod. The rod carries a little alcohol into the soup and makes DNA come out of solution onto the rod. Keep moving the rod through the alcohol into the cell soup, and more DNA will appear. *Do not totally mix the two layers.*

Observe and draw the tube. Label the different substances in the tube. Answer the questions.

Procedure 2

1. Obtain from your teacher 4 ml of *E. coli* cells and 4 ml of medium in test tubes. Label the tubes. Shake your *E. coli* culture gently to resuspend the cells. Add to each labeled tube 3 ml of a 50% solution of dishwashing detergent in water. (Your teacher may substitute some other detergent.) Shake each tube to ensure complete mixing.

2. Your teacher will provide a 60 to 70°C water bath. Place each tube into the water bath for

15 min. *Note:* Maintain the water bath temperature above 60°C but below 70°C. A temperature higher than 60°C is needed to destroy enzymes that degrade DNA.

3. Cool the tubes to room temperature (on ice if you have it).

4. For the DNA to be visible, it must be taken out of solution, or precipitated. Watch your teacher demonstrate the following technique. Use a dropper to carefully layer 3 ml of 95% ethanol on top of the suspension in each tube. The alcohol should float on top and not mix with the cell lysate. (It *will* mix if you stir or squirt it in forcefully, so be careful.) Water-soluble DNA is insoluble in alcohol and precipitates when it comes in contact with it.

5. A weblike mass (precipitate) of DNA will float at the junction of the two layers (the interface). Push a rod through the alcohol into the soup and turn the rod. The rod carries a little alcohol into the soup and makes DNA come out of solution onto the rod. Keep moving the rod through the alcohol into the cell soup, and more DNA will appear. *Do not totally mix the two layers.*

Observe and draw both tubes. Indicate the substances in each tube. Answer the questions.

Questions

1. What information storage advantage(s) lies in DNA's double helix structure?

2. Why does the alcohol stay on top of the cell suspension and the broth in step 3?

Manipulation and Analysis of DNA

In the last 4 decades, our knowledge of DNA structure and function and of the biochemical processes cells use to modify the structure and carry out the functions has grown explosively. This extensive new knowledge has led to the ability to manipulate and analyze DNA outside the cellular environment. For example, we can cut DNA into specific pieces, separate and isolate those pieces, join pieces, copy DNA, and determine its base sequence. We are using our skills at manipulating genes to gather even more knowledge about how basic life processes are carried out and to make useful products for our daily lives.

The activities in this section deal with methods of manipulating and analyzing DNA: restriction digestion, ligation, gel electrophoresis, hybridization analysis, DNA sequencing, and the polymerase chain reaction. Many of these activities are suitable for 9th graders; they can also be used with adults.

The first activity uses paper models to illustrate restriction digestion and gel electrophoresis. Two wet laboratories then let students perform restriction digests and carry out agarose gel electrophoresis for themselves. Next, students simulate restriction digestion and ligation to construct a recombinant paper plasmid. Worksheets then challenge students to apply what they have learned about restriction digestion, ligation, and electrophoresis to realistic problems. Finally, the more sophisticated techniques of hybridization analysis, DNA sequencing, and the polymerase chain reaction are illustrated through paper simulations.

Puzzles, problems, and information are included throughout this section to illustrate how these techniques are applied to answer various kinds of questions.

DNA Scissors: Introduction to Restriction Enzymes

10

About This Activity

In this activity, students are introduced to restriction enzymes, and they simulate the activity of restriction enzymes with scissors. They are also introduced to restriction maps and asked to make simple predictions based on a map.

Class periods required: 1/2-1

Introduction

Restriction enzymes

Restriction enzymes, or restriction endonucleases, are proteins that recognize and bind to specific DNA sequences and cut the DNA at or near the recognition site. A nuclease is any enzyme that cuts the phosphodiester bonds of the DNA backbone, and an endonuclease is an enzyme that cuts somewhere within a DNA molecule. In contrast, an exonuclease cuts phosphodiester bonds by starting from a free end of the DNA and working inward.

Restriction enzymes were originally discovered through their ability to break down, or restrict, foreign DNA. Restriction enzymes can distinguish between the DNA normally present in the cell and foreign DNA, such as infecting bacteriophage DNA. They defend the cell from invasion by cutting foreign DNA into pieces and thereby rendering it nonfunctional. Restriction enzymes appear to be made exclusively by prokaryotes.

The restriction enzymes commonly used in laboratories generally recognize specific DNA sequences of 4 or 6 base pairs. These recognition sites are palindromic in that the 5′-to-3′ base sequence on each of the two strands is the same. Most of the enzymes make a cut in the phosphodiester backbone of DNA at a specific position within the recognition site, resulting in a break in the DNA. These recognition-cleavage sites are called restriction sites. Below are some examples of restriction enzymes (their names

are combinations of italics and roman numerals) and their recognition sequences, with arrows indicating cut sites.

```
            ↓                              ↓
EcoRI:  5′  GAATTC  3′    HindIII: 5′  AAGCTT  3′
        3′  CTTAAG  5′             3′  TTCGAA  5′
                ↑                              ↑


            ↓                              ↓
BamHI: 5′  GGATCC  3′     AluI:   5′  AGCT  3′
       3′  CCTAGG  5′             3′  TCGA  5′
               ↑                            ↑


            ↓                              ↓
SmaI:  5′  CCCGGG  3′     HhaI:   5′  GCGC  3′
       3′  GGGCCC  5′             3′  CGCG  5′
               ↑                            ↑
```

Notice that the "top" and "bottom" strands read the same from 5′ to 3′; this characteristic defines a DNA palindrome. Also notice that some of the enzymes introduce two staggered cuts in the DNA, while others cut each strand at the same place. Enzymes like *Sma*I that cut both strands at the same place are said to produce blunt ends. Enzymes like *Eco*RI leave two identical DNA ends with single-stranded protrusions:

```
5′ G            AATTC 3′
3′ CTTAA            G 5′
```

Under appropriate conditions (salt concentration, pH, and temperature), a given restriction enzyme will cleave a piece of DNA into a series of fragments. The number and sizes of the fragments depend on the number and location of restriction sites for that enzyme in the given DNA. A specific combination of 4 bases will occur at random only once every few hundred bases, while a specific sequence of 6 will occur randomly only once every few thousand bases. It is possible that a DNA molecule will contain no restriction site for a given enzyme. For example, bacteriophage T7 DNA (approximately 40,000 base pairs) contains no *Eco*RI sites. The action of restriction enzymes is introduced and modeled in this activity.

Rejoining restriction fragments

DNA fragments generated by restriction digestion can be put back together with the enzyme DNA ligase, which forms phosphodiester bonds between the 5′ and 3′ ends of nucleotides. As you might expect, any blunt-ended DNA can be ligated to any other blunt-ended DNA without regard to the sequence of the two molecules. Restriction fragments with single-stranded protrusions, as the *Eco*RI products shown above, are pickier. For efficient ligation, the single-stranded regions must be able to hybridize to a complementary single-stranded region. The idea of rejoining restriction fragments and the need for complementarity in the single-stranded "tails" is introduced in this activity.

This requirement for complementarity may sound limiting, but an examination of the *Eco*RI digestion products shown above reveals that two *Eco*RI ends are perfectly complementary. Any two DNA fragments produced by *Eco*RI digestion can be ligated together, because their single-stranded protrusions are complementary. In fact, fragments with complementary single-stranded protrusions can be ligated much more readily than blunt-ended fragments, presumably because hybridization between the single-stranded regions holds the fragments together in the proper position for ligation. Because these single-stranded protrusions actually facilitate the joining of DNA segments with matching protrusions, they are often called sticky ends.

Restriction enzymes and DNA ligase play starring roles in DNA cloning. To a molecular biologist, cloning a piece of DNA means adding that piece to a plasmid or other vector and then putting the plasmid (or other vector) back into a host cell. One of the simplest methods of cloning is to ligate a restriction fragment into a plasmid that has been cut once with the same restriction enzyme(s). The restriction fragment becomes part of the plasmid when DNA ligase forms phosphodiester bonds between the two formerly separate DNA molecules. This type of cloning is modeled in *Recombinant Paper Plasmids*.

Restriction enzymes and genetic engineering

We often read that the discovery of restriction enzymes made genetic engineering possible. Why is that so? Because restriction enzymes first made it possible to work with small, defined pieces of DNA. Chromosomes are huge molecules that usually contain many genes. Before restriction enzymes were discovered, a scientist might be able to tell that a chro-

mosome contained a gene for an enzyme required to ferment lactose because he knew that the bacterium could ferment lactose and he could purify the protein from bacterial cells. He could use genetic analysis to tell what other genes were close to "his" gene. But he could neither physically locate the gene on the chromosome nor manipulate that gene.

The scientist could purify the chromosome from the bacterium, but then he had a huge piece of DNA containing thousands of genes. The only way to break the chromosome into smaller segments was to use physical force and break it randomly. Then what would he have? A tube full of random fragments. Could they be cloned? Not by themselves. If you introduce a simple linear fragment of DNA (like those produced by shearing) into most bacteria, it will rapidly be degraded by cellular nucleases. Cloning usually requires a vector to introduce and maintain the new DNA. Could our scientist use a vector such as a virus or plasmid to clone his DNA fragments? No. In order to clone DNA into a vector, you have to cut the vector DNA to insert the new piece. Could he simply study the random fragments? No. Every single chromosome from each bacterial cell would give different fragments, preventing systematic analysis. So for many years, physical manipulation of DNA was virtually impossible.

The discovery of restriction enzymes gave scientists a way to cut DNA into defined pieces. Every time a given piece of DNA was cut with a given enzyme, the same fragments were produced. These defined pieces could be put back together in new ways. A new phrase was coined to describe a DNA molecule that had been assembled from different starting molecules: recombinant DNA.

The seemingly simple achievement of cutting DNA molecules in a reproducible way opened a whole new world of experimental possibilities. Now scientists could study specific small regions of chromosomes, clone segments of DNA into plasmids and viruses, and otherwise manipulate specific pieces of DNA. The science of molecular biology literally exploded with the new information that became available. And genetic engineering, which essentially is the directed manipulation of specific pieces of DNA, became possible.

Separating restriction fragments

After restriction digestion, the fragments of DNA are often separated by gel electrophoresis. Background information about electrophoresis, a paper simulation of the process, and two wet laboratories follow this chapter.

Objectives

At the end of this activity, students should be able to

1. Describe a typical restriction site as a 4- or 6-base-pair palindrome;
2. Describe what a restriction enzyme does (recognize and cut at its restriction site);
3. Use a restriction map to predict how many fragments will be produced in a given restriction digest.

Materials

- Photocopies of the *Student Activity* pages, if needed
- Scissors

Preparation

Photocopy the *Student Activity* pages and DNA model sheet for your class, if necessary.

Tips

As students use the paper models, remind them that real DNA is three-dimensional and has no "back" and "front," nor does it matter if the letters representing the bases are upside down.

Answers to Exercise Questions

The numbers are the item numbers in *Exercises and Questions* in the *Student Activity*. Two of the numbered items, 1 and 5, contain only instructions and no questions and so are not represented below.

2. The DNA should be cut so that 5′ AATT protrudes from each end. The ends are sticky.

3. The DNA should be cut straight across between the C and G in the middle of the *Sma*I site. The ends are blunt.

4. The DNA should be cut so that 5′ AGCT protrudes from each end. The ends are sticky.

6. The two tails are 5′ AATT 3′ (*Eco*RI) and 5′ AGCT 3′ (*Hin*dIII). They are not complementary.

7. Each single-stranded tail has the sequence 5′ AATT 3′. They are complementary. Remember that to look for complementarity, you compare the 5′-to-3′ sequence of one strand with the 3′-to-5′ sequence on the other strand.

8. If the fragments were generated in an *Eco*RI digest, then all of them will have single-stranded 5′ AATT 3′ extensions on the ends. The ends of all of the fragments will be complementary.

9. Answers may vary. It is easier for DNA ligase to form a phosphodiester bond between two *Eco*RI fragments because of the complementary single-stranded tails. Hybridization between the bases in the tails brings the backbones into just the right position for resealing. With noncomplementary tails (*Hin*dIII and *Eco*RI), the noncomplementary base pairs keep the nucleotide backbones from coming into proper position for bond formation. It is very difficult to get two fragments with noncomplementary sticky ends to reseal with DNA ligase. Two fragments with blunt ends of any sequence can be connected by DNA ligase. Blunt-ended fragments are harder to connect than fragments with complementary sticky ends but much easier to connect than fragments with noncomplementary sticky ends (which hardly ever occurs). In a sense, "sticky ends" is a poor name for the ends with single-stranded tails, because these ends are sticky only with respect to complementary sticky ends and are very unsticky with respect to noncomplementary sticky ends.

10. Two linear fragments of 942 and 4,599 base pairs (5,541 − 942 = 4,599).

11. Two linear fragments of 2,003 (2,035 − 32) and 3,538 (5,541 − 2,003) base pairs.

12. Three linear fragments of 2,003, 2,881 (4,916 − 2,035), and 657 [5,541 − (2,003 + 2,881)] base pairs.

13. The 942-base-pair fragment contains no *Pvu*II sites and would not be cut. The 4,599-base-pair fragment would be cleaved into two fragments of 2,305 (3,247 − 942) and 2,294 (4,599 − 2,305) base pairs, giving three total fragments.

DNA Scissors

Background Reading

Genetic engineering is possible because of special enzymes that cut DNA. These enzymes are called **restriction enzymes** or **restriction endonucleases.** Restriction enzymes are proteins produced by bacteria to prevent or restrict invasion by foreign DNA. They act as DNA scissors, cutting the foreign DNA into pieces so that it cannot function.

Restriction enzymes recognize and cut at specific places along the DNA molecule called restriction sites. Each different restriction enzyme (and there are hundreds, made by many different bacteria) has its own type of site. In general, a restriction site is a 4- or 6-base-pair sequence that is a palindrome. A DNA palindrome is a sequence in which the "top" strand read from 5′ to 3′ is the same as the "bottom" strand read from 5′ to 3′. For example,

5′ GAATTC 3′
3′ CTTAAG 5′

is a DNA palindrome. To verify this, read the sequences of the top strand and the bottom strand from the 5′ ends to the 3′ ends. This sequence is also a restriction site for the restriction enzyme called *Eco*RI. The name *Eco*RI comes from the bacterium in which it was discovered, *Escherichia coli* RY 13 (*Eco*R), and I, because it was the first restriction enzyme found in this organism.

*Eco*RI makes one cut between the G and A in each of the DNA strands (see below). After the cuts are made, the DNA is held together only by the hydrogen bonds between the four bases in the middle. Hydrogen bonds are weak, and the DNA comes apart.

```
                 ↓
Cut sites:   5′  GAATTC 3′
             3′  CTTAAG 5′
                      ↑

Cut DNA:     5′  G        AATTC 3′
             3′  CTTAA        G 5′
```

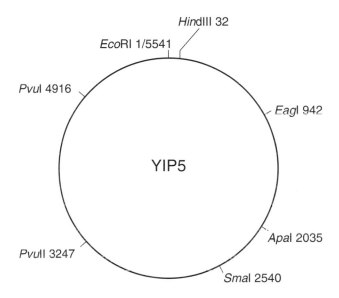

Figure 10.1   Restriction map of YIP5, a 5,541-base-pair plasmid. The number after each restriction enzyme name indicates at which base pair the DNA is cut by that enzyme.

The *Eco*RI cut sites are not directly across from each other on the DNA molecule. When *Eco*RI cuts a DNA molecule, it therefore leaves single-stranded "tails" on the new ends (see the example just given). This type of end has been called a "sticky end" because it is easy to rejoin it to complementary sticky ends. Not all restriction enzymes make sticky ends; some cut the two strands of DNA directly across from one another, producing a blunt end.

When scientists study a DNA molecule, one of the first things they do is figure out where many restriction sites are. They then create a restriction map, showing the locations of cleavage sites for many different enzymes. These maps are used like road maps to the DNA molecule. A restriction map of a plasmid is shown in Fig. 10.1.

The restriction sites of several different restriction enzymes, with their cut sites, are shown on the next page.

```
         ↓                          ↓
EcoRI:   5′ GAATTC 3′    HindIII: 5′ AAGCTT 3′
         3′ CTTAAG 5′             3′ TTCGAA 5′
                 ↑                          ↑


         ↓                          ↓
BamHI: 5′ GGATCC 3′      AluI:   5′ AGCT 3′
       3′ CCTAGG 5′              3′ TCGA 5′
               ↑                        ↑


         ↓                          ↓
SmaI:  5′ CCCGGG 3′      HhaI:   5′ GCGC 3′
       3′ GGGCCC 5′              3′ CGCG 5′
               ↑                        ↑
```

Which ones of these enzymes would leave blunt ends? Which ones would leave sticky ends? Refer to this list of enzyme cut sites as you do the activity.

Exercises and Questions

Exercise 1

Cut the DNA sequence strips (Appendix A) along their borders. These strips represent double-stranded DNA molecules. Each chain of letters represents the phosphodiester backbone, and the vertical lines between base pairs represent hydrogen bonds between the bases.

1. You will now simulate the activity of *Eco*RI. Scan along the DNA sequence of strip 1 until you find the *Eco*RI site (refer to the list above for the sequence). Make cuts through the phosphodiester backbone by cutting just between the G and the first A of the restriction site on both strands. Do not cut all the way through the strip. Remember that *Eco*RI cuts the backbone of each DNA strand separately.

2. Now separate the hydrogen bonds between the cut sites by cutting through the vertical lines. Separate the two pieces of DNA. Look at the new DNA ends produced by *Eco*RI. Are they sticky or blunt? Write *Eco*RI on the cut ends. Keep the cut fragments on your desk.

3. Repeat the procedure with strip 2, this time simulating the activity of *Sma*I. Find the *Sma*I site, and cut through the phosphodiester backbones at the cut sites indicated above. Are there any hydrogen bonds between the cut sites? Are the new ends sticky or blunt? Label the new ends *Sma*I, and keep the DNA fragments on your desk.

4. Simulate the activity of *Hin*dIII with strip 3. Are these ends sticky or blunt? Label the new ends *Hin*dIII, and keep the fragments.

5. Repeat the procedure once more with strip 4, again simulating *Eco*RI.

6. Pick up the "front-end" DNA fragment from strip 4 (an *Eco*RI fragment) and the "back end" *Hin*dIII fragment from strip 3. Both fragments have single-stranded tails of 4 bases. Write down the base sequences of the two tails, and label them *Eco*RI and *Hin*dIII. Label the 5′ and 3′ ends. Are the base sequences of the *Hin*dIII and *Eco*RI tails complementary?

7. Put down the *Hin*dIII fragment, and pick up the back-end DNA fragment from strip 1 (cut with *Eco*RI). Compare the single-stranded tails of the *Eco*RI fragment from strip 1 and the *Eco*RI fragment from strip 4. Write down the base sequences of the single-stranded tails, and label the 3′ and 5′ ends. Are they complementary?

8. Imagine that you have cut a completely unknown DNA fragment with *Eco*RI. Do you think that the single-stranded tails of these fragments would be complementary to the single-stranded tails of the fragments from strip 1 and strip 4?

9. An enzyme called **DNA ligase** re-forms phosphodiester bonds between nucleotides. For DNA ligase to work, two nucleotides must come close together in the proper orientation for a bond (the 5′ side of one must be next to the 3′ side of the other). Do you think it would be easier for DNA ligase to reconnect two fragments cut by *Eco*RI or one fragment cut by *Eco*RI with one cut by *Hin*dIII? What is your reason?

Exercise 2

Figure 10.1 is a restriction map of the circular plasmid YIP5. This plasmid contains 5,541 base pairs. There is an *Eco*RI site at base pair 1. The locations of other restriction sites are shown on the map. The numbers after the enzyme names tell at which base pair that enzyme cleaves the DNA. If you digest YIP5 with *Eco*RI, you will get a linear piece of DNA that is 5,541 base pairs long.

10. What would be the products of a digestion with the two enzymes *Eco*RI and *Eag*I?

11. What would be the products of a digestion with the two enzymes *Hin*dIII and *Apa*I?

12. What would be the products of a digestion with the three enzymes *Hin*dIII, *Apa*I, and *Pvu*I?

13. If you took the digestion products from question 10 and digested them with *Pvu*II, what would the products be?

Introduction to Gel Electrophoresis Laboratory Activities

About the Activities

Following this introduction are three activities that focus on gel electrophoretic separation of DNA fragments. The first, *DNA Goes to the Races,* is a paper-and-pencil simulation to introduce students to the concepts of electrophoresis before they deal with the procedure itself. The second, *Gel Electrophoresis of Precut Lambda DNA,* has students cast a gel and load precut DNA. The third, *Restriction Analysis of Lambda DNA,* has students perform restriction digests and then cast, load, and run gels. The last laboratory takes an extra day to perform.

Agarose Gel Electrophoresis

The standard method for separating DNA fragments is electrophoresis through agarose gels. Agarose is a polysaccharide like agar or pectin that dissolves in boiling water and then gels as it cools. In electrophoresis, DNA is applied to a slab of gelled agarose, and then an electric current is applied across the gel. Because DNA is negatively charged, it migrates through the gel toward the positive electrode.

The rate of migration of DNA through agarose depends on the size of the fragment. The smaller the DNA fragment, the more quickly it can progress through the agarose. The rate of migration of linear fragments through agarose is inversely proportional to the $\log_{10}$ of their molecular weights. Not surprisingly, the rate of migration is also affected by the shape of the DNA molecule; circular molecules (like plasmids) migrate differently from linear fragments with the same molecular weight.

Another important factor in electrophoresis is the concentration of agarose in the gel. The higher the concentration of agarose, the more it retards the movement of all DNA fragments. Therefore, it is better to use relatively high concentrations of agarose to separate and see small DNA fragments and low concentrations if you are interested in large fragments. How large and what concentrations? Examples of agarose concentrations and the sizes of DNA molecules that they efficiently separate are given in Table 1. The sizes of the DNA molecules are expressed in thousands of base pairs (kilobase pairs).

Agarose gels must be prepared and run in a buffer. The buffer is necessary because ions that would otherwise cause the anode to become alkaline and the cathode acidic are generated during electrophoresis. If a gel is accidentally prepared and run with water, the DNA bands will look bad. Recall that a buffer is a mixture either of a weak acid and its salt or a weak base and its salt. The usual electrophoresis buffer is made from the weak base Tris [tris(hydroxymethyl) aminomethane], and the Tris salt is made by adding boric acid. The metal chelator ethylenediamine-tetraacetic acid (EDTA) is also added. The resulting buffer is called TBE, for Tris-borate-EDTA. Other buffers sometimes used for agarose gel electrophoresis are TEA (Tris-acetate-EDTA; made with acetic acid) and TPE (Tris-phosphate-EDTA; made with phosphoric acid).

The voltage applied to the gel affects how quickly the DNA migrates. The higher the voltage, the more quickly the gel runs. There is a trade-off, however. Gels run at high voltages do not separate DNA fragments as efficiently as gels run slowly do. For good separation, gels should be run at no more than 5 V/cm of gel length. You will have to decide what voltage you want your students to use. Specific recommendations are made in the wet laboratory procedures in chapters 12 and 13.

Table 1

% Agarose in gel	Range of efficient separation of linear DNA molecules (kb)
0.3	60–5
0.6	20–1
0.9	7–0.5
1.5	4–0.2
2.0[a]	3–0.1

[a]A 2.0% agarose gel is a very hard gel.

Classroom Activities

Staining Gels

To make DNA fragments visible after electrophoresis, they must be stained. The favorite DNA stain of researchers is ethidium bromide. When ethidium bromide is bound to DNA, it fluoresces under ultraviolet (UV) light. It is a sensitive stain but has several drawbacks for high school use. The first drawback is that to see it, a UV light must be used. The second is that ethidium bromide is a mutagen and requires very careful handling and disposal.

Methylene blue is an alternative stain for DNA gels that is recommended for classroom use by the National Association of Biology Teachers. It is not as sensitive as ethidium bromide, so more DNA must be loaded into the gels. All laboratory protocols in this manual have been prepared with methylene blue staining. You can use either of two procedures to stain with methylene blue, depending on your time frame.

Method 1: methylene blue

Flood the gel with 0.025% methylene blue. Let it stand for 20 to 30 min. Use a funnel to return as much stain as possible to a container. Rinse the gel in tap water. Let it soak for several minutes in several changes of fresh water. DNA bands will become increasingly distinct as the gel destains. It will be about 2 h before DNA is clearly visible. For best results, continue to destain the gel overnight in a small volume of water. View over a light box or on an overhead projector but not for too long, because the dye will fade.

Method 2: methylene blue

Stain the gel with 0.25% methylene blue for 15 min. (*Note:* This concentration is ten times higher than that used in method 1.) Destain the gel in several changes of water, and then leave it overnight in a small volume of water (enough to cover the gel in a reasonably small container).

Staining with ethidium bromide

If you wish to use ethidium bromide, confine its use to a small area of your classroom. Wear gloves when you prepare the stain, handle stained gels, or dispose of any waste. Do not let students use the stain. The greatest risk is to inhale ethidium bromide powder when dissolving it. The safest route is to purchase a ready-mixed stock solution.

To stain with ethidium bromide, dilute the stock to 1 µg/ml. Soak the gel in this solution for 5–10 min. Use a funnel to put as much stain as possible back into a storage container (ideally, a brown glass bottle). If desired, soak the stained gel in water for 5 min or more to clear background ethidium bromide from the gel. View the gel under a UV light source or on a UV transilluminator.

To dispose of ethidium bromide, follow this procedure. If necessary, add sufficient water to reduce the concentration of ethidium bromide to less than 0.5 mg/ml. Add 1 volume of 0.5 M $KMnO_4$ (potassium permanganate), and mix carefully. Add 1 volume of 2.5 N HCl, and mix carefully. Let the mixture stand at room temperature for several hours (or overnight). Add 1 volume of 2.5 N NaOH, and mix carefully. Discard the disabled solution down the sink drain. Drain the disabled gels, and discard them in the trash. Caution: $KMnO_4$ is an irritant and is explosive. It should be handled in a chemical hood.

Commercial DNA stains

Methylene blue staining works, but it is not ideal. It requires high DNA concentrations and a significant amount of time. Ethidium bromide is sensitive and fast but requires UV light for visualization and presents chemical hazards. Some companies that market biotechnology equipment and supplies to educators offer proprietary visual DNA stains that they state are improvements over conventional methylene blue.

At this time we are aware of two such stains, one offered by Carolina Biological Supply Co. and the other offered by Edvotek, Inc. Both stains are more expensive than methylene blue. If you decide to try either one (or any additional stains that become available), be sure to follow the manufacturer's instructions carefully.

The Carolina Biological Supply Co. Carolina BLU DNA stain is added to the agarose gel and buffer. Faint DNA bands can be seen without additional staining. A few minutes of additional staining improves visibility. The company states that this stain can be used with half the amount of DNA required for methylene blue staining.

Edvotek offers a special formulation of methylene blue designed to optimize sensitivity when used with Edvotek equipment.

Recording Data
Photography

One way to record data from electrophoresis gels is to photograph the gels. Biotechnology supply compa-

Classroom Activities

nies sell cameras and camera systems designed for gel photography. These are convenient but can be very expensive.

To photograph a methylene blue-stained gel, use a Polaroid camera with film type 667, an aperture of f/8, and a shutter speed of 1/125 s. For UV light photography of ethidium bromide-stained gels, use Polaroid film type 667 (ASA 3,000). Set the camera aperture to f/8 and the shutter speed to B. Depress the shutter for 2 to 3 s. Type 667 film can be purchased at photography supply stores.

A related film type, 665, is used in the same way but generates a positive black-and-white picture plus a negative. The negative must be soaked in a developing solution (inquire at a photography store). Having a negative is important only if you wish to make enlarged prints of the photographs (e.g., for a publication or a display).

Copying gel patterns

The simplest method for copying the band pattern in a stained gel is to lay a piece of clear plastic (such as an overhead transparency sheet) on the gel and carefully trace the bands.

Data from a wet gel can also be preserved in the following manner. Tape a piece of graph paper (centimeter ruled, if possible) on a table. Cover the paper smoothly with plastic wrap, and tape the plastic down. Lay the wet gel on the plastic wrap, and line it up so that the wells are even with a ruled line. Use a needle or pin to carefully pierce through the gel and onto the graph paper at the center of each band's leading edge. When all the bands are marked, remove the gel and plastic wrap. Draw lines to represent the bands in the gel at each pinprick. Mark the ruled line where the wells were. The distance migrated by each fragment can be read directly off the graph paper by counting squares.

If you would like to pursue the mathematical relationship between DNA fragment size and migration, there is an excellent exercise on pages 269–270 of *DNA Science* (see Appendix H). If you don't have this book, try plotting the migration distance (the distance from the front of the well to the leading edge of each band) on the x axis and the $\log_{10}$ of the size of the fragment in base pairs (substitute for molecular weight, since they are directly proportional) on the y axis. Talk to the mathematics teacher in your school about ways to use the migration of DNA fragments as an illustration of the use of logarithms or of semilog paper in plotting data.

Preserving Gels

Wet methylene blue-stained gels can be kept in sealed containers or sealed plastic bags in the refrigerator for a long time. Include a small amount (a few milliliters) of destaining solution with the gel. This is enough to keep the gel moist but not to allow additional destaining.

Polyacrylamide Gel Electrophoresis

The other gel material used in electrophoresis of DNA is polyacrylamide. Polyacrylamide forms a tighter mesh than does agarose, so polyacrylamide gels can separate smaller molecules. Polyacrylamide also has a higher resolving power, meaning a polyacrylamide gel can separate two molecules whose molecular weights differ by only a small amount. For example, long polyacrylamide gels are used to separate DNA fragments that differ in length by only one nucleotide in DNA sequencing (see chapter 17). They are used in forensic DNA fingerprinting applications, in which it is critical to determine whether two DNA fragments are exactly the same size. Polyacrylamide gels are also used for protein electrophoresis.

Polyacrylamide is a polymer of the chemical acrylamide. To make a polyacrylamide gel, acrylamide is dissolved in buffer along with the cross-linking agent bisacrylamide. Catalysts are added to start polymerization. The liquid mixture is then quickly poured into a thin space between two glass or plastic plates. After the gel polymerizes, the plates and gel are clamped to an upright support and run vertically, rather than horizontally.

Polyacrylamide gels used to separate DNA are usually made with and run in TBE buffer, just like agarose gels. The chemical urea can be added to the gel to keep the DNA single stranded for applications like DNA sequencing. The most common buffer used for protein electrophoresis contains Tris base and the amino acid glycine. The detergent sodium dodecyl sulfate (SDS) can be added to this buffer to denature the proteins in the sample. As you can see, polyacrylamide gels can be made in several ways to accommodate different purposes.

Polyacrylamide Gel Electrophoresis in the Classroom

There are several points to consider if you are wondering about using polyacrylamide gels in your classes. First, they require vertical electrophoresis chambers rather than the horizontal ones used for agarose gels. Second, the casting procedure for polyacrylamide gels is significantly more technically demanding than that for agarose gels. Finally, the gel monomer acrylamide is toxic and must be handled carefully. Once acrylamide has polymerized into polyacrylamide it is no longer toxic, but the gels should still be handled with gloves since some free acrylamide could remain.

Because of the difficulty of casting the gels and the toxicity of acrylamide, we recommend that teachers use agarose gel electrophoresis in all but advanced high school or college classes. Even in these settings, agarose gel electrophoresis is sufficient for teaching so many concepts that there is no need to venture into polyacrylamide. However, if you emphasize proteins in your classes and find that agarose gels are not adequate for demonstrating your points, you may eventually want to incorporate this technology.

If you want to try polyacrylamide gel electrophoresis, we recommend that you purchase precast gels. These are ready to use. Simply clamp them to the electrophoresis chamber, add buffer and sample, and run. Precast polyacrylamide gels are sold by several biological supply companies. They can be quite expensive, so shop around for the best price. Be sure that you know which buffer system you need for your gels; you can purchase gels made with TBE, Tris-glycine, Tris-glycine-SDS, and even some other buffers.

DNA Goes to the Races

About This Activity

This is a reading and paper activity that introduces electrophoresis. Students should already be familiar with the activity of restriction enzymes through the activity *DNA Scissors*. This lesson provides enough background for students to continue with activities that require them to know something about the process (hybridization analysis, DNA sequencing, DNA fingerprinting, etc.). It is also an excellent introduction if you plan to conduct electrophoresis in your classroom. *Restriction Analysis Challenge Worksheets* (chapter 15) illustrate applications of electrophoresis in the laboratory. Students should complete this activity and *Recombinant Paper Plasmids* before using *Restriction Analysis Challenge Worksheets*.

Class periods required: *1*

Introduction

Please read *Introduction to Gel Electrophoresis Laboratory Activities* for general background information.

Objectives

At the end of this activity, students should be able to

1. Predict the number and arrangement of bands in a gel after digestion and electrophoresis if shown a simple restriction map;

2. Explain why electrophoresis causes DNA fragments of different sizes to separate.

Materials

DNA Goes to the Races student reading pages, restriction maps, and gel outline

Preparation

Photocopy the pages, if necessary.

Procedure

This activity is self-explanatory. Students can read the handout and do the exercises. Show them as much real material (or as many pictures) as you can: gel boxes, power supplies, an old gel or a photograph of one. Relate the items to their activity and reading.

Results of the Exercise

If possible, check the students' work as they do the exercise. Make sure they line the fragments up correctly between lanes as well as within an individual gel lane (for example, be sure that the 2,500-base-pair [bp] *Hind*III fragment is lined up between the 2,000- and 3,000-bp *Bam*HI fragments and even with the 2,500-bp *Eco*RI fragment). Use their own fragment size labels to assist you. When they are done, the fragments in the "gel" should be lined up as shown here.

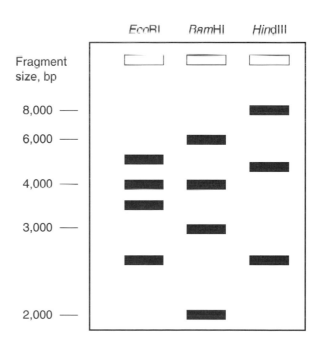

DNA Goes to the Races

You have already learned about restriction enzymes and how they cut DNA into fragments. You may have even looked at some DNA restriction maps and figured out how many pieces a particular enzyme would produce from that DNA. But when you actually perform a restriction digest, you put the DNA and the enzyme into a small tube and let the enzyme do its work. Before the reaction starts, the mixture in the tube looks like a clear fluid. Guess what? After the reaction is finished, it still looks like a clear fluid! Just by looking at it, you can't tell that anything has happened.

In order for restriction digestion to mean much, you have to be able somehow to see the different DNA fragments that are produced. There are chemical dyes that stain DNA, but obviously it doesn't do much good to add these dyes to the mixture in the test tube. In the laboratory, scientists use a process called **gel electrophoresis** to separate DNA fragments so that they can look at the results of restriction digests (and other procedures).

Gel electrophoresis takes advantage of the chemistry of DNA to separate fragments. Under normal circumstances, the phosphate groups in the backbone of DNA are negatively charged. In electrical society, opposites do attract, so DNA molecules are very much attracted to anything that is positively charged. In gel electrophoresis, DNA molecules are placed in an electric field (which has a positive and a negative pole) so that they will migrate toward the positive pole.

The electric field makes the DNA molecules move, but to cause the molecules to separate and be easy to look at later on, the whole process is carried out in a gel (obviously the source of the name *gel* electrophoresis). If you have ever eaten Jell-O, you have had experience with a gel. The gel material in Jell-O is gelatin; different gel materials are used to separate DNA. One gel material often used for electrophoresis of DNA is called *agarose,* and it behaves much like Jell-O but without the sugar and color. To make a gel for DNA (called *pouring* or *casting* a gel), you dis-

solve agarose powder in boiling buffer, pour it into the desired dish, and let it cool. As it cools, it hardens (sound familiar?).

Since the plan for agarose gels is usually to add DNA to them, scientists place a device called a *comb* in the liquid agarose after it has been poured into the desired dish and let the agarose harden around the comb. Imagine what would happen if you stuck the teeth of a comb into liquid Jell-O and let it harden. Afterwards, when you pulled the comb out, you would have a row of tiny holes in the solid Jell-O where the teeth had been. This is exactly what happens with laboratory combs. When the comb is removed from the hardened agarose gel, a row of holes in the gel remains (look at Fig. 11.1). The holes are called *sample wells.* DNA samples are placed into the wells before electrophoresis is begun.

For electrophoresis, the entire gel is placed in a tank of salt water (not table salt) called buffer. An electric current is applied across the tank so that it flows through the salt water and the gel. When the current is applied, the DNA molecules begin to migrate through the gel toward the positive pole of the electric field (Fig. 11.2). Figure 11.3 shows a scientist loading a DNA sample into an agarose gel sitting in an electrophoresis tank.

At this point, the gel does its most important work. All of the DNA in the gel migrates through the gel toward the positive pole, but the gel material makes it more difficult for larger DNA molecules to move than smaller ones. So in the same amount of time, a small DNA fragment can migrate much farther than a large one. You can therefore think of gel electrophoresis as a DNA footrace, where the "runners" (the molecules being separated) separate just like runners in a real race (Fig. 11.4). The smaller the molecule, the faster it runs. Two molecules the same size run exactly together.

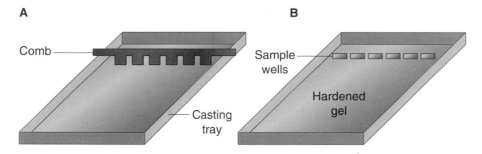

A

Comb

Casting tray

B

Sample wells

Hardened gel

Figure 11.1 Casting an agarose gel. (A) To make a gel, hot liquid agarose solution is poured into a casting tray (any shallow container), and the comb is put in place. (B) After the agarose cools and hardens, the comb is removed, leaving behind pits in the gel called sample wells. Samples are loaded into the wells prior to electrophoresis.

After a time, the electric current is turned off, and the entire gel is placed into a DNA staining solution. After staining, the DNA can be seen. The pattern looks like a series of stripes (bands) in the gel; each separate band is composed of one size of DNA molecule. There are millions of actual molecules in the band, but they are all the same size (or very close to it). At any rate, after a restriction digest, there should be one band in the gel for each different-size fragment produced in the digest. The smallest fragment will be the one that has migrated furthest from the sample well, and the largest will be closest to the well, as shown in Fig. 11.5.

Figure 11.2 In electrophoresis, the gel is placed in a tank of salt solution, and an electric current is applied. The DNA migrates toward the positive pole.

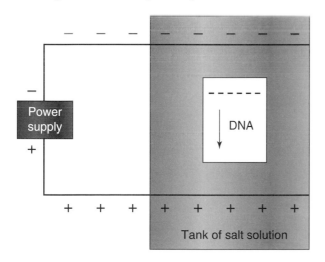

Activity

In Appendix A, you have three representations of a DNA molecule and the outline of an electrophoresis gel. The representations show the cut sites of three different restriction enzymes on the same DNA molecule. You will simulate the digestion of this DNA with each of the three enzymes and then simulate agarose gel electrophoresis of the restriction fragments.

Figure 11.3 A scientist is using a micropipette to load a DNA sample into an agarose gel. The gel is in an electrophoresis chamber full of buffer. The power supply for the chamber is on the laboratory bench behind the chamber.

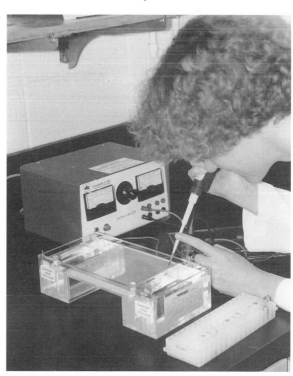

Figure 11.4   In electrophoresis races, the small DNA always wins!

1. Cut out the three pictures of the DNA molecule.

2. Simulate the activity of the restriction enzyme *Eco*RI on the DNA molecule that shows the *Eco*RI sites by cutting across the strip at the vertical lines representing *Eco*RI sites. You have now digested the molecule with *Eco*RI. Put your "restriction fragments" in a pile apart from the other two DNA strips.

3. "Digest" the second DNA strip with *Bam*HI. Put the *Bam*HI fragments in a separate pile.

4. Now "digest" the remaining DNA molecule with *Hin*dIII. Put these fragments in a third pile.

5. In our imaginary gel electrophoresis, you will separate the *Eco*RI, *Bam*HI, and *Hin*dIII fragments as if you had loaded the three sets of fragments into separate but adjacent sample wells. Arrange your fragments as they would be separated by agarose gel electrophoresis. Designate an area on your desk as the end of the gel with the sample wells. Starting with the *Eco*RI fragments, arrange them from longest to shortest, with the longest one closest to the well.

6. Next, separate the *Bam*HI fragments, and place them adjacent to the *Eco*RI fragments. Be sure to order each fragment correctly by size with re-

Figure 11.5   Gel electrophoresis is used to separate products of restriction digestion. (A) Restriction map, with fragment sizes in base pairs; (B) view of gel after electrophoresis.

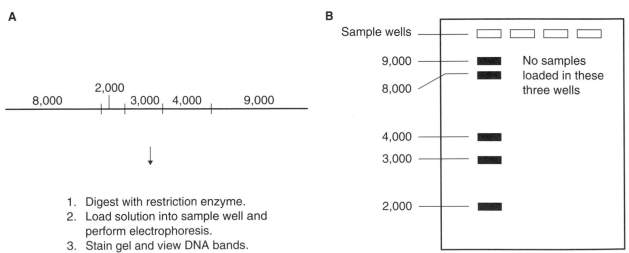

A

8,000   2,000   3,000   4,000   9,000

1. Digest with restriction enzyme.
2. Load solution into sample well and perform electrophoresis.
3. Stain gel and view DNA bands.

B

Sample wells

9,000   No samples
8,000   loaded in these
three wells

4,000

3,000

2,000

spect to other *Bam*HI fragments and to the *Eco*RI fragments you have already laid out.

7. Repeat the same procedure for the *Hin*dIII fragments. You should now have all three of your sets of fragments arranged in order in front of you.

8. Look at the outline of the electrophoresis gel provided in Appendix A. Notice that it has a size scale in base pairs on the left-hand side and that sample wells are drawn in. Using the outline and the size scale as a guide for where to draw your fragments, draw the pattern your restriction digest would make in the gel. Use the *Eco*RI sample well for the *Eco*RI fragments, and so on.

9. After you draw the bands representing the restriction fragments, use the size information on the paper DNA strips to label the bands on the gel with the sizes of the fragments in base pairs.

10. Use the fragment sizes as a check for your work.

Are all the smaller fragments across all the gel "lanes" in front of all the larger fragments? Did you notice that the size scale doesn't seem to have regular intervals? The size scale looks the way it does because agarose gels separate fragments that way.

Gel Electrophoresis of Precut Lambda DNA

12

About This Activity

This activity is a short version of the activity *Restriction Analysis of Lambda DNA*. Students perform electrophoresis using lambda DNA that has already been digested with restriction enzymes. Prior to this exercise, students should perform the activities *DNA Scissors* and *DNA Goes to the Races*.

Class periods required: *approximately 2*

Introduction

Please read the introduction to *DNA Scissors* and *Introduction to Gel Electrophoresis Laboratory Activities* for general background information.

Two samples of predigested DNA from the bacteriophage lambda are used in this exercise. Each sample has been cut by *Hin*dIII, *Eco*RI, *Pst*I, or *Bam*HI endonuclease. There are seven *Hin*dIII restriction sites on lambda DNA, resulting in eight DNA fragments (two of the fragments are small and may appear as one very faint band). The five *Eco*RI restriction sites of lambda DNA yield six fragments of DNA. You may not be able to resolve all of these fragments. *Bam*HI also cuts lambda DNA at five sites, yielding six DNA fragments. Since two pairs of these fragments are fairly similar in size, you may see only four bands in the gel. Finally, *Pst*I cuts lambda DNA at 28 sites. Many of the DNA fragments generated are too small to see in the gel, and many cannot be resolved. However, *Pst*I does produce a pattern of many bands and provides a good example of restriction enzyme cutting, if all you want to do is show cutting versus no cutting.

Objectives

At the end of this laboratory, students should be able to

1. Describe three steps involved in restriction analysis: restriction digestion, electrophoresis, and staining;
2. Analyze stained DNA fragments of lambda DNA in a gel and compare them to a restriction map.

Materials

Equipment

- Electrophoresis chamber and gel trays
- Power supply
- Digital micropipettes and tips (very dextrous individuals could use graduated microcapillary pipettes or similar alternatives.)
- Dishes for staining and destaining gels

Supplies (per laboratory team)

- 30 to 50 ml of melted 1% agarose solution (made with 1× Tris-borate-EDTA (TBE buffer [EDTA is ethylenediaminetetraacetic acid])

- 2 μg of lambda DNA digested with *Eco*RI, *Hin*dIII, *Bam*HI, or *Pst*I (>0.2 μg/μl), with loading dye added; students need to load about 2 μg of DNA per gel lane for it to be visible after methylene blue staining

- 2 μg of lambda DNA digested with a second enzyme from the list (>0.2 μg/μl), with loading dye added

- 50 ml of 0.025% methylene blue solution or other staining solution

- Distilled water

Resource Materials

Lambda DNA, restriction enzymes, concentrated TBE, and, in many cases, precut lambda DNA are available from many biotechnology supply companies (see Appendix I). In addition, the following kits are available:

- Edvotek Kit 112, "Analysis of *Eco*RI Cleavage Patterns of Lambda DNA"
- Fotodyne Safekit 102, "Analysis of Predigested Lambda DNA"

- Carolina Biological Supply Co. kit, "Restriction Enzyme Cleavage of DNA" (catalog no. 21-1149)
- Carolina Biological Supply Co. kit, "Introductory Gel Electrophoresis Kit," (catalog no. 21-1148), supplies materials for students to use in separating colored dyes on an agarose gel as an introduction to electrophoresis.

Preparation

You can buy uncut lambda DNA and restriction enzymes and cut some of the DNA yourself for this activity and for use as a precut standard in the next activity, *Restriction Analysis of Lambda DNA*. The enzymes and uncut DNA will also be used in the next activity. To make your own predigested lambda DNA, simply follow the directions for digests in the next activity (including diluting the DNA, if needed), but scale up the reaction size by multiplying everything by, say, 10, *except* the amount of enzyme. You can get great digestion with much less enzyme and thus avoid waste. For example, use 40 μl of uncut DNA (0.4 to 0.5 μg/μl), 10 μl of 10× buffer (supplied with enzymes), 45 μl of water, and 5 μl of enzyme. Let the digest incubate for several hours, and then add 15 to 20 μl of loading dye to stop the digestion. Store your precut DNA in the refrigerator or freezer. This is enough sample for 8 to 10 gel lanes.

1. If necessary, dilute concentrated TBE buffer to 1× concentration. For example, dilute 10× TBE to 1× by adding 9 volumes of distilled water to 1 volume of 10× TBE (90 ml water + 10 ml 10× TBE). The 1× buffer solution is used in the gel chamber and for dissolving the powdered agarose.

2. Prepare the agarose solution.
   - Determine how much 1% agarose solution will be needed to cast your gels. A single minigel requires 30 to 50 ml of agarose solution, depending on the size of the gel tray. For example, 1 g of agarose powder mixed with 100 ml of 1× TBE buffer (from step 1 above) will yield 100 ml of a 1% agarose solution. Weigh out the agarose powder, and mix it with the appropriate amount of 1× TBE buffer in a clean flask that holds at least twice (preferably more) the volume of buffer.

   - Heat the agarose and buffer, using a boiling-water bath (30 to 60 min) or microwave (3 to 6 min) to dissolve the agarose powder. Do not cap the flask during heating. The gel can be cast when the container feels very warm but not hot when you touch it to your cheek. If you pour gels when the agarose is too hot, you can warp the casting tray. Melted agarose can be kept in a 55°C (or warmer) water bath until time to pour gels.

Tips

1. DNA should be stored in the refrigerator. The 10× TBE is stable at room temperature.

2. Running the gels. Small gels can be run for 1 to 2 h at 90 V, but this high voltage does not give a pretty separation. If you can adjust the voltage on your power supply, a run at 40 to 50 V for 2 to 3 h will give a prettier gel with the lambda digests. You can also run the gels at 12 V overnight or at 130 V for 45 min.

3. As you destain the gels, it may seem at first that no DNA is present. Continue to destain, and the DNA will appear as the background stain diffuses out of the gel.

4. Hardened excess agarose can be remelted for subsequent use. Store the hardened agarose in a closed container at room temperature. (This is agarose you did not use, not used gels.)

5. For *Bam*HI-, *Eco*RI-, and *Hin*dIII-cut lambda DNA, 0.7 to 0.8% agarose gels resolve the DNA fragments better. To make these gels, use 0.7 or 0.8 g of agarose per 100 ml of 1× TBE buffer. These gels are more fragile than 1% gels and require more careful handling.

6. If students have never loaded a gel, have them mix loading dye and water in the appropriate amounts and practice (make an extra gel for practice loading; you can cut off the loaded wells and melt the rest of the gel for reuse later). If you pour the practice gel ahead of time, they can practice loading while their gels harden.

7. The book *DNA Science* by Micklos and Freyer (available from Carolina Biological Supply Co.) has a wonderful two-page spread of "bad gels" with the causes of each problem listed. There is a similar section in the Fotodyne catalog.

8. TBE buffer can be reused a few times.

Procedure

1. Review the activity of restriction enzymes, if necessary. Show students the restriction map of lambda. Have them predict the relative positions

of the bands from different digests as they might look in a gel.

2. Follow the procedure in the *Student Activity* pages for casting the gels. Load enough of the restriction digestion mixture plus loading dye to give 2 μg of precut DNA per lane. In the example under *Preparation,* students would load 12 to 15 μl per lane. Casting and loading the gels may take an entire class period. When the electrophoresis is finished, the students' gels can simply be left in the gel boxes with the power off until the next class, when the students stain them. Alternatively, you can stain them yourself and let the students destain them. If you follow this plan, take the gels out of stain and store them overnight in a volume of water that is large enough to cover them. Some destaining will occur.

3. Have students record their results and answer the questions.

Answers to Student Questions

1. Students should draw their gel patterns.

2. No, because the gels tell you only the sizes of the fragments and give no information on the relative locations of the restriction sites.

3. Please refer to the restriction map of bacteriophage lambda at the end of this lesson. There are seven *Hin*dIII sites and five *Eco*RI sites in lambda DNA. A *Hin*dIII-*Eco*RI digest of lambda DNA will cleave lambda DNA at 12 sites, resulting in 13 restriction fragments. These fragments would have the following sizes in base pairs according to the restriction map: 21,226, 1,904, 2,027, 947, 1,375, 4,268, 5,148, 564, 125, 584, 4,973, 831, and 3,530. If you wish to test the prediction by performing a double digest, mix 1 μl of each enzyme in a microcentrifuge tube, and then add 1 μl of the mixture to a digest. Incubate for 2 h if possible. For the clearest interpretation, load the gel with single digests of *Hin*dIII and *Eco*RI and the double digest. Some of the bands in the double digest will be the same as those in the single digest. The single-digest bands will provide a size scale for the double-digest bands. It could be a good exercise for students to predict the migration patterns of these three samples run side by side.

4. An experiment to test whether restriction enzymes cut DNA must include an experimental digest with buffer, DNA, and enzyme and a control digest with buffer, DNA, and no enzyme. The two tubes should be incubated under identical conditions and their contents then run side by side in a gel. A good second control sample would be simply to run some of the starting DNA without mixing it with buffer and without incubating it to make sure that the starting DNA is intact. After the gel has run, it should be stained and destained, and the DNA banding patterns should be compared.

Gel Electrophoresis of Precut Lambda DNA

Procedure

1. Casting the gel
   - Obtain an electrophoresis chamber and gel tray.

   - Seal the ends of the tray with masking tape. Place the comb at one end of the tray, making sure it does not touch the bottom of the tray.

   - Pour enough agarose into the gel tray to cover the lower third (about 6 mm) of the comb. Allow the agarose to cool.

   - Remove the tape from the gel tray without damaging the ends of the gel. Do not remove the comb at this point. Place the gel tray into the electrophoresis chamber with the wells nearest the negative (black leads) electrode end of the chamber.

2. Fill the electrophoresis chamber with 1× TBE (Tris-borate-EDTA [ethylenediaminetetraacetic acid]) buffer. The buffer must completely cover the gel and electrodes.

3. Carefully remove the comb from the gel, leaving the wells that you will fill with your DNA samples.

4. Load the precut DNA plus loading dye into the appropriate well (avoid puncturing the bottom of the well). Your instructor will tell you what volume of DNA sample to load. Using a fresh pipette tip, repeat this step with a sample of DNA cut with the other restriction enzyme. Depending on the number of wells in the gel, several student teams may use the same gel. Be sure to sketch the location and identity of the contents of each well.

Note: It is not necessary to put the tip of the micropipette into the well, because the loading dye increases the density of the DNA solution and will help DNA flow into the submerged wells.

5. Plug the electrophoresis chamber into the power supply. The electrode nearest the DNA samples in the wells must be connected to the negative pole of the power supply. Turn on the current (your teacher will tell you if you need to set the power supply to a particular voltage). Allow the slow marker dye to migrate 2 to 2.5 cm. Depending on the available voltage, this process will take as long as 24 h (12 V) or as little as 40 min (130 V).

6. The final step is to stain the DNA with methylene blue.
   - Flood the gel in a petri dish or other container with 0.025% methylene blue for 1 h.

   - Destain the gel by repeated soakings of tap water. Destaining should continue until the DNA bands become distinct (approximately 30 to 60 min).

   - Gels can be photocopied or photographed for a permanent record or refrigerated in sandwich bags for storage. Alternatively, place the gel on plastic wrap, and lay the plastic wrap and gel on a piece of graph paper. Stick a pin through each band in the gel and down through the graph paper so that you produce a pattern of holes in the graph paper that shows the spacing of the bands in the gel exactly. Using the pinholes as a guide, draw the banding pattern. Viewing the gels with a yellow filter helps make the bands more visible.

Questions

1. Draw your gel and label the lanes with the name of the enzyme used. Compare your fragment patterns to the ideal fragment patterns in Fig. 12.1 for a digest of lambda DNA by *Eco*RI, *Hin*dIII, and/or *Bam*HI. Account for differences in separation and band intensity between the experimental gel and the ideal pattern.

2. Can you predict the fragment pattern for a digest using *Eco*RI and *Hin*dIII at the same time with only the information (the location of the bands) from the gels you ran? Why or why not?

3. Use the lambda restriction map in Fig. 12.2 to predict the sizes of fragments that would be generated in a double digest using *Eco*RI and *Hin*dIII.

4. This laboratory used precut DNA. Propose an experiment to test whether the restriction enzymes actually cut DNA. Be sure to describe exactly what you would use for control and experimental treatments. Assume that you have access to the restriction enzymes you wish to test.

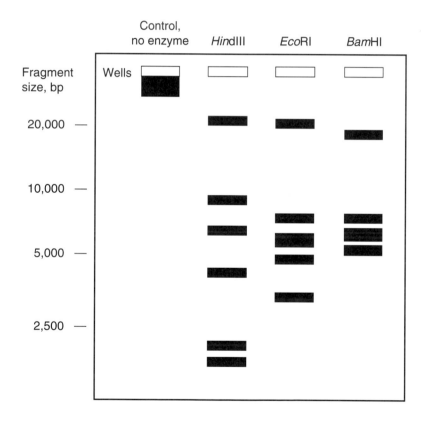

Figure 12.1 Predicted gel pattern. The 125-base-pair *Hin*dIII fragment will never be seen. The 564-base-pair *Hin*dIII fragment is also usually not visible. It is not shown in this example.

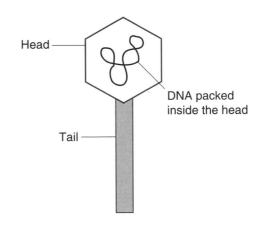

Head —

DNA packed inside the head

Tail —

*Hind*III restriction map, with fragment sizes in base pairs

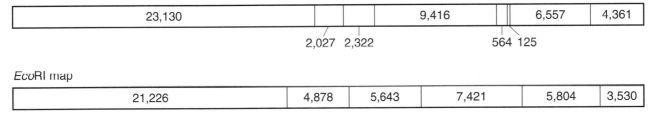

23,130			9,416		6,557	4,361

2,027 2,322   564 125

*Eco*RI map

21,226	4,878	5,643	7,421	5,804	3,530

*Bam*HI map

5,505	16,841	5,626	6,527	7,233	6,770

Figure 12.2 Bacteriophage lambda is the best known of the phages. It has a linear chromosome of approximately 48,500 base pairs and about 45 known genes. Its restriction map is shown.

Restriction Analysis of Lambda DNA

Classroom Activities

About This Activity

In this activity, students will digest lambda DNA with the restriction enzymes *Eco*RI, *Hin*dIII, and either *Bam*HI or *Pst*I. They will then prepare an agarose gel and separate the DNA fragments by agarose gel electrophoresis. Setting up the digests occupies the first day, preparing and loading a gel is done on the second day, and staining and destaining the gel is done on the third day. Depending on the destaining time, results may not be clearly visible until the fourth day, which is spent analyzing the gels.

Class periods required: 4

Introduction

Please read the introduction to *DNA Scissors* and *Introduction to Gel Electrophoresis Laboratory Activities* before you do this laboratory. The introduction to the lesson *Gel Electrophoresis of Precut Lambda DNA* contains information on the products of digestion of lambda DNA with different enzymes.

It is recommended that you do the paper activity *DNA Scissors* (which introduces students to the activity of restriction enzymes) and *DNA Goes to the Races* (which introduces students to gel electrophoresis) before you do this laboratory. In *DNA Scissors*, students use restriction maps to predict the sizes of DNA fragments after digestion. You could show them the restriction map of lambda DNA from chapter 12 and have them predict the sizes of the fragments their digests will generate.

Objectives

At the end of this laboratory, students should be able to

1. Describe three steps involved in restriction analysis: restriction digestion, gel electrophoresis, and staining;

2. Compare a pattern of restriction fragments of lambda DNA in a gel to a restriction map of lambda.

Materials

Equipment

- Electrophoresis chambers and gel trays
- Power supplies
- Micropipettes (0.5 to 10 µl)
- Trays for staining gels (any small, shallow dish will do)
- Microwave or hot plate for melting agarose
- Hot-water bath (optional for keeping agarose molten)
- Freezer (*not* frost free, if possible)
- Light box or table (optional)
- Polaroid camera and type 667 film (optional)
- Cooler or ice bucket
- 37°C water bath or dry incubator
- Microcentrifuge (optional)

Supplies

- Lambda DNA, 0.4 to 0.5 µg/µl; 4 µl per digest
- Restriction enzymes (*Eco*RI, *Hin*dIII, and *Pst*I or *Bam*HI)
- Restriction digest buffer
- Agarose
- TBE (Tris-borate-EDTA) buffer
- Loading dye
- Sterile water (for digests)
- Sterile tips for micropipettes
- Sterile microcentrifuge tubes
- Masking tape
- Waterproof marking pens (or other labeling system)
- Methylene blue solution
- Flask for melting agarose
- Centimeter-ruled graph paper (optional for recording data; students can bring this)
- Plastic wrap (optional for recording data)
- Distilled water (for preparing gels)
- One holder or rack for microcentrifuge tubes per student team (A Styrofoam cup full of crushed ice will work; the ice is simply to provide support.)

Resources

Lambda DNA, restriction enzymes, agarose, and concentrated TBE buffer can be purchased from many biotechnology supply companies (see Appendix I). Restriction enzymes are supplied with concentrated digestion buffer, usually 10×.

The following kits are also available:

- Fotodyne SafeKit 104, "Analysis of Restriction Enzyme Digestion"
- Carolina Biological Supply Co., "DNA Restriction Analysis Kit" (catalog no. 21-1151; refills available)
- Edvotek Kit 212, "Cleavage of Lambda DNA with *Eco*RI Endonuclease"

Preparation

When the reagents arrive, place the enzymes in the freezer immediately. If your freezer is frost free, it is a good idea to keep the enzymes in a Styrofoam box with an ice pack inside the freezer. Frost-free freezers go through cycles of heating to eliminate frost build-up. Keeping the enzymes in a Styrofoam container on an ice pack inside the freezer keeps them cold during these cycles.

Keep the concentrated restriction buffer frozen. Keep the DNA in the refrigerator. Agarose and 10× TBE may be stored at room temperature.

Before day 1 (restriction digest day)

1. To protect your supply, you may wish to aliquot the restriction enzyme into several small batches for student use. Since each student digest uses 1 μl of enzyme, it seems as though 1 μl of enzyme per student group should be enough, but it probably won't be. In fact, you should count on only eight groups using about 12 μl. There are probably two reasons for this: pipetting errors by students and (more important) the tendency for the viscous enzyme solution to stick to pipette tips. Keep the enzyme in the freezer or on ice.

2. Thaw the restriction buffers (they are 10×; do not dilute them). Put 5 to 6 μl of 10× restriction buffer in a sterile microcentrifuge tube for each student group. Keep the thawed buffer in the refrigerator or on ice until class.

3. Dilute the lambda DNA to approximately 0.5 μg/μl if it is not close to this concentration; 0.4 μg/μl is fine. To make the dilution, use the following formula:

$$(\text{starting concentration})(x) = (0.5 \ \mu g/\mu l)$$
$$(\text{desired volume})$$

where x is the amount of DNA you will start with.

Express the starting concentration in micrograms per microliter (for example, if your DNA is 1 mg/ml, use 1 μg/μl as the starting concentration). The desired volume is how much of the 0.5-μg/μl solution you want. So if you want 100 μl of 0.5-μg/μl lambda DNA and the solution you bought is 2.0 mg/ml, then

$$(2 \ \mu g/\mu l) \ (x) = (0.5 \ \mu g/\mu l) \ (100 \ \mu l)$$
$$x = 25 \ \mu l$$

Take x (in this case, 25) μl of the starting solution, and put it in a sterile microcentrifuge tube. Add (desired volume − x) μl of sterile Tris-EDTA buffer (TE). In this case, 100 μl is our desired volume, so $100 − x(25) = 75$ μl of TE. So in our example, 25 μl of starting solution plus 75 μl of TE gives 100 μl of 0.5-μg/μl DNA.

Put 20 μl of uncut, 0.5-μg/μl lambda DNA in a sterile microcentrifuge tube for each student group. If you have a precut standard, put the equivalent of 2 μg DNA in a sterile microcentrifuge tube for each student group.

4. Put 20 μl of sterile water in a sterile microcentrifuge tube.

5. Photocopy the *Student Activity* pages and lambda restriction map for your class if students do not have manuals.

Before day 2 (gel electrophoresis day)

1. Dilute the concentrated TBE to 1×; e.g., add 9 volumes of distilled water to 1 volume of 10× TBE (example: 90 ml of water + 10 ml of 10× TBE). The 1× buffer solution is used in the gel chamber and for dissolving the powdered agarose.

2. Prepare the agarose solution.
   - Determine how much 1% agarose solution will be needed to cast your gels. Each gel requires 30 to 50 ml of agarose solution, depending on the size of the gel tray. For example, 1 g of agarose powder mixed with 100 ml of 1× TBE (from step 1) will yield 100 ml of 1% agarose solution (enough for two or three gels). Weigh out the agarose powder, and mix it with the appropriate amount of 1× TBE in a clean flask. The flask should have a capacity at least twice the volume of the TBE-agarose.
   - Heat the agarose and buffer, using a boiling-water bath (30 to 60 min) or a microwave (3 to 6 min) to dissolve the agarose powder. Make sure the agarose is completely dissolved by swirling the flask and holding it up to the light (wear an insulated glove!) to look for ghostly

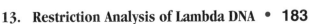

clear fragments of undissolved agarose. The gels can be cast when the container is just cool enough to hold comfortably. Melted agarose can be kept in a 55°C (or warmer) water bath until time to pour the gels.

Before day 3 (staining)

Prepare 0.025 or 0.25% methylene blue, depending on the staining method you choose (refer to *Introduction to Gel Electrophoresis Laboratory Activities*).

Before day 4 (analysis)

If you are photographing gels, make sure you have supplies on hand. If not, use the graph paper method for recording data (see *Introduction to Gel Electrophoresis Laboratory Activities*). Have students bring (or supply for them) centimeter-ruled graph paper. Bring plastic wrap to class, and make sure you have pins or dissecting needles for piercing the gels.

Tips

1. Before you begin the activity, go over the outline of the entire activity. Review restriction enzymes and electrophoresis if necessary.

2. Each day, have students read the procedure for the day very carefully before beginning. Success depends on their carrying out the directions accurately.

3. If students have not used micropipettes or electrophoresis apparatus or have not cast gels before, it is a good idea to let them practice pipetting, loading a practice gel, and plugging in the electrophoresis apparatus.

4. Most gel combs have at least eight wells. These will be enough to let two student groups share one gel if desired. Make sure the students do a good job of recording which sample went into which well.

5. The prettiest gels will result from running at 40 to 50 V for 2 to 3 h or 12 V overnight. After you turn the power supplies off, you may simply leave the gels in the boxes and covered with buffer until the next day.

6. Agarose gels can be prepared the day before use and stored covered with 1× TBE buffer at room temperature (right in the gel box, ready to load the next day).

7. Hardened agarose can be remelted and used again. Store hardened agarose in a closed container at room temperature. Do not reuse gels that have had DNA in them or have been stained.

8. Gels with 0.7 or 0.8% agarose will give better resolution of the *Hind*III, *Eco*RI, or *Bam*HI digests, but the gels are more fragile. If you decide to use the lower concentrations, add 0.7 or 0.8 g of agarose, respectively, to 100 ml of 1× TBE. The lower-concentration gels will also run a little faster; keep an eye on the marker dye, and stop it when the slow marker dye has moved 2.5 to 3.5 cm.

9. The book *DNA Science* by Micklos and Freyer (order from Carolina Biological Supply Co.) has a wonderful two-page spread of bad gels and their causes. The company Fotodyne has a similar display in the back of their education products catalog.

10. If you like, talk to the advanced-mathematics teacher about using semilog paper to plot the relationship between fragment migration and fragment size in base pairs (see *Introduction to Gel Electrophoresis Laboratory Activities*).

11. In a 3-h laboratory period, students can complete the activities for day 1 and day 2 and leave their gels running. You can shut the gels off yourself later, or, if you teach at the college level, you can have students come back at an appointed time to shut them off. Students can return to the laboratory later to stain their gels and leave them in a small amount of water overnight to finish destaining. They can record their data the next day, or you can store their gels until the following laboratory period.

Procedure

1. Review and practice as necessary. Show students the restriction map of lambda in chapter 12. Have them predict the relative positions of the bands as they might appear in a gel.

2. Have students study the procedure for the day at hand.

3. Students should follow the procedure carefully, keeping the enzymes on ice at all times.

4. After day 1 activities, you will need to put their digests in the freezer. Digests should incubate for 1 to several hours at 37°C.

5. After day 2 activities, you may need to stop the electrophoresis by turning off the power supplies (unless you run the gels for 24 h at 10 to 12 V). You may leave the gels in their boxes at room temperature overnight after turning off the power.

6. After the activity is completed, students should record their data (either by the graph paper method or a photograph) and answer questions.

Answers to Student Questions

1. The control tube is to show that there are no restriction fragments if the DNA is not treated with an enzyme.

2. The extra microliter of water in the control tube is to make up for not adding 1 μl of enzyme. By adding an extra microliter of water, the final volume in all of the tubes is the same.

3. Only five *Eco*RI fragments are seen (in the ideal student gel) because the 5,804-base-pair fragment and the 5,643-base-pair fragment did not resolve under these electrophoresis conditions. If *Bam*HI was used, the 6,770- and 6,527-base-pair fragments did not resolve, and the 5,626- and 5,505-base-pair fragments did not resolve.

To resolve these pairs of fragments, you could run the gel much longer (this might require casting a larger gel) or cast a lower-percentage agarose gel that more efficiently resolves large fragments (see *Introduction to Gel Electrophoresis Laboratory Activities*).

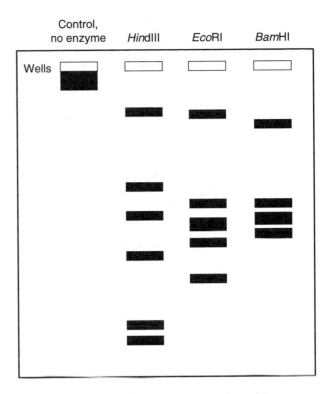

Predicted student gel. The 125-base-pair *Hind*III fragment will not be visible; the 564-base-pair *Hind*III fragment probably will not be visible and is not shown.

Predicted student gel, with fragment sizes in base pairs.

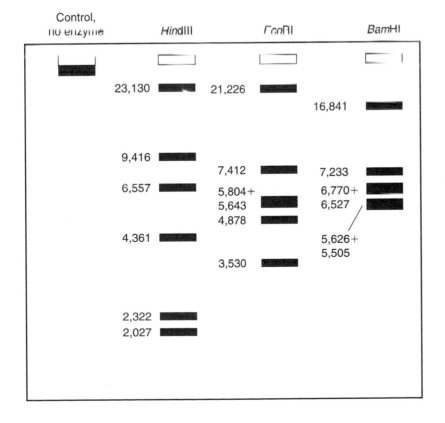

Restriction Analysis of Lambda DNA

13

Introduction

In this activity, you will be carrying out a series of steps that molecular biologists and biotechnologists use very frequently in their work. You will digest a DNA sample with different restriction enzymes, separate the resulting fragments by electrophoresis through an agarose gel, stain the fragments so that you can see them, and analyze the products of the digest. This whole procedure is called restriction analysis.

The procedure has been broken into several parts: setting up the restriction digests, preparing and loading the agarose gels, staining the gels, and analyzing the data. If you were in a laboratory all day (as scientists are), you could carry out the analysis in a single day. As it is, you will carry out a different part of the procedure each day for several days.

The DNA you will digest is from bacteriophage lambda. Lambda is a virus that infects *Escherichia coli* and destroys it. Lambda DNA contains about 48,500 base pairs. You will cut lambda DNA with the enzymes *Eco*RI, *Hin*dIII, and either *Pst*I or *Bam*HI. These enzymes will cut the DNA into a number of different-size pieces called restriction fragments.

You will separate the fragments by agarose gel electrophoresis. The mixture of fragments is loaded into a well in the gel, and then an electric current is applied. Because DNA is negatively charged, the fragments will migrate toward the positive electrode in the gel box. The shorter the fragment, the faster it can

progress through the agarose. When you stop the electrophoresis (by turning off the current), the smallest fragment will be closest to the positive electrode, and the largest will be furthest away. The intervening fragments will be sorted by their sizes. You will stain the DNA so that you can see the fragments and analyze your results by comparing them to a restriction map of lambda.

Day 1: Restriction Digests

Each laboratory team will need

- One micropipette (0.5 to 10 or 1 to 20 μl)
- One box of sterile tips for the micropipette
- Four sterile microcentrifuge tubes
- Waterproof marking pen or some other way to label the tubes
- One holder or rack for microcentrifuge tubes

Obtain from your teacher microcentrifuge tubes containing

- 10× restriction buffer
- Uncut lambda DNA
- Sterile water

1. Label your four sterile microcentrifuge tubes *Eco*RI, *Hin*dIII, *Pst*I or *Bam*HI (depending on the enzyme), and Control. Set up the restriction digests according to the following instructions. Use Table 13.1 below to check off each component of the reaction as you add it to each tube.

Table 13.1 Guide for setting up restriction digests

| Component | Amount (μl) in tube with: | | | |
	*Eco*RI	*Hin*dIII	*Bam*HI or *Pst*I	Control
Water	4	4	4	5
10× buffer	1	1	1	1
Uncut DNA	4	4	4	4
Enzyme	1	1	1	None

2. Set your micropipette to 4 μl, and carefully add 4 μl of sterile water (from the microcentrifuge tube you got from your teacher) to the *Eco*RI, the *Hin*dIII, and the *Pst*I or *Bam*HI tubes. When you add a droplet to the tube, touch the pipette tip to the side of the tube to deposit a small bead on the inside of the tube.

3. Change the micropipette to 5 μl, and add 5 μl of sterile water to the Control tube.

4. Change the micropipette to 1 μl, and get a fresh tip. Carefully add 1 μl of 10× restriction buffer to each of the four tubes, checking off in Table 13.1 as you go.

5. Change the micropipette to 4 μl, and get a fresh tip. Carefully add 4 μl of uncut lambda DNA to each of the four tubes, checking off as you go.

6. Your digests are ready for the enzymes. Your teacher has the enzymes on ice. Set your micropipette to 1 μl. When you are ready to add the enzymes, put on a fresh tip, and immediately and carefully add 1 μl of *Eco*RI to the *Eco*RI tube. Close the tube.

7. Change to a fresh tip. Immediately and carefully add 1 μl of *Hin*dIII to the *Hin*dIII tube. Close the tube.

8. Change to a fresh tip. Immediately and carefully add 1 μl of *Pst*I or *Bam*HI to the appropriate tube. Close the tube.

9. Mix the reagents in the closed tubes either by microcentrifuging them for a few seconds or by tapping the bottoms of the tubes gently on the top of your desk. Tap the bottoms of the tubes gently with your finger to ensure good mixing.

10. Place the tightly closed tubes in a 37°C water bath or dry incubator. Incubate for 1 to several hours. Your teacher will put the tubes in the freezer for you.

Day 2: Gel Electrophoresis

Each laboratory team will need

- 30 to 50 ml of agarose solution
- Gel-casting tray
- Gel electrophoresis chamber
- Electrophoresis power supply (one per two groups)
- Micropipette (0.5 to 10 μl)
- Sterile tips for micropipette
- Loading dye
- Precut lambda DNA standard
- 1× TBE (Tris-borate-EDTA [ethylenediaminetetraacetic acid]) buffer

1. Remove your frozen digests from the freezer, and let them thaw while you prepare the gel.

2. Cast the gel.
   - Seal the ends of the tray with masking tape. Place the comb at one end of the tray, making sure it does *not* touch the bottom of the tray but is close to it.
   - Pour enough agarose into the gel tray to cover the lower third of the comb. Allow the agarose to cool (it will become whitish and opaque).
   - Remove the tape from the gel tray without damaging the ends of the gel. Do not remove the comb at this point. Place the gel tray into the electrophoresis chamber with the comb nearest the *negative* electrode end (black leads) of the chamber.

3. Fill the electrophoresis chamber with 1× TBE buffer. The buffer must completely cover the gel.

4. Carefully remove the comb from the gel. The holes left in the gel are the wells that you will fill with your restriction digests.

5. Set your micropipette to 1.5 μl. Add 1.5 μl of loading dye to each of your four tubes. Mix the contents by spinning the tubes briefly or by tapping them on your desk.

6. Load each sample into a separate well. *Be sure to draw a diagram of the gel and label which well contains which sample.* Also load one lane with the precut standard (obtain this from your teacher). Record its position.

7. Plug in the electrophoresis chamber to the power supply by connecting the red (positive) lead to the red electrode and the black (negative) lead to the black electrode. Be sure that the end of the gel with the DNA is connected to the negative pole of the power supply (even if you have to do the colors backwards).

8. Turn on the current to the level indicated by your teacher. Allow the slow marker dye to migrate 2.5 to 3.5 cm into the gel. Depending on the voltage applied to the gel, this could take as little as 40 min (130 V) or as much as 24 h (10 to 12 V).

Day 3: Staining

Each laboratory team will need

- One gel-staining tray
- Methylene blue solution

1. Carefully remove your gel from the electrophoresis chamber, and place it in the staining tray.

2. Add enough methylene blue solution to the tray to just cover your gel. Soak the gel in the stain for the amount of time indicated by your teacher.

3. Using a funnel, return as much as possible of the stain solution to the stain solution container. Rinse the gel carefully with tap water.

4. Cover the gel with fresh tap water, and soak it for 3 to 5 min. Pour off the tap water.

5. Repeat step 4 until you can see the stained DNA bands clearly or until you run out of time. You will probably need to soak your gel overnight in a volume of tap water sufficient to cover it.

Day 4: Data Analysis

Each laboratory team will need either photography equipment or centimeter-ruled graph paper, plastic wrap, and a needle or pin.

Make a permanent record of your data in one of two ways.

1. Photograph the gel with Polaroid film, and fasten the picture in your laboratory notebook.
2. Use the graph paper method to make an accurate reproduction of your gel.

If you photograph the gel, follow your teacher's instructions exactly. If you use the graph paper method,

1. Tape a piece of centimeter-ruled graph paper to your desk.

2. Cover the graph paper smoothly with plastic wrap. Tape down the plastic wrap.

3. Lay your gel on the plastic wrap, and line up the leading edge of the wells with a line on the graph paper.

4. Use the needle or pin to pierce through the gel at the center of the leading edge of each band. Also pierce through the leading edge of each well you used. The goal is to generate a pattern of pinpricks on the graph paper that will mark exactly the locations of the bands in the gel.

5. Remove the gel and plastic wrap. Using the pinpricks as a guide, draw the band patterns on the graph paper. Keep the graph paper in your laboratory notebook as a permanent record of your data.

Look at the restriction map of lambda. List in descending order (from largest to smallest) the sizes of the DNA fragments that should be generated in the following digests: *Eco*RI, *Hin*dIII, *Bam*HI (do only if you used *Bam*HI). Now sketch the pattern of bands from your gel. Label each gel lane with the name of the enzyme used in that digest. Compare the pattern of bands in your *Eco*RI lane with the expected set of fragments you just listed. Starting with the largest band in your gel, label each band with the size of the fragment in base pairs. Remember that two fragments very similar in size may not resolve and that very small fragments may have run off the gel or may not be detectable. Do the same for the *Hin*dIII fragments and the *Bam*HI fragments, if you used that enzyme. Discuss your band assignments with your laboratory teammates and your teacher.

When you are satisfied that you have made the best possible assignment of sizes to the bands, resketch your diagram with the size labels and put it in your laboratory notebook with the permanent record of your gel.

Questions

1. What is the purpose of the control tube?

2. Why did the control tube receive 5 µl of water instead of 4 µl?

3. The lambda restriction map from chapter 12 indicates that six *Eco*RI fragments should be generated, yet you probably saw only five in your gel. Why? If you used *Bam*HI, you expected six fragments, yet you probably saw only four. Why? Can you suggest a way to see all of the expected fragments?

4. If you have not already done so, answer the questions from chapter 12.

Recombinant Paper Plasmids 14

About This Activity

This activity introduces students to some basic recombinant DNA techniques. Paper models are used to demonstrate how plasmid DNA is cut with restriction enzymes and then recombined to create a recombinant DNA molecule. Students should already be familiar with plasmids from the activity *Sizes of the* Escherichia coli *and Human Genomes* and from class discussion. They should also have done *DNA Scissors* as an introduction to restriction enzymes. This activity is very effective when used in combination with a laboratory on bacterial transformation, such as that in chapter 19. If students have completed *DNA Goes to the Races* or are comfortable with electrophoresis, *Restriction Analysis Challenge Worksheets* (chapter 15) makes a challenging follow-up assignment.

Class periods required: 1

Introduction

Recombinant DNA is simply a DNA molecule that has been assembled from pieces taken from more than one source of DNA. Making recombinant DNA became possible with the discovery of restriction enzymes and DNA ligase. Restriction enzymes are used to cut DNA into reproducible fragments, and ligase is used to form new phosphodiester bonds between them. When phosphodiester bonds are formed between DNA pieces from different sources, a recombinant DNA molecule has been created.

The most important application of recombinant DNA technology is gene cloning. To clone a gene, a fragment of DNA containing the gene must be isolated. The gene-containing DNA fragment is then added to a vector DNA molecule that will be replicated inside a cell. The most common type of vector is a plasmid, although other vectors are also used.

Plasmids are (relatively) small circular DNA molecules found in many bacteria and some yeasts. They are du-

plicated by the host cell's DNA replication enzymes and transmitted to daughter cells during cell division. Some plasmids maintain a high copy number within their hosts, with 50 or more copies per cell. Others may have a copy number as low as 1 per cell. Plasmid copy number is a characteristic of the plasmid itself.

In order to be duplicated by the host's replication enzymes, plasmids must contain an origin of replication (see the introduction to *DNA Replication* [chapter 6]). An origin of replication is a base sequence within a DNA molecule at which the DNA replication process begins. Most plasmids also contain one or a few genes whose products play a role in plasmid DNA replication. In addition, many plasmids contain genes that are useful to the host cell under certain environmental conditions. For example, some plasmids contain genes for toxic products that kill competing bacteria. These plasmids also contain genes that make their hosts immune to the toxin. However, the most familiar plasmid-borne genes are probably antibiotic resistance genes.

Antibiotics are natural substances (or manmade copies and derivatives of natural substances) that kill or suppress the growth of microorganisms. Antibiotics are also made by microorganisms in nature. Production of an antibiotic may confer a survival advantage on the producer by allowing it to kill nearby microbes that would normally compete for nutrients.

Since antibiotics arose in nature, it is not surprising that the capacity to resist them also arose naturally. The antibiotic resistance genes found on plasmids are presumed to have evolved in response to the presence of antibiotics in nature. This natural system of antibiotic and resistance gene has provided molecular biologists with a powerful tool for laboratory applications. One example is illustrated below.

After a DNA fragment has been ligated into a plasmid, the recombinant plasmid is introduced into a host cell. The host cell replicates the recombinant plasmid DNA as it divides. Plasmid DNA is often introduced into bacterial host cells by transformation (see *Transformation of* Escherichia coli [chapter 19]). In a typical transformation, billions of bacteria are treated and exposed to the plasmid DNA. Only a fraction (usually fewer than 1 in 1,000) will acquire the plasmid. Antibiotic resistance genes provide a means of finding these transformants among all the nontransformed cells.

After transformation by plasmids containing antibiotic resistance genes, bacteria that acquired the plasmid (transformants) can be detected by plating on media containing the antibiotic(s). Only bacteria expressing the new antibiotic resistance genes can form colonies on the antibiotic plates. The only bacteria in the mixture of transformed and nontransformed cells that have the antibiotic resistance genes are those that acquired the plasmid: the transformants. This method of selecting transformants allows scientists to quite easily find the 1 cell in 1,000 that has acquired the plasmid.

Because of the ease with which scientists can find cells transformed with plasmids containing antibiotic resistance genes, scientists usually clone their genes of interest into these plasmids. The gene of interest may not change the bacteria in any way that a scientist can readily detect, so it is impossible to tell whether a bacterium has been transformed with the gene. However, if the gene of interest is in a plasmid encoding antibiotic resistance, the scientist need only select those bacteria that become antibiotic resistant after transformation. Those bacteria should also contain the gene.

Students usually ask how you can be sure you will get the product you want from a ligation and not alternative products they see it is possible to form or why the "right" product forms instead of the "wrong" product. (They will undoubtedly form many alternative products with their paper plasmids.) Be sure that they understand that in any ligation, all possible products will be formed. The ligase enzyme is "blind"; it forms phosphodiester bonds between DNA molecules without regard to the base sequence of those molecules. So after a ligation, you will have a mixture of products in the test tube. How can you separate and identify them? Scientists usually use transformation.

The key to understanding how transformation can help identify ligation products is to realize that transformation is a very inefficient process. Very few DNA molecules actually get into cells, so essentially only one molecule gets into a given transformed cell. That cell now contains an isolated ligation product. If a batch of competent cells is transformed with a ligation reaction mixture, representatives of all the products will likely enter bacterial cells. Those products that contain replication origins will be replicated by their new hosts and transmitted to offspring.

So if a scientist performs a ligation with a vector containing an ampicillin resistance gene, transforms the ligated DNA into ampicillin-sensitive cells, and selects for transformants by plating on ampicillin-containing medium, she will get a collection of cells that contain a plasmid with a replication origin, an ampicillin resistance gene, and possibly other components, depending on the ligation reaction. In many cases, therefore, a scientist is faced with a collection of transformants which may or may not contain the plasmid she desires.

To determine whether a given transformant contains the desired recombinant plasmid (as opposed to an alternative ligation product), the scientist often must prepare plasmid DNA from that transformant. The plasmid DNA is then tested (by restriction analysis, DNA sequencing, or other means) to determine whether it has the desired structure.

In this activity, students will assemble plasmids carrying genes for ampicillin and kanamycin resistance. Starting with two parent plasmids, one encoding ampicillin resistance and the other encoding kanamycin resistance, students will simulate digestion of each plasmid with *Bam*HI and *Hin*dIII. Next, they will ligate pieces together to create a recombinant plasmid containing both resistance genes.

It is very effective to follow this activity with the wet laboratory *Transformation of* Escherichia coli. The EZ Gene Splicer kit available from Carolina Biological Supply Company (catalog no. 21-1162) lets students perform the ligation and transformation modeled in this activity.

Objectives

At the end of this activity, students should be able to

1. Describe the activity of DNA ligase;
2. Define recombinant DNA;
3. Describe how to make a recombinant plasmid from two starting plasmids.

Materials

- *Student Activity* pages
- Scissors, paste, Scotch tape

Preparation

Make photocopies of the *Student Activity* pages if necessary.

Tips

As students use the paper models, remind them that DNA is three-dimensional and has no "front" and "back"; nor does it matter if the letters are upside down. The thing that does matter is complementarity of base pairs and the correct antiparallel orientation of the strands.

Procedure

1. Have students read the background information and/or discuss the background material.

2. Have them read the instructions, and make sure they understand them. You may want to review the activities of restriction enzymes and DNA ligase.

3. Perform the indicated activities.

4. Answer questions. These questions provide a good lead-in to the transformation laboratory.

Answers to Student Questions

1. Please refer to the diagram. The restriction fragments (shown underneath the plasmid diagrams) are labeled. There are six possible combinations. (Remember that DNA is three-dimensional, and it doesn't matter if the letters in the paper model are upside down.) If the two fragments of pAMP are called 1A and 1B (see diagram), and the two from pKAN are 2A and 2B (see diagram), then the combinations are 1A + 1B, 1A + 2B, 2A + 1B, 2A + 2B, 1A + 2A, and 1B + 2B. You could detect only four of these: 1A + 2A, 1A +1B, 1A + 2B, and 2A + 2B. The combination 1B + 2B has no selectable marker (antibiotic resistance), and 2A + 1B has no origin of replication. Both of these molecules could form and be transformed into bacteria, but 1B + 2B could not be detected on antibiotic media. The 2A + 1B molecule would not be copied once it entered the cell and would not be transmitted to the daughter cells that form the colony.

2. The essential elements of the experiment are as follows. Make two samples of bacteria from one culture. Transform one of the samples with the DNA mixture (experimental). Take the other sample through the transformation procedure,

Diagram of the paper plasmids and their restriction products.

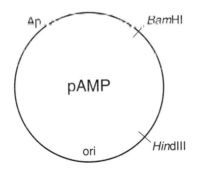

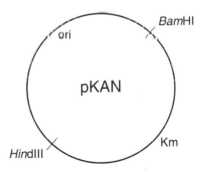

Both plasmids are digested with *Bam*HI and *Hind*III, producing the following fragments:

pAMP

H   ori   Ap   B (1A)

H   B (1B)

pKAN

(2A) B   Km   H

(2B) B   ori   H

Ap: ampicillin resistance
Km: kanamycin resistance
ori: origin of replication
B: *Bam*HI end
H: *Hind*III end

but do not add DNA (control). Plate some bacteria from the experimental and control samples on the following media: no antibiotics (are any of the bacteria alive?), ampicillin added (none of the control organisms should grow [this is a test of whether the ampicillin is effective]; some of the experimental organisms should form colonies); kanamycin added (similar to ampicillin plate); ampicillin and kanamycin added (only cells transformed with the ampicillin-kanamycin recombinant plasmid should grow).

3. The critical thing about these bacteria is that they will survive on one antibiotic plate but not on the other or on the two-antibiotic plate. To find the transformed bacteria, plate some of the experimental mixture on an ampicillin plate and some on a kanamycin plate. After colonies grow on the plates, transfer some of the cells from each colony to a plate containing the other antibiotic (or both antibiotics). Those that survive on one antibiotic but fail to produce a colony on plates containing the other antibiotic are the bacteria you seek.

Two of the plasmids predicted in question 1 will grow only on ampicillin; one will grow only on kanamycin.

4. The most important reason is that scientists need to be able to identify bacteria that have been transformed with the gene they desire. These bacteria may be very rare among the experimental bacteria. Associating the desired gene with an antibiotic resistance gene gives a tremendous advantage: by plating on an antibiotic medium, a scientist can select only those organisms that have been transformed with the antibiotic resistance gene and, by association, the gene of interest. Another advantage to having the antibiotic gene associated with the desired gene is that this association provides a mechanism for ensuring that the desired gene is still there. By propagating the transformed bacteria on antibiotic media, the scientist can continuously select for only those bacteria that contain the antibiotic resistance gene and, by association, the desired gene.

Recombinant Paper Plasmids

Background Reading

Some of the most important techniques used in biotechnology today involve making recombinant DNA molecules. A recombinant object has been re-assembled from parts taken from more than one source. You could make a recombinant bicycle by disassembling two bicycles and reassembling them in a new way: putting the wheels of one on the frame of the other, for example. Your genome is recombinant in that part of it came from your mother and part came from your father. Recombinant DNA molecules are pieces of DNA that have been reassembled from pieces taken from more than one source of DNA. Often, one of these DNA sources is a plasmid.

Plasmids are small, circular DNA molecules that can reside in cells. Plasmids are copied by the cell's DNA replication enzymes because they contain a special sequence of DNA bases called an **origin of replication.** DNA replication enzymes assemble at this special sequence to begin synthesizing a new DNA molecule. As you might expect, bacterial and other chromosomes have replication origins, too. Replication origins are essential to heredity; if a DNA molecule does not have a replication origin, it cannot be copied by the cell and will not be transmitted to future generations.

Plasmids often contain genes for resistance to antibiotics. Antibiotics are natural substances produced mostly by soil microorganisms. Antibiotic production allows these microorganisms to kill off competing microbes. Antibiotic resistance is also a natural phenomenon; at the very least, the antibiotic producers must be resistant to the antibiotics they make! We will be working with genes for resistance to the antibiotics ampicillin and kanamycin.

In this activity, we will assemble plasmids carrying genes for ampicillin and kanamycin resistance and then recombine the two plasmids. We call the plasmid with ampicillin resistance pAMP, the plasmid with kanamycin resistance pKAN, and the recombinant plasmid pAMP/KAN. We will use paper plasmid DNA models to go through the process that scientists use when making recombinant DNA. Scissors will substitute for restriction enzymes. The enzyme DNA ligase, which forms phosphodiester bonds between pieces of DNA, is represented by Scotch tape. Our result will be a model of a recombinant DNA molecule. Scientists place real recombinant plasmids back into bacteria where they multiply. The bacteria also multiply, making millions of copies of the recombinant DNA molecule and the proteins it encodes.

Construction of the pAMP and pKAN Plasmids

Locate the three strips of DNA code on the worksheet (Appendix A) marked "Paper pAMP plasmid model." On each strip, the two rows of letters indicate the nucleotide bases, and the solid horizontal lines indicate the sugar-phosphate backbone of the DNA molecule. The hydrogen bonds between the base pairs are located in the white space between the rows of letters.

1. Use your scissors to cut carefully along the *solid* lines. Cut out each strip, leaving the solid lines intact. Make a vertical cut to connect the open end of the box formed by the solid lines. This cut will remove the 5′ and 3′ labels from the strip.

2. After all three strips are cut out, glue or tape the "1" end to the "paste 1" area, covering the vertical lines. Connect "2" to "paste 2" and "3" to "paste 3" until you complete the circular model of the pAMP plasmid.

3. Using the page (Appendix A) marked "Paper pKAN plasmid model," cut out and paste together a plasmid containing a kanamycin resistance gene. The procedure is exactly the same as for the pAMP plasmid.

Constructing a Recombinant pAMP/KAN Plasmid

You have now prepared a pAMP plasmid and a pKAN plasmid. In this part of the exercise, you will use them as starting materials to make a recombinant plasmid. You will cut pAMP and pKAN with two specific enzymes, *Bam*HI, and *Hin*dIII. You will ligate together fragments that come from each plasmid, creating a pAMP/KAN plasmid.

1. First, simulate the activity of the restriction enzyme *Bam*HI. Reading from 5′ to 3′ (left to right) along the top row of your pAMP plasmid, find the base sequence GGATCC. This is the *Bam*HI restriction site. Notice that it is a palindrome. Cut through the sugar-phosphate backbone between the two G's, stopping at the center of the white area containing the hydrogen bonds (don't cut all the way through the paper). Do the same on the opposite strand. Cut through the hydrogen bonds between the two cut sites, and open the plasmid into a strip. Each end of the strip should have a single-stranded protrusion with the sequence 3′ CTAG 5′. Mark the ends of the strip *Bam*HI.

2. Next, simulate the activity of the restriction enzyme *Hin*dIII. Reading from 5′ to 3′ (left to right) along the top row of the pAMP strip, find the sequence AAGCTT. This is the *Hin*dIII restriction site. It is also a palindrome. Cut the sugar-phosphate backbone between the two A's, stopping at the center of the white space containing the hydrogen bonds. Repeat the cut between the two A's on the opposite strand of the restriction site. Cut the intervening hydrogen bonds. This time you should have two pieces with single-stranded protrusions. One protrusion on each piece is the *Bam*HI end; the other should have the sequence 3′ TCGA 5′. Mark each of these two new ends *Hin*dIII. You now have two pieces of pAMP DNA. Set them aside.

3. Using your pKAN plasmid model, repeat steps 1 and 2. The pKAN plasmid is now in two pieces with labeled ends. Including the two pieces of pAMP, you now have four pieces of plasmid.

4. Take the piece of plasmid with the ampicillin resistance gene in it, and connect it to the piece containing the kanamycin resistance gene by using DNA ligase (Scotch tape). Be sure that complementary bases are paired where you make the ligations. Notice that the *Bam*HI end will not pair with the *Hin*dIII end but will pair with another *Bam*HI end. Likewise, the *Hin*dIII end must pair with another *Hin*dIII end.

Remember that DNA is three-dimensional. In our model, the letters representing the bases can look upside down, but in real DNA, it doesn't matter. So in this paper simulation, the letters representing the bases do not need to be right side up. All that matters is that the 5′-to-3′ directions match within a strand and that the base pairs are correct.

You have now created a recombinant pAMP/KAN plasmid.

Transformation

It is possible to introduce plasmids into bacterial cells through the process of **transformation.** Bacteria that can be transformed (can take up DNA) are called competent. Some bacteria are naturally competent. Others can be made competent by chemical and physical treatment. After the bacteria absorb the plasmid DNA, they express the new antibiotic resistance genes as instructed by the new DNA. Bacteria that express new proteins in this way are said to be transformed. New copies of the plasmid are synthesized by the cell's DNA replication enzymes and passed to daughter cells as the bacteria multiply. Because many identical copies of the new genes are generated in this process, you are said to have **cloned** the genes.

After transformation by plasmids containing antibiotic resistance genes, transformed bacteria can be detected by plating on media containing the antibiotic(s). Only bacteria expressing the new antibiotic resistance genes (the transformed bacteria) can form colonies on the antibiotic plates. This method of selecting transformants (because they are the only ones that can grow on the media) is a big advantage, because transformation is usually very inefficient. In a typical experiment, less than 1 cell in 1,000 will be transformed.

Questions

1. A *Bam*HI cut site will ligate only to the matching end of another *Bam*HI site. The same is true for *Hin*dIII cut sites and most other restriction enzyme cut sites. If you use two fragments at a time, how many possible different combinations could be formed from the four fragments you made in exercise 3? How many of these could you detect in colonies of transformants on antibiotic-containing media?

2. Assume you did the activity with real DNA and attempted to transform bacteria with your new pAMP/KAN plasmid. Describe an experimental procedure for growing the transformed bacteria on plates to find out if the bacteria have actually been transformed by pAMP/KAN. Include controls that would tell you whether any of the experimental bacteria are alive and whether the antibiotics are effective.

3. Referring to question 2, describe an experimental procedure for finding experimental bacteria that were transformed by the other DNA molecules that formed during the ligation step (you described these molecules in question 1).

4. Scientists often combine an antibiotic resistance gene with whatever gene they are trying to clone. The desired gene is then associated with the antibiotic resistance gene. Any bacterium that contains the desired gene is then resistant to an antibiotic. In most cases, the scientists have no use for the protein that destroys the antibiotic. Why would scientists combine genes in this way?

Restriction Analysis Challenge Worksheets

15

About These Worksheets

The restriction analysis challenge worksheets contain three problems that illustrate actual uses of restriction analysis in the laboratory. The problems, particularly the second and third ones, require multistep analysis. They are most appropriate for advanced classes or for students who enjoy analytical puzzles. Students must be familiar with restriction enzymes, construction of recombinant DNA molecules with DNA ligase, and gel electrophoresis. The required background is found in *DNA Scissors, Recombinant Paper Plasmids,* and *DNA Goes to the Races.*

These worksheets require students to apply their knowledge of construction of recombinant DNA molecules and restriction analysis to three problems like those encountered every day in research laboratories. The first two problems involve analysis of the molecules produced during a ligation, and the third one requires students to generate a restriction map from restriction analysis data.

Students may require assistance getting started with the first two problems, because the problems require that they integrate and use information from several lessons. You may find it helpful to do the first problem together as a class and then let students work on the remaining problems individually or in small groups.

Discussion

When a scientist sets out to make a recombinant molecule, he must plan how he will recognize that molecule when he gets it. He usually looks at the structure of the DNA molecule he wants to make and the structures of other products that will be present in the ligation (such as the starting plasmid) and then designs a restriction analysis that will let him identify the molecule he wants. Students will be going through this same process as they do challenges I and II.

In challenge I, the described ligation will produce a regenerated vector (where the ligase has simply re-

sealed the *Eco*RI site in the plasmid vector) as well as the desired recombinant molecule. The *Eco*RI fragment can be inserted in either of two orientations, but in this problem, there is no way to distinguish between the two. It is also possible that more than one *Eco*RI fragment could be inserted into the same vector molecule. In reality, the resealed vector would be the most common product, followed by recombinant plasmids containing one insert. In the laboratory, it is usually uncommon to get plasmids with multiple copies of an insert.

Students may want to know how it is possible to analyze individual products of a ligation when the reaction is done in a single tube. You can explain to them that the process of transformation separates the ligation products (see the introduction to chapter 14). Transformation efficiency is very low: about 1 in 1,000 to 1 in 1,000,000 cells may be transformed in a given experiment. Each transformed cell then contains one plasmid molecule, which can replicate into many copies and is propagated as the cell divides. A transformed cell divides to make a colony on the appropriate antibiotic medium, and every cell in that colony contains a plasmid that is a copy of one of the original product molecules. By analyzing the plasmids present in many different colonies of transformants, we can look individually at the different products of one ligation reaction. Scientists often analyze the plasmids from many transformants when they are looking for a single product from a ligation that could produce many different products.

Challenge II is an amplification of the ideas introduced in challenge I. Here the vector is ligated in the presence of two different inserts: the one that is to be removed and the one that is desired. It is possible that either insert could be ligated into the vector (in either orientation), and it is also possible that the vector will simply be sealed without an insert. Notice that the reaction products to be drawn are restricted to those with a single replication origin (one plasmid molecule) and are less than 7,000 base pairs (bp) in size.

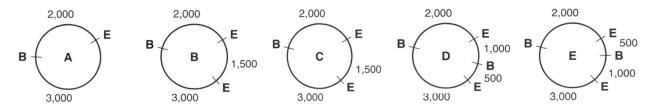

Diagram 15.1 Challenge II: possible products with one replication origin and a total size of less than 7,000 bp.

Use this problem to help your students see that constructing a recombinant molecule can be a complicated task and that it requires the use of analytical skills to plan for and identify the various products. These exercises should let your students do some analytical thinking and help them see how restriction analysis is used in real-life problems.

Challenge III should stand on its own. There is no magic method for solving the third problem; students simply have to work with it. If your students enjoy figuring out the restriction map, challenge them to create similar problems for their classmates. They must imagine a DNA molecule with a restriction map and generate the data (have them draw a gel!). Let their classmates construct the maps from the data. If there are problems or ambiguities, the other students will find them. The exercise of imagining a restriction map and drawing the data from it will reinforce all the lessons about electrophoresis.

Answers to Challenges

I. The plasmid vector can be distinguished from the recombinant molecule with either an EcoRI digest or a BamHI digest. The vector produces a single 5,000-bp EcoRI fragment, while the recombinant produces that fragment plus the 1,500-bp insert fragment. The vector produces a single 5,000-bp BamHI fragment, while the recombinant produces a single 6,500-bp fragment. There is no way to distinguish the two possible orientations of the insert with the information given. If students have trouble, have them draw and compare the structures of the starting vector and the

recombinant plasmid. Check their gel drawings to make sure they have their fragments labeled properly and drawn in a reasonable manner.

IIA. There are five potential products, given the constraints of the problem (see Diagram 15.1). The two plasmids with the original insert in different orientations (plasmids B and C) are indistinguishable, given the available information.

IIB. The best way to distinguish the products, given the available information, is to perform a digest with BamHI. Product A (regenerated vector with no insert) will give a single 5,000-bp fragment. Products B and C (vector with original insert in either orientation) will give a single 6,500-bp fragment. Product D will give two fragments of 3,500 and 3,000 bp. Product E will give two fragments of 4,000 and 2,500 bp. An EcoRI digest will not distinguish between the old and new inserts, since both are 1,500 bp long. An EcoRI-BamHI digest will not distinguish between products D and E. Each will give fragments of 2,000, 3,000, 500, and 1,000 bp.

IIC. If students have good answers for IIB, it should be easy for them to draw the gel. Check their gels to make sure they have the fragments correctly labeled and drawn properly with respect to the size scale.

III. See Diagram 15.2. Either orientation is correct. E, EcoRI site; B, BamHI site. The size scale is in kilobase pairs (1,000 bp).

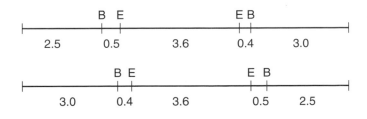

Diagram 15.2 Challenge III: restriction map of the linear DNA fragment.

Restriction Analysis Challenge

The combination of restriction digestion and gel electrophoresis is often called restriction analysis, since the information obtained from these procedures can be used to analyze the structure of a DNA molecule. Restriction analysis is especially important in checking the structure of recombinant DNA molecules and in analyzing an unknown DNA. The following examples show you some typical restriction analysis problems encountered by scientists in the laboratory.

Restriction Analysis Challenge I

You wish to insert a 1,500-base-pair (bp) *Eco*RI restriction fragment into the plasmid vector shown in Fig. 15.1. You digest the plasmid DNA with *Eco*RI, stop the digest, add the 1,500-bp fragment, and ligate. You know that this procedure will give you a mixture of regenerated starting plasmid and the recombinant

molecule you desire. Outline a restriction analysis procedure you could use to distinguish between regenerated vector and the desired product. State the predicted fragment sizes. Use the gel outline provided to show the predicted products from your analysis (one product type per lane). Label each lane with the DNA molecule type (vector or recombinant) and the expected fragment sizes.

Restriction Analysis Challenge II

You wish to remove the 1,500-bp *Eco*RI fragment from the starting plasmid shown in Fig. 15.2 and replace it with a *different* 1,500-bp *Eco*RI fragment (also shown). You digest the starting plasmid with *Eco*RI, stop the digest, add the new fragment to the mixture, and religate.

Figure 15.1 Information and gel outline for restriction analysis challenge I.

Gel outline

Starting plasmid vector

Distance between restriction sites is shown in base pairs.

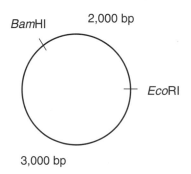

Fragment to be inserted into vector

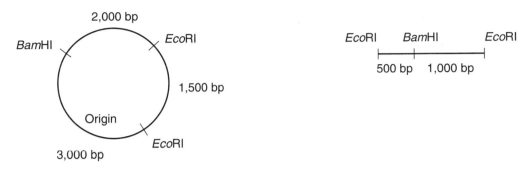

Starting plasmid   ***Eco*RI fragment to be inserted**

Distances between restriction sites are shown in base pairs.

Figure 15.2   Information for restriction analysis challenge II.

A. Draw the products you could generate that would have *one replication origin* and be *less than 7,000 bp in size*. (Hint: what are all the fragments present in the ligation reaction?) Label the products A, B, etc., and indicate distances between restriction sites.

B. Design a restriction analysis procedure that will let you identify as many of your proposed products as possible after a single digest. You may use more than one restriction enzyme in the digest. State the predicted fragment sizes for each one of your products from IIA.

C. Use the gel outline in Fig. 15.3 to draw the predicted gel pattern from your proposed analysis of each of the products shown in IIA. Label the gel lanes with the product molecule (A, B, etc.). Label the bands with fragment sizes. (A size scale has been provided.)

Figure 15.3   Gel outline for restriction analysis challenge IIC.

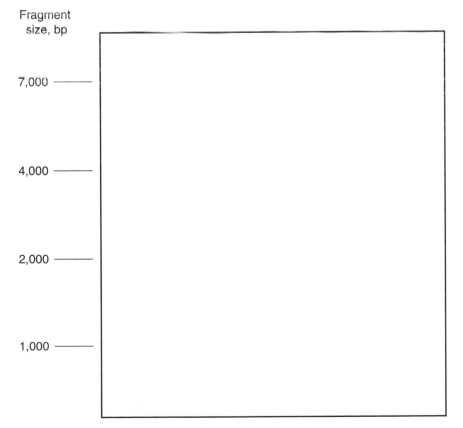

Fragment size, bp

7,000 —

4,000 —

2,000 —

1,000 —

Restriction Analysis Challenge III

When scientists study a new piece of DNA, often one of the first things they do is put together a restriction map of that piece. The restriction map is then used as a road map when they study the genes encoded by the DNA. How is a restriction map put together? Digests with single enzymes and combinations of enzymes are performed. Observing the sizes of fragments produced, scientists determine restriction site locations that would give them the patterns they see. Here is one such puzzle to try for yourself.

A *linear* (not circular) piece of DNA is digested with the restriction endonuclease *Eco*RI and gives frag-ments of 3,000, 3,600, and 3,400 bp. When digested with *Bam*HI, the DNA molecule gives fragments of 4,500, 3,000, and 2,500 bp. A double digest with *Eco*RI and *Bam*HI gives fragments of 2,500, 500, 3,600, 3,000, and 400 bp.

Draw a restriction map of the starting piece of DNA, showing the locations of the *Eco*RI and *Bam*HI restriction sites. Indicate the distances between the restriction sites in base pairs, and label the sites *Eco*RI or *Bam*HI as appropriate. Also indicate the distance from each end of the starting piece to the nearest restriction site.

Classroom Activities

Detection of Specific DNA Sequences: Hybridization Analysis

16

About These Activities

The activities in this lesson plan use paper models to illustrate basic concepts of hybridization analysis. Hybridization is a technique that takes advantage of the specificity of DNA base pairing for the detection of specific DNA sequences in a mixed sample. It is one of the fundamental methods of analysis of DNA. The first activity (*Fishing for DNA*) requires only that students be familiar with the structure of DNA and the base-pairing rules; it is suitable for young students. The second and third activities assume that students are familiar with restriction enzymes and the process of electrophoresis. If you plan to use the subsequent lessons on DNA sequencing, the polymerase chain reaction, or the analysis of human DNA, it is important to introduce your class to hybridization.

Class periods required: 1-2

Introduction

Restriction digestion, electrophoresis, and staining allow us to cut DNA molecules into reproducible fragments and to look at the sizes of those fragments. However, it does not give us much information about the DNA base sequences within the fragments. We know that the restriction site was present at the end of each fragment, but that is all. Often it is important to know whether a specific DNA base sequence is present in a sample and where it is located with respect to restriction sites. These kinds of questions come up in the screening of DNA libraries (see chapter 4), in testing to determine whether a certain gene has been introduced into an organism, and in analyzing the structure of a certain region of DNA (as in DNA fingerprinting). The technique most commonly used to answer these questions is hybridization analysis.

In brief, hybridization analysis involves separating the strands of (denaturing) the DNA molecules to be analyzed and then mixing those separated strands with many copies of a single-stranded DNA or RNA mole-

cule. This single-stranded DNA or RNA molecule has the complement of the base sequence of interest and is labeled for detection (often with radioactive isotopes). It is called a **probe.**

When a probe is mixed with single-stranded sample DNA under the right conditions, hydrogen bonds form between the probe and its complementary sequence in the sample DNA. The formation of hydrogen bonds between two complementary single strands to re-create a double helix is called **hybridization** or **annealing.** If the sample DNA does not contain the base sequence complementary to the probe, no probe molecules will anneal to the sample. After time for hybridization has been allowed, the sample is rinsed to remove unhybridized probe and then tested for the presence of hybridized probe (see below). Hybridized probe indicates the presence of the DNA sequence of interest.

Denaturation of DNA

To separate the two strands of a DNA molecule, the hydrogen bonds between the base pairs must be broken. Breaking these bonds can be accomplished physically with heat or chemically with base (acid works, too, but it destroys parts of the nucleotides in the process). The extremity of the conditions required to separate two strands depends on how stably the two strands are associated. Since G-C pairs are held together by three hydrogen bonds while A-T pairs are bound by only two, it takes a higher temperature or a higher pH to separate strands with more G-C pairs. The temperature required to denature a given DNA molecule is called its **melting temperature,** and the melting temperature of a particular molecule can be predicted on the basis of its G-C content. In hybridization applications, denaturing conditions are generally set so that any DNA molecule will melt (for example, 95 to 100°C). The point at which consideration of melting temperature may become important is when the probe is hybridized to the sample (see below).

Hybridization

If a DNA molecule in solution is boiled, it will denature. If the solution is then allowed to cool, the strands will reanneal, and the double helix will form again. Annealing will occur between any complementary single strands of DNA under the right conditions. No enzyme is needed. As might be imagined, short complementary molecules tend to hybridize much faster than long ones, because it is easier for them to achieve proper alignment for base pairing through random contact. To facilitate hybridization and subsequent detection of the probe, the DNA to be analyzed is commonly transferred to a membrane, usually nylon or nitrocellulose.

Probes

Theoretically, any DNA or RNA molecule can be used as a probe. Short synthetic oligonucleotides of 15 to 30 bases are often employed because they are easy to obtain (in research environments!) and because the scientist has complete control over the base sequence. Purified restriction fragments, whole linearized plasmids, or products of in vitro enzymatic DNA or RNA synthesis are also used. Probes can be radioactively labeled by any of several methods, including synthesis in the presence of radioactive nucleotides or attachment of a radioactive terminal phosphate group by the enzyme polynucleotide kinase. Nonradioactive labeling methods are also becoming increasingly popular. In these methods, the probe DNA is chemically modified with a substituent that results in a colored or luminescent product after the DNA is put through a detection reaction (usually an enzymatic one).

The critical aspect in probe selection is to be certain that the probe will hybridize only to the DNA of interest. The chance of a random occurrence of any particular DNA sequence is 1 in 4^n, where n is the number of bases in the sequence. Therefore, the chances of random hybridization decrease dramatically as the probe becomes longer. However, it is possible for two single-stranded molecules that are somewhat mismatched to hybridize to each other. Because of this possibility, the choices of both probe and hybridization conditions are important.

When two slightly mismatched molecules hybridize, the double helix they form is less stable than a perfectly base-paired helix. (Consider that there are no hydrogen bonds between mismatched bases, plus a mismatch can cause structural disturbance.) The imperfectly base-paired helix will denature at a lower temperature than that required for its perfectly paired counterpart. Thus, at certain temperatures, perfectly paired helices will form and be stable, but mismatched helices will not. By adjusting the hybridization temperature, scientists can either permit or block the mismatched annealing of a probe.

Would it ever be desirable to hybridize under conditions that permit annealing of the probe to an imperfectly matched sequence? Yes, often. When scientists "fish" for genes, one technique they often use is to employ a known gene, perhaps a yeast gene, to probe for its counterpart in a distantly related organism such as a human. These scientists do not expect the yeast gene to be a perfect match to the human gene, but they hope that the two sequences will be similar enough to permit hybridization under mild conditions. This type of approach is often successful. In other applications, hybridization may be carried out under such demanding conditions that only perfect matches will occur.

Two examples of applications of hybridization analysis, with transfer processes, are outlined below.

Screening DNA libraries

A DNA library is a collection of clones that together represent the entire genome of an organism. A plasmid library can be made by digesting the DNA of the organism with a restriction enzyme and then ligating the resulting fragments en masse into plasmid DNA molecules that were previously cleaved with the same enzyme (see chapter 4). The result is a large number of plasmids containing different restriction fragments of the organism's DNA, or a DNA library. There are many ways of making and screening libraries, but the main idea is the same. The example discussed below describes a method for screening plasmids in bacterial colonies.

To find a plasmid containing a specific DNA sequence (for example, from a particular gene), the entire library is transformed into a host such as *Escherichia coli*. The transformants are plated on agar to produce individual colonies, each of which contains one particular plasmid (with a unique insert from the organism's DNA). Next, cells from each colony are transferred to a membrane by gently pressing the membrane down onto the agar plate (most of each colony is left on the plate). In this way, the pattern of the colonies is preserved on the membrane (Fig. 16.1).

Now the cells must be lysed to expose their DNA, and their DNA must be denatured. This can be accomplished in a single step by soaking the membrane in a solution of detergent (to lyse the cell membranes) and base (to denature the DNA). The membrane is then rinsed, and the single-stranded DNA molecules may be fixed to it through heating or exposure to ultraviolet light.

The membrane with the fixed single-stranded DNA is now ready for the hybridization step. The membrane is immersed in a special hybridization solution, and radioactive single-stranded probe is added. The composition of the solution and the conditions (particularly temperature) of hybridization are varied to suit the experiment. After the hybridization period has elapsed, the membrane is rinsed repeatedly under conditions that will remove unhybridized probe but will not disrupt hydrogen bonds between any hybridized probe and sample DNA (washing conditions can also be adjusted to remove imperfectly hybridized probe).

The membrane is allowed to dry and then placed against a piece of photographic film. If radioactive probe is hybridized to the DNA from any of the colonies transferred to the membrane, the film next to that region of the membrane will be exposed and will turn dark upon developing. In this manner, any colony that contains a plasmid with DNA sequences complementary to the probe will show up as a black spot on the film. The process of exposing film by placing it next to a radioactive sample is called autoradiography, and the resulting "picture" is called an autoradiogram, or "autorad" for short (Fig. 16.1).

In the last step of the process, the autoradiogram is lined up with the original petri plate from which the colonies were transferred. The black spots on the autoradiogram (representing hybridized probe) are matched to specific colonies. The cells in these colonies should contain plasmids with the sequence of interest. To verify that they do, the indicated colonies are transferred to separate cultures. Plasmid DNA is prepared from them and usually is sequenced.

Southern blot hybridization

Another major application of hybridization analysis is the testing of the products of a restriction digest to determine which fragments, if any, contain a certain DNA sequence or to determine the size of a fragment containing a particular sequence. The procedure most commonly used for this purpose involves a process called Southern blotting that is followed by hybridization analysis. Specific applications of Southern blot hybridization can be found in the lessons on genetic diseases and DNA fingerprinting in this book.

To begin the analysis, the sample DNA is digested with restriction enzymes, and the resulting fragments are separated by agarose gel electrophoresis in the standard manner. Unfortunately, hybridization performed on DNA molecules embedded in agarose is not very effective. In 1975, a scientist named Southern published a method for transferring the fragments from an agarose gel to a membrane in a way that preserved the arrangement of the fragments in the gel. This transfer method is called *Southern transfer* or **Southern blotting.**

In Southern transfer, the agarose gel is soaked in base to denature the DNA fragments. Then the gel is placed on a long piece of blotting paper whose ends are suspended in a reservoir of salt solution. The membrane is placed directly on the gel, and a stack of dry absorbent paper (such as paper towels) is placed on the membrane (Fig. 16.2). The blotting paper acts like a wick, drawing fluid from the reservoir up through the gel into the stack of dry paper. This process is driven by capillary action, which is what you see when you put the corner of a paper towel into a glass of water.

As the fluid migrates upward through the gel, it carries the denatured DNA fragments with it. When the

Figure 16.1 Colony hybridization.

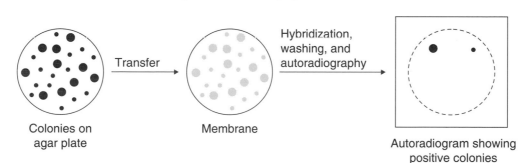

Colonies on agar plate   Membrane   Autoradiogram showing positive colonies

Hybridization, washing, and autoradiography

Transfer

A

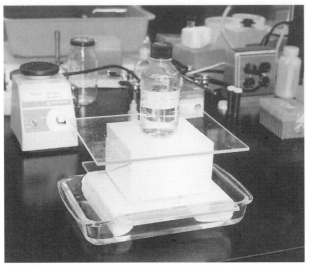

B

Figure 16.2 Assembling a Southern blot. (A) The student is placing the membrane on top of the gel. (B) The complete transfer setup. The bottle on top is simply a weight.

fragments reach the membrane on top of the gel, they stick to it and remain there. The assembly of wick, gel, membrane, and paper is put together so that the DNA fragments are transferred straight up in exactly the pattern they were found in the gel. They stick to the membrane in exactly this pattern, duplicating the pattern in the gel itself.

Once the denatured DNA fragments have been transferred to the membrane, the hybridization procedure is exactly as described previously for library screening. Autoradiography of a Southern blot shows dark bands corresponding to the bands in the gel that contained DNA sequences complementary to the probe (Fig. 16.3).

Other uses of hybridization

Hybridization is not always used as an analytical method per se. It plays a critical role in other laboratory procedures such as DNA sequencing and the polymerase chain reaction (illustrated in subsequent lessons). Both of these procedures involve in vitro DNA synthesis by the enzyme DNA polymerase. DNA polymerase must have a primer at which to begin synthesis (see chapter 6, *DNA Replication*). In vitro, the primer is usually a short synthetic oligonucleotide annealed to the template DNA molecule. Very often, the template molecule starts out as double-stranded DNA. An excess of the short primer is added to it in solution, and then the solution is heated to nearly 100°C to denature the template. As the solution cools, the short primers hybridize to the template strands.

This process of annealing primers is the first step in both DNA sequencing and the polymerase chain reaction. As described in a subsequent lesson, the polymerase chain reaction itself can be used as an analytical technique, and its specificity is actually the specificity of hybridization between the primers and the template DNA (see chapter 18, *The Polymerase Chain Reaction: Paper PCR*).

Objectives

After completing part I of the lesson *(Fishing for DNA)*, students should be able to

1. Define hybridization;

2. Describe how single-stranded DNA can be used to detect a DNA sequence of interest.

After completing parts II and III, students should be able to

3. Describe how to use a DNA probe and restriction analysis to determine which restriction fragment of a given molecule contains a sequence of interest;

4. If shown a probe sequence and the sequence of a target molecule, predict which restriction fragments of the target will hybridize to the probe.

Materials

Part I, fishing for DNA

- The student reading *Fishing for DNA* and the accompanying DNA sequence sheet (If you use photocopies, copy the DNA sequence sheet and the reading onto separate pieces of paper.)

- Scissors

A

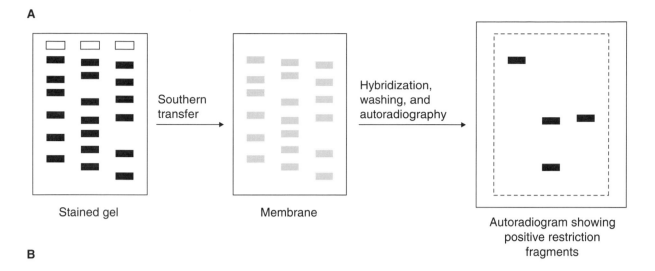

Stained gel → Southern transfer → Membrane → Hybridization, washing, and autoradiography → Autoradiogram showing positive restriction fragments

B

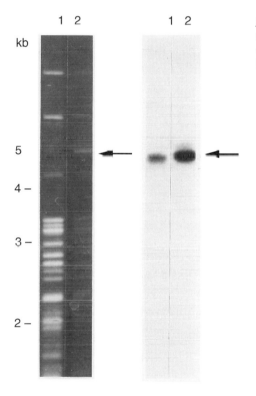

Figure 16.3 Southern hybridization analysis. (A) Diagrammatic representation; (B) two stained gel lanes and the results of a hybridization to those lanes. kb, DNA fragment size in kilobase pairs.

• Optional demonstration model: two small glass fishbowls or other glass containers (large beakers are fine), some string, paper, tape, a refrigerator magnet, a few paper clips, and some kind of colored paper or objects

Part II, combining restriction and hybridization analysis

• The student reading *Southern Hybridization* and gel outlines
• Tape

Part III, Southern hybridization

Southern hybridization worksheet

Preparation

• Optional demonstration model. Make a few copies of the DNA sequence sheet for yourself, and cut out the sequence strips and two copies of the probe. Tape a paper clip to the back of one of the strips containing the target sequence. Put half of the sequence strips into one glass fishbowl (or other glass container) and the rest into another.

(The target sequence with paper clip will be in only one bowl.) Make two "hook" assemblies by fixing the probe to a slightly larger piece of paper. Glue the probe to a magnet (such as a refrigerator magnet.) Tie a string to the magnet-probe assembly. To the other end of the string, tie or fasten anything colored to symbolize the detection method that allows the location of the probe to be seen.

• Make any necessary photocopies.

Procedure

Part I, fishing for DNA

1. Remind your students how large the DNA molecule of any organism is. If you did the activity *Sizes of the* Escherichia coli *and Human Genomes,* get out the scale models of the *E. coli* chromosome and gene. Show the class the models of the chromosome and the tiny gene, and tell the students that for this activity, they would like to know whether a particular DNA sequence, such as a certain gene, is present in a huge molecule like the *E. coli* chromosome. Ask them if they can think of a way to find out. (Sequencing the entire chromosome would give the answer but is not practical; in all the years of research on *E. coli,* its chromosome is still not completely sequenced as of mid-1995.) If you made the fishbowl prop, show it to your class, and tell them that they are about to do an activity that will show them how scientists fish for specific DNA sequences.

2. Have students read the background reading in *Fishing for DNA* and do the activity. It may be helpful to get out your DNA model to show the class single-stranded DNA. Give them as much additional information about hybridization as you feel is appropriate. Talk through the questions.

3. Optional demonstration for summary. Tell the class that the sequences in the bowls are DNA from two different samples, and we want to know whether the gene for X (whatever you like) is in the samples. Show them the probes. Lower the probes into the two bowls. Tell them that if the probe can hybridize to a sequence within the test DNA, it will stick when you try to rinse it out. Fish in the target-containing bowl until the magnet-probe sticks to the target sequence with the paper clip. Fish in the other bowl, too (you won't "catch" anything). Ask students what it means that the probe stuck to the DNA in one bowl but not in the other. Get them to tell you that the gene of

interest is present in the sample that the probe stuck to but not in the other. Be sure that they relate the answer to base pairing between the probe and the sample DNA. Depending on your class, you may want to do this as part of the introduction to the activity.

The activity Fishing for DNA *is a sufficient introduction to hybridization for younger students.*

Part II, restriction and hybridization

After you complete *Fishing for DNA,* have students read *Combining Restriction and Hybridization Analysis* and do the activity. Give them any additional information about Southern blotting that you deem appropriate. An advanced class should be able to complete parts I and II within a single class period.

Part III, Southern hybridization

The Southern hybridization worksheet can be done in class or for homework. It shows how Southern hybridization can be used to map the location of a gene.

Answers to Student Questions

Part I, fishing for DNA

1. DNA hybridization is complementary base pairing between two single-stranded DNA molecules to form a double helix.

2. Students should describe the use of probes to find a specific DNA sequence in a sample.

3. Be sure that students have the correct 5′-to-3′ orientation: since the written sequence is 5′ to 3′, the answer sequence should be 3′ to 5′ (omit this requirement for younger students if you think it is too detailed).

Part II, restriction and hybridization

The DNA strip is cut into eight pieces; the probe hybridizes to the fourth smallest.

Perform Southern hybridization analysis on a sample of virus DNA, using the probe for the desired gene. This analysis will show which restriction fragment contains the gene of interest. Digest a second sample of virus DNA, perform electrophoresis and staining, and then cut the correct fragment out of the gel and purify it for cloning.

Part III, Southern hybridization

1. Check students' pictures for correct labels.

2. The DNA polymerase gene is located in the region where the 10,500-base-pair (bp) *Bam*HI fragment and the 5,500-bp *Eco*RI fragments overlap.

3. Since the entire DNA polymerase gene was used as a probe and did not hybridize to either the 7,500-bp *Bam*HI fragment or the 4,000-bp *Eco*RI fragment, the entire virus X DNA polymerase gene is probably inside the indicated area, neither to the right of the *Eco*RI site nor to the left of the *Bam*HI site. Of course, there is still some uncertainty, since we do not know how similar the two genes actually are, but it is not necessary to discuss this fine point in class unless students bring it up.

4. The point of this question is for students to realize that proteins that carry out the same function in different organisms often have very similar amino acid and DNA base sequences. The similarity is usually more pronounced the more closely related the organisms are. It is therefore often effective to use a known gene to search for a gene encoding the same kind of protein in a different organism.

For the answer to IIIB, see the diagram below.

Hybridization Wet Laboratories

Hybridization analysis is not hard to do, but it requires quite a few more steps (extra laboratory periods) than restriction analysis, and you have to be careful. The extra steps require extra materials and solutions, too.

If you are familiar with hybridization analysis and know what you are doing, we recommend a labeling system such as the Boehringer Mannheim "Genius" kit (available from Boehringer Mannheim and Carolina Biological Supply Co.) for nonradioactive labeling and detection of your probe. If you buy this kit, you will still have to make quite a few solutions to use with it.

If you are inexperienced with hybridization analysis but are determined to try it in your classroom, we strongly recommend you get a prepared kit for the entire activity. At this time, Edvotek sells a DNA fingerprinting kit (no. 311) that involves Southern hybridization analysis. In addition to electrophoresis equipment, the kit requires that you have access to an oven that reaches 80°C. We have not tested this kit. If you decide to try it, follow all the instructions very carefully.

Classroom Activities

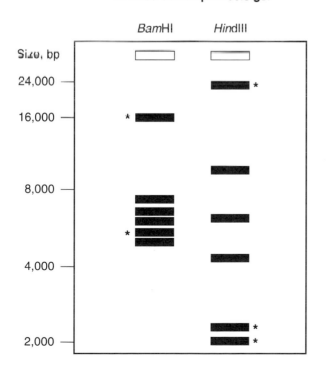

Stained electrophoresis gel

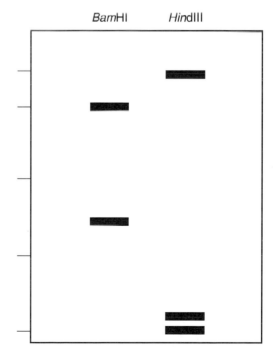

Results of hybridization analysis

Detection of Specific DNA Sequences Part I. Fishing for DNA

16

Imagine this: you are a scientist staring at a test tube full of DNA that you just prepared. You want very much to know whether that DNA contains the gene for cystic fibrosis. You know a little about the cystic fibrosis gene, so how can you tell if it is in your sample?

Scientists often need to fish for particular pieces of DNA such as the cystic fibrosis gene. How do they do it? Once they know a little about the DNA they are looking for, it is not very hard.

When a scientist fishes for a particular piece of DNA, she uses a special DNA "hook" called a **probe.** A probe is usually a form of DNA called single-stranded DNA. Think of pictures of DNA or of models of DNA you have made. The base pairs are in the middle, and the two backbones are on the outside. Single-stranded DNA is only half of this structure: one backbone with bases sticking out from it (rather like a messenger RNA molecule; look at the picture below). If a piece of single-stranded DNA finds another piece of single-stranded DNA with the right base sequence, it will pair with it to make a regular DNA helix. Scientists use a special term for the pairing of two single DNA strands to make a double helix: **hybridization** (see Diagram 16.I).

To see whether a gene is present in a DNA sample, our scientist uses a probe that has the same sequence as part of the gene. She adds it to the DNA sample to see whether it will hybridize with anything. How does she know when it hybridizes? If the probe hybridizes to the sample, it will stick to the DNA in the sample. If not, it can be easily rinsed away. After the scientist rinses her sample, she determines whether the probe has stuck to it. If it has, she knows that the sequence of DNA in the probe is also present in her sample. Since the probe has the sequence of part of a gene, that gene is most likely in the sample DNA.

Activity

On Worksheet 16.I (see Appendix A), you have a long DNA sequence representing your sample. You also have a short sequence that is a probe for the cystic fibrosis gene. You will use the probe to determine whether the cystic fibrosis gene is in your sample DNA.

Do you notice anything funny about your long DNA sequence? It is only one strand. To test whether a probe will hybridize to a DNA sample, it is necessary to separate the strands of the sample DNA so that the probe can find complementary base pairs. Your sample DNA has already been prepared.

1. Cut out the probe sequence.

2. Scan the sample DNA sequence to see whether the probe will hybridize to it.

3. Is the cystic fibrosis gene present in your DNA sample? If you found a place where the probe could hybridize, mark it on the DNA sequence.

Diagram 16.I Hybridization of a probe to a sequence within single-stranded sample DNA.

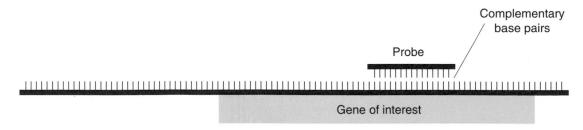

Questions

1. What is DNA hybridization?

2. How can you use hybridization to tell you whether a certain DNA sequence is present in a DNA sample?

3. Below is a DNA sequence. Write out the sequence of a 10-base-pair (bp) probe that would hybridize to it.

5′ AATGCAGGCCCTATATGCCTTAACGGCATATGCAATGTACAATGCAAGTCCAACCGG 3′

Detection of Specific DNA Sequences Part II. Combining Restriction and Hybridization Analysis

Introduction

The hybridization analysis you learned about in *Fishing for DNA* can indicate whether a given DNA sequence is present in a sample. Sometimes this simple piece of information is all that is needed. However, a positive hybridization test does not give any information about where the sequence of interest is located within the sample DNA molecule.

To get information about the presence and location of a particular DNA sequence, scientists combine restriction analysis with hybridization analysis. Basically, the sample DNA is digested with a restriction enzyme (or enzymes), and the fragments are separated by electrophoresis. After electrophoresis, hybridization analysis is performed on the fragments. Only the fragment or the fragments to which the probe can base pair will be detected (Diagram 16.II).

There are a few important technical details about this whole process. First, it doesn't work very well just to soak an agarose gel in a probe solution. In 1975, a scientist named Southern figured out a way to transfer DNA fragments from a gel directly to a membrane so that the exact pattern of the fragments in the gel was preserved. After the fragments were stuck on the membrane, they could be tested for hybridization to a probe. This method of transfer is called Southern blotting, and the combination of restriction analysis, transfer to membrane, and hybridization to probe is called Southern hybridization analysis.

Activity: Southern Hybridization Analysis

1. Take the sample DNA sheet from the previous exercise, and cut out the DNA strips by cutting along the dotted lines. Tape strip 1 to strip 2 to strip 3 through strip 10 to form one long linear molecule. This molecule could be a chromosome like one of yours.

2. Now simulate the activity of the restriction enzyme *Sma*I. This enzyme cuts the DNA sequence 5′ CCCGGG 3′ between the C and G in the middle of the sequence. Digest your chromosome by cleaving at every *Sma*I site. (Your fragments are single stranded in anticipation of the hybridization. In reality, the restriction digestion and electrophoresis are performed on double-stranded DNA, and then the two strands of the molecules are separated.)

3. Next, simulate electrophoresis of your fragments. Sort them by size, and lay them on your desk top as if they were in a gel.

4. Using the outlines provided on Worksheet 16.II (see Appendix A), draw the arrangement of fragments in your gel.

5. On your drawing of your gel, mark with an asterisk any band(s) that would hybridize to the probe.

6. Keep in mind that the pattern of DNA fragments in your gel will be exactly transferred to a membrane. On the membrane, the fragments will be

Diagram 16.II Southern hybridization analysis shows which restriction fragments hybridize to a probe.

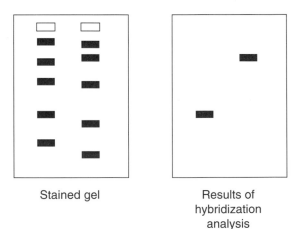

Stained gel

Results of hybridization analysis

tested for hybridization with the probe. Afterward, only the fragment(s) that hybridizes to the probe will be detected (as in the accompanying diagram of Southern hybridization analysis). Draw what will be detected after hybridization in the box marked "Results of hybridization analysis."

Question

1. It is possible to cut a slice from an electrophoresis gel, purify the DNA in that slice, and clone it. Suppose that you have a probe for a viral gene that you would like to clone, but you do not know where it is within the 40,000-bp virus chromosome. How could you use Southern hybridization analysis to help you clone your gene?

Detection of Specific DNA Sequences
Part III. Southern Hybridization

Problem A

You are analyzing the chromosome of a newly discovered virus, X. You have already constructed a restriction map of the 26,500-base-pair (bp) linear chromosome, shown in Worksheet 16.IIIA (see Appendix A). Since you are interested in DNA polymerase enzymes, you conducted a hybridization analysis on virus X DNA using as a probe a DNA polymerase gene from another virus you believe is closely related to X. To your delight, the probe hybridized to the DNA from virus X. You believe the virus X DNA to which the probe hybridized is the virus X DNA polymerase gene.

Now you would like to determine the location of the virus X DNA polymerase gene, so you perform a Southern hybridization analysis. You digest viral DNA with *Eco*RI and *Bam*HI separately, separate the fragments in a gel, transfer them to a membrane, and hybridize with the same DNA polymerase gene probe. Shown on Worksheet 16.IIIA is a picture of the fragments in the electrophoresis gel and the pattern seen after the hybridized probe is detected on the membrane.

Problem B

On Worksheet 16.IIIB (see Appendix A) is a restriction map of bacteriophage lambda. You digest some lambda DNA with the enzymes *Bam*HI and *Hin*dIII separately, load the fragments into an agarose gel, and perform electrophoresis.

Next, you transfer the fragments to a membrane and carry out hybridization analysis using the 4,878-bp *Eco*RI lambda fragment (refer to the map) as a probe.

1. Draw a picture of the electrophoresis gel, using the stained electrophoresis gel outline in Worksheet 16.IIIB (the two smallest *Hin*dIII fragments will run off the gel).

2. Indicate which fragments will hybridize to the 4,878-bp probe. (Hint: Even if there is only a partial overlap of the probe with a restriction fragment, the probe will still hybridize to that fragment.)

3. In the second outline, draw what would be seen after detection of the probe on the membrane.

Questions (Problem A)

1. Label the fragments on the gels with their sizes in base pairs.

2. Indicate on the restriction maps the region of the virus X chromosome in which the DNA polymerase gene is located.

3. How far to the left in terms of restriction sites could the gene lie? How far to the right?

4. Why would a DNA polymerase gene from another virus hybridize to DNA fragments from virus X?

DNA Sequencing

17

About This Activity

In this lesson, a paper-and-paper clip simulation is used to illustrate the most common method of DNA sequencing. If you have pop beads for DNA models, they can be substituted for the paper and paper clips.

Class periods required: 1–2

Introduction

The methods most commonly used for sequencing DNA on a small scale employ compounds called *chain terminators,* which are chemicals that specifically stop the elongation of a new DNA strand by DNA polymerase. For background information on how chain termination methods of DNA sequencing work, please read the introduction to the *Student Activity* pages.

The dideoxy sequencing method is easily illustrated in the following activity. If you are uncertain as to how the method works, try the activity yourself first. In the activity, your class will "sequence" a piece of DNA. They will synthesize a DNA strand based on a paper template, using colored paper clips to represent the new nucleotides. Dideoxynucleotides are represented by a special color. The results of the entire class will be pooled and symbolically "run" in a sequencing gel. This activity will not only illustrate the DNA sequencing method but will also reinforce students' knowledge of base-pairing rules, the mechanism of DNA synthesis by DNA polymerase, and the principle of electrophoretic separation by size.

Several chain terminators are also used as antiviral drugs. The reading at the end of this chapter explains how terminators work to fight human immunodeficiency virus and herpesviruses.

Objectives

After this lesson, students should be able to

1. Explain what a chain terminator is and why it causes DNA synthesis to stop;
2. Explain how chain terminators are used to reveal the base sequence of a DNA molecule;
3. Read a DNA sequence from a sequencing gel.

Materials

- Enlarged photocopies of template-primer (four "molecules" per student). The long strand is the template; the short strand base paired to it is the primer.

- Colored paper clips in five colors: 20 each of four colors and 4 of the "stop" color per student. (Colored pop beads can be substituted for the paper clips and the paper template-primers.)

- Small paper cups (or other container), four per student

- Scissors

- Large corkboard or paper sheet on classroom wall, pins, large letters (A, G, C, and T; one each), and labels P + 1, P + 2, etc., up to P + 16 (one each). These materials are for setting up the "sequencing gel" (see below).

Resource Materials

Colored pop beads designed for use in DNA models are available from Carolina Biological Supply Co. in bulk (catalog no. 17-1041 to 17-1048) or in special kits.

Preparation

- Photocopy and enlarge the template-primers. If possible, laminate them, punch a hole at the end of the primer (to attach the first paper clip), and reinforce the hole. Save these for future use.

- Sort the paper clips into colors.

- Set up your sequencing gel on the corkboard, or draw it on the blackboard like this (also see below):

	A	G	C	T
P + 16				
P + 15				
•				
•				
•				
P + 4				
P + 3				
P + 2				
P + 1				

Procedure

1. Review the mechanism of DNA synthesis by DNA polymerase. Be sure to emphasize that the 3′ hydroxyl group is necessary for forming a new bond in the growing DNA chain (see Fig. 17.2). Show them the structure of the dideoxynucleotides (see Fig. 17.1 in the *Student Activity* pages), and explain that these molecules can be incorporated like regular deoxynucleotides, but that when they are, DNA synthesis is stopped.

2. Hand out four DNA template-primers per student. Tell the students they will sequence the template by the dideoxynucleotide method, with paper clips representing the radioactive new nucleotides.

3. Have the class count off as A, G, C, and T, as if they were numbering off. Each student will conduct only one type of reaction on the four template molecules. The reason for having each student "synthesize" four molecules is to be sure that a stop paper clip is incorporated by someone at every possible position. There should be at least 8 (preferably 10 to 12) new strands synthesized for each type of reaction. You may adjust the number of strands each student synthesizes to fit your class size.

4. Assign one color of paper clip to be A, one to be G, and so on. The fifth color will be the dideoxynucleotide, that is, the stop nucleotide.

5. Each student should get four paper cups and label them A, G, C, and T. The A students put 20 paper clips of the correct colors in the G, C, and T cups. In the A cup, they put 16 paper clips of the A color and 4 of the stop color and mix them thoroughly. The stop color represents dideoxyadenosine (ddA) molecules.

The G students get 20 A, C, and T paper clips and put 16 G and 4 stop paper clips in their G cups (mixed well). This time the stop paper clips represent dideoxyguanosine (ddG) molecules.

The C and T students should set their "reaction mixtures" up appropriately with their paper clips.

6. Students should line up their four template molecules on their desks, with the paper cups in easy reach. Be sure the "spiked" paper clips (the ones with dideoxynucleotides) are well mixed. Now have the class begin to synthesize new DNA from the 3′ end of the paper primer. Since the first template nucleotide is an A, they should draw a complementary nucleotide (a T) from the correct paper cup and bond it to the paper primer (by piercing the paper). The students must do this four times, one for each template molecule. The drawings must be done at random (eyes closed). If any student with the T reaction pulls out a stop paper clip, that student should add the stop paper clip to the primer and then set that template aside. No more nucleotides can be added to it.

Figure 17.5 Reaction products from the C reaction. The four shorter products were produced when a student drew a stop paper clip at a G in the template sequence. The long product resulted when a student happened not to draw a stop paper clip at any of the template G's.

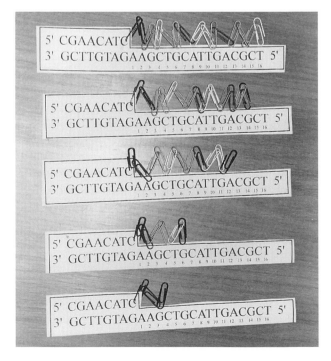

7. Repeat step 6 for the next base in the template (the second A), linking the second T paper clip to the first. Again, if a student randomly selects a stop paper clip, that paper clip should be added to the growing DNA strand, but then that template must be set aside, and no more nucleotides can be added to it. Remind the students that the stop paper clips are dideoxynucleotides that lack the 3′ hydroxyl group for forming a new bond.

8. Continue down the template to the end. Figure 17.5 shows all possible products for the C reaction. Notice that some products stop at each G in the template, and for one full-length product, the student never drew a stop paper clip from the C paper clips.

9. Now it is time to load the gel and perform electrophoresis. DNA sequencing reactions are denatured before loading, so the students should "denature" their molecules by cutting the paper primer away from the paper template but leaving the paper primer attached to the paper clips.

Point to your gel set-up and tell the students that all the A reaction products will be loaded and run in the A lane, the G-reaction products will be run in the G lane, and so forth. The students should imagine that they have loaded their products into the appropriate lane and that electrophoresis has occurred.

10. To see the banding pattern, focus on one lane at a time. For example, ask the A-reaction students if anyone has a product molecule with just one new nucleotide (paper clip) added to the primer. (No one should, since the first nucleotide added is a T.) Ask for chains with two, three, four, and five new nucleotides. Someone in the A group should have a molecule with exactly five new nucleotides, since there is a T at that position in the template. If you are using corkboard, have that student come forward and pin the molecule in the A lane right at the P+5 level (there may be several such molecules; hang them on pins on top of each other). If you are using a blackboard, simply draw a band in the A lane at the P+5 level.

Continue asking for molecules up to P+16. You should get products at P+5, +9, +10, and +16 from the A-reaction students. Repeat for the G-reaction students (you should get P+4, +7, +13, +15, and +16), the C-reaction students (P+3, +6,

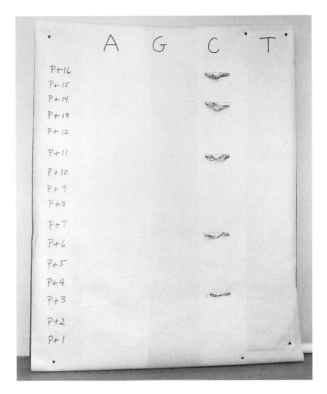

Figure 17.6 The C-reaction products have been "loaded" and "separated" on the "gel." Their lengths are P+3, P+6, P+11, P+14, and P+16 (refer to Fig. 17.5).

+11, +14, and +16), and the T-reaction students (P+1, +2, +8, +12, and +16).

Figure 17.6 shows the C lane of the gel loaded with the reaction products pictured in Fig. 17.5. This figure also shows how to draw the gel outline. Figure 17.7 shows what your completed "gel" will look like.

11. Now read the sequencing gel. Starting at P+1, identify the lane the band is in (T). Write T on the blackboard. Continue with P+2, P+3, etc., writing the sequence as you go. Compare the base sequence you write on the board with the sequence of the newly synthesized DNA strand (they should match). Compare it with the sequence of the paper template (they should be perfectly complementary).

12. (Optional) Project a transparency of the photograph of the sequencing gel in Fig. 17.4, and have students read some of the DNA sequence from it.

13. (Optional) Read and discuss the extra reading *Chain Terminators as Antiviral Drugs*.

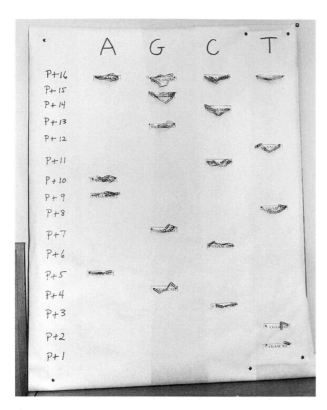

Figure 17.7 The completed sequencing gel. The sequence is read from the bottom up: TTCGACGTAACTGCG. This is a perfect complement to the template sequence.

Answers to Student Questions

1. The template molecules are present in the gel (they are loaded along with the rest of the reaction mixture), but you cannot see them because they are not radioactive.

2. DNA replication

3. Dideoxycytosine (ddC)

DNA Sequencing: the Terminators

Determining the base sequence of a piece of DNA is a critical step in many applications of biotechnology. How is it done? The method most commonly used today employs compounds called chain terminators, a term that describes the lethal effect these compounds have on DNA synthesis, and nature's sequence readers, the DNA polymerase enzymes. These enzymes read the sequence of a single template DNA strand and synthesize a complementary strand. The chain terminators allow us to "look over the shoulder" of the DNA polymerase enzyme: we can see the order in which it adds bases to a new DNA strand and therefore deduce the sequence of the template strand.

What are chain terminators, and how do they block DNA synthesis? Chain terminators are molecules that closely resemble normal nucleotides but lack the essential 3′ hydroxyl (OH) group (Fig. 17.1). In DNA replication, a nucleotide complementary to the tem-

plate base is brought into position. The DNA polymerase then adds this nucleotide to the growing DNA strand by forming a bond between the 5′ phosphate group of the new nucleotide and the 3′ OH group of the previous nucleotide (see Fig. 17.2). DNA polymerases cannot synthesize DNA without a preexisting 3′ OH group to use as a starting point (this is referred to as a requirement for a primer; see chapter 6). If a DNA polymerase mistakenly adds a chain terminator instead of a normal deoxynucleotide, no further nucleotides can be added, and DNA synthesis is terminated.

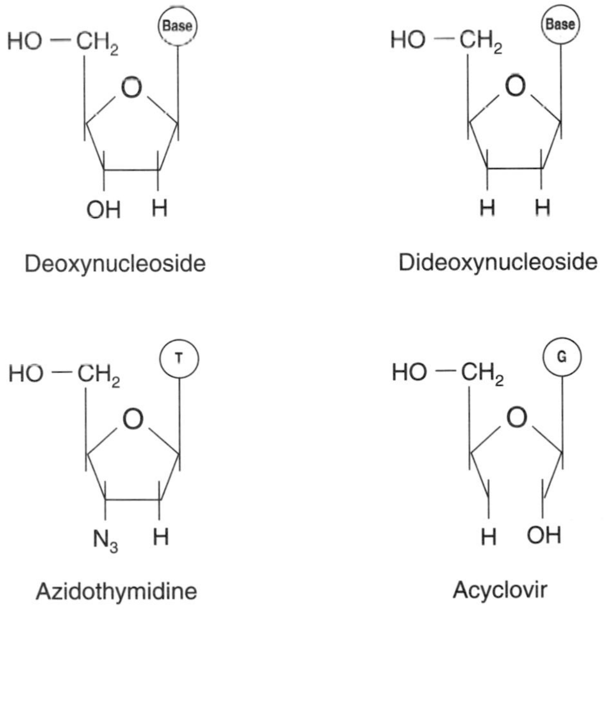

The chain terminators used in DNA sequencing are the dideoxynucleotides (Fig. 17.1). How do dideoxynucleotides let us see the base sequence of a DNA molecule? To set up a sequencing reaction, the template DNA molecule to be sequenced is mixed with primers and radioactive normal deoxynucleotides (deoxyadenosine [dA], dG, dC, dT). This master mix is divided into four batches, and each batch is then "spiked" with a different dideoxynucleotide: dideoxyadenosine (ddA), ddG, ddC, or ddT. DNA polymerase enzyme is added, and it synthesizes new DNA strands on the template molecules. Occasionally, however, a dideoxynucleotide chain terminator is inserted in place of the analogous normal deoxynucleotide, and the synthesis of that DNA molecule is terminated. In the reaction mixture containing ddA (the A reaction), some percentage of the new molecules will get a ddA at each place there is a T in the template. The result is a set of new DNA molecules in which some of the molecules terminate at each T in the template. Similarly, in the G reaction, some of the new molecules will terminate at each C in the template. In the C reaction, some of the new molecules terminate at each G in the template, and in the T reaction, molecules terminate at each template A.

After the synthesis reactions are complete, the A-, G-, C-, and T-reaction mixtures are denatured by heating and loaded separately into adjacent lanes of a gel, and a current is applied (Fig. 17.3). The newly synthesized molecules are separated by size and then visual-

Figure 17.1 Normal deoxynucleoside shown with chain terminators. All are incorporated into DNA from their triphosphate forms.

Deoxynucleoside

Dideoxynucleoside

Azidothymidine

Acyclovir

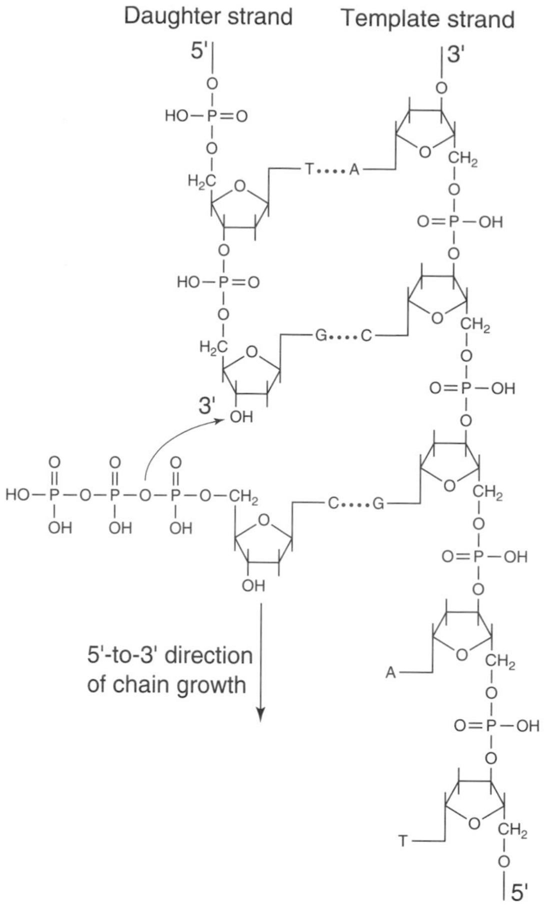

Daughter strand   Template strand

5'-to-3' direction of chain growth

Figure 17.2   DNA replication. Base pairing between an incoming deoxynucleoside triphosphate and the template strand of DNA guides the formation of a new complementary strand.

Figure 17.3   A student loading a sequencing gel. A sequencing gel is a vertical gel: the DNA runs from top to bottom. It is tall and quite thin (less than 1 mm thick). Sequencing gels are made with acrylamide instead of agarose. Acrylamide forms a much tighter mesh than agarose, making it possible to separate DNA molecules.

ized by exposing the gel to photographic film (remember that the new nucleotides are radioactive). The A-reaction lane will show bands that correspond in length to the sites of the T's in the template. The G-reaction lane shows bands whose lengths correspond to the sites of template C's, and so on. The sequence of the new DNA strand is "read" from the sequencing gel by starting at the bottom (the shortest new molecule) and reading upward (see below). Figure 17.4 shows an autoradiogram of a sequencing gel.

If this written explanation seems confusing, don't worry. The sequencing simulation you will do shows how the whole approach works. After you complete the simulation, go back and reread the introductory material, and then answer the questions.

Figure 17.4   Autoradiogram of a sequencing gel. The scientist who did this sequencing procedure was screening several plasmids from transformants to see if she had gotten the clone she was trying to construct. She did. In fact, every one of the plasmids had the sequence she wanted.

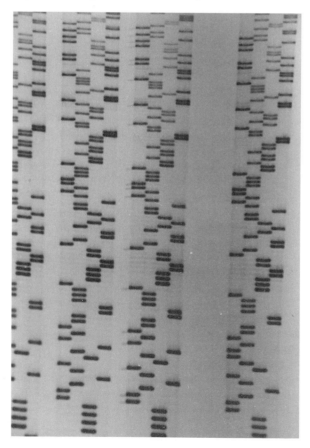

Questions

1. Would the long template molecule (represented by the paper DNA strand) show up in the sequencing gel?

2. If a cell culture were "fed" some dideoxynucleotides in media, what cellular process, if any, would be most affected?

3. (Optional: refer to the additional reading.) What chain terminator is currently used in DNA sequencing and as an anti-AIDS drug?

Chain Terminators as Antiviral Drugs

Chemicals that were first used as investigative tools in research laboratories often find their way into the pharmacy. Compounds that affect fundamental biological processes are important to basic research and sometimes turn out to be therapeutically useful as well. The chain terminators are a good example of such compounds. As shown in the previous activity, chain terminators are essential to the now-favorite method of DNA sequencing. In addition, three of the currently (1995) available anti-AIDS drugs and the best available antiherpes drug belong to this class of compounds.

How are chain terminators used in fighting AIDS and herpesvirus infections? They fight viral infections in the same way they assist in DNA sequencing: by terminating DNA synthesis. For a virus to establish an infection, it must replicate its nucleic acid. Blocking this replication is potentially a very effective way to fight the spread of a virus, but blocking DNA synthesis could be as harmful to the patient as to the virus invader. Herpesviruses and the AIDS virus are good candidates for antireplication drugs because they encode their own DNA-synthesizing enzymes that have unique properties that can be exploited.

The AIDS virus (human immunodeficiency virus [HIV]) is an RNA virus. It encodes a special enzyme, reverse transcriptase, that synthesizes DNA by using the viral RNA as a template. This step is essential to HIV infection. Chain terminators fight the spread of HIV in the body by interfering at this stage. Reverse transcriptase is a good target for chain terminator drugs because it is a sloppy enzyme: it is much more likely to incorporate an incorrect nucleotide or a chain terminator than is the human polymerase. Furthermore, reverse transcriptase lacks the ability to proofread its work, so it cannot remove an incorrect nucleotide once it is incorporated.

There are more issues involved in developing a successful chain terminator drug than simply the sensitivity of the viral replication enzymes. The drug must survive in the body and be absorbed by the proper cells. The active triphosphate form of the chain terminators (shown in Fig. 17.2 for the nucleotide being added to the growing chain) is not absorbed by cells, so the drugs are given in unphosphorylated form (as shown in Fig. 17.1). Once these compounds enter the cell, the human enzymes must add the phosphate groups (phosphorylate the molecules) to activate them. Different compounds are absorbed and phosphorylated with different efficiencies. So the development of a new drug depends not only on its toxicity to the virus and harm to the host but also on how it is metabolized in the human system.

The chain terminators currently employed against the AIDS virus are azidothymidine (AZT), dideoxycytosine (ddC; the same compound used in DNA sequencing), and dideoxyinosine (ddI) (Fig. 17.1). AZT, an analog of thymidine, is incorporated into a growing DNA molecule in place of normal thymidine. Likewise, ddC is incorporated in place of cytosine. The compound inosine is a nucleoside analog that is identical to adenosine except for the absence of one amino group. Cells synthesize inosine and convert it directly to adenosine. When ddI is given as a drug, it enters cells and is rapidly converted to ddA. The ddA is incorporated by reverse transcriptase in place of normal adenosine.

These drugs have toxic side effects in patients. Although human DNA polymerase is less sensitive to the drugs than reverse transcriptase is, the drugs do affect DNA replication in normal human cells. AZT is particularly toxic to the bone marrow, while ddC and ddI affect the peripheral nerves. Current AIDS therapy seems to be moving toward alternating the use of these drugs to minimize harm to the patient. Unfortunately, the rapidly mutating HIV can acquire resistance to these compounds. It is hoped that more research will lead to better treatments and an actual cure for AIDS.

The herpesviruses are DNA viruses with relatively large genomes. They encode many of their own DNA replication enzymes, including a DNA polymerase.

After the initial infection, herpesviruses remain in the body in an inactive (latent) state from which they can be activated and can then cause outbreaks of disease. Herpes simplex viruses 1 and 2 cause fever blisters and genital herpes, respectively. The herpesvirus varicella-zoster causes chicken pox when it first infects a person and shingles in subsequent outbreaks.

Herpesvirus diseases can now be treated with the chain terminator acyclovir (Fig. 17.1). Acyclovir (marketed under the name Zovirax) is relatively nontoxic to the human host because human cells take up acyclovir but do not phosphorylate it, so the drug remains inactive. Herpesviruses, however, encode an enzyme that does phosphorylate acyclovir. Therefore, acyclovir becomes an active chain terminator only in herpesvirus-infected cells. Acyclovir is an analog of guanidine and is incorporated opposite cytosine residues in the template. Once incorporated, it terminates further DNA synthesis and inhibits the activity of the herpes polymerase.

Classroom Activities

The Polymerase Chain Reaction: Paper PCR

18

About This Activity

The polymerase chain reaction (PCR) has become a key tool in molecular biology research and biotechnology applications. PCR is not conceptually difficult if students have sufficient preparation (see below). However, most people, whether school-age students or adult scientists, will not really grasp what is going on until they have gone through all the steps themselves, either by drawing pictures or by manipulating models. In this lesson, students will use paper models to simulate the steps of the PCR. The model exercise demonstrates how DNA polymerase can be used to make multiple copies of a specific DNA fragment and shows how the technique can be used to detect a specific DNA molecule (such as the chromosome of a disease-causing microorganism) in a sample. Students should have performed both activities in *DNA Replication* and have been introduced to hybridization before performing this activity.

Figures 18.1 and 18.2 are shown in the *Student Activity* pages.

Class periods required: 1–2

Introduction

The PCR has become very popular with both researchers and clinicians. This clever technique allows many copies of a specific DNA segment to be produced from a single copy and is extremely useful as a detection method and a cloning tool. For example, PCR can be used to identify viruses, to analyze scarce DNA (such as from small bloodstains), or even to probe antique DNA from museum specimens.

PCR is a simple technique that combines in vitro DNA synthesis by DNA polymerase and hybridization. It requires a solution containing a starting sample, two primers (many molecules of each), a DNA polymerase enzyme, and free nucleotides. The primers are short single-stranded pieces of DNA, typically 16 to 20 bases long. One primer of the pair hybridizes to each

of the two strands of the sample DNA at either end of the segment to be amplified. The primers are mixed in great excess with the sample DNA. To start the reaction, the double-stranded sample DNA in the mixture is denatured into single strands by heating (usually to 95°C). The mixture is then cooled so that hybridization of complementary strands can occur. Because the primers are in such great excess, one primer molecule will anneal to each single strand before the strands of sample DNA can find one another (Fig. 18.1).

In the next step, the DNA polymerase enzyme synthesizes a second DNA strand on each of the two original strands, using the free nucleotides in the solution. The annealed primer serves as a starting point (Fig. 18.1). New DNA is synthesized from the 3′ end of each primer and extends in only one direction. The result is two helices where before there was only one. (DNA polymerase enzymes must have a 3′ end from which to start synthesizing, and they can add nucleotides in only one direction; see *DNA Replication* [chapter 6].)

The process of denaturation, hybridization, and synthesis is then repeated, giving four helices (Fig. 18.1). A further round yields 8, then 16, then 32, and so on. Typical PCR procedures call for 25 to 40 rounds of amplification and yield huge numbers of molecules. Nearly all of the new DNA molecules have the primers as their ends and are the same length.

In early PCR methods, new DNA polymerase had to be added after each denaturation step because the high heat necessary for denaturation destroyed the enzyme. Now, however, scientists have purified a DNA polymerase enzyme from a bacterium that inhabits hot springs *(Thermus aquaticus)*. This enzyme, the *Taq* polymerase, remains active after being heated to 95°C and does not need to be added after each denaturation step. The *Taq* polymerase has made it simple to automate PCR: all that is needed is a heater with programmable temperature cycles. "PCR machines," called thermal cyclers, have become a fairly standard piece of laboratory equipment (Fig. 18.2).

PCR can produce enough product DNA, even from a minute amount of template, to be visible in a gel after electrophoresis. Biotechnologists often use this technique to produce many copies of an interesting DNA segment for cloning. PCR is also used in DNA typing when only minimal samples are available (see chapter 24). And, since the PCR process depends on the specific base pairing of primers to the template DNA, PCR can be used as a diagnostic tool.

In PCR-based diagnosis, the primers are chosen so that they hybridize only to a specific portion of the target organism's DNA or so that the distance between the primer hybridization sites is unique to the target. To determine whether the target organism is present, PCR is performed on the sample. If no DNA in the sample can hybridize to the primers, no unique PCR product will be synthesized. On the other hand, if the desired target DNA is present, there will be a unique product representing the segment between the primer hybridization sites. Because PCR produces so many copies of the segment, the product DNA can easily be detected by a variety of methods, including agarose gel electrophoresis and simple staining.

The great advantage of PCR as a diagnostic tool is its ability to amplify rare DNA. The amplification allows tiny amounts of specific DNA to be detected. In many clinical specimens, the disease-causing organism is present in very low numbers. PCR can reveal the presence of these rare organisms without taking time for culturing. PCR also permits the use of very small specimens. This advantage has made PCR an invaluable tool in studying ancient DNA samples. Researchers can use minute samples of museum specimens to produce enough DNA for analysis, thus preserving the original specimen essentially intact. Because the products of PCR are not usually designed to be longer than 2,000 base pairs or so, PCR also works well on samples in which the DNA has undergone extensive fragmentation. (For more information on the use of PCR to study ancient samples, see the reference below.)

Below is a paper PCR activity you can do with your students. Although real PCR primers are at least 16 nucleotides long (to provide greater specificity), the paper primers are only 5 nucleotides long. In addition, the segments amplified by PCR are usually hundreds of nucleotides long, whereas this activity amplifies a short segment. However, this paper model very accurately demonstrates the steps of PCR and shows how a specific DNA segment can be amplified from a single copy. The second part of the activity illustrates how PCR is used as a diagnostic tool.

Objectives

After this lesson, students should be able to

1. Describe the steps in the PCR and explain how these steps can generate multiple copies of a specific DNA fragment;
2. Describe how PCR can be used in disease diagnosis.

Materials

- Strips of light- and dark-colored paper (at least eight of each per student group)
- Removable tape (one roll per student group)
- Scissors (one pair per group)
- At least eight light and eight dark primers per student group (Appendix K)
- One "double-stranded" parental DNA molecule per group (Appendix K)
- One "sample 1" and "sample 2" (Appendix K) page per group
- One copy of student questions per student

Teaching Resource

Carolina Biological Supply Co. sells an instructional video (catalog no. 21-2734) in which the model presented in this activity is used to explain how PCR works. The video also demonstrates a PCR laboratory activity that does not require a thermal cycler (see *PCR in the Classroom*, below).

Preparation

Cut the light and dark paper strips to about the width of the primer and template models.

If students do not have manuals, photocopy the page with the parental DNA sequence and the primers as many times as needed for your class. For the part of the activity that illustrates the use of PCR in disease diagnosis, make enough photocopies of the "Sample" pages for your class (or make transparencies).

Procedure
PCR
Refer to Fig. 18.3 through 18.8.

Each student group has a parental DNA molecule, primers, and strips of colored paper that will represent newly synthesized DNA (these strips are somewhat analogous to the free nucleotides in the

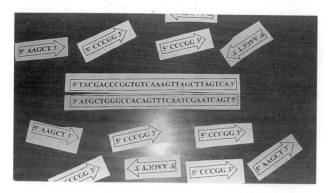

Figure 18.3 The starting template and primers for paper PCR. The colored paper strips are not shown.

solution). The two parental strands should be taped together to form a double-stranded molecule. There is no representative of the DNA polymerase in this model; the DNA "synthesis" will be performed by the students (Fig. 18.3).

Step 1. Denaturation

This is the 95°C step. Students should denature their double-stranded parental DNA into two single strands by removing the tape holding the strands together.

Step 2. Hybridization

The temperature is reduced, and hybridization can occur. Because there are so many primers in solution, they will hybridize to the parental DNA before the two parental strands can find each other (ideally, students should have many more than eight primers of each color on their desks). Students should "hybridize" the primers to the long DNA strands by matching the primer sequences to their complementary sequences and taping the primer in place. Be sure that the students hybridize the primers in the correct orientation: if the parental strand is oriented with the 5′ end on the left, then the 3′ end of the primer should be on the left. It doesn't matter if the letters are upside down. Real DNA strands are twisted

into a helix, so there is no upside down. Because of the way the paper primers are designed, a dark primer should always hybridize to a light DNA strand (Fig. 18.4).

Step 3. DNA Synthesis

Students will "synthesize" DNA by taping one end of a strip of light-colored paper to the 3′ end of the light primer that is hybridized to the dark single strand. The light-colored strip of paper (the new DNA strand) should be extended to the end of the parental DNA "template," and then any excess paper should be cut off. The new light-colored DNA strand should be taped to its complementary dark strand to represent the new double-stranded DNA molecule. Finally, students should write the correct DNA sequence on the new strand. Go through the same procedure with the dark primer that is hybridized to the light-colored strand, and synthesize a new dark strand.

After one round of synthesis, there will be two double-stranded DNA molecules (one end of each will be uneven; Fig. 18.5).

Now, repeat steps 1 through 3 with the two DNA molecules. *Note:* The primers that started the synthesis of strands in the previous round are now part of the new DNA strands. They must be used as part of the new templates, too (Fig. 18.6). The products will be four double-stranded molecules (Fig. 18.7). Notice how the length of the new strands is changing.

Repeat steps 1 through 3. This repetition produces eight double-stranded molecules, of which two stretch only from one primer sequence to the other (Fig. 18.8; note the two products with even ends stretching between the primer hybridization sites).

Figure 18.5 Paper PCR: products of the first round of DNA synthesis. Note that the primers are part of the product DNA strands. The base sequences of the product strands have been written on the paper strips.

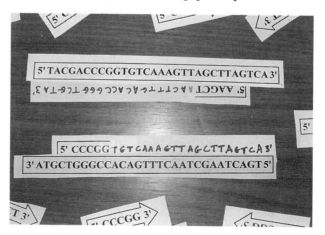

Figure 18.4 Paper PCR: primers are hybridized to the denatured template strands.

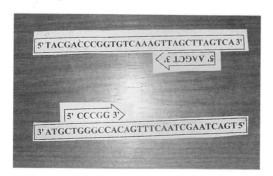

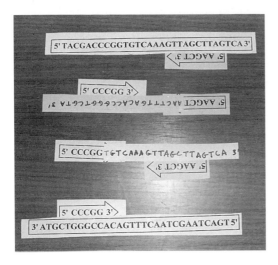

Figure 18.6 Paper PCR: hybridization for the second round of DNA synthesis. It does not matter that the letters on some of the strands are upside down.

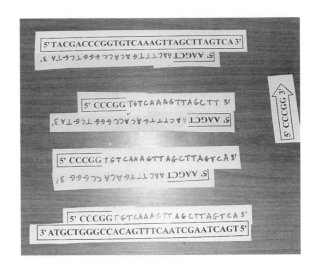

Figure 18.7 Paper PCR: products of the second round of DNA synthesis.

Ask your students to predict the products of another round of synthesis. Predicted products can be drawn on paper or the blackboard or done with paper materials. (If you have students perform another round of paper synthesis, they will need additional light and dark paper strips [eight of each] as well as additional light and dark primers [eight of each].) There will be 16 molecules, and 8 of them will be the short species. Yet another round will give 32 molecules, of which 22 will be the short species. A sixth round of synthesis yields 64 helices, of which 52 are short. The bottom line is that the number of molecules with primer sequences at their ends increases dramatically, forming the vast majority of the products.

It is important for students to see that the products of PCR are almost all identical and that they are double-stranded DNA segments that begin and end with the two primer sequences.

DNA analysis with PCR

Hand out the sample 1 and 2 DNA and primer sequences to your students if students do not have manuals. Have them cut out the primers. Tell the students that the two sample sheets represent DNA prepared from two different sources. The students are now laboratory technicians, and they will perform PCR on both samples, using the primers provided (they are to assume that they have large quantities of the primers).

Have students predict the DNA sequences of the major products from the two separate reactions.

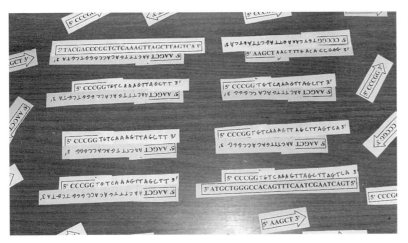

Figure 18.8 Paper PCR: products of the third round of DNA synthesis. These products include the first molecules of what will be the major product of many rounds of synthesis: the short products extending exactly from one primer sequence to the other. These molecules are the third one down the left-hand column and the second one down the right-hand column.

The product from sample 1 (primer sequences underlined) is

<u>TTCCAGCC</u>AGAGTCTCGGAACTAGCCTTATG
AAGGTCGGTCTCAGAGCCTTGAT<u>CGGAATAC</u>

There is no product from sample 2: primers do not hybridize.

The students will notice that "something is wrong" with sample 2 and that the primers do not hybridize anywhere. Get them to talk about this. Ask them what they would see if they simply performed the reactions with the two samples, loaded the results into a gel, electrophoresed, and stained. (Answer: A product band in the sample 1 lane but none in the sample 2 lane.) Ask them if they think it would be possible to use PCR to determine whether a specific microorganism was present in a sample or whether a specific gene was present in an organism.

Explain to the class that PCR can be used to detect the presence of disease-causing microorganisms in medical specimens. DNA is prepared from a sample taken from the patient (blood, tissue, sputum, etc.), and PCR is performed with primers that hybridize *only* to the DNA of the microorganism of interest. Laboratory workers can then tell if the microorganism was present in the sample by whether or not any DNA product is produced. This kind of test can be used in disease diagnosis.

Have students answer the questions (in class or for homework).

Answers to Student Questions

1. She would have to know enough of the DNA sequence of the virus to design primers that would hybridize to it. She would have to test to make sure that those primers did not hybridize to human DNA or to samples prepared from healthy patients (which might contain small amounts of DNA from other microorganisms). She could use computerized DNA sequence comparisons for some of her primer designing (to make sure the primer sequence was not present in any known organisms other than virus X).

2. Make the primers long. The longer the primer, the less likely that it will accidentally hybridize to other DNA molecules, since the odds of a given base sequence occurring randomly are 1 in 4^n, where n is the number of bases in the sequence (see chapter 16).

3. Answer: 2^n

4. Answer: $2n$

5. No. With only one primer, you would be limited to synthesizing one strand over and over and could never synthesize the complementary strand. After one round, you would have one new single strand. For the second round, you would still have only one template for the primer, so you could generate only one more new single strand, and so on.

Selected Readings

Mullis, K.B. 1990. The unusual origin of the polymerase chain reaction. *Scientific American* 262(4):56–65.

Paabo, S. 1993. Ancient DNA. *Scientific American* 269(5): 86–92.

PCR in the Classroom

If you do not have a thermal cycler

Carolina Biological Supply Co. sells kits (catalog no. 21-1220, 21-1222, and 21-1224) for classroom PCR that you can use even if you do not have a thermal cycler. They require two water baths, one boiling and one set at 55°C. Teacher and student directions are included. You must supply electrophoresis equipment to separate the products. These demonstration kits have been specifically designed to work with the less than ideal conditions of two water baths. The water baths will not give detectable products with commercial kits that require a thermal cycler. Carolina Biological Supply Co. also sells a videotape (catalog no. 21-2734) in which this activity is demonstrated.

If you have a thermal cycler

The experiment kits listed above, which were designed for teaching, can be used with a thermal cycler.

You can also order research-grade kits from Perkin-Elmer Corp., 761 Main Ave., Norwalk, CT 06854 (800-762-4002). Call and ask for a catalog.

If you go on to the lesson about DNA typing (chapter 24), you can order and perform PCR-based DNA typing with an educational kit from Carolina Biological Supply Co. Their kit "Human DNA Fingerprinting by Polymerase Chain Reaction" (catalog no. 21-1226) provides reagents needed to amplify a VNTR region on chromosome 1. You must supply electrophoresis equipment to separate the PCR products.

The Polymerase Chain Reaction

Introduction

One of the difficulties scientists often face in the course of DNA-based analysis is a shortage of DNA. A forensic scientist may have only a tiny drop of blood or saliva to test. An evolutionary biologist may want to analyze DNA from a museum specimen without destroying the specimen. Even if ample amounts of tissue or numbers of sample cells are available, it takes some work to purify specific DNA fragments in large quantities.

In 1985, a new technique that changed the whole picture was introduced. This technique essentially allows a scientist to generate an unlimited number of copies of a specific DNA fragment. It was invented by biotechnology industry scientist Kary Mullis, who had the initial inspiration one night in 1983 as he was driving and thinking about a technical problem he faced at work.

The essence of Mullis's idea was this. If you could set up a test tube reaction in which DNA polymerase duplicated a single template DNA molecule into 2 molecules, then duplicated those into 4, then duplicated those into 8, then 16, then 32, etc., you would soon have a virtually infinite number of copies of the original molecule. Each round of DNA synthesis would yield twice as many molecules as the previous round: a chain reaction producing specific pieces of DNA. Mullis's new technique was called the **polymerase chain reaction,** or PCR.

Of course, Mullis did more than just realize that DNA polymerase can copy one DNA helix into two. You already knew that, too. What he did was figure out how to generate a chain reaction in a test tube and how to get the reaction to copy the DNA segment of the scientist's choosing.

Mullis's approach relies on the characteristics of DNA polymerase enzymes and on the process of hybridization. Recall that DNA polymerases must have a primer base paired to a template DNA strand so that they can synthesize the complement to the template strand. Also remember that hybridization is the spontaneous formation of base pairs between two complementary single strands: you can separate the two strands of a helix by heating, but if you then allow the mixture to cool, the base pairs between the strands will re-form.

Now, how did Mullis use DNA polymerase and hybridization to get a chain reaction of DNA synthesis? Refer to the diagram in Fig. 18.1 during this explanation.

First, you decide what DNA segment you wish to duplicate (scientists say *amplify* instead of duplicate, because they are making so many copies). Then you synthesize two short single-stranded DNA molecules that are complementary to the very ends of the segment. These two short molecules must have specific characteristics. Look at the first panel in Fig. 18.1 under round 1. It shows a double-stranded parental DNA molecule with two copies each of the two short, single-stranded DNAs. Each of the single-stranded molecules is complementary to only one strand of the parental DNA, and each one is complementary to only one end of the segment. Furthermore, if you imagine these short molecules base paired to the complementary regions in the duplex, their 3' ends would point toward each other. These short, single-stranded molecules are the **primers.**

To begin the chain reaction, a large number of primers are mixed with the template molecule in a test tube containing buffer and many deoxynucleoside triphosphates. (What are they for?) This mixture is heated almost to boiling, so that the two strands of the parental molecule denature (Denaturation in Fig. 18.1).

Next, the mixture is allowed to cool. Ordinarily, the two strands of the parental DNA molecule would eventually line up and re-form their base pairs. However, there are so many molecules of primers in the mixture that the short primers will find their complementary sites on the parental strands before the two

Round 1

Double-stranded parental DNA

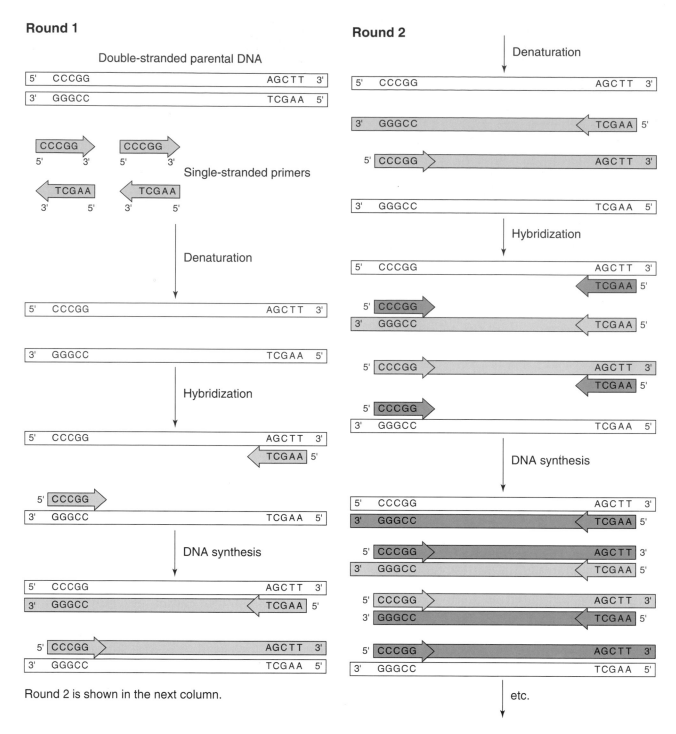

Round 2 is shown in the next column.

Round 2

Figure 18.1 PCR.

parental strands can line up correctly for base pairing. So a primer molecule hybridizes to each of the parental strands (Hybridization in Fig. 18.1).

Now DNA polymerase enzyme is added. The primers hybridized to the single-stranded parental molecules meet the requirements of the enzyme for DNA syn-

thesis. DNA polymerase begins adding the correct de-oxyribonucleotides to the 3′ ends of the primers, forming new complementary strands (DNA synthesis in Fig. 18.1).

After a short time, the mixture is heated up again. Now the two new double-stranded molecules denature, leaving four single strands (Denaturation, round 2). The mixture is cooled, and the abundant primer molecules hybridize to the single-stranded molecules (Hybridization, round 2). DNA polymerase is added again, and new deoxynucleotides are added to the 3' ends of the hybridized primers, yielding four double-stranded molecules (DNA synthesis, round 2). Notice that two of the newly synthesized strands begin and end at the primer hybridization sites.

This process of denaturation, hybridization, and DNA synthesis is repeated over and over, often 25 to 30 times, yielding huge numbers of molecules. The overwhelming majority of the newly synthesized molecules reach exactly from one primer hybridization site to the other. So by choosing the primers, a scientist controls which segment of the parental molecule is amplified. PCR is now used routinely for many different purposes: to amplify a specific fragment of DNA for cloning, to generate a DNA fingerprint from a minute sample, and even to diagnose diseases.

One technical improvement to the process outlined previously has made performing PCR even easier. Did you notice that we added DNA polymerase before each DNA synthesis step? That is because PCR was first carried out with *Escherichia coli* DNA polymerase, which is rendered inactive at the high denaturation temperature. So more enzyme had to be added for each round of synthesis. However, some organisms inhabit the very hot waters of hot springs and thermal ocean vents. The DNA polymerase enzymes from these organisms are not inactivated by the temperatures required for DNA denaturation. Now PCR is carried out with heat-resistant DNA polymerase, so the enzyme needs to be added only once at the beginning of the reaction cycles.

When PCR was first developed, scientists preset three water baths to the temperatures required for denaturation, hybridization, and DNA synthesis. They performed PCR by simply moving the reaction tube from one water bath to another. It wasn't long before enterprising biotechnology companies manufactured incubators that rapidly cycled between the desired temperatures, eliminating the need for manually moving tubes. These *thermal cyclers* have further simplified the performance of PCR. PCR can now be carried out by mixing parental DNA, primers, buffer, deoxynucleotides, and heat-resistant polymerase in a reaction tube; placing the tube in the thermal cycler; programming the thermal cycler to the desired time and temperature specifications; and waiting for the cycles to finish. It usually takes a few hours for many cycles of amplification. Figure 18.2 shows a scientist placing a

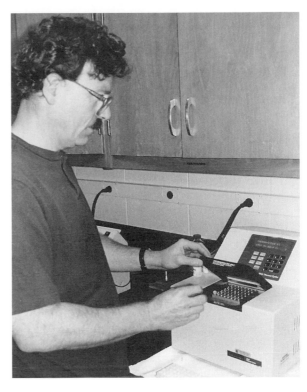

Figure 18.2 A scientist is loading a reaction tube into an automated thermal cycler (PCR machine). The digital keypad on the machine is for programming the desired temperatures, times, and number of reaction cycles.

PCR reaction tube into a thermal cycler in his laboratory.

When we (the authors writing this material) and other scientists first learned about PCR, the idea made perfect sense. However, we did not completely get it until we had worked through several reaction cycles ourselves, drawing the parental DNA molecules, the primers, the products, etc. Because we really learned PCR by working through it, we have provided a paper simulation for you to do the same thing. The template and primers are included in this chapter. Your instructor has directions, though you may be able to figure it out yourself. We have also included materials to simulate a PCR-based diagnostic test.

If you do not do the paper simulation, we highly recommend that you get some clean paper and draw through four rounds of PCR yourself. Figure 18.1 should get you started. We like using different colors of pens to keep track of the parental DNA, the primers, and the newly synthesized molecules (remember, primers will form part of the new molecules).

Here are some questions to think about.

Questions

1. What would a scientist have to know before she could design a PCR-based diagnostic test for virus X?

2. How could a scientist help ensure that her primers will hybridize only to the DNA she wishes to detect?

3. Write an expression that predicts the number of product molecules generated from a single double-stranded DNA molecule after n rounds of synthesis.

4. Predict the number of product DNA strands that are *not* the short primer to primer strands, that would be generated from one double-stranded DNA molecule after n rounds of synthesis.

5. Could you amplify DNA given only one primer? What would the products be after one round? Two rounds? Four rounds?

Transfer of Genetic Information

The previous set of classroom activities focused on the manipulation and analysis of DNA. Most of the methods we discussed depend on the availability of large numbers of identical DNA molecules. We get these identical molecules primarily through DNA cloning. The activities in the previous section illustrate part of the cloning process: the construction of a recombinant DNA molecule and its analysis by restriction and DNA sequencing. The other essential part of the cloning process is gene transfer: the introduction of new genetic information into an organism.

We use gene transfer in cloning as a means of propagating recombinant DNA molecules. The recombinant molecule is replicated inside the host cell as the cell multiplies, producing an essentially unlimited number of copies. We also use gene transfer to deliberately change the genotype of an organism so that the organism will have a new trait or make a useful product. For example, we have made corn resistant to certain caterpillar pests by introducing a gene for an insecticidal bacterial protein. We have induced *Escherichia coli* to make human insulin by introducing a complementary DNA copy of the human messenger RNA.

This section highlights the process of gene transfer. It begins with three wet laboratories that illustrate natural methods of gene transfer in *E. coli:* conjugation, transformation, and transduction. Although it is easy to think of these methods simply as tools of biotechnology (and they are important tools), the natural occurrence of gene transfer by each of these mechanisms is also significant. Introductory material at the beginning of each lesson gives examples of the medical importance of each of these natural gene transfer methods.

Following the *E. coli* laboratories is a reading about methods of plant genetic engineering and an exercise showing genetic alteration of a living plant by *Agrobacterium tumefaciens*.

Transformation of *Escherichia coli*

Classroom Activities

About This Activity

This laboratory is an economical and reliable protocol that teaches students one way to insert new genetic information into bacteria. Students demonstrate that *Escherichia coli* is unable to grow in the presence of the antibiotic ampicillin. They then insert a plasmid with an ampicillin resistance gene into the bacterium. After the transformation process, some of the treated *E. coli* cells express the antibiotic resistance gene and grow in the presence of the drug. Students observe phenotypic proof of a genotypic change in *E. coli*.

Two options for observing the transformed phenotype are given. The first calls for inoculating tubes containing liquid medium (broth versus broth plus ampicillin) with transformed and untransformed cells. Growth is observed as cloudiness in the medium. This method is very simple to carry out; it requires no agar plates or cell spreading by students. In the second method, students spread some of the transformed and untransformed cells on solid media with and without antibiotic. For this procedure, you must make or obtain the appropriate agar plates and supervise the students' use of alcohol and fire.

The liquid method of observing transformation is definitely the easier of the two and will demonstrate the phenomenon. The plating method is more interesting because it allows students to compare the number of cells in the transformed sample that will form colonies on the antibiotic with the number that will grow on the antibiotic-free medium (the percentage of antibiotic-resistant cells will be *very* low). The plating method is also the method of choice if you wish to go on and prepare plasmid DNA from the transformed cells (the plasmid minipreparation; a simple procedure is given in *DNA Science* by Micklos and Freyer [see Appendix H]). If you wish to perform minipreparations, start a liquid culture in antibiotic medium from one of the transformed colonies on the antibiotic plate.

Class periods required: 1–2 50-min periods or 1 90-min period plus observation and data recording on 1–2 subsequent days. There is an optional break-point in the procedure.

Introduction

Transformation occurs when cells take up free DNA molecules from the environment and express the information encoded in this DNA. This phenomenon is of great importance to experimental molecular biology because it provides a means of inserting new genes into cells. Transformation was first observed by Griffith, who found that when a living unencapsulated (rough) strain of pneumococcus (*Streptococcus pneumoniae,* formerly called *Diplococcus pneumoniae*) was mixed with dead cells of an encapsulated (smooth) strain, the rough-colony strain was transformed into the smooth-colony form. In 1944, Avery, McCarty, and McLeod demonstrated that purified DNA from the smooth cells is the substance that causes the transformation.

Bacterial cells in a state that allows them to be transformed are said to be **competent.** Only some bacterial strains, such as pneumococci, are naturally competent. Naturally competent cells contain proteins dedicated to the process of transformation. Proteins on the outsides of these organisms bind DNA and transport it into the cell. Internal proteins then compare the base sequence of the new DNA to the genome of the organism. If sufficient similarity (homology) is found, the new DNA is recombined into the genome and expressed. As part of the genome, this new DNA is replicated and passed on to daughter cells.

If the new DNA is not similar in sequence to the genome of the organism, it is not incorporated into the genome and is lost. In this way, naturally competent cells have access to genetic variability through transformation but do not waste their time expressing completely irrelevant genes. Natural transformation is believed to be the most important mechanism of genetic exchange for a number of species important to humans, notably *Streptococcus pneumoniae* (a causative agent of pneumonia) and *Neisseria gonorrhoeae* (the causative agent of gonorrhea).

Cells that are not naturally competent can often be artificially induced to take up DNA. In 1970, two scientists developed a method that enabled *E. coli* to take up and express DNA. Their method involved treating the cells with cold calcium chloride solution and then heat shocking them. They found that rapidly growing (log-phase) cells took up foreign DNA most efficiently after these treatments. To this day, no one knows exactly why these treatments cause *E. coli* to take up DNA. Obviously, some characteristic of the cell membrane must be altered to allow DNA molecules to enter.

Once inside *E. coli*, the DNA must be expressed for transformation to be achieved. This step can be a problem. *E. coli* contains enzymes called exonucleases that degrade linear pieces of DNA, starting at the ends and working inward. In most cases, new linear DNA fragments are destroyed before they even have a chance to recombine into the *E. coli* genome. This destruction is one reason scientists transform *E. coli* with plasmids. Plasmids are circular and are thus immune to exonuclease attack (because they have no ends).

Another reason scientists use plasmids is that plasmids contain their own origins of replication and so will be replicated and transmitted to daughter cells. It is not necessary for a plasmid to recombine into the *E. coli* genome to be maintained and expressed, so plasmids need not be similar in base sequence to the *E. coli* chromosome. Plasmids therefore make very convenient vehicles (or vectors) for introducing new genetic material into *E. coli*.

Where do plasmids come from? Plasmids have been found in many different species of bacteria and in yeasts. Plasmids seem to be extra pieces of DNA; they do not contain any genes that are essential to the life of the organism. However, they often (but not always) contain genetic material that gives the cell survival advantages under certain conditions, such as genes for resistance to antibiotics, resistance to normally poisonous heavy metals, degradation of unusual chemicals, or killing of other bacteria.

This laboratory exercise uses the calcium chloride-heat shock method to introduce a plasmid into *E. coli*. The plasmid carries a gene for ampicillin resistance that is used to detect its presence in transformed cells. The antibiotic ampicillin prevents the formation of cell walls in *E. coli* (and many other bacteria), thus preventing the bacterium from producing new cells. The ampicillin resistance gene encodes an enzyme called ß-lactamase that breaks down the ampicillin molecule, allowing cells to multiply in ampicillin-containing media.

You may notice an interesting phenomenon if you use the plating method of detecting transformants. After a period of incubation, tiny colonies begin to grow in a halo around the original ampicillin-resistant colonies of transformed bacteria on the ampicillin plates. These are called satellite colonies. They form because the ß-lactamase enzyme produced by the transformed cells diffuses into the medium, destroying the ampicillin in the vicinity of the colony. When enough of the antibiotic has been destroyed, nontransformed, ampicillin-sensitive cells that were on the plate all the time but couldn't form colonies because of the antibiotic begin to multiply successfully, forming the satellite colonies.

Note: This activity can also be done with other plasmids such as pKAN, pUC18, or pBR322. The pUC18 and pBR322 plasmids contain ampicillin resistance genes and can be substituted into the procedure without any changes. pKAN contains a kanamycin resistance gene instead. If you use pKAN, you must substitute kanamycin for ampicillin in the selective media (recipes are in Appendix E), but otherwise, the procedure is the same.

Kanamycin works differently from ampicillin; instead of interfering with cell wall construction, kanamycin blocks translation. Blocking translation is lethal for the cells. Several different kanamycin resistance genes exist, but all of them work in basically the same way: they all encode enzymes that modify the kanamycin molecule so that it cannot enter the *E. coli* cell. (The different versions of kanamycin resistance enzymes cause different modifications to the drug.) The kanamycin resistance enzymes stay in the periplasmic space (the area between the inner and outer membranes of *E. coli*) and modify the drug there so that it cannot cross the inner membrane. Because the enzymes stay inside the cell, you won't see satellite colonies on kanamycin plates.

Objectives

At the end of this laboratory, students should

1. Be able to discuss how the transformation procedure enabled some *E. coli* cells to form colonies on antibiotic-containing media;

2. Have reinforced their understanding of genotype and phenotype;

3. Understand the importance of sterile technique when handling bacteria.

Materials

Equipment

- Thermometers (Centigrade)
- 0.5- to 10-µl pipettors or other small-volume measuring devices
- 100- to 1,000-µl pipettors
- Refrigerator (optional)
- Autoclave or pressure cooker (optional for making plates and sterilizing test tubes)

Supplies

- *E. coli* streaked on plates (strain MM294, JM101, or other suitable strain)
- 0.1 M (100 mM) calcium chloride
- Plasmid DNA (pAMP or other)
- Tryptic soy broth (TSB) or other suitable medium such as Luria broth
- TSB plus ampicillin
- Styrofoam cups
- 15-ml sterile culture tubes
- Marking pencils
- Microcentrifuge tubes
- Tips for 0.5- to 10-µl pipettors
- Tips for 100- to 1,000-µl pipettors
- Ice
- Inoculating loop
- Lysol or bleach sterilizing solution
- Rack(s) for holding microfuge tubes

Optional

- Nutrient plates (such as TSB plates or L plates)
- Nutrient plates plus ampicillin

Recipes for media and solutions are in Appendix E.

Resource materials

- Carolina Biological Supply Co. "Colony Transformation Kit" (catalog no. 21-1142)

- Connecticut Valley Biological Supply "Genetic Engineering II: Plasmid Uptake Kit" (catalog no. AP821)

- Edvotek kit 201, "Transformation of *E. coli* Cells with Plasmid DNA"

- Plasmid DNA can be purchased from many biotechnology supply companies.

Preparation

Each laboratory team will need the following materials:

- Marking pencil
- 100- to 1,000-µl pipettor and sterile tips
- 0.5- to 10-µl pipettor and sterile tips
- One Styrofoam coffee cup (for a hot water bath)
- Centigrade thermometer
- One Styrofoam cup of ice
- 500 µl of 100 mM calcium chloride solution
- 0.25 µg of plasmid DNA
- Two 1.5-ml sterile microcentrifuge tubes
- Inoculating loop
- Access to flame or other sterilizer for loop
- 1 ml sterile TSB

For detecting transformants in liquid culture (option 1):

- Three sterile 15-ml culture tubes with 4 ml of TSB each
- Two sterile 15-ml culture tubes with 4 ml of TSB + ampicillin each
- 100- to 1,000-µl pipettor and sterile tips

For plating the transformation mixtures (option 2):

- Bacterial spreader (bent glass rod)
- Alcohol in a petri dish (enough to wet the spreader)
- Access to a flame (alcohol or Bunsen burner)
- Two nutrient medium plates (TSB agar or Luria agar or nutrient agar)
- Two nutrient medium plates with antibiotic

1. *E. coli.* Prepare one streak plate for every two laboratory teams the day before doing this laboratory. *This procedure works best with plates incubated for 1 day at 37°C or for 2 days at room temperature (20 to 25°C).* If you like, have each student team prepare one streak plate the day before the laboratory.

2. Plasmid DNA. The plasmid DNA should be at a concentration of about 0.05 µg/µl (µg/µl is equivalent to mg/ml). Most companies sell DNA in more concentrated solutions, so you may have to dilute the DNA before use. For example, if your DNA is at a concentration of 1 mg/ml (same as 1 µg/µl), you need to dilute it 20-fold. Add 5 µl of plasmid DNA to 95 µl of sterile water or Tris-EDTA (ethylenediaminetetraacetic acid) buffer. This yields 100 µl of DNA at 0.05 µg/µl, which is enough for 20 student experiments. Store the diluted DNA in the refrigerator if you need to keep it overnight.

3. Calcium chloride. The recipe is in Appendix E.

4. TSB plus antibiotic. Using a sterile pipette tip, add 0.1 ml of 100× antibiotic (ampicillin or kanamycin) solution to every 10 ml of broth. For example, add 0.4 ml (400 μl) of ampicillin to 40 ml of TSB (recipes are in Appendix E).

5. Ensure proper treatment of microbial waste. Glass- or plasticware and any other materials that contact cells should be placed into a 10% Lysol solution (or a 20% bleach solution) overnight. Autoclaving the materials at 15 lb for 15 min is an alternative method but is not necessary if you use the disinfectant solutions properly. Prepare appropriately designated waste containers for your class.

6. Make copies of the *Student Activity* pages for each student if necessary.

Tips

1. Students should study the protocol before beginning the laboratory. Success depends on accurate movement through the steps.

2. In step 4 of the procedure, it is important to use a sufficiently large cell mass. If the plates do not contain enough 3-mm-diameter colonies, instruct students to scrape up several smaller colonies. Plates incubated for 24 h at 37°C or for 48 h at room temperature should contain sufficiently large colonies.

3. In step 5 of the student procedure, it is very important that students achieve good resuspension of their cells. No visible clumps should remain. It is also important that they resuspend the cells in the + tube first, because these cells then have time to preincubate in the calcium chloride while the cells in the − tube are being resuspended. The preincubation improves transformation efficiency.

4. After students have heat shocked their cells and added broth to the tubes, the cells can incubate for up to several hours before being plated or inoculated into liquid media. You will achieve better results with pKAN if you let the cells sit in the broth for at least 30 min before plating; the newly transformed cells need time to express the kanamycin resistance enzyme that will protect them from the lethal effect of the antibiotic. This outgrowth period is not as important with ampicillin. Cells transformed to ampicillin resistance can be plated immediately. If you need to leave the cells

until the next day before plating, let them sit at room temperature or 37°C for a while, and then put them in the refrigerator overnight. If your class ends at the end of the school day, put the cells in the refrigerator then but get them out in the morning and allow them to incubate at room temperature for a while before class.

5. The cells in tubes B and D (in the liquid detection option) are not dependent on transformation, and they will cloud their tubes within 24 h. It may take the transformants longer to cloud their tube: there are far fewer of them to start with, and they must express the ampicillin resistance enzyme before they can multiply. Tube A cells will cloud within 48 h.

6. Expected results. Tubes A, B, and D: Cell growth, indicated by cloudiness, should occur. Tubes C and E: Cell growth should *not* occur; the fluid will be clear. Plates: There should be luxurious (confluent) growth on both tryptic soy agar (TSA) plates, some colonies on the TSA plate with ampicillin (TSA-AMP+), and no colonies on the TSA plate without ampicillin (TSA-AMP−).

Answers to Student Questions

1. The expected results are given above. If students get other results (such as growth on the TSA-AMP− plate or growth in tube C), they should explain.

2. Tube C (plate TSA-AMP−) is a control to make sure that the ampicillin inhibits the growth of nontransformed cells. There should be no growth. Tube E is a control for medium contamination (no equivalent plate control). Tube D is a control for cell viability of untransformed cells in nutrient medium (analogous to plate TSA−). Tube B is also a control for cell viability that uses the cells to which DNA was added (plate TSA+ is the analog). Tube A and plate TSA-AMP+ allow us to specifically ask whether any cells were transformed by the plasmid DNA. Only cells that express the plasmid-borne ampicillin resistance gene can grow here.

3. No cells grew in tube E (let's hope) because it was not inoculated and the broth was sterile. No cells grew in tube C (on plate TSA-AMP−) because the medium contained ampicillin and the bacteria are not resistant and were not given the opportunity to be transformed by an ampicillin resistance plasmid. Cells grew in tubes B and D (on plates TSA+ and TSA−) because these cul-

tures each contain simple nutrient medium that supports the growth of transformed or untransformed cells. Cells grew in tube A (on plate TSA-AMP+) because the cells had become ampicillin resistant by taking up the pAMP plasmid DNA and expressing the ampicillin resistance gene.

4. Tube A (plate TSA-AMP+) specifically selects transformants. The medium contains ampicillin, which prevents the growth of any cells that have not taken up the plasmid DNA and expressed its ampicillin resistance gene.

5. Answers will vary.

Technical Note

Medium plates and solutions containing antibiotics can be stored in the refrigerator for a month or two without loss of the antibiotic's activity. Antibiotics in media will lose their activity during long-term storage. *Do not keep antibiotic media from one year to another*. If you use antibiotic media that is too old, the experiment will appear not to have worked; untransformed cells will grow on the antibiotic-containing plates (because the antibiotic is no longer active). Stock solutions of antibiotics (see Appendix E) can be stored in the freezer for years. Make a stock and freeze it in small amounts. When you need antibiotic media, thaw one of the small tubes, and add the drug to some fresh media.

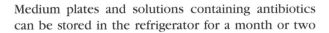

Classroom Activities

Transformation of *Escherichia coli*

Introduction

Transformation is the uptake and expression of foreign DNA by bacterial cells. This phenomenon occurs infrequently in nature: a few types of bacteria take up DNA from their environment, but most don't. *Escherichia coli* is one of the many bacterial strains that do not undergo transformation naturally. However, in 1970, a process for increasing the ability of *E. coli* cells to be transformed was developed. Rapidly growing cells were suspended in cold calcium chloride and exposed to high concentrations of plasmid DNA. The cells were then briefly incubated at a relatively high temperature. After this treatment, some of the cells expressed genes encoded on the plasmid: the cells had been transformed. This laboratory exercise is based on that procedure.

In this activity, you will cause *E. coli* to take up plasmid DNA. A plasmid is a small, circular DNA molecule that acts like a minichromosome in a cell. The cell's DNA replication enzymes duplicate the plasmid DNA just as they duplicate the regular chromosome, so plasmid DNA molecules are inherited by both daughter cells when the bacterium divides. This means that if a single bacterial cell takes up a plasmid molecule, all its descendants will contain the molecule, too. Their genetic makeup will include the genes encoded by the plasmid DNA.

Scientists use plasmids as convenient vehicles for introducing new genes into cells. It is easy to isolate plasmid DNA. New genes can be added to the purified plasmid DNA by using restriction enzymes and DNA ligase. Finally, the new recombinant plasmid can be introduced into host cells by transformation. When plasmids are used this way, they are called vectors. A vector is any DNA molecule used to deliver new genes to cells.

Here is a problem for you to think about. Even the most carefully conducted transformations of *E. coli* are very inefficient. Only one cell in thousands or millions takes up plasmid DNA. How can you find the few transformed cells among the many untransformed ones?

When scientists do recombinant DNA work, they usually use plasmids that carry marker genes. Marker genes are genes that produce an easily detected phenotype such as resistance to an antibiotic or a color change under certain conditions. The plasmid you will be using in this activity carries an antibiotic resistance marker gene. How could you take advantage of this marker to detect cells that take up your plasmid DNA?

Procedure

1. Label one sterile 1.5-ml microcentrifuge tube as + (with plasmid) and another as − (without plasmid). Plasmid DNA (pAMP) will be added to the + tube; none will be added to the − tube, which is our control.

2. Using a 100- to 1,000-μl pipettor and a sterile tip (or a sterile transfer pipette), add 250 μl of 0.1 M calcium chloride solution to each labeled tube (+ and −).

3. Place both tubes in an ice-filled cup.

4. Transfer one or two large (3-mm-diameter) colonies from an agar plate to the + tube as follows.
   - Sterilize an inoculating loop in a Bunsen burner flame until the loop glows red-hot. Continue to pass the lower third of the shaft through the flame.

   - Stab the loop into the agar several times to cool the loop. *Do not touch the bacterial colonies until you have cooled the loop.* If you touch the cells with a hot loop, you will kill them.

   - Scrape into the loop one or two 3-mm-diameter bacterial colonies, but be careful not to transfer any agar. Impurities in the agar can inhibit transformation and ruin your experiment.

- Immerse the filled loop in the calcium chloride solution in the + tube, and *vigorously* tap the loop against the tube's wall to dislodge the cells. Hold the tube up to the light to observe whether the cell mass fell off the loop.

  - Reflame the loop before setting it down. Flaming sterilizes the loop so that you won't contaminate your work area. (Remember, you just touched living *E. coli* cells with the loop.)

5. Immediately resuspend the cells in the + tube by repeatedly pipetting in and out with the 100- to 1,000-μl pipettor and sterile tip (or sterile transfer pipette). Hold the tube up to the light, and inspect it carefully to ensure that the suspension is homogeneous. You cannot hurt the bacterial cells by being too vigorous. *It is important that no visible clumps of cells remain.*

To avoid inhaling or splashing into your eyes any aerosol that might be created, keep your face away from the tip end of the pipette when you are pipetting the suspension culture.

6. Return the + tube to ice.

7. Transfer to the − tube and suspend a second mass of cells, as described in steps 4 and 5.

8. Return the − tube to the ice. Incubate both tubes on ice for 10 to 15 min.

9. Use the 0.5- to 10-μl micropipette and a sterile tip to add 5 μl of plasmid DNA to the + tube. Tap the tube with your finger to mix the contents. Avoid making bubbles in the suspension or splashing the suspension up the sides of the tube.

10. Return the + tube to the ice. Incubate both tubes on ice for an additional 15 min.

11. When the incubation period is nearly over, prepare the heat shock bath. Go to the sink and fill your second Styrofoam cup with water that is at 43°C. Use the thermometer to determine the temperature of the water. You will need the water to be at 42 to 43°C at the end of the incubation period.

12. When the incubation on ice is over, heat shock the bacteria. *It is essential that the cells receive a sharp and distinct shock.*
    - Make sure that your heat shock bath is 42 to 43°C.

- Remove both tubes from the ice bath, and immerse them in the hot water bath for exactly 90 s.

- Immediately return the tubes to the ice bath, and let them stand for at least one additional minute.

13. Set the tubes in a rack at room temperature.

14. Use a 100- to 1,000-μl pipettor to add 250 μl of tryptic soy broth (TSB) to each tube. Gently tap the tubes to mix the contents.

The cells can be plated immediately if you are looking for ampicillin resistance, or they can be left until the next day. If you are looking for kanamycin resistance, incubate the cells for at least 30 min before plating them or inoculating them into liquid media. If the cells are to be left overnight, incubate them for about 1 h at 37°C or for longer at room temperature, and then refrigerate them.

15. Clean up responsibly. Put all waste that has come in contact with bacterial cells in the designated biological waste containers.

Your instructor will inform you which detection option to use.

Option 1: detection of transformants in liquid culture

1. Obtain from your instructor three sterile culture tubes containing 4 ml of TSB each and two sterile culture tubes containing 4 ml of TSB with ampicillin each.

2. Take your + tube from the procedure discussed previously. Using a sterile device, add 100 μl of the mixture in the + tube to one of the two tubes containing TSB with ampicillin. Label this tube A.

3. Using a fresh sterile device (if you are using micropipettes, this means using a fresh sterile tip), add 100 μl of the mixture in microcentrifuge tube + to a culture tube containing TSB (no ampicillin). Label this tube B.

4. Using a fresh sterile device, add 100 μl of the mixture in microcentrifuge tube − to the other tube of TSB with ampicillin. Label this tube C.

5. Using a fresh sterile device, add 100 μl of the mixture in microcentrifuge tube − to a tube of TSB (no ampicillin). Label this tube D.

Option 1: detection of transformants in liquid media

Tube	Source of cells	Medium in detection tube	Expected results	Observed growth	
				24 h	48 h
A	+ Tube	Antibiotic			
B	+ Tube	No antibiotic			
C	− Tube	Antibiotic			
D	− Tube	No antibiotic			
E	None	No antibiotic			

Option 2: detection of transformants by plating on solid media

Plate	Source of cells	Expected results (indicate many, few, or no colonies)	Observed growth	
			24 h	48 h
TSA-AMP	+ Tube			
TSA-AMP	− Tube			
TSA	+ Tube			
TSA	− Tube			

6. Do not inoculate the last tube of TSB. Label this tube E.

7. Incubate all tubes at 37°C or room temperature.

8. Fill out the Expected Results table with what you expect to see in terms of growth in the five tubes.

9. Clean up responsibly. Put all waste that has come in contact with bacterial cells in the designated biological waste containers.

Next Day (Day 2 or 3)

Observe the five tubes by tapping them gently and smelling the contents. Do the tubes become cloudy when you tap them? Is there a new smell (apart from the smell of the medium)? Cloudiness and a new smell indicate growth of *E. coli*. Record your observations, and fill in the table under "Observed growth" at 24 h.

Day After (Day 3 or 4)

Repeat the observation procedure for day 3. Fill in the table under "Observed growth" at 48 h. Has any change occurred?

Option 2: detection of transformants by plating on solid media

Obtain from your teacher the following materials.

- Two plates of tryptic soy agar (TSA) or other nutrient medium
- Two plates of TSA with ampicillin (TSA-AMP)
- One spreader (bent glass rod)

1. Label one TSA plate and one TSA-AMP plate as +; label the other two plates as −.

2. Use the matrix given here as a checklist as you spread the + tube and the − tube cells on each type of plate.

	TSA plate	TSA-AMP plate
Control cells (− tube)	100 μl	100 μl
Transformed cells (+ tube)	100 μl	100 μl

3. Use a 100- to 1,000-μl micropipette and sterile tip (or a sterile transfer pipette) to add 100 μl of cell suspension from the − tube to the TSA− plate and another 100 μl to the TSA-AMP plate. Do not let the suspension sit on the plate too long before spreading it; if too much liquid is absorbed by the agar, the cells cannot be evenly distributed.

To avoid inhaling or splashing in your eyes any aerosol that might be created, keep your face away from the tip end of the pipette while you are pipetting the suspension culture.

4. Spread the cells with a sterile spreader. The object is to spread the cells out evenly and isolate

them from each other on the agar surface so that each cell can give rise to a distinct colony. Use the following procedure.

- Dip the spreader in ethanol, and then pass it through a Bunsen flame only long enough to ignite the alcohol. Remove the spreader from the flame (the spreading rod will become too hot if left in the flame).

- Allow the alcohol to burn off. Lift the lid from the TSA− plate, but do not set the lid down on the laboratory bench.

- Cool the spreader by touching it to the agar surface away from the 100 µl of cell suspension. It is essential to cool the spreader before touching the cells.

- Touch the spreader to the cell suspension, and then gently drag it back and forth across the surface of the agar.

- Rotate the plate a quarter turn, and repeat the spreading motion. Remember, the object is to spread the cells out as evenly as possible.

5. Repeat step 4 with the TSA-AMP− plate, and spread the cell suspension.

6. Use a new sterile tip to add 100 µl of cell suspension from the + tube to the TSA+ plate, and another 100 µl to the TSA-AMP+ plate.

7. Repeat step 4 to spread the cell suspensions on the TSA+ and the TSA-AMP+ plates.

8. *Reflame the spreader one last time.* Let it cool in the air a minute, and then put it down.

9. Place the four plates upside down in a 37°C incubator. Incubate them for 12 to 24 h. If an incubator is not available, incubate them upside down at room temperature for 2 days. After this incubation, move the plates to a refrigerator to preserve the culture. Plates are incubated upside down so that if any condensation takes place inside them, the liquid will not drip back onto the agar surface. If there is liquid on the surface of the agar, the bacterial cells will spread out and you will see a film of growth rather than isolated colonies.

10. Clean up responsibly. Put all waste that has come in contact with bacterial cells in the designated biological waste container.

Last Day
Sketch the four plates.

Questions

1. Compare the observed growth to what you expected. Account for the similarity or dissimilarity in the results.

2. What was the purpose of the TSA-AMP− plate? Of the TSA− plate? Of the TSA+ plate? Of the TSA-AMP+ plate? If you detected transformants in liquid, indicate the purpose of each tube.

3. Explain why growth occurred or did not occur on each of the plates or in the tubes.

4. On which plate(s) did you specifically detect transformants? Justify your answer.

5. Using the results of the laboratory, discuss the relationship of genotype to phenotype.

Conjugative Transfer of Antibiotic Resistance in *Escherichia coli*

20

About This Activity

This quick, easy activity allows students to observe conjugational transfer of an antibiotic resistance marker from one strain of *Escherichia coli* to another. It has been used successfully in high school and college classes.

Class periods required: *1 for the experiment; part of another to record results*

Introduction

In 1946, Lederburg and Tatum discovered that genes could be exchanged between *E. coli* cells in a process that required direct contact between the cells and a special fertility (F) factor in the donor cell. This process was named conjugation and is also referred to as bacterial sexuality because of the direct donation of genetic material. Since 1946, other conjugation systems fairly similar to the F system have been discovered.

The basic form of the F factor is the F plasmid, a very large plasmid that contains several genes required for its conjugational transfer. Any cell that contains the F plasmid can synthesize all the proteins needed for conjugation (from the F genes), and so is considered fertile. Fertile cells synthesize a special structure called a pilus, a tubelike appendage that protrudes from the outer membrane. The pilus binds to a recipient cell (a cell lacking F) and brings the pair together. In a process that is not completely understood, the F plasmid is then copied from a special replication-transfer origin by the fertile cell, and one copy is transferred to the recipient cell. The recipient thus becomes fertile, and the donor remains so.

Additional genes can be incorporated into the F plasmid by natural recombination processes or through laboratory manipulation. F plasmids that contain other genes are often called F′ (pronounced F prime) plasmids. These bacterial genes are then transferred during conjugation. Finally, the F plasmid occasionally recombines into the *E. coli* chromosome. When these fertile cells begin conjugation, they attempt to replicate and transmit the entire circular chromosome! (They can actually transmit the whole thing if they remain paired with the recipient long enough.)

Many plasmids besides F encode proteins that allow them to be transmitted by conjugation. These conjugative plasmids also occasionally allow the transmission of nonconjugative plasmids, so virtually any plasmid can be transferred at some frequency. Some plasmids can replicate in only a few types of host and promote conjugation only between those hosts. Others, however, have a very broad host range and promote conjugation in hundreds of bacterial species.

The medical implications of conjugation are profound. Virtually every clinically important antibiotic resistance gene is carried on a plasmid. Conjugation allows the spread of plasmids not only between different individuals of the same bacterial species but also between species and even between genera. Conjugation is believed to be the most important route of transmission of antibiotic resistance in most disease-causing bacteria.

In this activity, students will observe the conjugative transmission of ampicillin resistance to a cell that is already resistant to streptomycin. The antibiotic streptomycin acts by binding to a ribosomal protein and preventing protein synthesis. Streptomycin resistance is conferred by a mutation in the gene for that ribosomal protein. This specific mutation alters the shape of the protein so that streptomycin can no longer bind to it but leaves the protein functional. Since streptomycin resistance is a chromosomal mutation, it is not normally transmitted by conjugation. Ampicillin resistance, however, is conferred by a plasmid-borne gene and can readily be transmitted on a conjugative plasmid. Conjugation experiments are usually set up so that cells that receive the desired information, the recipient cells, can be easily detected. In this experiment, the donor cells are resistant to ampicillin, and the recipient cells are resistant to streptomycin. Only the desired conjugation products are resistant to both. Thus, by plating on medium containing both antibiotics, we permit only the conjugation products to grow.

Objectives

At the end of this activity, students should be able to

1. Define the term "conjugation";

2. Describe how to set up a bacterial conjugation experiment and explain how to detect the conjugation products;

3. Give an example of one way in which bacterial conjugation is important to people.

Materials

- Frozen or stab culture of *E. coli* cI (streptomycin-resistant recipient cells)
- Frozen or stab culture of *E. coli* cII (ampicillin-resistant donor cells)
- Tryptic soy broth, Luria broth, nutrient broth, or other medium
- Ampicillin solution
- Streptomycin solution
- Tryptic soy (or other medium) agar plates, one per student group
- Tryptic soy agar plates with ampicillin, one per student group
- Tryptic soy agar plates with streptomycin, one per student group
- Tryptic soy agar plates with ampicillin and streptomycin, one per student group
- Micropipettes and sterile tips
- Optional: other sterile inoculators (sterile bacterial loops, sterile cotton swabs, etc.)
- Small sterile containers: microcentrifuge tubes, glass test tubes, or other containers
- Marking pens for writing on petri plates

Note: Recipes for media and antibiotic solutions are given in Appendix E.

Resources

- Connecticut Valley Biological Supply "Introduction to Natural Genetic Engineering Kit" (catalog no. AP8200)

- Carolina Biological Supply Co. "Introductory Bacterial Conjugation Kit" (catalog no. 21-1125) and "Advanced Bacterial Conjugation Kit" (catalog no. 21-1127). Individual materials are also available. The kit refill contains bacterial strains cI and cII.

- The Carolina Biological Supply Co. "Genetic Construction Kit" (catalog no. 17-1915) is a bacterial conjugation kit in which the marker transferred is the β-galactosidase gene.

Preparation

Ahead of time

On the day of the experiment, you will need to have freshly grown overnight cultures of cI and of cII. Each student group will use 200 to 400 μl of each culture, so plan a convenient volume (10 ml is more than enough for 20 student groups). Broth medium can be made, sterilized ahead of time, and then stored at room temperature in a closed container until you need it. The day before you plan to do the experiment, add ampicillin to one culture flask, and inoculate the culture with cII. Add streptomycin to the other culture flask, and inoculate the culture with cI. Grow the cells without shaking (or with gentle shaking) at 37°C or room temperature.

If you prefer, you can have each student group inoculate small overnight cultures of cI and cII for their own use the next day. Depending on your time and resources, you can have the groups streak out each strain on appropriate antibiotic media and then use an isolated colony to inoculate an overnight culture for the experiment.

Make and pour the plates ahead of time. If you have sterile containers, you can make one batch of agar, sterilize it, and then divide it among four sterile containers while it is liquid, using sterile technique. Add antibiotics to three of the containers, and pour your plates. Alternatively, prepare four batches of agar, sterilize them all, and add antibiotics to three of them. Antibiotic-containing agar plates will keep in the refrigerator for a few weeks.

Sterilize anything else you will need.

Photocopy the *Student Activity* pages, if necessary.

Day of experiment

Set up containers of 1:5 dilute bleach or Lysol for disposal of tips and other materials that come in contact with bacterial cells.

Each student group will need

- One each of the four types of plates
- Three small, sterile containers
- Access to a micropipette and sterile tips

- Optional: sterile inoculating tool (such as cotton swabs or inoculating loops)
- A marker
- Photocopies of the *Student Activity* pages, if necessary

Procedure

1. Review the process of conjugation. Have students look at the diagram in the *Student Activity* section and suggest a method for detecting only successful conjugation products. Discuss appropriate controls. We are told that cI is streptomycin resistant and ampicillin sensitive, but what do we need to do to make sure? What about cII?

2. Have students follow the procedure in the *Student Activity* pages. Since the cells must be left together for 20 min (approximately) to conjugate, you can use the time to continue your discussion. You may want to do question 1 as a class activity at this time.

   Ask students what they think happens to their own intestinal bacteria when the doctor prescribes ampicillin to them for an ear infection. They should eventually realize that the ampicillin can kill sensitive intestinal bacteria, leaving any resistant forms to multiply more freely. These resistant forms could then transmit their resistance genes. The spread of antibiotic resistance among disease-causing microbes is a significant and growing medical problem.

3. On the next day, have students observe and record their results. If the cells have not grown sufficiently, incubate them for another day.

Tips

Your students can use any sterile volumetric device to measure out the cI and cII cells. The exact volume is not important (although the procedure says 100 or 200 μl); just have the students mix equal volumes of the two cell cultures.

It is also not important that the indicator plates be inoculated with a micropipette. Any sterile inoculation device will do. For example, sterile cotton swabs purchased at the pharmacy work admirably. Have students use a new swab for each inoculation. Plates may also be "spread" by using bent glass rods flamed in alcohol as in the transformation activity (please refer to that lesson plan [chapter 19] for instructions). If you spread cells on the indicator plates, you will need three plain, three ampicillin-containing, three streptomycin-containing, and three ampicillin-streptomycin-containing plates per student group, since only one sample can be spread on a plate.

Optional Follow-Up

Invite someone from your hospital laboratory (or a knowledgeable physician) to talk to your class about the spread of antibiotic resistance among disease-causing organisms.

Answers to Student Questions

1. Make sure that students fill in Table 20.1 correctly (see next page) and understand which cells they are looking at in each case. Go over the table in class, and refer to it when you discuss the observed results of the experiment.

2. Only the conjugation products (recipient cells that received the ampicillin resistance gene from the donor) can grow on both antibiotics. Both parental strains will fail to grow.

3. Having the recipient be streptomycin resistant makes it possible to distinguish the conjugation products from the parental strains. The conjugation products have a unique combination of resistances that neither parent has. If the recipient strain were streptomycin sensitive (like the donor strain cII), it would not be possible to tell the conjugation products (ampicillin resistant, streptomycin sensitive) from the ampicillin-resistant, streptomycin-sensitive donor cells.

Selected Reading

Amábile-Cuevas, C., M. Cárdenas-Garcia, and M. Ludgar. 1995. Antibiotic resistance. *American Scientist* 83:320–329.

Classroom Activities

Table 20.1   Expected answers on student copies of Table 20.1

Medium	Inoculated with:		
	cI	**cII**	**cI + cII**
Plain agar			
Do you expect growth?	Yes	Yes	Yes
Reason?	No antibiotic in medium	No antibiotic in medium	No antibiotic in medium
Amp agar			
Do you expect growth?	No	Yes	Yes
Reason?	cI is ampicillin sensitive.	cII is ampicillin resistant.	The mixture contains cII, which is ampicillin resistant, and conjugation products, which are resistant to ampicillin and streptomycin.
Strep agar			
Do you expect growth?	Yes	No	Yes
Reason?	cI is streptomycin resistant.	cII is streptomycin sensitive.	The mixture contains cI, which is streptomycin resistant, and conjugation products, which are resistant to ampicillin and streptomycin.
Amp-Strep			
Do you expect growth?	No	No	Yes
Reason?	cI is ampicillin sensitive.	cII is streptomycin sensitive.	Only the conjugation products will grow. They are resistant to both ampicillin and streptomycin.

Classroom Activities

Conjugational Transfer of Antibiotic Resistance in *Escherichia coli*

20

Introduction

Conjugation refers to the transfer of genetic information from one bacterial cell to another in a process that requires contact between the cells. In conjugation, one cell always donates genetic information to the other; the donor cell is called the male cell or fertile cell, and the recipient is called the female cell.

What makes a bacterial cell fertile? Fertile (male) cells contain special plasmids called conjugative plasmids. The first conjugative plasmid discovered was called the F (for fertility) plasmid. Other conjugative plasmids work in a manner similar to F.

The F plasmid encodes proteins that enable the host cell to donate DNA. In fact, the DNA that fertile cells donate is the F DNA. F encodes proteins that form a special structure on the outside of the bacterial cell called a pilus. The pilus is a long tubelike appendage. It binds to a recipient cell (a cell without the F plasmid) and draws the mating pair together. The fertile cell then copies the F plasmid and simultaneously transfers the copy to the recipient cell. The recipient becomes fertile, too (Fig. 20.1).

Conjugative plasmids can contain extra genes, such as genes for antibiotic resistance. In the experiment you will conduct today, the fertile cell (cII) will transfer a gene for ampicillin resistance on a conjugative plasmid to the recipient cell (cI). The recipient is already resistant to streptomycin but cannot transmit this characteristic to the donor.

Classroom Activities

Figure 20.1   Conjugative transfer of an ampicillin resistance plasmid.

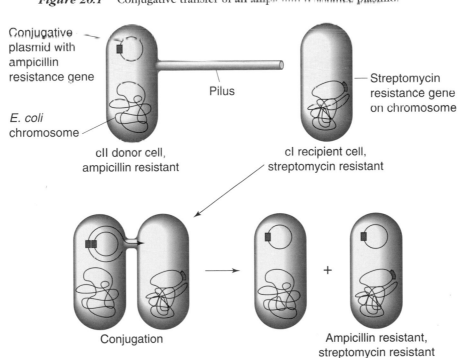

Procedure

Obtain the following materials from your instructor:

- Agar medium plate (no antibiotics)
- One ampicillin agar plate (label this plate Amp)
- One streptomycin agar plate (label this plate Strep)
- One ampicillin-streptomycin agar plate (label this plate Amp-Strep)
- Three small sterile containers
- Micropipette and sterile tips or other sterile measurement device
- Marker
- Optional: sterile inoculation tools

1. Label the three sterile containers cI, cII, and cI + cII

2. Following your teacher's instructions, add 100 or 200 µl of cI culture to the tubes marked cI and cI + cII. Use sterile technique. Change micropipette tips between additions. Dispose of used tips in the containers provided by your teacher.

3. Add the same volume of cII to the tubes marked cII and cI + cII. Change tips between additions.

4. Let the tubes stand at room temperature for approximately 20 min. During this time, the cI and cII cells will conjugate.

5. While you wait, mark the back of each of your agar plates with a large Y, so that the plate is divided into three approximately equal areas. Label one area cI, another cII, and the third cI + cII.

6. When the 20 min are up, use a sterile micropipette tip to pipette 5 µl of the cells in tube cI to the area of each agar plate marked cI. Change tips between plates. Dispose of used tips properly.

7. Repeat step 6, adding 5 µl of the cells in tube cII to the four plate areas marked cII.

8. Repeat step 6 once more, using the cells in tube cI + cII and the plate areas marked cI + cII.

Note: Your instructor may want you to use an alternative method to inoculate your plates. Listen for directions.

9. Incubate the plates overnight at room temperature or 37°C.

Day 2

Record your results by sketching any growth on the plates in Fig. 20.2.

Day 2

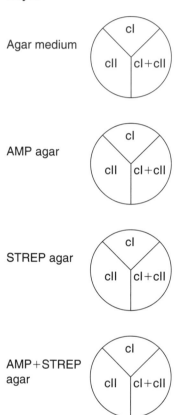

Figure 20.2 Diagram for recording results of conjugation experiment.

Questions

1. In Table 20.1, indicate where you expect to see growth on your plates. Write why you expect to see growth or no growth. For example, on the Amp plate, you expect to see growth of the cells from tube cII because strain cII is supposed to be ampicillin resistant. On the other hand, you do not expect to see growth of cII on the Strep plate because cII is streptomycin sensitive.

2. In this experiment, which cells grow on the Amp-Strep plate?

3. In this experiment, how does having the recipient cell be streptomycin resistant make it possible to detect conjugation products? What would happen in this experiment if the recipient cells were not streptomycin resistant?

Table 20.1 Expected results

Medium	Inoculated with:		
	cI	cII	cI + cII
Plain agar Do you expect growth? Reason?			
Amp agar Do you expect growth? Reason?			
Strep agar Do you expect growth? Reason?			
Amp-Strep agar Do you expect growth? Reason?			

Transduction of an Antibiotic Resistance Gene

21

About This Activity

This extremely simple wet-laboratory activity shows students the transmission of a bacterial antibiotic resistance gene by a bacterial virus. The laboratory is technically easy enough for students in grade 9 to perform if they can be prepared well enough that they understand what they are doing and seeing. With the more sophisticated discussion and optional modifications offered in the instructor's introduction, this activity could be incorporated into college-level classes in microbiology, virology, or genetics.

Class periods required: *1, plus observation and data recording on another day*

Introduction

In the process of transduction, a bacterial virus (bacteriophage) carries bacterial genes from one cell to another. Many different bacteriophages are capable of transduction; the details of transduction by any one of them depends on that phage's life cycle.

Viruses recognize their host cells through molecular interactions on the cell and virus surfaces. The proper interaction triggers changes that cause the viral genetic material to be injected into the host cell. At this point, essentially two different types of infection may take place, depending on the specific virus.

Lytic and latent viral infections

After a lytic virus infects its host cell, the virus takes over the cellular machinery. Normal cellular metabolism slows or stops. Cellular enzymes are diverted to make many new copies of the viral genetic material and many viral proteins. As the cell is filled with viral components, new virus particles assemble. Finally, the host cell dies, releasing the progeny virus into the environment. Sometimes this release of progeny is a gradual process; in other cases, the infected cell bursts (lyses), releasing all of the new virus particles at once.

The course of a latent infection is very different. After a latent virus injects its genetic material into the host cell, the virus does not hijack the cellular metabolism. Instead, a few viral proteins that direct the incorporation of viral DNA into the host chromosome are produced. If the host cell is a bacterium, the latently infected host is called a lysogen. (Bacteriophages that set up latent infections are called lysogenic bacteriophages; lambda is the best known of this group.) The viral DNA lies dormant in the host chromosome until a signal directs it to begin an active (often lytic) infection cycle. In most cases, the nature of that signal is unknown. During the active cycle, viral genetic material is reproduced, and viral proteins are made. New virus particles are assembled and released.

There are variations on the lytic- and latent-infection themes. For example, the varicella virus infects its human host and causes the disease chicken pox. During the disease, the virus goes through active infection cycles. However, when the patient recovers, the virus is not gone. Copies of the viral DNA remain integrated into the chromosomes of certain cells as a latent infection. This viral DNA may remain dormant for the rest of the patient's life, or it may reactivate, causing a second disease known as shingles.

Transduction

In general, transduction is the result of an error in bacteriophage reproduction. As bacteriophage reproduce, they replicate their genetic material and also produce new virus coats. The coats themselves (properly called capsids) are assemblies of viral proteins. At some point in the construction of the new virus, the newly replicated viral genetic material must be packaged into the capsids to create a virus particle. Each bacteriophage has á mechanism for packaging its genome into a capsid. Some bacteriophage occasionally make an error and package a piece of the host cell's DNA instead. This event is usually random, so any bacterial gene could end up inside a virus particle.

Virus particles that contain bacterial DNA instead of viral DNA are completely capable of attaching to a new host cell and injecting DNA (those functions are carried out by the protein capsid and are independent of its contents). Once inside the new cell, the bacterial DNA can recombine with the resident genome and be expressed and transmitted to future generations. When this happens, transduction has occurred.

Sometimes bacterial genes actually become part of a bacteriophage chromosome. These events happen in lysogenic infections in which the phage genetic material recombines with the host genome for a time. Apparently, when the phage DNA pops back out of the host chromosome to reproduce and package itself, it occasionally brings a piece of host DNA with it. This host DNA then acts like part of the viral genome and is replicated and transmitted along with it. Newly infected cells receive that particular fragment of bacterial DNA instead of the random fragments described previously.

Medical importance of transduction

Is transduction important to people other than scientists? Yes, several transduction events have great medical significance. The ones that we know about involve the second form of transduction described previously, in which a bacterial gene has apparently become part of a bacteriophage chromosome and is transferred to other hosts by the phage.

One example of a phage-borne disease is the usually fatal food poisoning called botulism. This disease is associated with the bacterium *Clostridium botulinum*, but the disease itself is caused by only one particular protein made by that organism. This protein is the botulism toxin (a toxin is any protein that has a poisonous effect on an organism). The gene for the botulism toxin is carried by a bacteriophage that infects *Clostridium botulinum* and is thought to have been transduced from another bacterium. Without the phage, there would be no botulism.

Similarly, *Staphylococcus aureus* food poisoning is associated with a bacterial virus. The most common toxin involved in this disease is encoded by a gene on a lysogenic phage. Finally, the disease diphtheria is also caused by a phage-borne gene, in this case in the bacterium *Corynebacterium diphtheriae*. This gene encodes the diphtheria toxin, the protein responsible for essentially all the symptoms and effects of this severe disease. Strains of *Corynebacterium diphtheriae* without the lysogenic phage are fairly harmless.

Applications of transduction

Viruses are used in research laboratories as another means of introducing foreign DNA into cells. The genomes of small viruses are easy to work with, and scientists often use such viruses as cloning vectors. Viruses are especially important in working with mammalian cells. Researchers clone genes of interest into the viral DNA (often after removing viral genes to inactivate the virus) and then package the recombinant DNA into virus particles. The particles inject the DNA into the cells. Some viruses recombine their DNA into the mammalian genome, while others replicate in the cytoplasm, much as bacterial plasmids do. Using viruses to introduce foreign DNA into mammalian cells is called **transfection** (as opposed to transduction, which refers to gene transfer between bacteria by their viruses).

For example, one avenue for cystic fibrosis gene therapy being tested by researchers involves having a respiratory virus deliver a healthy copy of the defective gene to the patient's lungs. The virus in question, adenovirus, normally infects cells lining the respiratory tract. The virus injects its DNA into the cells, and the DNA then replicates in the cytoplasm of the infected cells, rather like a plasmid. Scientists have altered adenovirus DNA to render it incapable of causing disease, and they have added a healthy copy of the defective gene. They have shown that adenovirus particles containing the altered DNA can deliver it to airway cells and that the cystic fibrosis protein is then expressed. At the time of this writing, the efficiency of the process is low, and several important technical issues remain to be worked out, so it is not clear whether adenovirus transfection will become part of cystic fibrosis gene therapy.

Demonstration of random transduction

This laboratory activity demonstrates transmission of a gene for ampicillin resistance to *E. coli* by the bacteriophage called T4. Random transduction actually involves two infection steps: in the first infection, fragments of the host genome are packaged by mistake; in the second infection, new host cells receive the bacterial genes. In this exercise, however, you and your students will perform only the second step. The bacteriophage lysate (a suspension of phage particles) you will work with already contains virus particles with packaged bacterial DNA. The suspension of virus particles is called a lysate because the phage are harvested from an infected culture of *E. coli* after they lyse (break open) the cells.

The transducing lysate was produced by growing a special strain (see below) of bacteriophage T4 on host cells that contained the plasmid pKK061. This plasmid contains a gene for resistance to ampicillin but also has (in addition to its normal replication origin) an origin of replication recognized by bacteriophage T4 DNA replication proteins. When T4 infected host cells containing pKK061, the viral proteins replicated the plasmid along with its own DNA. The result was a lot of plasmid DNA lying around in the host cell for T4 to package by mistake. For every 10,000 virus particles produced in that infection, we can observe one transduction event involving plasmid DNA.

How will transduction be detected? Transfer of the plasmid DNA to a new host will render that host resistant to ampicillin, so transductants (cells that have been successfully transduced) can be selected by plating on ampicillin-containing media. As we just said, the transducing particles represent only 0.01% of the viruses in the lysate. What about the other bacteriophage particles, which normally infect and kill *E. coli*? These normal viruses would usually make it difficult to detect the transductants. To avoid this problem, a special strain of T4 was used to make the transducing lysate.

Amber mutations and amber suppressors

The transducing T4 strain carries mutations in two different genes that change an amino acid codon to the stop codon UAG (see Table 3.1 in chapter 3). This particular stop codon has the casual name amber codon, acquired at the time of its discovery, and the mutations are called amber mutations. Amber mutations terminate protein synthesis and usually result in loss of function of the protein (if the amber mutation is very near the end of the protein-encoding sequence, it may not have much effect). In the transducing T4 strain, the two amber mutations block production of two essential proteins and are therefore lethal.

Amber mutations in bacteriophages are very useful to scientists because in combination with special host strains, they provide an "on-off" switch for phage growth. Here is how the on-off system works. Amber codons (and the other two stop codons) stop translation because no transfer RNAs (tRNA) have matching anticodons. When an amber codon in messenger RNA reaches the ribosome, no new amino acid is added to the growing peptide chain; instead, protein synthesis is terminated. There is nothing magic about the base sequence of the terminator codons except that they have no matching tRNAs that will add amino acids to a peptide chain in response to them.

However, tRNA molecules are also encoded in DNA (to make tRNA, the DNA base sequence is simply transcribed into RNA by a special RNA polymerase). The base sequence of any tRNA can be changed by mutations in the gene encoding it. In some bacterial strains, mutations change the anticodon of a tRNA so that the tRNA can recognize what used to be one of the stop codons. In these strains, that codon is no longer a stop codon but instead encodes whatever amino acid the mutant tRNA carries. If the mutant tRNA recognizes the amber stop codon and inserts an amino acid at its position, an amber codon no longer terminates protein synthesis. A bacterial strain that makes a tRNA that recognizes the amber codon is called an amber suppressor.

In amber suppressor strains, amber mutations are essentially erased. The effect of an amber mutation comes from its ability to terminate protein synthesis; in amber-suppressing strains, protein synthesis is not terminated. Since the mutant tRNA does not usually insert the original amino acid at the site of the amber mutation, the protein may be impaired in function, but often, it can still perform adequately. So bacteriophage carrying normally lethal amber mutations may be able to reproduce in bacterial hosts that suppress the amber mutations through a mutant tRNA.

Our transducing T4 strain fits this description. In bacterial hosts with normal tRNAs, the phage is "dead" (unable to produce new virus particles) as a result of its amber mutations. However, in strains that make a mutant tRNA that suppresses the amber stop codon, the phage can reproduce quite well. Thus, laboratory personnel can control whether or not the phage reproduces through choice of the host bacterial strain. The amber mutations in our bacteriophage T4 are an example of **conditional lethal mutations:** under some conditions, they are lethal; under other conditions, they are not.

For this activity, the transducing lysate was produced by infection of an amber-suppressing strain carrying pKK061 (see above). The suppressing strain allowed the bacteriophage to reproduce. To detect transductants without interference, the lysate is allowed to infect *E. coli* strain B_E, which does not suppress amber codons. The T4 strain cannot grow in B_E, so no bacteriophage plaques will be observed. However, any transducing particles that were made during the first infection can transduce B_E to ampicillin resistance. So when the B_E-T4 mixture is plated on ampicillin-containing medium, transductants can be detected without interference from cell killing by the virus.

If you would like your students to see cell lysis of *E. coli* by T4 or to observe a conditional lethal mutation at work, use the amber suppressor strain CR63 as described below in part B. This strain allows transducing T4 to grow by suppressing the amber mutations in the two essential genes. If you mix the lysate with CR63 cells and then spread the mixture on a plate, you will be able to see that the phage kills the cells.

Objectives

After completing this activity, students should be able to

1. Describe how random transduction of bacterial genes occurs;

2. State what happens in the first infection of a transduction and what happens in the second;

3. (If you include the material in the lesson) explain what an amber mutation is, why it can be very harmful to the organism, what an amber-suppressing host is, and how it negates the effects of amber mutations.

Materials

Part A: demonstration of transduction

- Photocopies of *Student Activity* pages, if necessary

- Overnight culture of *E. coli* B_E

- Transducing T4 lysate. *Never freeze a T4 lysate.*

- Sterile tubes. Each student group will need four for mixing cells and phage (0.5-ml total volume); for dilutions, you will need at least three for the entire class (one must be able to hold 10 ml comfortably).

- Micropipettors and sterile tips for measuring 10 µl

- Micropipettors and sterile tips for measuring 100 and 500 µl

- Sterile pipettes for measuring 1 and 10 ml

- Plates with tryptic soy agar and ampicillin, four per student group (or L agar or nutrient agar; recipes in Appendix E)

- Racks, beakers, or ice buckets for holding test tubes upright

- Glass spreader (glass rod bent into an L shape), one per student group *or* sterile cotton swabs

- Bunsen or alcohol burner (not necessary if you use sterile swabs)

- Beaker with methanol or ethanol; must be large enough for the glass spreader to fit (not necessary if you use sterile swabs)

Optional additional materials for detecting cell lysis (part B)

- Overnight culture of *E. coli* CR63
- Four sterile test tubes per student group
- Four tryptic soy agar plates without antibiotic

Resources

At this time, the only commercial source of the transducing lysate is Carolina Biological Supply Co. You can also purchase the bacterial strains from them. Their "Advanced Transduction Kit" takes students through both infection cycles: the original one in which the bacterial DNA is packaged and the second one in which B_E is transduced.

Preparation

Advance preparation

Prepare the agar plates (directions are in Appendix E). Sterilize test tubes and any other necessary equipment. You will also need sterile tryptic soy broth or L broth for the day of the experiment.

Day before

Start a culture of *E. coli* B_E. If you plan to use it, also start a culture of strain CR63.

Day of class

Set up some containers with 1:5 diluted bleach or Lysol for biological waste. The T4 phage lysate is not dangerous and can be treated just like *E. coli* for disposal.

Aliquot small amounts (about 100 to 200 µl) of phage lysate into sterile tubes marked "undilute" for student use. To dilute the transducing lysate, you will need (depending on the final volume of diluted lysate needed) at least one sterile test tube containing 900 µl (0.9 ml) of sterile broth and at least one tube containing 1 ml of broth. These tubes can be set up the day before and refrigerated overnight.

Make the phage dilutions (referred to in the *Student Activity* pages). Add 100 µl of phage to the tube containing 900 µl of sterile broth (use a sterile tip). Thump the tube gently to mix the contents. Mark this tube "1 to 10." With a fresh tip, add 10 µl of undilute phage lysate to the test tube containing 1 ml of broth. Thump gently, and label "1 to 100." As you draw the phage suspension into the pipette tip, do it gently. If phage suspension splashes up onto the plunger of the pipettor, the pipettor can become contaminated. For this reason, we do not recommend that students be allowed to prepare phage dilutions.

If you make the dilutions ahead of time, place them in the refrigerator or on ice for storage.

Dilution Mathematics

1-to-10 dilution

Add 100 µl of phage to 900 µl of broth.
Initial volume of phage = 100 µl
Final volume of phage = 1,000 µl (broth + phage)
Ratio of initial/final volume = 100 µl/1,000 µl = 1/10

1-to-100 dilution

Add 10 µl of phage to 1 ml (1,000 µl) of broth.
Initial volume of phage = 10 µl
Final volume of phage = 1,010 µl
Ratio of initial/final volume = 10 µl/1,010 µl = 1/101, or approximately 1/100

If we wanted an exact 1-to-100 dilution, we would add 10 µl of phage to 990 µl of broth. (Do this if you prefer; it won't affect the results significantly.)

Procedure

Hand out the *Student Activity* pages, if necessary. Discuss the process of transduction with your class. Make sure they understand what happens. When they add the diluted virus lysate to the host cells, there will be many, many more bacterial cells than virus particles, so if a cell is infected, it will most likely be infected by only one virus. If the virus is a normal virus particle, it will inject its DNA but not be able to reproduce on account of the mutations discussed previously. If the virus is a transducing particle, it will inject host DNA into the cell, and that cell may be transduced. The class will be looking for host cells that have been transduced with plasmid DNA. Ask the students how transduction of pKK061 DNA might be detected. They should be able to suggest testing for ampicillin resistance.

The picture shows replicated plasmid DNA as a linear molecule. This is correct. When T4 replicates the plas-

mid DNA, it produces long linear molecules containing many copies of the plasmid DNA: imagine paper towels being pulled off a roll. When this linear DNA enters the new host cell, it recombines with itself and recircularizes. It is possible to isolate circular plasmid DNA from the transductants generated in this experiment. If you wish to, simply use the minipreparation procedure from *DNA Science* (see Appendix H).

Follow the procedure on the activity sheet. The lysate is diluted so that a reasonable number of transductants will be produced. By testing different dilutions of the lysate, you should get at least one or two plates with a good number of transductants, plus you can take the opportunity to talk to your students about dilutions if you wish.

If you wish to look at cell lysis by the bacteriophage, have students do part B on the *Student Activity* pages, too. The plates to which the most phage were added may look "trashed" after incubation because of lysis of cells. If you use soft top agar to plate the cells, the results will be prettier (but this procedure isn't necessary; don't worry about it if you do not know how). Explain only as much about the amber mutations as you wish. You can simply tell your students that strain B_E does not permit the T4 strain to grow, while strain CR63 does.

Tips

You may want to have students read the *Student Activity* pages the day before. Talk through the science, and have students make up a flow chart based on the options of the activity you have chosen. Students could answer several of the questions before doing the experiment.

The activity can be modified for advanced students by having them make their own transducing lysates or by having them recover plasmid DNA from the transductants by using the minipreparation procedure described in *DNA Science*.

Answers to Questions

Part A

1. The phage suspension was diluted so that a reasonable number of transductants could be detected on one or two plates. The starting suspension was too concentrated and would have given too many transductants.

2. There should be no colonies on the no-phage plate. This plate has ampicillin in it, and the no-

phage tube has only *E. coli* B$_E$ in it. This strain is not resistant to ampicillin and should not grow on the plate. If you see isolated colonies, there has been contamination. Either the phage was accidentally added, giving some ampicillin-resistant transductants, or the culture was contaminated with ampicillin-resistant organisms. The no-phage plate was intended to act as a control. This plate should verify that without the addition of transducing phage, there are no ampicillin-resistant cells in the *E. coli* culture. If the no-phage plate fails to show colonies, we can conclude that the ampicillin-resistant colonies seen on the phage plates are there because of the phage.

3. The colonies growing on the plates to which phage were added are transductants (unless there were colonies on the no-phage plate; then we can't be sure what any of the colonies are). These colonies contain plasmid DNA transduced by the phage. There should be approximately 10 times more colonies on the undilute plate than on the 1-to-10 plate, because the undilute phage suspension should have 10 times more phage in it than the 1-to-10 phage dilution. The 1-to-10 plate should have 10 times more colonies than the 1-to-100 plate, because 10 times more phage suspension (and therefore 10 times more phage) was added to it.

4. Five hundred colonies would be expected on the undilute plate because 10 times more phage sus-pension was added to it. Five colonies would be expected on the 1-to-100 plate (see question 3 above).

Part B

The no-phage plate should have a smooth lawn of cells covering it. The plates to which phage were added may look anywhere from empty (all the cells were lysed) to trashy (some of the cells were lysed) to splotchy. The plates may all look similar, depending on how well the phage was able to spread.

1. The phage plates should be different from the no-phage plate. The difference is that the phage lyses CR63, and new phage may then infect neighbor cells. Perhaps no cells at all will be growing on the phage plates because of the spread of reproducing phage. On the other hand, some growth may be visible, depending on the efficiency of spreading.

2. The purpose of the no-phage plate was to let you see what a normal plate of CR63 would look like so that you could compare it to the plates to which phage were added. The differences in the no-phage and phage plates (which were treated identically except for the phage) are attributable to the phage. The no-phage plate is a control.

Classroom Activities

Transduction of an Antibiotic Resistance Gene

Transduction is a natural method of gene transfer that occurs in bacteria. The key player in transduction is a bacterial virus, or bacteriophage (phage for short). Many different bacteriophages infect many different bacteria. You may have already met one of the phages: lambda. Today you will meet a different phage, one that also infects *Escherichia coli*. It is called T4.

How does T4 transfer genetic material between *E. coli* cells? The answer is found in its life cycle. T4 infects *E. coli* by attaching to its outer membrane and injecting its DNA into the bacterial cell. Once inside the cell, the phage DNA takes over. The *E. coli* cell becomes a factory for producing many copies of the T4 genome and large amounts of viral proteins. Some of these proteins help replicate the T4 DNA; others are assembled into new T4 heads and tails. After many copies of the T4 genome have been made and many new heads and tails are floating around in the cytoplasm, still other T4 proteins begin to put together new virus particles. These proteins fill the empty phage heads with T4 DNA and then attach the tails. After many new viruses are assembled, the *E. coli* cell bursts, releasing the virus progeny.

What does this have to do with transferring *E. coli* genes? The critical step is the point at which the new virus particles are assembled. Once in a while, the T4 assembly proteins make a mistake. Instead of filling a phage head with T4 DNA, they fill it with a piece of host DNA. The filled head gets a tail and becomes a virus particle fully capable of injecting DNA into a new bacterial cell. However, when it does so, the new host cell receives that bacterial DNA instead of dangerous viral DNA. When the new host expresses the bacterial DNA it received, it is said to have been transduced. (Remember, no virus infection took place, since the virus particle was like a "dummy warhead" filled with harmless bacterial DNA.)

This Activity

In this activity, you will observe the transmission of an antibiotic resistance gene by phage T4. The T4 virus particles you will work with were grown on a plasmid-containing host strain, and some of the virus particles produced from that infection contain plasmid DNA. Your job is to detect some of these plasmid DNA-containing particles by their ability to transduce antibiotic-sensitive *E. coli*. Figure 21.1 summarizes the transduction process.

How do you think you could detect transductants (*E. coli* that have received plasmid DNA)?

Procedure

Part A: transduction of E. coli

Obtain the following materials from your instructor:

- Four sterile test tubes
- Two micropipettes, one for small volumes and the other for larger volumes
- Sterile micropipette tips for both
- Four ampicillin medium plates
- Marking pen

1. Label all the plates "Amp." In addition, label one plate "no phage," one plate "undilute," one plate "1 to 10," and the last one "1 to 100."

2. Label the tubes the same way (it is not necessary to write Amp on them).

3. When you have labeled the tubes, use the large micropipette to place 0.5 ml of *E. coli* B_E cells in each tube. It is not necessary to change tips for each addition unless you touch the outside of the tube or some other nonsterile surface with the tip.

To avoid inhaling or splashing in your eyes any aerosol that might be created, keep your face away from the tip end of the pipette while you are pipetting the culture.

Your teacher is making 1-to-10 and 1-to-100 dilutions of the phage suspension for you to use. Why do you think the phage suspension is being diluted?

1. Infection of plasmid-containing host cell

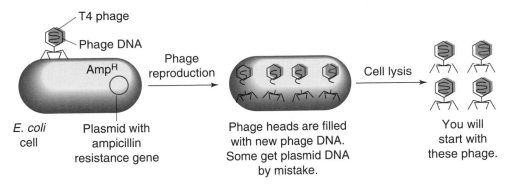

2. Second infection: transfer of plasmid DNA (today's activity)

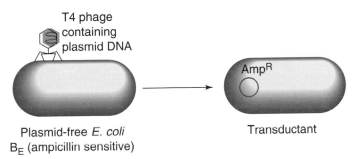

Figure 21.1   Transduction of plasmid DNA by bacteriophage T4.

4. Take your tube of B$_E$ cells labeled "undilute" and add 10 μl of the undilute phage lysate to it, using a sterile tip. Thump the tube gently to mix the phage and cells. To the tube marked "1 to 10," add 10 μl of the 1-to-10 dilution. Mix gently. Finally, to the tube marked "1 to 100," add 10 μl of the 1-to-100 dilution, using a fresh tip. Mix gently.

5. Let all of the tubes stand at room temperature for about 15 min.

Two methods for spreading the phage-cell mixtures on the ampicillin plates are described below. Your instructor will tell you which option you will use.

Spreading Option 1

6. Obtain four wrapped sterile cotton swabs.

7. Take the tube labeled "no phage," and pour its contents on the appropriately labeled plate. Open one sterile cotton swab. Use the swab to spread the cells over the entire surface of the agar. Do the spreading gently so as not to tear up the agar. Place the used swab and the tube in the biological waste container provided for you.

8. Repeat for the other tubes and plates. Use a fresh swab each time.

Spreading Option 2

6. You will need a bent glass rod, a large beaker containing alcohol, and access to a flame.

7. Take the tube labeled "no phage," and pour its contents on the appropriately labeled plate. Replace the plate lid.

8. Dip the spreader in the alcohol, and let the excess alcohol drip off. Pass the spreader through the flame only long enough to ignite the alcohol (do not let the spreader heat up in the flame). Remove the spreader from the flame, and allow the alcohol to burn off completely. Repeat.

9. Remove the lid from the plate, and cool the spreader by touching it to the agar away from the cells. It is essential that the spreader not be hot, or it can kill the cells.

10. Spread the cells as evenly as possible over the surface of the agar. To do this, drag the spreader

back and forth across the plate, turn the plate a quarter turn, and repeat. Do this until you have turned the plate a full turn.

11. Repeat steps 7, 8, 9, and 10 with the other tubes and plates.

After the plates have sat for a few minutes and the liquid is absorbed, invert the plates. Incubate the plates overnight at 37°C or at room temperature.

What do you expect to see on the no-phage plate?

Part B: detection of cell lysis by bacteriophage T4

In part A, you set up an experiment to detect transductants. The *E. coli* B_E strain used in this experiment allows the transducing T4 strain to inject its genome but does not permit the virus to multiply inside it. Therefore, you will not see cell lysis (bursting) by the phage. Other *E. coli* strains such as CR63 will permit the transducing T4 strain to reproduce and will therefore be lysed by the phage. In part B of this activity, you will add T4 phage to CR63 cells to observe the lethal action of the bacteriophage.

To do this optional segment of the activity, you will need the following:

- Four more sterile test tubes
- Four agar plates *without* antibiotic

1. On each of these tubes, write "CR63." In addition, write "no phage" on the first tube, "undilute" on the second tube, "1 to 10" on the third tube, and "1 to 100" on the fourth.

2. Label the agar plates in the same manner, writing "CR63" and the rest of the information on each plate.

3. To each of the four test tubes, add 0.5 ml of the culture of *E. coli* CR63. It is not necessary to change tips unless you contaminate one by touching some nonsterile surface.

4. Take your tube of CR63 cells labeled "undilute," and add 10 μl of the undilute phage lysate to it, using a sterile tip. Thump the tube gently to mix

the phage and cells. To the tube marked "1 to 10," add 10 μl of the 1-to-10 dilution. Mix gently. Finally, to the tube marked "1 to 100," add 10 μl of the 1-to-100 dilution, using a fresh tip. Mix gently.

5. Let all of the tubes stand at room temperature for 5 to 15 min.

Spread the cells on the appropriately labeled plates as you did in part A. After the liquid is absorbed, invert the plates, and incubate them overnight.

Next day

Examine your plates. Are there colonies? If the plates were incubated at room temperature, you may need to let them grow another day. Record your results.

Results

Part A

How many colonies are on the no-phage plate?

How many colonies are on the undilute plate?

The 1-to-10 plate?

The 1-to-100 plate?

If there is a reasonable number on the plate, count them and record the number. If there are very many colonies on the plate, divide the plate evenly into four quadrants by making a large + on the back of the plate, and count the colonies in one quadrant. Estimate the total number on the plate by multiplying the number in the quadrant by four.

Part B

Record your results (drawing the plates may be helpful).

Describe the no-phage CR63 plate.

Describe the 1-to-10 plate.

Describe the 1-to-100 plate.

Questions

Part A

1. Why was the phage suspension diluted?

2. Did you see colonies on the no-phage plate? Was this what you expected? Why or why not? What was the purpose of this plate?

3. What are the colonies growing on the plates to which you added phage? Are there more of them on the undilute plate than on the 1-to-10 or 1-to-100 plate? Why?

4. Suppose you looked first at the plate to which you added the cells containing 10 μl of the 1-to-10 phage dilution and counted 50 colonies. How many would you expect to see on the plate with the cells to which you added 10 μl of the undilute lysate? On the plate with the cells to which you added 10 μl of the 1-to-100 dilution?

Part B

1. Do the plates to which phage were added look the same as the no-phage plate? If they are different from the no-phage plate, what is causing the difference? If they are the same, why do you think they are?

2. What was the purpose of the no-phage plate?

Classroom Activities

Agrobacterium tumefaciens: **22**
Nature's Plant Genetic Engineer

About This Activity

In this simple activity, students inoculate plants with the bacterium *Agrobacterium tumefaciens* and observe the subsequent formation of a plant tumor. Middle-school students have performed the exercise successfully, though the science behind it is sophisticated enough for college students. *A. tumefaciens* causes tumor formation because it transfers genes to the plant. This property makes it very useful in plant genetic engineering. The lesson includes information on how *A. tumefaciens* is used in genetic engineering. A reading on other methods of plant genetic engineering accompanies this lesson.

Class periods required: *1 to inoculate the plants and then a few minutes in several later classes for observation*

Introduction

The common soil bacterium *A. tumefaciens* causes crown gall disease in many dicotyledonous plants. Crown gall is a plant tumor, the result of a massive proliferation of plant cells. Virulent strains of *A. tumefaciens* contain genes that cause the plant cells to divide. *A. tumefaciens* causes crown gall disease by inserting these and a few other genes into a host plant's genome. *A. tumefaciens* is therefore a natural genetic engineer. Interestingly enough, *A. tumefaciens,* like many human molecular biologists, uses a plasmid to modify the genome of plants. The plasmid used by *A. tumefaciens* is the tumor-inducing, or Ti, plasmid.

Chemicals secreted from freshly wounded plant tissue attract *A. tumefaciens* to the wound site. The bacterium binds to the walls of the broken cells and enters these now-dead cells. From there, it injects a segment of Ti plasmid DNA into the adjacent living plant cells.

The injected bacterial DNA diverts the plant cell's machinery to tasks that support the growth and reproduction of *A. tumefaciens.* Living cells surrounding the wound site proliferate into a tumor, or gall, to house and protect the bacteria. These rapidly dividing tumor cells also synthesize novel chemical compounds, called opines, that provide nutrients critical to the bacteria but useless to the plant. Opines consist of amino acids bound to common metabolic intermediates such as pyruvate.

Different strains of *A. tumefaciens* induce the production of different opines. The opine synthesized by a particular tumor can be catabolized (broken down) only by the strain that caused the tumor. The strain-specific opine also promotes the conjugational transfer of tumor-inducing plasmids from that virulent strain to avirulent (plasmid-free) *Agrobacterium* strains.

The gall induced by *A. tumefaciens* is located at soil level, where the roots join the stem (the crown). Strain-specific opines produced by the gall seep into the soil, thus helping that strain of *A. tumefaciens* compete with other strains as its plasmid is transferred at a higher rate. Are *A. tumefaciens* and the plant tumor it induces simply a plasmid's way of making more plasmids?

The Ti plasmid

The Ti plasmid encodes proteins that enable *A. tumefaciens* to infect plants and induce gall formation. The portion of the Ti plasmid that is inserted into the plant's chromosome is called transferred DNA, or T-DNA. This DNA segment contains genes that code for opine synthesis as well as genes that encourage plant cell proliferation by increasing the production of plant hormones such as cytokinins and auxins. None of the T-DNA genes are involved in the transfer process.

Located to the left of the T-DNA segment are virulence *(vir)* genes, which encode proteins that control the steps in the infective process: binding of the bacterium to the plant cell wall and transfer of the T-DNA into the plant cell. Thus, *vir* genes are essential for gene transfer but are not themselves transferred.

Wound juices secreted by the plant appear to induce the expression of the *vir* genes.

Genes for opine catabolism and conjugational plasmid transfer are located elsewhere on the Ti plasmid.

The disarmed Ti plasmid

The unique biology of *A. tumefaciens* provides important tools for plant genetic engineers. Before researchers can exploit the natural genetic engineering capabilities of *A. tumefaciens,* however, they must first "disarm" the Ti plasmid by removing its tumor-causing genes. The *vir* genes that control gene transfer ability remain intact. Researchers then replace the T-DNA genes with the foreign gene(s) they wish to transfer to the plant (see Fig. 22.1). The disarmed plasmid containing the new gene(s) is returned to *A. tumefaciens,* which is grown in culture so that many engineered bacteria are produced. Pieces of plants to be engineered are then placed in solution with the genetically altered *A. tumefaciens.*

Figure 22.1   (A) Transfer of *Agrobacterium* T-DNA to plant cell. (B) The Ti plasmid.

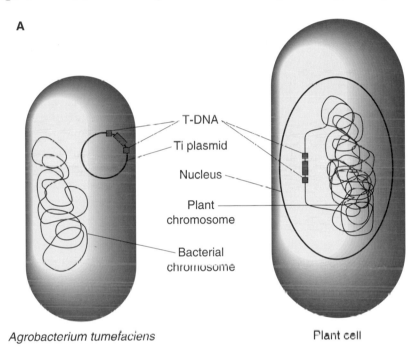

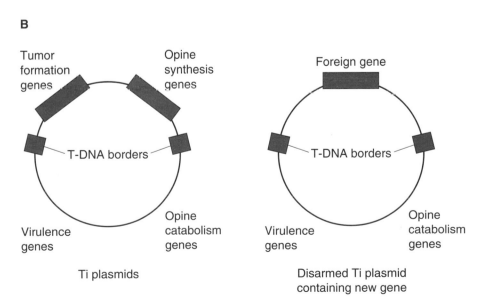

To date, plant biologists have used disarmed *A. tumefaciens* plasmids as vectors in the genetic transformation of a wide variety of dicots. More than 20 important agricultural plants have been successfully engineered by using *A. tumefaciens.*

Laboratory activity

In this laboratory, students will wound the stems and leaves of a kalanchoe plant and inoculate them with *A. tumefaciens.* Several different kalanchoe plants are available. One, *Kalanchoe diagremontiana,* is the ornamental succulent known as the pregnant plant and is available from local florists. Other varieties can be ordered from biological supply companies.

Objectives

After completing this activity, students should be able to

1. Explain the molecular biology underlying crown gall disease and tumor formation;

2. Describe how *A. tumefaciens* can be used to genetically engineer plants.

Materials

Each laboratory team will need the following:

- Sterile toothpicks
- Test tube rack
- Two squares of Parafilm, 1 by 1 cm
- One test tube containing 5 to 10 ml of sterile water
- Two small kalanchoe plants or other dicotyledenous potted plants (but not brassica)
- Access to an agar slant culture of *A. tumefaciens*
- Lysol water (30 ml of Lysol in 4 liters of water) or other disinfectant solution
- One inoculating needle
- Alcohol or Bunsen burner
- Rubbing alcohol and cotton swabs or cotton balls for wiping plant surface

Resources

- *Kalanchoe* spp. and *A. tumefaciens* can be ordered from Carolina Biological Supply Co. *A. tumefaciens* can also be ordered from other biological supply houses.

- The "Plant Cancer Kit" from Carolina Biological Supply Co. can be used instead of this procedure. In this kit, the experimental organism is sunflower.

Preparation

- Photocopy the *Student Activity* pages, if necessary.

- Make overhead transparencies of the diagrams if you wish to use them in discussing Ti plasmid biology.

- Assemble the materials, and prepare the disinfectant.

Procedure

Have students read the introductory material on their *Student Activity* pages. Make sure they understand how the Ti plasmid works, and how it can be used as a genetic engineering vector. Have them inoculate their plants, following the procedure given.

Tips

- Be sure to review safety procedures with your students before proceeding. Remember that rubbing alcohol is flammable; be sure that it has been removed from the work area before burners are lit. The entire procedure should be demonstrated by the instructor before students begin.

- Show the students a plant infected with a crown gall if one is available. It may be weeks before they see galls form on their plants.

- Different species of plants can be tested for susceptibility. Monocots and plants in the brassica family (e.g., broccoli, cauliflower, and Wisconsin Fast Plants) appear resistant to infection.

- Different portions of different plants can be wounded and tested for susceptibility to infection.

Selected Readings

Chilton, Mary-Dell. 1983. A vector for introducing new genes into plants. *Scientific American* 48(6):50-59.

Drexler, Edward, Maud A. W. Hinchee, Douglas T. Lundberg, Laurence B. McCullough, Joseph D. McInerney, Jeffrey Murray, and Richard Storey. 1989. *Advances in Genetic Technology: Biological Sciences Curriculum Study,* p. 1-11. D.C. Heath and Co., Lexington, Mass.

Classroom Activities

Agrobacterium tumefaciens: Nature's Plant Genetic Engineer

Would you believe that there exists in nature a genetic engineer that inserts new genes into plant cells by using a plasmid as its vector? Would you believe that this natural genetic engineer is a common inhabitant of the soil? It's all true, and the natural genetic engineer is the soil bacterium *Agrobacterium tumefaciens.*

A. tumefaciens infects certain types of plants (most dicots but not monocots) at wound sites. Once in the wound, the bacterium injects a segment of its plasmid, called Ti, for tumor-inducing, into the adjacent living plant cells. This piece of DNA, called T-DNA for transferred DNA, is only one region of the plasmid. The T-DNA inserts itself into the plant's genome, where it goes to work hijacking the plant's machinery to support the reproduction of *A. tumefaciens.*

Around the infected wound, living cells proliferate into a tumor, or gall, to house and protect the bacterium. The tumor cells synthesize new chemicals that provide critical nourishment to the bacterium but are useless to the plant. Both of these effects are driven by genes on the T-DNA.

In the late 1970s, plant scientists realized they might be able to take advantage of *A. tumefaciens*' natural genetic engineering abilities. They developed an important method for plant genetic engineering based on this organism and its Ti plasmid.

In this method, the T-DNA genes that induce tumor formation and nourish the bacterium are removed from the Ti plasmid and replaced with any gene of interest. This is accomplished through the use of restriction enzymes, DNA ligase, and other recombinant DNA techniques. The new plasmid is returned to *A. tumefaciens,* which is grown in culture so that many bacteria carrying the engineered plasmid are produced.

Plants to be engineered are then infected with the bacterium carrying the "designer" Ti plasmid. *A. tumefaciens* injects the engineered T-DNA into the plant.

Instead of receiving genes for tumor formation, the plant gets the genes inserted into the Ti plasmid by the scientist. This method has been used to genetically transform more than 20 important agricultural plants. For example, tobacco has been genetically engineered to produce medically important proteins such as hemoglobin. In a more lighthearted experiment, scientists have produced plants that synthesize the protein luciferase, the enzyme that causes the light of fireflies. These plants glow in the dark, though not as brightly as fireflies flash.

Today's Activity

In this laboratory, you will infect a plant with *A. tumefaciens* to start the tumor formation process. Because gall formation is slow, you will need to observe the plant for several weeks to see development of the tumor.

Procedure

1. Observe your teacher's demonstration of the proper inoculation technique.

2. Wipe down the laboratory table with Lysol water or other disinfectant.

3. Use an alcohol swab to wipe the plant surface to be wounded (stem or upper leaf). Put the alcohol away.

4. Pick up a sterile toothpick by touching only one end. Pierce the stem of the plant with the other end; the toothpick should go all the way through the stem.

5. Remove the cap from the test tube of water, and flame the tube's mouth. Use a sterile inoculating needle to apply a small amount of sterile water to the wound on the plant. Flame the needle, and reflame the mouth of the test tube before replacing its cap.

6. Remove the cap from the slant tube of *A. tumefaciens,* and flame the tube's mouth. Flame an inoculating needle, and then cool the needle. Use the sterile needle to apply a small amount of *A. tumefaciens* from the slant culture to the plant's wound. Reflame the needle. Reflame the mouth of the test tube before replacing its cap.

7. Remove the paper from the sterile side of a piece of Parafilm (don't touch that side). Apply the sterile side of the Parafilm to the plant's wound. It does not have to adhere tightly to the plant surface, but it should remain in place for a day or so.

8. With a second plant, repeat these steps but without *A. tumefaciens.*

9. Wipe down all laboratory surfaces with disinfectant.

10. Water the plants as you normally would.

Observe the plants regularly for gall formation. Evidence of gall formation may appear in 1 to 2 weeks. If this does not occur, however, do not discard the plants. Gall formation sometimes requires several months.

Record your observations on the log provided below. Supplement the log with sketches of the plant. Label each sketch with the date you make it.

Plant Observation Log

Date	Observations
Week 1	
Week 2	
Week 3	
Week 4	
Week 5	
Week 6	
Week 7	
Week 8	

Methods of Plant Genetic Engineering

It is humbling but true: most plants are far more genetically complex than humans and animals. This fact partly explains why plants have eluded genetic engineers longer than animals have. The plant kingdom is so genetically complex and diverse that our understanding of plant molecular genetics is more limited than our knowledge of animal and microbial molecular genetics. Factor in the federal government's relatively low funding of plant research versus medical research, and it's even clearer why plant engineering lags behind microbial engineering by 10 years or more.

Only recently have scientists delved deeply enough into plant molecular biology to make genetically engineered plants commercially feasible. Scientists now know enough about plant molecular biology to give certain crop plants single-gene traits such as insect resistance, herbicide tolerance, and better nutritional quality. Biotechnology gives scientists the tools they need to engineer plants with these and other desirable traits with far greater speed, precision, and certainty than is possible with traditional crossbreeding.

Engineering a desirable trait into plants usually involves two different biotechnologies. Recombinant DNA technology is used to isolate the desirable gene, prepare it for moving into its new plant host, and move it into individual plant cells. Getting DNA into plant cells proved to be challenging; techniques used for this purpose are described below. However, engineered plant cells aren't much use to most farmers and horticulturists; these people need whole plants. Once individual plant cells have received new genetic information, the problem becomes how to regenerate whole plants from these cells.

The technology used to regenerate whole plants from individual cells or tissues is called plant tissue culture. Plant cells have the unique property of totipotency, the potential to re-create a whole, multicellular organism from a single cell. It seems as though this property should be no big deal—after all, nearly all animal body cells contain the organism's entire genome—but it is. It has thus far proven impossible to regenerate animals from differentiated cells, but through the use of special media and hormone treatments, the feat is routinely accomplished with many plants. Plant tissue culture thus makes it possible to genetically engineer single plant cells and then re-create whole, engineered plants. Now let's get back to how to get foreign DNA into plant cells.

Bacterial Vectors

The first break in plant genetic engineering came in the late 1970s, when scientists found an effective way to insert foreign DNA into plant cells by using *A. tumefaciens*. Once they had identified a desirable gene from another organism, they could splice it into the plasmid DNA of *A. tumefaciens* for depositing into the plant's genome. Typically, pieces of leaves are bathed in a suspension of the bacterium. The pieces are then moved to a growth medium that permits only those cells that have received plasmid DNA to grow. Tissue culture technology is then employed to regenerate the engineered cells into plants.

A. tumefaciens is an effective vector for tobacco, petunias, tomatoes, and other dicots (plants with two seed leaves), but *A. tumefaciens* can't penetrate the cells of most monocots. This is a major limitation, because monocots include most of the important cereal food crops, such as corn, wheat, barley, rice, and rye. This shortcoming forced scientists to look beyond *A. tumefaciens* for ways to deliver foreign DNA into monocots.

Microinjection

Microinjection is a new twist on an old idea. Biologists first used fine glass microtools in the late 1800s to dissect animal tissues. Today, scientists use compound microscopes and micromanipulators fitted with tiny glass pipettes to inject DNA directly into the nuclei of plant cells. Most plant cell walls are so tough, however, that they usually must be stripped with enzymes before DNA can be injected. These wall-less cells, called protoplasts, can then be cultured into whole plants in vitro, that is, in glass dishes.

Electroporation

When scientists use electroporation, they shock protoplasts with electricity until the cells become receptive to foreign DNA. A high-voltage electrical pulse temporarily opens small holes in the protoplast membranes, allowing the foreign DNA to slip through before the membranes reseal themselves. The protoplasts can then be cultured into whole plants.

Biolistics

When bacterial vectors, microinjection, electroporation, or other techniques aren't suitable for a particular plant, scientists increasingly turn to biolistics, a blending of ballistics and biology. With a 0.22-caliber gun nicknamed the "bioblaster," plant engineers bombard cells with metallic microprojectiles coated with DNA. Scientists have already used this gun to transform yeast, algal, higher-plant, animal, and human cells, and they predict it will become one of the more useful and versatile tools for plant genetic engineers.

One advantage of using the gene gun with plants is that the device can be used on intact cells, while the other techniques are generally limited to protoplasts. Intact cells are far easier to culture into whole plants than are protoplasts.

Using the above techniques, scientists have engineered many new traits into plants. Among the potential new crops are corn made genetically resistant to the European corn borer, squash engineered to resist viruses, tobacco plants that produce the active ingredient in hepatitis B vaccine, and rice that resists fungal disease. One biotechnology company is even engineering the genes for the blue dye indigo (used to dye blue jeans) into cotton. The plants would produce blue cotton!

Analyzing Genetic Variation

One exciting way we can apply our ability to analyze DNA is to look at genetic variation. In years past, we could study differences among individuals by looking at phenotypic traits. Now we can look directly at DNA. We can use this ability to determine the amount of genetic variation within populations or between species. We can analyze an individual's genotype and try to predict its phenotype.

Ecologists use DNA typing to determine the genetic diversity still available in small threatened populations such as whooping cranes and cheetahs. Conservation biologists at zoos and wildlife preserves use measures of genetic variation to guide decisions about mating males and females from endangered or highly inbred species. Evolutionary biologists are looking at the differences in DNA between species to determine how closely related the species are. Forensic scientists use human DNA typing to exclude suspects in criminal and paternity cases. Medical scientists analyze genotypes to predict genetic disease.

This section of activities relates the techniques of DNA analysis to the analysis of genetic variation, focusing on human genetics. The first lesson, *Generating Genetic Variation: the Meiosis Game*, sets the stage for the following activities by reviewing the basics of inheritance and how genetic diversity is generated. The second lesson, *Analyzing Genetic Variation: DNA Typing*, uses examples to show how DNA typing is currently performed. The contents of several previous chapters are drawn upon. Following this lesson are three short puzzles in which students use DNA data to solve real-life mysteries.

The next activity is a reading and discussion of questions about human genetic diseases. It is designed to start students thinking about the molecular basis of these diseases. In one of the activities, students design a DNA test for sickle-cell disease. Finally, a short reading on the molecular genetics of cancer is included.

Generating Genetic Variation: the Meiosis Game

23

About This Activity

Meiosis is one of the most difficult aspects of general biology for teachers to teach and students to learn. And yet, a solid grasp of both the specific details and the overall process of meiosis is requisite for an accurate and complete understanding of biological topics as varied as Mendelian inheritance, evolution, and the diagnosis of genetic disorders.

Because meiosis is such a crucial concept in biology, we believe an accurate picture of its precise details and overall function is of fundamental importance. So we are momentarily shifting our focus from the techniques of molecular genetics to return to a fundamental topic in basic biology: meiosis—its primary function, specific details, and secondary result, the generation of genetic variation.

This lesson contains two activities. The first is a simple simulation of meiotic chromosome segregation and fertilization, the two components of sexual reproduction. Students make 23 simply constructed paper chromosome pairs, randomly generate gametes by selecting one member of each pair, and combine gametes to produce offspring. Although this activity mimics meiosis and fertilization in humans, the same principles can be illustrated with far fewer chromosomes.

The second activity is a paper simulation of chromosome behavior during meiosis; crossing over between chromatids is included. Only three pairs of chromosomes are used. This activity provides a very concrete illustration of how human heredity works on the chromosome level and why everyone is different.

These exercises are very versatile. They can be used to introduce middle-school students to the basics of human genetics (by performing the simple segregation exercise and discussing heredity in terms of families), to illustrate the principles of Mendelian genetics, or to prepare students for a molecular approach to human genetics. With respect to the last goal, stu-

dents will not successfully master the concepts of DNA paternity determination or DNA fingerprinting unless they understand *why* everyone's DNA is different. Genetic disease determination and gene mapping will also be easier to understand if they have a good grasp of these fundamentals.

Class periods required: *1 or more, depending on the number of variations and discussion topics*

Introduction

Evolution depends upon a constant source of genetic diversity. The ultimate source of new and different genes is mutation. Some mutations are beneficial, but by far most are neutral or harmful.

Some people question how the evolution of life on Earth could be based on a source of genetic variation that is fairly rare and often harmful or neutral. They wonder how the evolution of all of the Earth's species could be based on those relatively rare events in which a beneficial mutation occurs in an appropriate environment.

It is true that mutation alone introduces genetic variability into a population relatively slowly. But nature has provided another method of continually and reliably generating lots of genetic variation in populations: sexual reproduction. The great majority of the Earth's species reproduce sexually at some point in their life cycles. And, as you have learned in a previous section, those that don't have evolved other methods of producing genetically variable offspring (see chapter 2).

The genetic variation produced through sexual reproduction is derived from reassorting existing genes into novel combinations and not from creating new genes. This reassorting and recombining of genes occur during both steps in sexual reproduction: the production of haploid gametes through meiosis and the fusion of these gametes in fertilization.

Meiosis

The overall goal of meiosis is to produce haploid gametes for sexual reproduction. The behavior of chromosomes during meiosis ensures that genetic variability among offspring is achieved. Two meiotic processes create genetically variable gametes: first, independent assortment and segregation of chromosomes, and second, crossing over.

Independent Assortment and Segregation

You will recall from chapter 2 that individual maternal (or paternal) chromosomes segregate independently of each other. It is equally likely that any maternal chromosome will find itself in a gamete with any other nonhomologous chromosome. So genes on different chromosomes segregate independently of each other. This independence creates genetic variation in the gametes, because any one gamete will contain a random assortment of maternal and paternal chromosomes. For assistance in visualizing how segregation and independent assortment create genetic variation, see Fig. 2.1, 2.2, and 23.2. (*The first activity in this lesson plan demonstrates these processes.*)

Crossing Over

Even though reassortment of whole chromosomes during meiosis produces an enormous amount of genetic variation, nature has devised a way of increasing variability still further by creating genetic variability *within* a chromosome.

During meiosis I, the maternal and paternal homologous chromosomes line up opposite each other prior to segregation. You will recall that homologous chromosomes have the same genes and that early in meiosis, after DNA replication, each chromosome consists of two sister chromatids. Genetic variation within a chromosome is created when chromatids of the two homologs exchange equal segments of DNA by breaking and rejoining at precisely the same point. Some genes from the maternal chromosome come to reside on the paternal chromosome and vice versa.

This amazing feat of exchanging chromosome segments gene for gene is called **crossing over.** Sometimes only one of the sister chromatids is involved in an exchange. At other times, both are involved in what is termed a double crossover. In a double crossover, however, the sister chromatids can break and rejoin with their homologs at different places. Thus, all four chromatids differ genetically. (*Crossing over is demonstrated in the second activity of this lesson.*)

If homologous chromosomes have the same genes, how can all of the chromatids be different after cross-

ing over? When we say that homologous chromosomes have the same genes, we mean that they have genes for the same traits in the same order with the same general pattern of noncoding sequences between them. They are not genetically identical, however, because these genes occur in different forms called **alleles.** So homologs contain slightly different versions of the same gene.

How different are the chromosomes in a homologous pair? It has been estimated that the DNA of two individuals varies once every 300 to 1,000 base pairs. Since the two homologs of a chromosome pair came from two different individuals (the mother and the father), this estimate applies to homologous pairs.

Are the differences distributed evenly along the chromosomes? No, according to current analyses of human chromosomes (and also by logical prediction). Certain regions of chromosomes display much more variability between individuals than do other regions. These hypervariable regions seem, in general, to be in noncoding areas of the genome. Upon reflection, this distribution makes sense. Noncoding DNA would be freer to accumulate mutations, because changes in these regions would presumably have no effect on the survival of the individual. On the other hand, most mutations in coding regions would have harmful consequences. Natural selection would tend to eliminate these mutations from the population.

Teaching meiosis

Why is meiosis so difficult to teach and to understand? Partly because it occurs on a molecular level, so it's difficult to "see." But more important, there are just too many two's to keep straight.

- Chromosomes are present in two's, one from the mother and one from the father.
- At some stages in meiosis, each chromosome is composed of two strands called sister chromatids.
- Chromosomes are made of DNA, and DNA is a two-stranded molecule.
- The cells divide in two.

In addition to too many two's, there are too many strands. Chromosomes have strands called chromatids. Chromatids are composed of strands of DNA, which is a double-stranded molecule. All of these strands keep doubling and dividing in two. No wonder meiosis is so difficult to understand!

Another point students often miss is that the random process of chromosome segregation can (and does) produce results that don't come out even. Many families have four children of the same sex; two heterozygotes can produce a single homozygous recessive child, though the odds against it are 3 to 1. We included a coin toss in our simulation so that the results would really be random. If you ask students to choose one of each chromosome pair at random, they usually make selections so that half of the chromosomes in their gametes are of maternal origin and half are of paternal origin (which of course does happen in nature, but not as often as students think).

Before students can begin to understand how meiosis generates genetic variability, they must have a clear grasp of the meiotic process, a most difficult task for both teacher and student. Having attempted to teach meiosis many times, we are all too aware of the difficulties. We also are aware of some of the points at which the floor opens up and the students appear to fall through and get completely lost. To assist you, here are some points to keep in mind and to continually reinforce with your students.

Points to remember

- The overall function of meiosis is to produce haploid gametes.

- The maternal and paternal chromosomes do not fuse together when gametes fuse to produce an organism. The chromosomes retain their distinct identities. Students must keep this in mind when visualizing the diploid cell that will give rise to haploid gametes. Even when portions of the maternal and paternal chromosomes are exchanged during crossing over, a chromosome's identity remains either maternal or paternal.

- Meiosis is composed of three main events: first, the DNA is replicated; then the nucleus and cells undergo two divisions.

- The maternal and paternal chromosomes of a chromosome pair (a homologous pair) separate from each other during the *first* meiotic division.

- The second meiotic division is like mitosis. Replicated strands of DNA separate from each other.

Helpful hints

- To count chromosomes, count centromeres, not chromatids.

- It is helpful to think of the number of chromosomes and the amount of DNA separately. In a diploid cell prior to meiosis, the number of chromosomes is $2n$. Think of the amount of DNA at that point as being $2n$ also. When meiosis begins, the DNA replicates. At that point, the chromosome number is still $2n$, but the amount of DNA is $2 \times 2n$ ($4n$). At the end of meiosis I, the chromosome number is $1n$, but the amount of DNA is $2 \times 1n$ ($2n$). Finally at the end of meiosis II, both the chromosome number and the amount of DNA are $1n$.

Diagrammatic representations

Figures 23.1 and 23.2 (see the *Student Activity*) provide schematic descriptions of the *highlights* of meiosis: one at the chromosome to molecular level and one at the chromosome to cellular level. For a more complete description of meiotic events, see a general biology textbook.

It helps to describe what is happening to the chromosomes during meiosis one level at a time. Make sure the students understand what is going on at the molecular level before you tackle the behavior of the chromosomes. Then, when discussing chromosomal behavior, stress the independent assortment of non-homologous maternal and paternal chromosomes and the subsequent segregation of homologous chromosomes. In other words, first have them focus on what is going on between chromosomes 1, 2, 3, etc. Then have them focus on what goes on between members of a homologous pair, that is, chromosomes 1A and 1B, 2A and 2B, etc. Take everything one step at a time before putting it all together.

Only when the students are clear on both the molecular events and the behavior of the entire set of chromosomes should you even begin to discuss crossing over and linkage.

Resources

Carolina Biological Supply Co. "Chromosome Simulation BioKit" (catalog no. 17-1100)

The best biological systems for demonstrating segregation and crossing over at the phenotypic level are fungi. In fact, most of the observations that guide current thinking about molecular mechanisms of genetic recombination were made with fungi. Carolina Biological Supply Co. sells kits for conducting classroom experiments or demonstrations with the fungus *Sordaria fimicola* (for example, "*Sordaria* Genetics BioKit" [catalog no. 15-5847]). The kits contain appropriate strains along with other needed materials and directions. *Sordaria* strains can also be ordered separately. Make sure your students understand the process of meiosis before they perform the fungus experiments. To interpret the results of the crosses, they must be familiar with chromosome behavior during meiosis.

Activity I: Chromosome Segregation and Fertilization

Objectives

After completing this exercise, students should be able to

1. State that humans have 23 pairs of chromosomes;

2. Define homologous chromosomes as chromosomes that have the same genes in the same arrangement and understand that the genes can be in different versions (called alleles);

3. Explain that half a person's DNA is a copy of part of her mother's DNA and the other half is a copy of part of her father's DNA;

4. Explain how independent assortment and chromosome segregation during meiosis generate a genetically diverse human population.

Materials

- Four colors of construction paper (we suggest two sets of two similar shades each, such as red and pink for one and light and dark blue for the other) *or* photocopy the chromosome templates onto colored paper

- One penny per student team

- Markers (optional)

- Two pairs of scissors per student team

Preparation

Make a poster of parental chromosomes to use for comparison with the gametes and the offspring chromosomes your class generates. Make a set of maternal and paternal chromosomes, following the directions below (make Y small, if you wish). Divide a poster board or some other suitable surface in half. Glue the maternal chromosomes in pairs on one side and the paternal chromosomes on the other side. (A felt board and felt chromosomes would be useful for a class demonstration.)

Optional: Laminate colored paper with the chromosome patterns, cut out the laminated chromosomes, and save them to use in the whole chromosome segregation exercise.

Tips

After your class does the simulation, collect a few sets of offspring chromosomes. Paste them on a poster board or other surface, and save them as examples of offspring from the parental chromosome poster just discussed. These two chromosome posters will make it very easy for you to explain DNA-based paternity determination.

Chromosome segregation in meiosis and the random recombining of alleles in sexual reproduction provide excellent opportunities to introduce your students to basic concepts in probability and statistics.

Procedure

1. Set the stage appropriately for the simulation by reminding the class of whatever they have learned about heredity that is relevant to the points you wish to make (for example, meiosis, gametogenesis, sex determination, Mendel's laws, or whatever you deem useful). One way to look at this exercise is that the class is simulating possibilities for the genotypes of different offspring of the same parents, since all the students form gametes starting from the same parental chromosomes. Some suggestions for possible discussion topics are listed below.

2. Divide your students into playing groups. Each playing group consists of two teams. Each team will simulate chromosome segregation to produce a gamete, and then the two teams will combine the gametes' chromosomes to make an offspring's chromosomes. A playing group can be as small as two students (one per gamete). Since there is a lot of handwork to perform in cutting out the chromosomes and conducting the crossing-over simulation, a playing group of four (two students per team) may work better, depending on your class. Designate one team per group as the female gamete team and one team as the male gamete team.

3. Pass out materials to each playing group. Each group gets the following:
   - One red sheet and one pink sheet of construction paper for the female team
   - One light blue and one dark blue sheet of construction paper for the male team
   - One pair of scissors per team (two per group)
   - One marker per team (optional)
   - One penny per team (for coin toss)

4. The male and female gamete teams make their parental chromosomes. Each team should mark off their sheets of paper into 24 rectangles as shown (one rectangle will not be used). The rectangles should be numbered 1 through 22 on all the sheets. The female team marks the 23rd rectangle X on both sheets. The male team marks the 23rd rectangle X on one sheet and Y on the other sheet. After the rectangles are drawn and marked, each team should cut them out. Two differently colored rectangles with the same number represent a homologous pair of chromosomes. Each gamete team should have 23 marked pairs of chromosomes, and the two members of each pair should be different colors. Introduce the term "homologous." Be sure that the students understand that the two members of the pair are essentially alike but different in many small ways. The two homologs have the same types of genes in the same arrangement, but the exact version of the genes might be different, and the noncoding DNA (if you are talking about that) would also be different.

5. For simulation of chromosome assortment and segregation, have the students line up their homologous pairs of chromosomes. Each gamete team should assign one member of the pair to be heads and the other to be tails. Then use a coin toss to determine which member of the homologous pair is contributed to the gamete.

6. When the gamete teams have segregated their chromosomes, have the two teams in a playing group combine their gametes to make an offspring. Have the teams line up the new chromosome pairs.

Analysis

At this point, discuss the exercise at a level appropriate to your class. Make the following points.

No two teams produced the same offspring, even though the starting material was the same. One way to emphasize this is to make a table on the blackboard with 1 through 23 listed vertically and the playing groups listed horizontally. Have each playing group report the colors of the two chromosomes of each pair; record the information on the board. It will be clear that every group has different results. Have the students speculate on how many different combinations are possible.

If you are couching the discussion in terms of generating siblings from the same parents, you could now have students think about their own families or families they know with several children. Ask them how

similar and how different they are from their brothers and sisters. Ask them why they think they are somewhat like their siblings and somewhat different from them (this would be a particularly good discussion for very young students).

Exactly one half of the offspring's genetic material came from its mother and half from its father. All of the offspring's DNA has a match in either its mother's or its father's DNA. Ask your class about the mother's and father's chromosomes. Where did they come from? (The class should realize that the parents' chromosomes came from the grandparents.) If you like, you could identify (for example) the pink chromosomes as being from the maternal grandmother and the red chromosomes as being from the grandfather, and examine their distribution in the grandchildren. Ask the students if they have any traits they know are like those of their grandparents.

Activity II: Simulation of Chromosome Behavior during Meiosis

In this exercise, students simulate the behavior of chromosomes during meiosis, using a system with three pairs of chromosomes. The activity illustrates the formation of four haploid cells from the starting diploid, the process of crossing over, random segregation, and independent assortment. It can be used after the simple simulation of chromosome assortment (activity I) or on its own.

If the three pairs of chromosomes are marked with genes, the exercise can be used as a concrete illustration of the inheritance of traits. You can make this exercise as complex as you wish: mark only one gene, or put several genes on each chromosome. No matter what you decide to do, we suggest you make sure that students understand the chromosome behavior before you begin discussing genes, genotypes, and phenotypes. If students are confused about what is happening to the chromosomes, they will also be confused about the genetics. You may choose to do the chromosome simulation once without discussing genes and then repeat it with marked alleles.

Materials

Each student or student team will need the following:

- Three pairs of chromosomes prepared as described under *Preparation*
- Scissors
- Tape
- Penny for coin tosses

For the preparation, you will need

- Chromosome patterns
- Removable tape
- Colored and white paper for photocopies

Preparation

The chromosome pairs can be made during class by the students or before class.

To prepare the three pairs of chromosomes, photocopy the patterns as many times as you need for your class plus once for you. Use double-sided copying for the shaded chromosomes so that the shading will be on both the front and the back. If possible, copy half of the maternal and paternal patterns on colored paper for fertilization exercise. Note that the patterns for paternal chromosomes are separate from the patterns for maternal chromosomes. Each homologous pair of chromosomes contains one paternal and one maternal chromosome. Match the shapes of the chromosomes to make the pairs.

- Cut out the chromosome patterns. Fold each chromosome longitudinally in half (along the dotted line), and use removable tape to keep it folded. This is the starting chromosome.

- If you intend to tie this exercise to a discussion of genetics, mark genes on the starting chromosomes. For example, put a capital A on one arm of a maternal chromosome, front and back (so the A will be on both chromatids after the chromosome is unfolded for "replication"). At the same position on the homologous paternal chromosome, put a lowercase a front and back. On the other arm of the maternal chromosome, mark a B allele. Put a lowercase b allele on the paternal chromosome at the same "locus." Depending on your plans, you may want to mark all three pairs of chromosomes with genes. If you intend to discuss linkage, mark two genes that are fairly close together on the same chromosome arm.

- Attach one set of starting chromosomes to a poster board or other material, and save it for reference.

- Hand out three pairs of folded chromosomes to each student or student group. Some students should get three white pairs; others should get three colored pairs.

Procedure

1. Explain to your students that they are about to simulate chromosome behavior during meiosis.

The model cell has three pairs of chromosomes. The homologs from the father of this individual are the unshaded chromosomes; the homologs from the mother are the shaded chromosomes. Review the fact that each chromosome is one long DNA molecule associated with proteins.

2. DNA replication. Before meiosis begins, the DNA molecules replicate. The two replicated molecules remain attached at the centromeres. To simulate DNA replication, have students remove the tape and unfold the paper chromosomes so that they look like X's. Each arm of the X is a chromatid.

3. Homologous pairing and crossing over. After replication, the homologous replicated chromosomes pair closely with one another in what is called a synaptonemal complex. At this time, the homologous chromatids exchange segments in the process of crossing over. Microscopic observations of many meiotic cells show that on average, human cells undergo slightly more than one exchange per chromosome pair during meiosis.

Have students place the two homologs of each pair directly on top of each other. This represents the synaptonemal complex. They should then perform one crossover event per pair by choosing one of the four arms of the X, cutting the paired chromosomes at the same place on that arm, and exchanging the two segments. The new segments should be taped in place. The students may perform a second crossover event on one chromosome pair if they wish. The second event may be on the same arm or a different one. Crossing over does not occur exactly at the centromere.

At this point, you may wish to remind students of the idea of recombinant DNA. Ask them if their own chromosomes are an example of recombinant DNA. (They are, in the sense that their chromosomes are products of recombination of their parents' chromosomes.) Students should realize that cutting and pasting DNA together is an activity that cells mastered long ago. Ask your students to speculate on what kinds of enzymes might be involved in crossing over. (Examples: some kind of cutting enzyme to break the chromosome, a ligase to reseal it, an enzyme or enzymes that could read DNA sequences to make sure the cutting and resealing happens at matching sites along the chromosome.)

4. Segregation of replicated chromosomes. After completing the crossing over activity, the repli-

cated chromosomes separate from their homologs and migrate to opposite ends of the cell before cell division. Each daughter cell will get one of the X-shaped chromosomes of each pair. Have students designate heads or tails for the maternal and paternal chromosomes. They should then use a coin toss to determine which member of each homologous pair migrates to the right side of the desk and which migrates to the left. When this part of the activity is complete, each student or student team should have separated their chromosomes into two sets of three chromosomes, with one chromosome of each type.

5. Cell division. This is the end of meiosis I. Students can simply imagine the cell dividing, or they can put some kind of divider across their desks to separate the chromosome sets. Examine the sets, and compare them to the starting cell. How many centromeres were there in the starting cell? (Refer to your poster; there should be six.) How many centromeres are in the two cells produced in meiosis I? (Three.) These two cells are haploid, but they still have $2n$ DNA. For gametes, we need haploid cells with $1n$ DNA. Meiosis II reduces the amount of DNA to $1n$.

6. Meiosis II. Meiosis II is much like mitosis. The DNA in the chromosomes is already replicated, so the chromosomes line up along the spindle and split down the centromeres. The separated chromatids migrate to opposite poles, and cell division separates them into two daughter cells.

   Have students line up their three chromosomes and cut the two sides of the X apart down the fold line. Now they should use the coin toss as in step 4 to decide which daughter chromosomes will migrate to the right and which will migrate to the left. After chromosome migration, the cell divides. Repeat this step for the second cell generated in step 5.

   At the end of this part of the activity, students should have four sets of three chromosomes, each with $1n$ DNA, in four "cells." These are the four gametes produced by meiosis.

7. Analysis. Make a chart on the blackboard to record the genetic contents of the gametes. Put 1 through 3 vertically and students group numbers horizontally. Record the contents of the gametes, using M for the shaded maternal chromosome and P for the unshaded paternal chromosome. If you marked each chromosome with genes, you could record the genetic content of each gamete (for example, AB, cd, eF). Put an asterisk by any chromosome that has undergone crossing over (for example, M* or eF* in the example above). To determine whether a chromosome is maternal or paternal, use the pattern at the centromere. (If you have a large class, you may not need to record all the gamete genotypes to make the point about genetic variability.)

It should be evident from the summary chart that meiosis generates a great deal of variation. The variation from random segregation and independent assortment is reflected in the different distributions of M's and P's among the gametes. Ask your class to figure out how many different possible combinations there are of M's and P's from the three chromosomes. In other words, if there were no crossing over, how many genetically different gametes could you generate? The answer is eight. How many gametes could they get with no crossing over and two chromosome pairs? (Four.) Four chromosome pairs? (Sixteen.) Perhaps your students will realize that there is a pattern to the answers: 2^n, where n is the number of chromosome pairs. Then how many combinations could a human cell generate? (2^{23} = 8,388,608.) This is a lot of variability, but crossing over generates even more. With crossing over, the ability of cells to rearrange their genetic material is essentially unlimited.

8. Fertilization. Have students with white chromosomes combine one "gamete" with one "gamete" of students with colored chromosomes (the colors identify which parent the chromosomes came from). You may want to tape some of these sets of offspring chromosomes on the blackboard for discussion. Note that exactly one half of the offspring's genetic material came from its mother's gamete and half from its father's gamete. All of the offspring's DNA has a match in either its mother's or its father's DNA. The shaded chromosomes represent contributions from the grandmothers; the unshaded chromosomes are contributions from the grandfathers. Remind students that humans have 23 pairs of chromosomes, but the process is the same.

After the simulation is complete, add one set (or more) of offspring chromosomes to your poster of the parental chromosomes. You may want to refer to it during discussions of DNA typing.

Discussion topics
Mendelian Genetics

The simple chromosome segregation exercise (activity I) provides a useful way of looking at Mendelian inheritance. As an example, ABO blood group genetics is a topic that often appears in biology texts. The ABO blood group gene is found on chromosome 9 (the location of a particular gene is called its genetic locus). Choose a blood group genotype for each parent. For example, let the mother be AO (a blood type A genotype, since the O gene is recessive), and let the father be BO (a blood type B genotype, for the same reason).

Have students mark pink chromosome 9 with a straight line across it (to represent the gene) and an A, mark red chromosome 9 with an identically placed line and O, mark light blue chromosome 9 with the line and B, and mark dark blue chromosome 9 with the line and O. Then observe the blood group genotypes of the offspring. Relate them to phenotypes (observed blood type). Ask your students if they can always tell the genotype from the phenotype (the answer is no: blood type A could be AA or AO; type B could be BB or BO). Show them how Punnett squares work to show the possible products of a genetic cross. The chromosome simulation should make it easy for students to understand how Punnett squares work.

Obviously, you can take this exercise as far as you want, including other traits on other chromosomes, etc. One teacher who field tested this activity prepared the whole set of chromosomes ahead of time and marked some with genetic-disease alleles on the back. After the segregation and mating exercise, the students were asked to look at their chromosomes to see who had inherited a disease.

The second activity can also be used as a tool for looking at Mendelian genetics if the chromosomes are marked with genes as described in the activity. Make up phenotypes for the genes to make discussion more meaningful. You could use Mendel's pea traits, human blood groups, or anything else.

Sex Determination and Sex-Linked Characteristics

Talk about the X and Y chromosomes (activity I). Have the male gamete team cut the Y chromosome extra small to simulate its actual small size with respect to the X chromosome. Use the chromosome segregation exercise to show younger students how the paternal gamete determines the sex of the offspring. For more advanced classes, use the large X chromosome and the small Y chromosome as props to talk about the fact that most genes on the X chromosome are missing on the Y and the effect that this difference has on the inheritance of diseases like hemophilia A and red-green color blindness (both are X linked). You could mark one of the mother's X chromosomes with a genetic defect and observe the inheritance.

Genetic Linkage (Only If You Do Activity II)

When two traits are inherited together more frequently than they should be on the basis of random segregation, the two traits are said to be linked. The term was coined before it was known that genes are located on chromosomes and can, in fact, be physically linked together. Use this lesson (with the crossing-over exercise) to show how linkage really works. The closer together two genes are, the more likely it is that they will be transmitted together. In fact, the frequency with which two genes on the same chromosome are inherited together is often used as an indicator of the physical distance between the genes.

The ABO blood group locus is linked to a gene encoding the enzyme adenylate kinase-1 and to another gene responsible for a genetic defect called nail patella syndrome (deformities of the nails and knees). If you want to illustrate linkage, have students make three marks very close together across one pair of their chromosomes and think about the likelihood of the marks being separated in a given round of crossing over. What if the marks were further apart? What if they were at opposite ends of the chromosome?

The crossing-over activity should make it clear to students that two genes on opposite ends of a chromosome are not linked, although they are on the same chromosome. You can tell them that physicians and scientists can sometimes use linkage to make predictions about inherited diseases. If they do not know how to detect a disease gene but can detect another gene (or other DNA marker) that is consistently inherited with the gene (tightly linked), they can use that marker as a probable indicator of the disease gene. (This method is not 100% foolproof, since crossing over could occur between the marker and the disease gene. If it ever did, the people who inherited the separated disease gene would not have the marker as an indicator.) Analysis of a linked marker was used in the original genetic tests for cystic fibrosis before the actual gene responsible for the disease was identified. The marker used was not another gene but a restriction fragment length polymorphism (see *Analyzing Genetic Variation: DNA Typing* [chapter 24]).

Evolution

Genetic variation provides the raw material for evolution. Environmental factors put selective pressure on the members of a population, and those members best suited to survive and reproduce in the given environment produce more offspring. If the traits that lead to more offspring have a genetic basis, the characteristics of the population as a whole change over time.

This mechanism absolutely depends on the presence of genetically different members of a population. Mutation is the original source of genetic variation. After mutations happen, genetic recombination through sexual reproduction shuffles genes, creating new combinations of traits. The more variation there is in a population, the better the chances are that some member will be successful in any given set of environmental circumstances.

Discuss the evolutionary role of mutation and recombination with your class. You could ask them to think about what might happen if no recombination occurred. If the only source of genetic variation were mutations passed along directly to offspring through asexual reproduction, what might be different in the way a population could respond to drastic environmental changes? What if a succession of different changes occurred close together in time?

You could also discuss the evolutionary role of selective pressure. Remember that in large populations experiencing no selective pressure, gene frequencies do not change. This statement is reflected in the Hardy-Weinberg equilibrium equation. If you teach Hardy-Weinberg to your classes, you could use this activity as a way to remind them of it or introduce it. Evolution is defined as a change in gene frequency. For such a change to occur in a large population, there must be selective pressure.

Generating Genetic Variation: the Meiosis Game

Introduction

Evolution depends on a constant source of genetic diversity. The ultimate source of new and different genes is mutation. Some mutations are beneficial, but by far most are neutral or harmful. Some people question how the evolution of life on Earth could be based on a source of genetic diversity that is fairly rare and often harmful or neutral. They wonder how the evolution of all of the Earth's species could be based on those rare times when a beneficial mutation occurs in an appropriate environment.

It is true that mutation alone introduces genetic variability into a population relatively slowly, but nature has provided another method of continually and reliably generating lots of genetic variation in populations: sexual reproduction. The great majority of the Earth's species reproduce sexually at some point in their life cycles. As you have learned in a previous section, those that don't reproduce sexually have evolved other methods of producing genetically variable offspring (see chapter 2).

The genetic variation produced through sexual reproduction is derived from the reassortment of existing genes (alleles) into novel combinations and not from the creation of new genes (alleles). This reassorting and recombining of genes occurs during both steps in sexual reproduction: the production of haploid gametes through meiosis and the fusion of these gametes in fertilization.

Meiosis and Genetic Variation

The overall goal of meiosis is to produce gametes for sexual reproduction. Each gamete contains one half the number of chromosomes that the parent's somatic cells contain and consequently is haploid. The behavior of chromosomes during meiosis ensures that genetic variability among gametes, and ultimately offspring, is achieved. Two meiotic processes create genetically variable gametes: first, independent assort-

ment and segregation of chromosomes, and second, crossing over.

Independent assortment and segregation

One way to generate genetically variable gametes is by distributing different chromosomes to different gametes, much as one deals different hands to a table of poker players. You will recall from chapter 2 that individual maternal (or paternal) chromosomes segregate independently of each other. All chromosomes nonhomologous to a maternal chromosome are equally likely to find themselves in a gamete with that maternal chromosome, just as all cards in a deck are equally likely to be dealt to any player (provided the dealer is not crooked). It follows that genes on different chromosomes also segregate independently of each other. This creates genetic variation in the gametes, because any one gamete will contain a random assortment of maternal and paternal chromosomes and therefore genes. For assistance in visualizing how segregation and independent assortment create genetic variation, see Fig. 2.1, 2.2, and 23.2. The first activity in this chapter demonstrates the relationship between independent assortment, segregation, and the production of genetically variable gametes.

Crossing over

Even though distributing chromosomes into different groupings produces an enormous amount of genetic variation, nature has devised a way of increasing variability still further by creating genetic variability *within* a single chromosome. During meiosis I, the maternal and paternal homologous chromosomes line up opposite each other prior to segregation. You will recall that homologous chromosomes have the same genes and that early in meiosis, after DNA replication, each chromosome consists of two sister chromatids. Genetic variation within a chromosome is created when chromatids of the two homologs exchange equal segments of DNA by breaking and rejoining at the same point. Some genes from the maternal chromosome come to reside on the paternal chromosome and vice versa.

This amazing feat of exchanging chromosome segments gene for gene is called crossing over. Sometimes only one of the sister chromatids is involved in an exchange with a chromatid from a homologous chromosome. When only one chromatid is involved, one chromatid per chromosome is no longer genetically identical to its sister chromatid or to the parental chromosome prior to replication. At other times, both chromatids in a chromosome are involved in what is termed a double crossover. In a double crossover, however, the sister chromatids can break and rejoin with their homologs at different places. Thus, all four chromatids differ genetically from each other and from the original chromosomes that gave rise to them.

If homologous chromosomes have the same genes, how can the chromatids be different after crossing over? When we say homologous chromosomes have the same genes, we mean they have genes for the same traits in the same order with the same general pattern of noncoding sequences between them. They are not genetically identical, however, because, as you will recall from chapter 2, these genes occur in different forms called alleles. So homologs actually contain slightly different versions of the same genes.

How different are the chromosomes in a homologous pair? It has been estimated that the DNA of two individuals varies once every 300 to 1,000 base pairs. Since the two homologs of a chromosome pair came from two different individuals (the mother and the father), this estimate applies to homologous pairs.

Are the differences distributed evenly along the chromosomes? No, according to current analyses of human chromosomes (and also by logical prediction). Certain regions of chromosomes display much more variability between individuals than do other regions. These hypervariable regions seem, in general, to be in noncoding areas of the genome. Upon reflection, this distribution makes sense. Noncoding DNA is freer to accumulate mutations because changes in these regions presumably have no effect on the survival of the individual. On the other hand, most mutations in coding regions have harmful consequences. Natural selection tends to eliminate these mutations from the population.

Understanding meiosis

Understanding the details of meiosis and differentiating between meiosis and mitosis can be difficult.

Meiosis is hard to grasp for a number of reasons. It occurs on a molecular level, so it's difficult to "see." But more important, there are just too many two's to keep straight.

- Chromosomes are present in two's, one from the mother and one from the father.
- At some stages in meiosis, each chromosome is composed of two strands called sister chromatids.
- Chromosomes are made of DNA, and DNA is a two-stranded molecule.
- The cells divide in two.

In addition to too many two's, the process involves too many strands. Chromosomes have strands called chromatids. Chromatids are composed of strands of DNA, which is a double-stranded molecule. All of these strands keep doubling and dividing in two. No wonder meiosis is so difficult to understand!

Nonetheless, having a good grasp of meiosis is requisite if you want to understand biological topics as varied as Mendelian inheritance, evolution, and the diagnosis of genetic disorders. To assist you in acquiring this understanding, here are some points to keep in mind and some helpful hints.

Points To Remember

- The overall function of meiosis is to produce haploid gametes, that is, cells with one half the number of chromosomes that the parental somatic cells contain.

- The maternal and paternal chromosomes do not fuse together when gametes fuse to produce an organism. The chromosomes retain their distinct identities. Even when portions of the maternal and paternal chromosomes are exchanged during crossing over, a chromosome's identity remains either maternal or paternal.

- Meiosis is composed of three main events: first, the DNA is replicated; then the nucleus or cell undergoes two successive divisions.

- The maternal and paternal chromosomes of a chromosome pair (homologs) separate from each other during the first meiotic division.

- The second meiotic division is like mitosis. Replicated strands of DNA separate from each other.

Helpful Hints

- To count chromosomes, count centromeres, *not* chromatids.

- Think of the number of chromosomes and the amount of DNA separately (see Fig. 23.2). In a diploid cell prior to meiosis, the number of chromosomes is 2n. Think of the amount of DNA at that point as also being 2n. When meiosis begins, the DNA replicates. At that point, the chromosome number is still 2n, but the amount of DNA is 2 × 2n (4n). At the end of meiosis I, the chromosome number is 1n, but the amount of DNA is 2 × 1n (2n). Finally, at the end of meiosis II, both the chromosome number and the amount of DNA are 1n.

Figures 23.1 and 23.2 provide schematic descriptions of the highlights of meiosis: one at the chromosome to molecular level and one at the chromosome to cellular level. (For a complete description of meiotic events, see a general biology textbook.) When studying meiosis, first make sure you understand what is going on at the molecular level before you tackle the behavior of the chromosomes in relationship to one another.

Activity I: Independent Assortment and Segregation

In this exercise, you will work in a group that is divided into two teams. One team is the female gamete team, and the other is the male gamete team.

Each team will simulate the independent assortment and segregation of chromosomes in gamete production. The two teams in a group will combine male and female chromosomes to produce the chromosomal content of an offspring.

Your group will be given the following materials:

- One red and one pink sheet of construction paper for the female team
- One light blue and one dark blue sheet of construction paper for the male team
- One penny per team (for a coin toss)

1. Mark off your sheets of paper into 24 rectangles (one rectangle will not be used; see Appendix A). Number the rectangles 1 through 22 on all the sheets. The female team marks the 23rd rectangle X on both sheets. The male team marks the 23rd rectangle X on one sheet and Y on the other sheet. After the rectangles are drawn and marked, cut them out. Different-colored rectangles with the same number represent a homologous pair of chromosomes. Each team should have 23 marked pairs of chromosomes, and the two members of each pair will be different colors. For example, for

the female team, each chromosome pair is represented by one pink and one red chromosome.

2. Simulation of chromosome assortment and segregation. Each team lines up its homologous pairs of chromosomes and assigns one member of the chromosome pair to be heads and the other to be tails (for example, red = heads and pink = tails). Toss a coin to determine which member of the homologous pair to contribute the gamete.

3. When your team has determined which chromosome will be in the gamete, the two teams in a playing group should combine their gametes to make an offspring. Line up the new chromosome pairs found in the offspring.

Activity II: Simulation of Chromosome Behavior during Meiosis

In this exercise, you will simulate the behavior of chromosomes during meiosis, using a system with three pairs of chromosomes. The activity illustrates the formation of four haploid cells from the starting diploid cell, the process of crossing over, random segregation, and independent assortment.

Each student or student team will receive

- Three pairs of paper chromosomes
- A penny for coin tosses

1. The model cell has three pairs of chromosomes (see Appendix A). The homologs from the father of this individual are the unshaded chromosomes; the homologs from the mother are the shaded chromosomes.

2. DNA replication. Before meiosis begins, the DNA molecules replicate. The two replicated molecules remain attached at the centromeres. To simulate DNA replication, remove the tape, and unfold the paper chromosomes so that they look like X's. Each arm of the X is a chromatid.

3. Homologous pairing and crossing over. After replication, the homologous replicated chromosomes pair closely in what is called a synaptonemal complex. At this time, the homologous chromatids exchange segments in the process of crossing over. Microscopic observations of many meiotic cells show that on average, human cells undergo slightly more than one exchange per chromosome pair during meiosis.

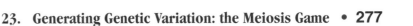

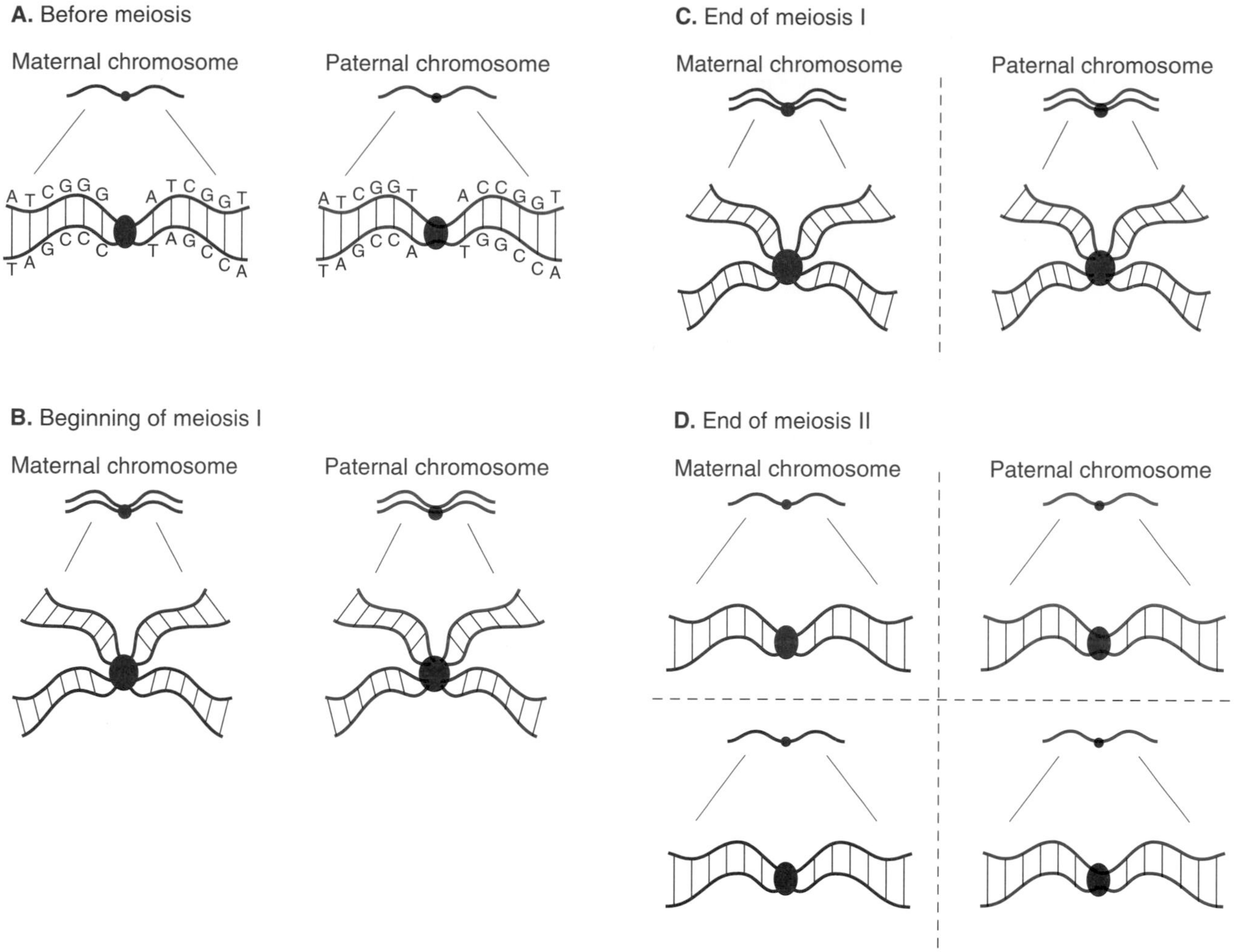

A. Before meiosis

Maternal chromosome   Paternal chromosome

B. Beginning of meiosis I

Maternal chromosome   Paternal chromosome

C. End of meiosis I

Maternal chromosome   Paternal chromosome

D. End of meiosis II

Maternal chromosome   Paternal chromosome

Figure 23.1   A view of chromosomes at the molecular level during meiosis. (A) Before meiosis. Before meiosis begins, a chromosome consists of a single molecule of DNA (a chromatid); but remember, DNA is a double-stranded molecule. (B) Beginning of meiosis I. At the beginning of meiosis, the DNA molecules are replicated. The result is a chromosome that consists of two strands of DNA held together by one centromere. (C) End of meiosis I. During the first meiotic division, the maternal and paternal chromosomes separate from one another, and the cell divides in two. (D) End of meiosis II. During the second meiotic division, the two chromatids separate from one another, and the cells divide in two.

Place the two homologs of each pair directly on top of each other. This represents the synaptonemal complex. Perform one crossover event per pair by choosing one of the four arms of the X, cutting the paired chromosomes at the same place on that arm, and exchanging the two segments. Tape the new segments in place. You may perform a second crossover event on one chromosome pair if you wish. The second event may be on the same arm or a different one. Crossing over does not occur exactly at the centromere.

4. Segregation of replicated chromosomes. After completing the crossing-over activity, separate

the homologs. Each daughter cell will get one of the X-shaped chromosomes of each pair. Designate heads and tails for the maternal and paternal chromosomes. Toss a coin to determine which member of each homologous pair migrates to the right side of the desk and which migrates to the left. When this part of the activity is complete, you will have separated your chromosomes into two sets of three chromosomes; each set will contain one chromosome of each type.

5. Cell division. Cell division marks the end of meiosis I. Imagine the cell dividing, or put some kind of divider across your desk to separate the chro-

A. Before meiosis begins

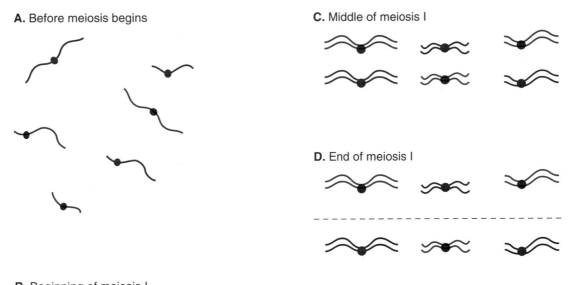

C. Middle of meiosis I

D. End of meiosis I

B. Beginning of meiosis I

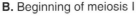

E. End of meiosis II

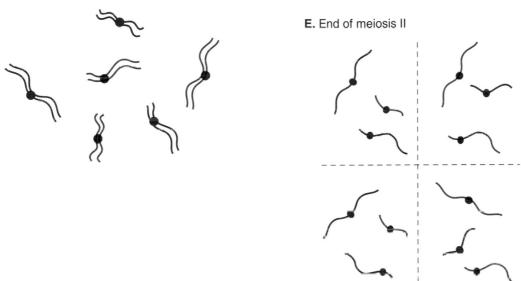

Figure 23.2 Chromosome behavior during meiosis. (A) Before meiosis begins. Chromosomes occur as homologous pairs, one from the mother and one from the father. This cell contains three chromosome pairs; its diploid number is six. (B) Beginning of meiosis I. As meiosis begins, the DNA in each chromosome is replicated, producing two-chromatid chromosomes. The cell is still diploid. (C) Middle of meiosis I. The homologous chromosomes align themselves opposite one another before separating. (D) End of meiosis I. Homologous chromosomes have separated, and the cell has divided. Each cell is haploid and contains maternal and paternal chromosomes. (E) End of meiosis II. The chromatids of a chromosome have separated, and each cell has divided to yield four haploid gametes.

mosome sets. Examine the sets and compare them to the starting cell. How many centromeres were there in the starting cell? How many centromeres are in the two cells produced in meiosis I? These two cells are haploid, but each one still has 2n DNA. For gametes, we need haploid cells with 1n DNA. Meiosis II reduces the amount of DNA to 1n.

6. Meiosis II. Meiosis II is much like mitosis. The DNA in the chromosomes is already replicated, so the chromosomes line up along the spindle and split through the centromeres. The separated chromatids migrate to opposite poles, and cell division separates them into two daughter cells. Line up your three chromosomes, and cut the two sides of the X apart down the fold line. Use the

coin toss as in step 4 to decide which chromatid will migrate to the right and which will migrate to the left. After chromatid migration, the cell divides. Repeat for the second cell generated in step 5. At the end of this part of the activity, you should have four sets of three chromosomes ($1n$), each with $1n$ DNA, in four cells. These are the four gametes produced by meiosis.

If there were no crossing over, how many genetically different gametes could you generate? How many gametes could you get with no crossing over and two chromosome pairs? Four chromosome pairs?

How many combinations could a human cell generate?

Analyzing Genetic Variation: DNA Typing

24

About This Activity

This chapter contains two activities. The first is a reading and paper activity that provides the foundation for understanding how human DNA is analyzed to solve paternity cases, to yield evidence in criminal cases, and to diagnose genetic disease. Students will need to have completed the Southern hybridization worksheets and the polymerase chain reaction (PCR) lessons before doing this activity, since those two lessons introduce the laboratory techniques used in DNA typing. The meiosis game (in chapter 23) provides an excellent background for this activity, because it is a concrete demonstration of the individuality of each person's genetic makeup.

The second activity is a method of demonstrating the probability calculations used in analyzing forensic DNA data.

Following this chapter are three short activities that illustrate specific applications of DNA typing in the form of puzzles: *A Mix-Up at the Hospital, A Paternity Case,* and *The Case of the Bloody Knife.*

Class periods required: 1–2

Introduction

DNA-based identification methods focus on the highly variable regions of the human genome that we described in *Generating Genetic Variation: the Meiosis Game.* Because the genome is so large, we cannot look at these regions by restriction digestion, electrophoresis, and staining. A restriction digest of human DNA looks like one giant smear down the gel lane because so many fragments are generated from 3 billion base pairs.

However, the techniques of Southern hybridization and PCR allow us to look at specific regions and ignore the rest. In Southern analysis, the region you look at is determined by where the probe hybridizes

to the sample DNA. In PCR, the specificity is determined by the base sequence of the primers; the DNA sequence between the two sites on the chromosome at which the primers hybridize is the region that is amplified in the reaction. If we use one of these techniques to characterize several highly variable regions from one person's genome, we generate a set of data referred to as that person's DNA profile, DNA type, or DNA fingerprint. Since the term "DNA fingerprinting" is now a registered trademark of the biotechnology company Cellmark, we will use the term "DNA typing" or "DNA profiling" to refer to the process.

To conduct DNA typing, a sample must be obtained. In paternity cases, blood is drawn from the child, its mother, and the alleged father, and DNA is extracted from the white cells. However, any DNA-containing body fluid or tissue can be used: hair follicles, skin, or semen, for example. DNA is extracted and then analyzed by one of the two approaches outlined previously. The results from a DNA typing analysis are compared to the results from identical analysis of other samples. In criminal cases, the DNA profile of crime scene evidence (such as blood or semen) is compared to the DNA profile of the suspect. In paternity cases, the DNA profile of the child is compared to that of its mother and alleged father.

The exact procedure for DNA typing depends on the laboratory. Different biotechnology companies have patented their own probes and primers. Law enforcement agencies such as the Federal Bureau of Investigation (FBI) have also developed probes and procedures. The first DNA typing methods used Southern hybridization; now PCR-based approaches are increasingly popular. PCR has the advantage of greater sensitivity. It can be used with much smaller samples, since it amplifies the DNA to be detected. This advantage is particularly important in criminal cases, where available samples (such as skin fragments under a victim's fingernails) may be exceedingly small. The disadvantage of PCR-based typing is that fewer loci are currently available for analysis.

How accurate is DNA typing? This question has been the subject of controversy with respect to forensic cases. No one disputes the underlying theory of DNA typing: that everyone's DNA is unique and therefore provides a definite means of identification (with the exception of identical siblings). The problems have arisen over two issues, both of which apply largely to criminal cases.

The first issue is technical. In the past, court cases have revealed some instances of sloppiness in laboratory procedure and data analysis. Accurate laboratory procedure is particularly important in criminal cases. A person's fate may be riding on the outcome of a DNA analysis, and the sample at the crime scene may have been contaminated with other substances such as fabric dyes, subjected to temperature extremes, or otherwise made difficult to analyze. Perhaps more important, forensic samples are often vanishingly small. If a result from a DNA analysis is somewhat questionable, it may be impossible to repeat the tests. High-quality laboratory work in these cases is of the utmost importance.

In response to the technical criticisms, the National Academy of Science issued a call in 1992 for laboratory certification, quality controls, and standardization of procedure. An FBI-organized consortium of scientists from academia, crime laboratories, and private industry, called the Technical Working Group on DNA Analysis Methods, issued a set of guidelines that included laboratory protocols, quality assurance guidelines, and standards for education and training of personnel. Testing laboratories that meet these standards can be certified by professional groups such as the American Association of Blood Banks and the College of American Pathologists. Reputable testing laboratories adhere to these standards and seek certification. Although forensic laboratories are not required to follow the guidelines, they have become a sort of "gold standard" in the field of forensic DNA analysis.

The second issue concerning the reliability of DNA typing questions the point at which it can be concluded that two samples came from the same person. This question is more fundamental than the technical issues. Here's an example to illustrate the problem. Forensic laboratories have compiled large databases of DNA profiles generated with the probes they use. Suppose a crime scene sample and a suspect's blood are tested with probe A and show the same profile. Suppose that particular profile occurs in 1% of all the profiles in a database. There is then a 1% chance that a sample from any random person would also match the crime scene sample. Suppose that the samples are tested with probes B and C and again reveal identical patterns. Let's say that the probe B pattern is present

in 5% of the database profiles, and the probe C pattern is present in 10%. What are the chances that the DNA profile of a random person would match all three profiles?

Forensic laboratories use a multiplication rule to calculate this probability. They assume that a person's DNA profile at one locus (for example, the one shown by probe A) is independent of his or her profile at a second locus (such as the one shown by probe B). In this case, the probability of finding all three patterns in a random person can be calculated simply by multiplying the individual probabilities: 0.01 (A) $\times$ 0.05 (B) $\times$ 0.1 (C) = 0.00005. This result says that 5 people in 100,000 would have the same profile with all three probes. If the samples matched using a fourth probe, revealing a pattern that showed up in 2% of the database profiles, the odds of a random match would decrease to 1 in 1,000,000. Obviously, the larger the number of probes used to test two samples, the less likely it becomes that a perfect match for all probes could be a random event. (An activity you can use to demonstrate this kind of probability calculation is described in this chapter immediately following the solutions to the student exercises.)

Here is where the controversy comes in. Critics say the chance of a random match might be much higher than the odds calculated by the multiplication rule. They argue that a database containing profiles from Caucasians and African-Americans might not be a reasonable standard for comparison if the relevant ethnic group in a particular case was Polynesian.

The National Academy addressed this problem, too. They recommended a very conservative method of calculating the chances that a particular DNA profile could be duplicated randomly in another person. The Academy also recommended a specific procedure for characterizing DNA profiles from different ethnic groups, so that interpretation of future DNA profiles could be based on a wider sampling of the population. A 1993 report on an analysis of more than 70 worldwide databases concluded that there were no meaningful differences in genetic profiles within or between racial groups.

Analysis of mitochondrial DNA

Mitochondria are the cellular organelles in which respiration occurs. Each mitochondrion contains a small circular DNA molecule that is completely different from any nuclear DNA. DNA typing can be applied to mitochondrial DNA, too. Conveniently, mitochondrial DNA has a highly variable region near its origin of replication, providing a good target for DNA typing.

Because mitochondria are inherited differently from nuclear DNA, mitochondrial DNA analysis can be particularly useful in some circumstances.

When a zygote is formed by fertilization, the sperm donates the paternal chromosomes. However, the sperm does not donate cytoplasm or cytoplasmic organelles (with possible rare exceptions). The zygote's cytoplasm and organelles are furnished by the ovum. A child's mitochondria are thus all descended from mitochondria that were present in the maternal egg.

As a result of this inheritance pattern, a child's mitochondrial DNA is exactly like its mother's, its mother's mother's, etc. In addition, brothers and sisters have identical mitochondrial DNA because they have the same mother. A child's mitochondrial DNA also matches the mitochondrial DNA of the brothers and sisters of its mother, since all their mitochondria came from the grandmother. As you can imagine, an analysis of mitochondrial DNA could show very clearly whether two people were related through the female line. Analysis of mitochondrial DNA can be particularly helpful in reuniting families (see the introduction to *A Paternity Case* [chapter 26] for an example).

Other applications of DNA typing

When we hear about DNA typing, we tend to think of crime cases. However, measuring genetic similarity and analyzing kinship are important in many areas of biology. DNA typing is finding applications in such varied areas as conservation biology, evolutionary biology, behavioral biology, and taxonomy.

Conservation Biology

Measuring genetic similarity and analyzing kinship can be very important in captive breeding programs for endangered species. Because the captive and wild populations of species in these programs are usually quite small, substantial inbreeding (both in the wild and in captivity) has often taken place, leading to individuals who are very similar genetically. Genetic similarity caused by inbreeding presents problems for populations and individuals. Populations of genetically identical (or nearly identical) individuals are usually less resilient in the face of environmental change. Highly inbred individuals are also frequently less healthy and less reproductively successful than outbred ones. Therefore, for the good of the individuals and the species, conservation biologists try to breed unrelated individuals. However, in a substantially inbred population such as that of the whooping crane or cheetah, the fact that two animals live in different zoos doesn't mean they are genetically different. Conservation biologists have been conducting DNA typing of whooping cranes to determine how genetically similar the individuals are. They use the results of the typing to choose the most genetically different birds to form breeding pairs.

Evolutionary Biology

Analysis of genetic variation is used as a clue in deciphering the history of evolution as well. The reasoning behind this application is as follows. If the accumulation of genetic changes is responsible for phenotypic differences between species, genera, families, orders, and so on, then the amount of evolutionary distance between two groups of organisms should be reflected in the amount of genetic difference. In other words, two groups that diverged long ago will have more genetic differences between them than two groups that diverged recently. Furthermore, an analysis of the nature of the differences might give clues as to what the ancestral genetic pattern was like.

In keeping with this thinking, evolutionary biologists with a molecular bent analyze DNA samples from related organisms and compare the amount of difference. Using computer programs, they analyze the nature of the difference and produce hypothetical family trees. The trees show how, with a minimum of total genetic changes, the modern patterns could be produced from a single ancestral pattern (similar to the evolutionary tree for the protein cytochrome *c* in chapter 4). In fact, a whole research journal (the *Journal of Molecular Evolution*) is now dedicated to this sort of study, and similar studies appear in other research publications as well. At present, there is no way to prove whether trees produced by molecular analysis are correct, but they give evolutionary biologists another line of evidence to consider.

Behavioral Biology

In behavioral biology, DNA typing can help biologists determine kinship between members of a group. Determining the effect (if any) of kinship on relationships between animals is a major endeavor in behavioral biology. However, determining kinship isn't always easy. Depending on the mating behavior of the species, observation may reveal which individual fathers a particular offspring or brood, but observation doesn't always yield the desired information. For example, chimpanzee males and females mate promiscuously, making it impossible to tell by watching which individual is the father of an offspring. This problem long frustrated primatologists who were trying to understand the role of kinship in chimpanzee social interactions. Using blood typing to determine paternity was not a desirable alternative, since it involved tranquilizing the animals, potentially traumatizing them and disrupting their social lives.

In 1994, a group of researchers published a solution to the problem. They developed a PCR-based method of DNA typing that used small numbers of hairs. Chimpanzees make impromptu beds in trees at night, abandon them in the morning, and make a new bed the next night. The scientists could wait for a chimp to get up in the morning, climb to its abandoned bed, and collect shed hairs for DNA typing without ever disturbing the animal. The scientists are using their new kinship information in analyses of chimpanzee social behavior, and they hope that wildlife biologists studying other species can also take advantage of their DNA typing method.

Taxonomy

Who gets excited about taxonomy? Actually, quite a few biologists, and DNA typing is giving them new material to argue about. Traditionally, taxonomists depended on structural, physiological, and behavioral differences to distinguish species. When techniques for protein electrophoresis were developed, taxonomists began including information on enzyme patterns, too. Some molecular scientists now argue that genetic distance should also be grounds for defining species. For example, in the chimpanzee study described previously, the researchers analyzed three populations of the chimpanzee *Pan troglodytes*. They compared DNA fingerprints and mitochondrial DNA sequences. Their results showed that the mitochondrial DNA of the western population of *P. troglodytes* was significantly different from that of the other two populations: the western sequences were not present at all in the eastern and central groups. Scientists calculated that it would take 1.6 million years for so many differences to accumulate.

Some of the scientists suggested that this amount of genetic difference should make the western chimps a separate species. This is a controversial suggestion; other primatologists point to the similar appearance and many similar behaviors of the two groups and argue that DNA sequences alone do not a species make. There is deep disagreement over what defines a species and what sort of criteria should be used in doing so.

Another species dilemma sparked by DNA analysis involves an endangered species, the red wolf. Red wolves were extinct in their former habitats along the East Coast. In the past few years, efforts have been made to reintroduce captive-bred wolves in certain areas. Scientists who were interested in studying the relationship of red wolves to gray wolves and coyotes analyzed the mitochondrial DNA from samples from living animals and from small pieces of pelts found in various museums. They expected to determine the

degree of DNA sequence divergence between the three groups and, presumably, to reach conclusions about the evolution of the three species.

To their surprise, the mitochondrial DNA of every red wolf sample they analyzed matched that of either coyote mitochondria or gray wolf mitochondria. The scientists concluded that the red wolf was probably nothing but a hybrid between wolves and coyotes! If this is true, the endangered species isn't even a species. Others were not convinced, arguing that because red wolves had become so scarce in recent history, they might have begun breeding with wolves and coyotes because mates of their own species were not available. The issue is not settled. This kind of taxonomic argument is going on in many areas of biology as scientists begin to use the new molecular tools to study differences between individuals.

Objectives

After completing this activity, students should be able to

1. Explain the statement "There are more differences among people at the DNA level than at the protein level";

2. Explain that DNA typing is based on an examination of variable regions of the human genome;

3. Describe the two technical methods used in DNA typing (PCR and Southern hybridization);

4. Tell what a restriction fragment length polymorphism (RFLP) is and give one explanation for how an RFLP could arise;

5. Explain why it is necessary to use either hybridization or PCR in DNA typing.

Materials

Photocopies of the *Student Activity* pages (reading, exercises, and worksheets), if necessary.

Preparation

Make photocopies, if necessary.

Procedure

Review Southern hybridization analysis and PCR, if necessary. The *Student Activity* pages assume that students already understand these procedures.

This activity is self-explanatory. Students read the material and do the exercises. If you made posters of human chromosomes for the meiosis game, they could be useful here. Students often forget that each person has two copies of every chromosome and that the results of any analysis of a person's DNA will be a combination of the results from both chromosomes. Also, since a person's two chromosomes came from different individuals, the chromosomes are likely to be different in their highly variable regions. Your posters could help make these points.

Remind students that the exercises show only a very tiny region of a single chromosome (the examples are completely imaginary). In real life, the entire human genome is digested, generating a huge number of fragments. Without the use of hybridization or PCR, every person's DNA would look the same in a gel: a smear.

Hybridization allows laboratory workers to look selectively at certain areas of the genome. PCR specifically amplifies certain regions, producing millions of copies of specific regions of the genome. PCR products can be detected by hybridization or can sometimes be seen by simple staining if the original sample DNA is present at a concentration low enough not to interfere (by causing a strong background smear).

This exercise is the foundation for understanding DNA-based paternity tests, forensic DNA typing, and DNA-based diagnosis of genetic disease. Use class discussion and questioning to make sure your students understand the concepts presented in this lesson before going on to *A Paternity Case, A Mix-Up at the Hospital,* or *The Case of the Bloody Knife.*

Answers to Questions

Exercise 1

1. 20, 15, and 10 base pairs
2. 7, 13, and 25 base pairs
3. See column 2.
4. See column 2.
5. 7 and 38 base pairs
6. 7, 28, and 10 base pairs
7. See column 2.
8. See column 2.
9. The stained pattern of a digest of Bob's or Mary's DNA would be a smear, because so many fragments would be generated from the 3-billion-base-pair genome. It is necessary to use hybridization to look at a particular region of the genome.

Answers to questions 3 and 7:

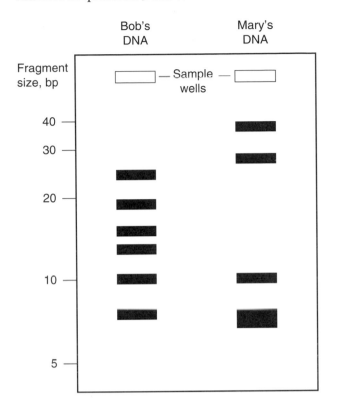

Answers to questions 4 and 8:

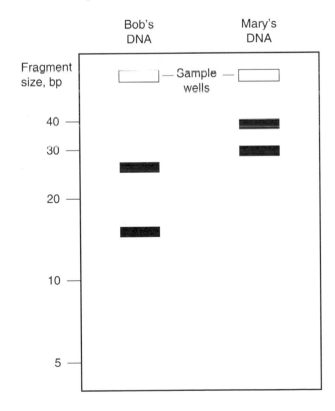

Exercise 2

1. AGCT or a permutation such as TAGC

2. Bob's maternal chromosome gives a fragment of 41 base pairs (also 3 and 12, but these are the same in every case).

   Bob's paternal chromosome gives a fragment of 53 base pairs (the difference is the three extra copies of the repeat in the paternal chromosome).

   Mary's maternal chromosome gives a fragment of 61 base pairs; her paternal chromosome gives a 41-base-pair fragment.

3. Students must work through several steps to arrive at the solution of this question. They must determine which fragments will be produced in the digest (which they have actually done for the previous question), then determine which of those fragments will hybridize to the probe, and then draw what the hybridization results would look like. You may need to help them through some of these steps.

Exercise 3

Note the orientation of the primers. The first primer hybridizes to the complement of the strand that is shown, and the second primer hybridizes to the strand whose sequence is shown. Remember to take into account the 5'-to-3' direction.

1. Bob's major maternal chromosome PCR product (53 base pairs long):
   5' GGCCTCTAGGACATGTAAAGCTAAAGCTAGCT AGCTAGCTAGCTAAGGCCTAGGTGCG 3'

2. Bob's major paternal chromosome PCR product (65 base pairs long):
   5' GGCTCTAGGACATGCTAAAGCTAGCTAGCTAG CTAGCTAGCTAGCTAGCTAAGGCCTAGGTGCG 3'

3. Mary's major maternal chromosome PCR product (73 base pairs long):
   5' GGCCTCTAGGACATGCTAAAGCTAGCTAGCTA GCTAGCTAGCTAGCTAGCTAGCTAGCTAAGGCC TAGGTGCG 3'

   Major product from Mary's paternal chromosome (53 base pairs long):
   5' GGCCTCTAGGACATGCTAAAGCTAGCTAGCTA GCTAGCTAAGGCCTAGGTGCG 3'

4. PCR products can often be seen by staining. If a tiny amount of sample DNA is used, there is very little background in the gel lanes to interfere with seeing the products. Note that Bob and Mary share a product.

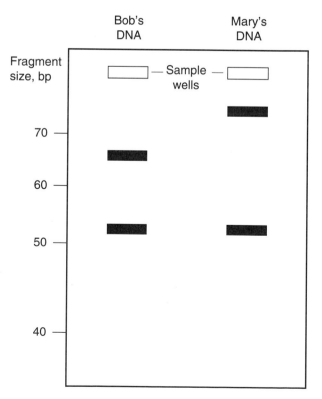

DNA Typing in the Classroom

Several companies market DNA fingerprinting kits that are simulations (often a "crime scene" with "suspects") in which students digest different samples of DNA with restriction enzymes, run the products on a gel, and then compare the stained banding patterns to "identify" a "criminal." We are not comfortable with these simulations because they convey a huge misimpression about human DNA typing: that you can do it by simply looking at stained digests of human DNA. This neglects the enormous complexity of the genome and the problem of looking selectively at regions that are informative.

Two simple simulated fingerprinting wet laboratories that we are comfortable with are the "PCR Forensics Simulation Kit" from Carolina Biological Supply Co. and the "DNA Fingerprinting Simulation Kit" from Stratagene. In the Carolina kit, samples that represent the products of PCR reactions from various suspects and a crime scene are provided. The "products" are separated on an agarose gel and examined. The Stratagene kit contains six sets of precut DNA; students digest each set with a second enzyme to simulate PCR products for "suspects" and an "evidence" sample. The "products" are also separated by agarose gel electrophoresis. You must provide the electrophoresis equipment. These are realistic simulations, since PCR products can be seen on stained gels.

A simulation of fingerprinting by Southern hybridization analysis is available from Edvotek (kit #11). This kit contains samples (no human DNA) to be separated on an agarose gel along with supplies for Southern transfer and a nonradioactive probe. If you purchase this kit, be sure to follow the instructions very carefully. Southern hybridization is not difficult, but it must be done correctly to yield good results. You must provide the electrophoresis equipment.

If you have access to a thermal cycler (PCR machine), real DNA fingerprinting kits can be ordered from Perkin-Elmer Corp., 761 Main Avenue, Norwalk, CT 06854 (1-800-762-4002). The two loci most commonly used are D1S80 and DQ-alpha; kits for both are available. Carolina Biological Supply Co. sells a classroom kit, "Human DNA Fingerprinting by PCR" (catalog no. 21-1226), that uses the D1S80 locus and also requires access to a thermal cycler. Their kit contains student and teacher instructions. All these kits use cheek cells as a source of DNA.

Calculating the Odds: a Demonstration Activity

Part 1: shuffling the genetic deck

This exercise provides an excellent opportunity to remind your students that mathematics plays a key role in forensic DNA identification. DNA typing is very much a numbers game. Because absolute certainty in DNA identification is not possible in practice, the next best thing is to claim virtual certainty because of the extremely small probabilities of a coincidental match. This activity is designed to teach students how to calculate allele frequencies and the frequency of a set of alleles simply by dealing playing cards from well-shuffled decks. By virtue of these card games, students should begin to understand and appreciate the strategy forensic scientists adopt to lower the probability of a chance match between evidence and suspect.

Objectives

After completing this lesson, students should be able to

1. Determine the predicted frequency of an allele;
2. Apply the multiplication rule to calculate the frequency of a set of alleles occurring together;
3. Explain why probabilities are influenced by the makeup of the database;
4. Discuss whether DNA evidence alone is sufficient to convict a suspect of murder in the absence of supporting evidence.

Materials

- Decks of playing cards. If possible, have everyone in the class bring in their own deck; in this way, everyone is involved, and the games go much more quickly.

- Calculator(s)

Procedure

Tell your students they will use decks of cards to demonstrate the multiplication rule as it is applied to calculating probabilities in DNA profiling. Each card dealt represents a locus (a specific site on a chromosome). The color of the card (red or black) *or* the suit (spade, diamond, heart, club) *or* a specific value (e.g., a king) *or* a combination (e.g., queen of hearts) represents alleles (specific DNA sequences or profiles that are found at a particular locus; see below).

Here's how the game works. Display a predetermined card or group of cards. Announce to the class that this card (or group of cards) represents the DNA profile of evidence found at a crime scene. Tell your students that they all are suspects in the crime. The question is, how many of them have DNA profiles that match the evidence? Have your students calculate the probability of a match before they deal. Then, have each student deal the same number of cards from their decks that you did. Count how many "suspect" profiles match the evidence profile. If matches occur, compare the experimental results to their calculated probability.

Start with high-probability events and work toward lower-probability events. Here is a possible sequence.

1. Turn over a red card. The single card is one locus analyzed by a single probe. The color red represents the allele revealed at that locus by the probe. How many students should also turn over a red card? (One in two. Half the cards in the deck are red, so the allele frequency for red is 0.5 = 1 in 2.)

   At this point, ask the students how they know the odds are 1 in 2. They will naturally respond that they know half the cards in the deck are red. Explain that criminologists use DNA profile databases so that they can know how frequently a particular pattern turns up at a given locus. If students were using a collection of cards whose makeup they knew nothing about, they couldn't calculate the odds of getting a red card. Similarly, criminologists must rely on their databases to make their calculations.

2. Turn over a club. Again, the single card represents a single locus; this locus has four alleles, represented by the four suits. How many students should match the club allele? (One in four. The allele frequency for clubs is 0.25 = 1 in 4.)

   So far, too many "suspects" match the evidence by chance, so let's lower the probability of a coincidental match. You could ask students how to lower the probability; they are bound to suggest using more cards or using the numbers on them.

3. Turn over four red cards in succession. This represents using four different probes to look at four loci, each of which has two alleles (red or black). How many students should match four red cards? (1 in 16. $0.5 \times 0.5 \times 0.5 \times 0.5 = 0.0625 = 1$ in 16.)

4. Turn over four red cards in succession, with the first card being an ace. How many "suspects" should match that sequence? (1 in 213.) The prob-

ability of getting the three red cards is $0.5 \times 0.5 \times 0.5$. The probability of drawing a red ace is 1 in 2 for the red and 1 in 13 for the ace, thus 1/26, or 0.0038 total. The probability of getting the four cards described is thus $0.038 \times 0.5 \times 0.5 \times 0.5 = 0.0047 = 1$ in 213. Note how the inclusion of the relatively low-frequency allele red ace makes the probability of this match much lower than the one posed in question 3.

Do a few more examples with more low-frequency alleles. The odds of getting a specific card like the queen of hearts is 1 in 52 (1 in 4 for the suit $\times$ 1 in 13 for the queen). The odds of getting four specific cards would be approximately 1 in 52 $\times$ 1 in 52 $\times$ 1 in 52 $\times$ 1 in 52.

We say approximately because a clever student may realize that if you draw a specific card from the deck (like the queen of hearts) and then want to know the odds of drawing the ace of spades, only 51 cards remain in the deck. The odds of drawing the ace of spades from the remaining cards is 1 in 51. However, in DNA testing, you are not removing anything from the "deck" when you use a probe to look at a particular locus. To be perfectly correct in this simulation, each card should be drawn separately, its identity should be recorded, and then it should be returned to the deck before the second card is drawn. If your students don't bring up this problem, we don't recommend that you do!

5. Remove all of the hearts from each deck. Now answer question 3. (1 in 84. $0.33 \times 0.33 \times 0.33 \times 0.33 = 0.01186 = 1$ in 84.) Note how altering the database has significantly changed the probability of this match compared to that of the allele set posed in question 3.

Tips

- In the case of low-probability sequences, such as the one posed in question 4, students should deal a hand, score it, and then reshuffle those cards back into the deck before dealing another hand. However, the exercise should work fine, especially for high-probability alleles, if you choose to skip the reshuffling step for the sake of time or convenience.

- Sometimes an unlikely combination is dealt in the first two or three hands. When this occurs, have your students continue to deal hands in order to validate the observed frequency. It should become obvious that observed probabilities only match calculated probabilities when sufficient trials are made.

This strategy will help your students understand that lower probabilities of a chance match result from choosing low-frequency alleles and including more cards (loci) in the game. Point out to your students that including more cards in the game is similar to using more than one DNA probe in a RFLP experiment.

Part 2: Discussion

The "Prosecutor's Fallacy"

There are two important questions with respect to the interpretation of DNA evidence. First, what is the probability that an innocent individual will match? Second, what is the probability that an individual who does match is innocent? It is the second question that is of direct interest to courts. The two questions are quite different. A common error, the "prosecutor's fallacy," consists of giving the answer to the first question in response to the second, thus confusing the match probability with the probability of innocence. The probability of innocence depends on the totality of evidence.

For example, consider a suspect whose DNA profile matches the evidence with a match probability of 1 in 700,000 and yet reliable eyewitnesses swear that he was 1,000 mi from the scene when the crime oc-

curred. Discuss such an example with your students. It is important that you impress upon your students that a small match probability may not, in itself, establish guilt.

Selected Readings

Balding, D., and P. Donnelly. How convincing is DNA evidence? *Nature* (London) 368:285–286.

Chen, L. 1994. O. J. Simpson case sparks a flurry of interest in DNA fingerprinting methods. *Genetic Engineering News* 14: 1, 30.

Morell, V. 1994. Decoding chimp genes and lives. *Science* 265:1172–1173. A nontechnical overview of the chimpanzee studies.

National Research Council. 1992. *DNA Technology in Forensic Science.* National Research Council, National Academy Press, Washington, D.C. This small book clearly outlines the complexities of generating and interpreting forensic DNA typing data and puts forth the National Academy's recommendations.

Nowak, R. 1994. Forensic DNA goes to court with O. J. *Science* 265:1352–1354.

Neufeld, P., and N. Collman. 1990. When science takes the witness stand. *Scientific American* 262(5):46–53. Written for a general audience, this article explains DNA typing and lays out the troublesome issues surrounding the technology. Many of the issues have been addressed since the article was published.

Zurer, P. 1994. DNA profiling fast becoming accepted tool for identification. *Chemical & Engineering News* 72:8–15.

Classroom Activities

Analyzing Genetic Variation: DNA Typing

Introduction: Studying Variation

We all recognize that everyone is different. Every day, we rely on differences in appearance, sound, or gait to recognize our friends and family. At work, school, and play, we observe people and note their great variety. In an informal way, we all study human diversity.

Some people literally make a career out of studying human diversity and similarity. These people may be scientists seeking to understand human evolution, physicians looking at the prevalence of genetic diseases, social workers trying to identify the father of a child, or law enforcement agents trying to determine the identity of a violent criminal. All of them need precise ways to measure the differences between individuals. How do they do it? Over the years, the techniques they use have evolved from a reliance on general appearance to measurement of specific phenotypes to, finally, examining the genotype itself. Let's look at what that means.

In the past, investigators of human diversity had to rely completely on outward traits such as eye color, ear shape, and the like. As you can imagine, information like this would not usually be precise enough to prove that a particular man fathered a child or to predict the occurrence of genetic diseases and would not even be available in the case of an unwitnessed murder.

As we learned more about biology, it became possible to get closer views of individuals by looking at specific proteins rather than overall appearance. For example, we learned that although everyone has blood, not everyone's blood is the same. Certain proteins on blood cells are different in different people, giving rise to blood types, and these blood types can be determined precisely. In recent years, an examination of proteins such as those that cause blood types has been very helpful in studying the amount of similarity in different groups of people, in ruling out suspects in crime cases, and in ruling out potential fathers in paternity cases.

Now we know that individuals have even more differences in their DNA than show up in their proteins. For example, an individual with blood type A could have the genes AA or AO. Even more important, we have learned that most of the differences in people's DNA may not even be in their protein-encoding genes but seems to be in the noncoding regions between their genes. Two individuals with blood type A might seem the same on that criterion, but an examination of their blood group genes and the surrounding DNA could reveal many differences.

It is very clear that the bottom line on inherited differences is in DNA. As a result of developments in the field of biotechnology, we can now look directly at DNA to study differences in various populations, to screen for disease genes, to determine paternity, and to identify individuals.

Analyzing differences at the DNA level

Everyone's DNA is different (with the exception of identical siblings). Therefore, it is easy to talk about identifying people by looking at their DNA, but when you begin to think about actually doing that, many questions arise. For example, humans have 3 billion base pairs of DNA. Obviously, we cannot look at all of them. Which ones should we look at? Are differences between people scattered uniformly over the chromosomes? Are some sections of the human genome likely to be more dissimilar from person to person than others? How much DNA do we need to look at to be sure that two samples are really the same? Just how different is the DNA from different people anyway?

Based on comparisons of many DNA sequences from many individuals, the average number of differences is estimated to be one for every 300 to 1,000 base pairs. Using the conservative end of this estimate, we can determine that a genome of 3 billion base pairs would show 3 million differences between two unrelated individuals.

Where can the differences be found? Again, studies show that the differences in people's DNA are not evenly distributed. Some regions of the genome are very similar between individuals, and other regions are highly variable. In general, these highly variable regions seem to be in noncoding DNA.

What is the nature of the differences? One type of difference is simply base changes, such as C to T or A to G. These can be used in DNA-based identification, as described below. A second kind of difference is also widely used in DNA-based identification. Human DNA contains end-to-end repeats of different short DNA sequences (from as few as 2 bases up to 30+ bases long). In some regions of the genome, the number of repeats varies highly from individual to individual. The abbreviation VNTR (variable number tandem repeat) is sometimes used to refer to these areas. DNA-based identification, or DNA typing, focuses on highly variable regions of the genome, whether the variety comes from many base changes or from a VNTR. In DNA typing, several of these highly variable regions from an individual are characterized, and a DNA "profile" is generated. The DNA profile can then be compared to other profiles (such as from potential fathers or from crime scene evidence). Let's see how the profiles are generated.

DNA typing

How do you examine DNA for typing? The immediate answer might seem to be, "Determine the DNA sequence." At this time, however, DNA sequencing is not a practical approach because of the time it takes and the cost involved. Instead, scientists use other methods to look at differences in people's DNA. You have already learned about the laboratory techniques involved. One of the approaches uses restriction enzymes and Southern hybridization; the other uses the polymerase chain reaction (PCR).

The basic approach to DNA typing by Southern hybridization analysis is to identify regions of the human genome that when cut with a particular restriction enzyme generate varying sizes of fragments from individual to individual. When different sizes of restriction fragments are generated from the same region of the genome in different people, we say we have found a **restriction fragment length polymorphism,** or **RFLP** ("poly" = many; "morph" = form; therefore, "polymorphism" = many forms).

An RFLP can arise in one of two ways. First, in a chromosome region with many changes in the base sequence, restriction sites can be created or destroyed by some of those changes. For example, the enzyme

*Hae*III cuts the sequence GGCC between GG and CC. Any base change in the sequence would destroy the site. Conversely, if the A in the sequence GACC were mutated to a G, a site would be created. Gain or loss of restriction sites would obviously change both the number and the sizes of restriction fragments generated in a digest. Exercise 1 shows an example of an RFLP created by sequence changes.

The second way that different-sized fragments can arise from the same region of a chromosome (from different people) is through a VNTR. Imagine a VNTR region with a *Hae*III site on either side of it. Obviously, the length of the *Hae*III fragment containing the VNTR depends on how many repeats are there. The more repeats, the longer the fragment. Exercise 2 shows an example of an RFLP created by a VNTR.

DNA typing by Southern hybridization is based on RFLPs, whether they are caused by sequence changes or by VNTRs. To do the typing, a laboratory worker digests the DNA sample with restriction enzymes and then performs Southern hybridization with a DNA probe that hybridizes to the highly variable region. The probe reveals the pattern of restriction fragments from that region. Often, the analysis is performed with multiple probes that reveal RFLPs from several different highly variable regions. The result is a DNA profile. If many highly variable regions of the genome are analyzed per sample, it is highly unlikely that two profiles from unrelated individuals will ever match by chance (see below).

Why not just look at the band pattern revealed by staining and skip the hybridization? Remember the size of the genome: 3 billion base pairs. If an average restriction fragment were 1,000 base pairs long, there would be 3 million of them. There is no way to look at individual bands by staining: you would only see, all the way down the gel lane, a smear caused by hundreds, thousands, or millions of overlapping fragments. DNA typing depends on the use of methods like hybridization and PCR, which allow you to look only at specific regions of the genome.

DNA typing by PCR is based on VNTRs. PCR primers that hybridize to the DNA sequence on each side of a VNTR are used to amplify the VNTR area. The PCR reaction cycle then produces millions of copies of the VNTR region. The PCR products are separated in an electrophoresis gel and visualized by staining or hybridization. (If the amount of background sample DNA is small, PCR products can sometimes be seen by simple staining, since they are present at such a high concentration.) The lengths of the PCR products are determined by the number of repeats present, so the sizes of products vary from person to person. Several sets of primers can be used to look at several VNTR regions, generating a unique profile for the individual.

Applications of DNA typing
Forensics
When most people think of DNA typing, they think of solving crimes or paternity cases. If a blood or tissue sample not belonging to the victim is found at the scene of a violent crime, it can be used to link a suspect to the crime scene through blood analysis and/or DNA typing. Most important, DNA typing can clear suspects who are innocent. In about 30% of DNA typing cases thus far, the results have proven that the prime suspect could not have left the incriminating sample. In some cases, men who were convicted of terrible crimes have been released after DNA testing of old crime scene samples proved these men could not have left these samples.

DNA typing is used to exclude suspects. If a suspect's DNA profile does not match the crime scene sample, he or she is cleared. If a suspect's DNA profile does match the crime scene sample, the question becomes how many other people in the population might have this same DNA profile just by chance. Studies of profiles of large numbers of people have made it possible to calculate the frequency of patterns seen with different probes.

For example, if a suspect's DNA profile with probe A matches the crime scene sample but 10% of the population also has that profile, then 1 in 10 people could also have left the sample. As a forensic scientist, what do you do? You use another probe. Suppose the suspect's profile with probe B also matches the crime scene sample profile. Let's say that particular profile is found in 5% of the general population. If the profiles from probe A and probe B are independent, we now have a $(0.1)(0.05) = 0.005$ or 0.5% probability of finding the two matches in a random person. Another way of looking at it is that 1 person in 200 would be expected to have both of these profiles.

In real crime scene analyses, eight or more probes may be used. The odds of two individuals matching by chance with this many probes are very small, though people argue about exactly how small. Originally, many scientists were concerned that individuals belonging to distinct ethnic groups might be short-changed by this type of calculation. Specifically, they worried that the DNA profiles of all the members of a particular ethnic group might be more similar to each other than the DNA profiles of the general population

would be (just as Scandinavians have blond hair and blue eyes more frequently than the general population). However, studies of DNA profiles from more than 70 ethnic groups have not shown great differences between the profiles in terms of frequencies. If an individual's DNA profile matches a crime scene profile exactly after analysis with several different probes, the odds that a different person left the sample are extremely small.

Biological Applications
Forensics is only one area in which DNA typing has become an important tool. Determining genetic similarity and analyzing kinship through DNA typing is useful in many areas of biological research such as conservation biology, evolutionary biology, taxonomy, and behavioral biology. For example, primatologists would like to understand how kinship affects the social behavior of chimpanzees. However, it is impossible to establish which chimpanzee is the father of an offspring simply by watching the troop, because both male and female chimpanzees mate promiscuously. If primatologists wanted to determine the father of a chimpanzee, they had to tranquilize the animals, take blood samples, and analyze them. This procedure could traumatize the animals and disrupt the very behavior the primatologists wanted to study.

DNA typing using PCR has provided a solution to this problem. The PCR method is so sensitive that it requires only a few hairs (there is DNA in the follicle cells) for analysis. Chimpanzees sleep alone at night and change sleeping sites each night. In the morning, without disturbing the animals at all, researchers can collect shed hairs from the sleeping sites after the chimps have left them. Now that researchers can determine kinships among the chimps, the scientists can analyze whether the degree of relatedness affects the social behavior of the chimps.

Exercise 1: Changes in Base Sequence Can Cause an RFLP

Use with worksheet 1.

Suppose that we are taking a close-up look at a highly variable region of chromosome 19 from Bob and Mary. Bob and Mary each have two chromosome 19s, one from their mothers and one from their fathers. The relevant DNA sequences are shown on worksheet 1. We are going to examine this region of chromosome 19 by Southern hybridization, using the probe shown on worksheet 1 and the restriction enzyme *Hae*III. This enzyme cuts the sequence GGCC between the G and the C.

Questions

1. What sizes of fragments would Bob's maternal chromosome 19 yield after digestion with *Hae*III? (Use the ends of the fragment as shown as actual ends.)

2. What sizes of fragments would Bob's paternal chromosome 19 yield after digestion with *Hae*III?

3. Use the gel outline below to draw the band pattern of Bob's chromosomes in the indicated lane. (Remember, both of Bob's chromosome 19s are present in the DNA sample prepared from his white blood cells.)

4. Determine which restriction fragments from the digest of Bob's chromosomes will hybridize to the probe shown on worksheet 1. (Any fragment that partially overlaps the probe can hybridize to it.) Draw the results of the hybridization analysis of Bob's DNA on the appropriate outline.

5. Repeat the procedure for Mary's DNA sample. What sizes of fragments would her maternal chromosome yield after digestion with *Hae*III?

6. What sizes of fragments would Mary's paternal chromosome yield?

7. Draw the gel pattern from these fragments of Mary's DNA in the indicated gel lane.

8. Draw the results of hybridization analysis on the indicated outline.

9. If you were analyzing DNA prepared from Bob and Mary's white blood cells, why couldn't you simply look at the stained gel pattern and skip the hybridization step?

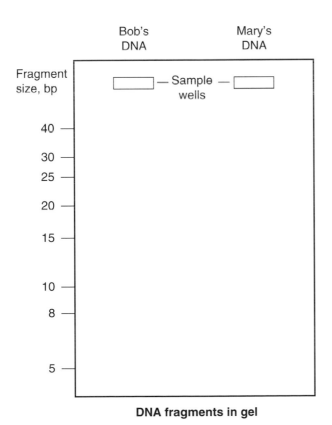

DNA fragments in gel

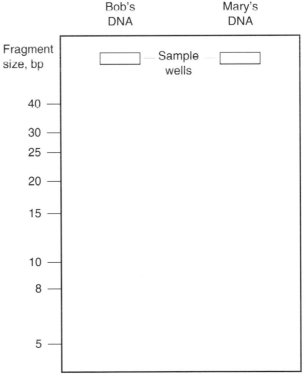

Results of hybridization analysis

Exercise 2: VNTRs Can Cause an RFLP

Use with worksheet 2.

Another laboratory also wishes to look at differences in Bob's and Mary's DNA. Workers in the second laboratory use a different approach: they focus on a region of chromosome 8 in which they have found a VNTR. Shown on worksheet 2 are the sequences from that region of Bob's and Mary's chromosome 8s. The laboratory will perform Southern hybridization analysis using the restriction enzyme *Hae*III and the probe shown on worksheet 2.

Questions

1. What is the tandemly repeated sequence in this chromosome region?

2. What size fragments would be generated from the region containing the repeats by digesting Bob's maternal chromosome with *Hae*III? His paternal chromosome? How large would the fragments from Mary's chromosomes be?

3. Suppose you digested samples of Mary's and Bob's DNA with *Hae*III and performed a Southern analysis using the probe shown on worksheet 2. Draw the results on the outline provided.

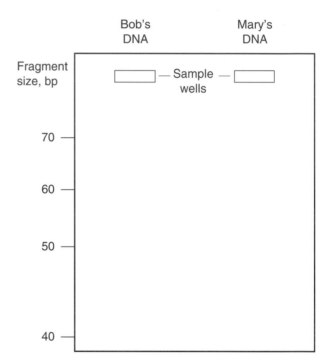

Exercise 3: PCR Can Reveal Differences between Chromosomes

Use with worksheet 3.

A third laboratory wishes to look at differences between Bob's and Mary's DNA. Workers there will also look at the highly variable region of chromosome 8 shown on worksheet 3 (the same region is also shown on worksheet 2), but they will use PCR instead of Southern hybridization analysis.

Laboratory workers prepare DNA from samples of Bob's and Mary's blood and then mix a tiny sample of DNA from each person with DNA polymerase enzyme, deoxynucleotides, and the primers shown on worksheet 3. The PCR is allowed to proceed through 30 replication cycles.

Questions

1. Write the sequence of the major PCR product from Bob's maternal chromosome.

2. Write the sequence of the major PCR product from Bob's paternal chromosome.

3. Write the sequence of the major PCR products from Mary's maternal and paternal chromosomes (label them "maternal" and "paternal").

4. Draw what the products would look like in a stained gel on the outline provided.

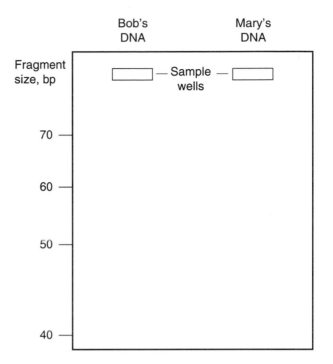

Because PCR yields so many copies of the major product molecule, it is possible to see the major products by simple staining even when there is too little original sample DNA to see in a stained gel. PCR products can also be visualized by hybridization to a probe, as in Southern analysis.

Worksheet 1

Use with Exercise 1: Changes in Base Sequence Can Cause an RFLP

Shown below are the probe sequence and the highly variable regions from Bob's and Mary's chromosome 19 homologs. Use the sequence information to answer the questions in exercise 1.

The restriction enzyme *Hae*III cuts the sequence 5′ GGCC 3′ between the G and the C.

Probe sequence: 3′ TTATATCGTAACC 5′

Highly variable region of Bob's two chromosome 19s:

Maternal chromosome:

5′-ATCTTGACCAATGTGAAAGGCCAATATAGCATTGGCCAAGCTCTC-3′

3′-TAGAACTGGTTACACTTTCCGGTTATATCGTAACCGGTTCGAGAG-5′

Paternal chromosome:

5′-ATCTTGGCCAATGTGAAGGGCCAATATAGCATTGGTCAAGTTCTC-3′

3′-TAGAACCGGTTACACTTCCCGGTTATATCGTAACCAGTTCAAGAG-5′

Highly variable region of Mary's two chromosome 19s:

Maternal chromosome:

5′-ATGTTGGCCAATGTGAGGACCAATATAGCATTGGCAAAGCTCTC-3′

3′-TACAACCGGTTACACTCCTGGTTATATCGTAACCGTTTCGAGAG-5′

Paternal chromosome:

5′-ATCTTGGCCAATATCAAAGACCAATATAGCATTGGCCAAGCTCTC-3′

3′-TAGAACCGGTTATACTTTCTGGTTATATCGTAACCGGTTCGAGAG-5′

Worksheet 2

Use with Exercise 2: VNTRs Can Cause an RFLP

Shown below are the probe sequence and the DNA sequence from the highly variable region of Bob's and Mary's chromosome 8 homologs. Use the sequence information to answer the questions in exercise 2.

The restriction enzyme *Hae*III recognizes the sequence 5' GGCC 3' and cuts between the G and the C.

Probe sequence: 3'-GGAGATCCTGTACGATTT-5'

Highly variable region of Bob's chromosome 8s:

Maternal chromosome:

5'-AGGCCTCTAGGACATGCTAAAGCTAGCTAGCTAGCTAGCTAAGGCCTAGGTGCGAT-3'

3'-TCCGGAGATCCTCTACGATTTCGATCGATCGATCGATCGATTCCGGATCCACGCTA-5'

Paternal chromosome:

5'-AGGCCTCTAGGACATGCTAAAGCTAGCTAGCTAGCTAGCTAGCTAGCTAGCTAAGGCCTAGGTGCGAT-3'

3'-TCCGGAGATCCTGTACGATTTCGATCGATCGATCGATCGATCGATCGATTCCGGATCCACGCTA-5'

Highly variable region of Mary's chromosome 8s:

Maternal chromosome:

5'-AGGCCTCTAGGACATGCTAAAGCTAGCTAGCTAGCTAGCTAGCTAGCTAGCTAGCTAAGGCCTAGGTGCGAT-3'

3'-TCCGGAGATCCTGTACGATTTCGATCGATCGATCGATCGATCGATCGATCGATCGATTCCGGATCCACGCTA-5'

Paternal chromosome:

5'-AGGCCTCTAGGACATGCTAAAGCTAGCTAGCTAGCTAGCTAAGGCCTAGGTGCGAT-3'

3'-TCCGGAGATCCTGTACGATTTCGATCGATCGATCGATCGATTCCGGATCCACGCTA-5'

Worksheet 3

Use with Exercise 3: PCR Can Reveal Differences between Chromosomes

Shown below are the sequences of the PCR primers and the highly variable regions from Bob's and Mary's chromosome 8 homologs. Use the information to answer the questions in exercise 3.

PCR primers:

5'-GGCCTCTAGGACATGTAAAGC-3' and 3'-TCGATTCCGGATCCACGC-5'

Remember to hybridize the 5'-to-3' primer to the 3'-to-5' DNA strand and vice versa.

Highly variable region of Bob's chromosome 8s:

Maternal chromosome:

5'-AGGCCTCTAGGACATGCTAAAGCTAGCTAGCTAGCTAGCTAAGGCCTAGGTGCGAT-3'

3'-TCCGGAGATCCTCTACGATTTCGATCGATCGATCGATCGATTCCGGATCCACGCTA-5'

Paternal chromosome:

5'-AGGCCTCTAGGACATGCTAAAGCTAGCTAGCTAGCTAGCTAGCTAGCTAGCTAAGGCCTAGGTGCGAT-3'

3'-TCCGGAGATCCTGTACGATTTCGATCGATCGATCGATCGATCGATCGATTCCGGATCCACGCTA-5'

Highly variable region of Mary's chromosome 8s:

Maternal chromosome:

5'-AGGCCTCTAGGACATGCTAAAGCTAGCTAGCTAGCTAGCTAGCTAGCTAGCTAGCTAGCTAAGGCCTAGGTGCGAT-3'

3'-TCCGGAGATCCTGTACGATTTCGATCGATCGATCGATCGATCGATCGATCGATCGATTCCGGATCCACGCTA-5'

Paternal chromosome:

5'-AGGCCTCTAGGACATGCTAAAGCTAGCTAGCTAGCTAGCTAAGGCCTAGGTGCGAT-3'

3'-TCCGGAGATCCTGTACGATTTCGATCGATCGATCGATCGATTCCGGATCCACGCTA-5'

A Mix-Up at the Hospital

25

About This Activity

This paper activity illustrates an application of DNA typing. Students assign babies to the correct pair of parents on the basis of DNA profiles. This activity can be done with some discussion by students who have completed *Analyzing Genetic Variation: DNA Typing* (chapter 24) or with considerable discussion by younger students who have completed *Generating Genetic Variation: the Meiosis Game* (chapter 23). Younger students should also have completed *DNA Scissors* (chapter 10, the introduction to restriction enzymes) and *DNA Goes to the Races* (chapter 11, the introduction to electrophoresis) before doing this activity.

Class periods required: 1/2–1

Introduction

Every now and then, we read of babies being switched in the hospital shortly after birth. These cases can come to light as a result of blood tests during medical procedures. Blood typing can be a relatively quick and simple way of determining whether a baby can be the offspring of two given parents. Likewise, blood typing can show that two blood samples did not come from the same person. However, the limited variety of blood types makes the chance of two unrelated people having the same type relatively high. Also, blood typing cannot distinguish between people with different genotypes (such as AO and AA) but the same phenotype (blood group A). When blood typing is inadequate to distinguish between people or to establish family groups, DNA typing can be used.

The DNA sequence contains much more variety than is seen at the phenotypic level of proteins or outward traits. This variety can be detected by restriction fragment length polymorphism (RFLP) analysis, by polymerase chain reaction (PCR) with variable tandem repeat regions, or even by the sequencing of short regions of DNA. Unlike a blood type, a person's DNA sequence is as individual as his or her fingerprint (with the exception of identical siblings, who share DNA sequence but have different fingerprints).

Although DNA typing is clearly more accurate in determining family relationships, blood tests can often give a simple "yes," "maybe," or "no" answer. Blood typing is less expensive to conduct than DNA typing, is a routine procedure of all major hospital laboratories, and is fast. Therefore, blood typing is usually the first test used in trying to establish family connection or to determine whether two blood samples came from the same person.

Objective

After completing this activity, students should be able to

1. Explain how to use DNA profiles to determine whether a couple are the parents of a particular child;

2. Give a clear description of what to look for in comparing the profiles.

Materials

- Photocopies of the *Student Activity* pages if students do not have manuals

- For younger students, the posters of human chromosomes (parents and offspring) from the meiosis game.

Preparation

Make photocopies, if necessary.

Procedure

For older students, the activity is self-explanatory. A relevant news story would be a good way to relate the activity to the real world. Depending on your class, you may want to suggest the analytical procedure described below. Your posters of human chromosomes will be helpful in reinforcing the idea that half of a person's chromosomes come from each parent and therefore would have counterpart patterns in the parents' DNA profiles. The posters will also show that not all of the DNA bands seen in the profile from either parent will show up in the child's profile. Review as much as you deem necessary for your class before beginning the activity.

To analyze the DNA profiles, students should carefully compare the babies' profiles to the profiles from each couple. For example, students can start with one baby's profile and compare it to the profile for the first couple. First, they should focus on one member of the couple and compare the baby's profile to that person's (for example, the woman's). Each band in the baby's profile should be checked to see if it matches a band in the first woman's profile. Any bands that match should be marked lightly in pencil. If no bands match, the woman cannot be the baby's mother. If some bands match, the student should compare the remaining bands to the man's DNA profile. Every remaining band in the baby's profile will match a band in the man's DNA profile if that man is the father. If some but not all bands match, then the couple are not the parents of that child. The student should go on to the next couple and compare the baby's profile to it in the same manner.

If you want to use this activity with younger students who have not completed the Southern hybridization, PCR, and DNA typing lessons, you will have to decide how much you want to tell them about how the DNA profiles are generated. They will need to be familiar with the activity of restriction enzymes and with gel electrophoresis of restriction fragments to understand the activity.

Start the lesson by showing them the posters of parental and offspring chromosomes from the meiosis game. Remind them that everyone is different because their chromosomes are different and that half a person's chromosomes come from the mother and half come from the father. Ask them if they think that by comparing chromosomes, they could tell whether a baby belonged to a particular couple. You might even ask them what they would look for.

Tell your class that scientists look at small regions of chromosomes when they want to tell whether people are related. These regions are different in nearly everyone.

The following explanation is intended only as an example of how you might prepare a younger class for the activity.

Ask your class a series of questions about what effect a highly variable DNA sequence would have on restriction sites. Continue asking questions until students come to the idea of RFLPs (the term is not necessary). It is not necessary to explain blotting and probing. Keep using your posters of chromosomes to generate examples. Make up a short sequence for a region of both of dad's chromosomes and both of mom's chromosomes, and then let the offspring have one of each. Let the class figure out the restriction fragments from mom, dad, and the offspring. (The chromosome 19s of Bob and Mary in the *Student Activity Analyzing Genetic Variation: DNA Typing* would be perfect examples.)

After the class has figured out the restriction patterns, draw these patterns in a gel outline on the blackboard. Make sure the class sees that each band in the child's profile has a counterpart in the parents' patterns but that not all of the parents' bands are present in the child's profile. Now have the class make up a different pattern for a second child of Bob and Mary (the second offspring could inherit the other chromosome 19s from the example). Ask the students what they think a profile from an unrelated child might look like (the important thing is that the bands would not match Bob's and Mary's).

When you feel your class is ready, give a short introduction, telling the class that they will be using DNA profiles to decide which baby belongs to which couple. (You could ask them if they have ever heard of babies being mixed up at a hospital.) Hand out the activity sheets. If you think it is necessary, go through one comparison with your class, following the procedure outlined previously. As your students work on the puzzle, circulate through the classroom and observe their work to make sure they understand what they are doing.

Answer

Baby 1 is the Stevenson baby, baby 2 is the Jones baby, and baby 3 is the Smith baby. Check each student's paper to see that the maternal and paternal bands have been correctly assigned.

Optional Follow-Ups

- Invite a speaker from a local hospital to talk about identification methods such as blood typing and DNA typing.

- Have students make up their own DNA typing puzzles.

A Mix-Up at the Hospital

25

On June 6 at approximately 1:00 p.m., Mrs. Smith, Mrs. Stevenson, and Mrs. Jones each delivered a healthy baby boy at Metropolitan General Hospital. At 1:20 p.m., the hospital's fire alarm sounded. Nurses and orderlies scrambled to evacuate patients, and the three new babies were rushed to safety. After the danger had passed, the hospital staff was distressed to find that in the confusion, they had forgotten which baby was which! Since the babies were rescued be-

fore receiving their identification bracelets, there was no easy way to identify them. Dr. Anne Robinson, head of pediatrics, ordered that DNA typing be performed on the babies and their parents.

The DNA typing laboratory looked at two different highly variable chromosome regions. The DNA profiles are shown in Fig. 25.1. Your job is to decide which baby belongs to which set of parents. To as-

Figure 25.1 DNA profile data from the Smith, Stevenson, and Jones parents and the three infants.

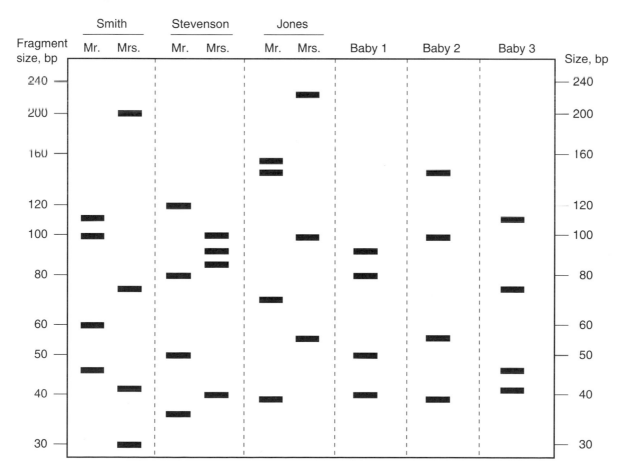

sign a baby to a set of parents, every band in the baby's profile should match a band from either the mother or the father. Not all of the bands in the mother's or father's profiles will have a counterpart in the baby's DNA profile. Hint: Use a ruler or a straight-edge to help you line up the bands.

Which baby belongs to which couple? Show which bands each baby inherited from its mother and from its father by marking the bands M and F.

Classroom Activities

A Paternity Case

26

About This Activity

This activity simulates a paternity case. Students analyze DNA typing data to determine whether I. M. Megabucks, a recently deceased megabillionaire, is actually the father of any of three children alleged to be his heirs. Students should have completed *Analyzing Genetic Variation: DNA Typing* (chapter 24) before doing this activity unless you are doing this lesson with younger students. For modifications for younger students, see below under Procedure. Younger students should have recently completed *Generating Genetic Variation: the Meiosis Game* (chapter 23; the crossing over exercise is not necessary). They also need to have completed *DNA Scissors* (chapter 10), so that they understand what a restriction enzyme does. Please refer to the introductions to *Generating Genetic Variation: the Meiosis Game* and *Analyzing Genetic Variation: DNA Typing* for background.

Class periods required: 1

Introduction

DNA-based paternity determinations are carried out by biotechnology companies such as Genetic Design, Inc., of Greensboro, N.C. The standard method is to analyze samples from the mother, the child, and the alleged father. Occasionally, however, the alleged father is deceased and did not (unlike Megabucks) prudently leave behind a blood sample for future DNA typing needs. In these cases, biotechnologists type DNA from known relatives of the alleged father, such as his acknowledged children or his parents. The DNA profiles are compared. Since two people who share the same father will have many more bands in common than two randomly selected individuals, it is possible, with careful analysis, to determine paternity on the basis of DNA profiles from a man's close relatives.

A particularly poignant application of DNA testing occurred in Argentina. During the Argentine military's brutal rule (1976 through 1983), many families were

torn apart. Often, parents were murdered, and their children were given away or sold. In other cases, parents were dragged away to prison, unwillingly leaving their babies to uncertain fates. Now that Argentina has a new government, the relatives of these kidnapped or disappeared children are trying desperately to find them. Many of the relatives are women whose children were murdered and who are now seeking their missing grandchildren. DNA testing has established the identities of dozens of formerly missing children, allowing them to be reunited with their relatives. "Genes of War" (*Discover,* October 1990, p. 46–52) is a story about the work of Dr. Mary-Claire King, the scientist who has been instrumental in helping the Argentinian families find their lost relatives.

Dr. King used mitochondrial DNA (see the introduction to chapter 24) in her analyses. Because the parents of the lost children had often been murdered, DNA from more distant suspected relatives was usually the only evidence available for comparison. Since mitochondrial DNA is passed on through the females in a family lineage, a child's mitochondrial DNA profile exactly matches the profile from her mother's mother, all of her mother's siblings, and the children of her mother's sisters.

Note: The activity in this chapter does not involve mitochondrial DNA.

Objective

After completing this activity, students should be able to explain the process of analyzing DNA data to determine whether a particular man is the father of a child.

Materials

Photocopies of the *Student Activity* pages if students do not have manuals

Preparation

Make photocopies, if necessary.

Procedure

This activity is self-explanatory. Students analyze the data on the worksheet. Your posters of human chromosomes from the meiosis game (chapter 23) will be helpful if you need to reinforce the idea that half of a person's chromosomes came from the mother and the other half came from the father. The posters will also illustrate that not all of the DNA bands seen in the typing data from the mother or the father would be in the child's DNA (fully half of each parent's DNA is not represented in the offspring's DNA).

Answer

Y's child could be Megabucks' child.

Question for Discussion

Why are the number of bands from the mother and the number from the father different in different children?

The difference is a consequence of variations in the DNA sequences. As restriction sites are created and destroyed by changes in base sequences, the number of restriction fragments in the region that hybridizes to a probe also changes. See the activity about Bob's and Mary's chromosome 19s in *Analyzing Genetic Variation: DNA Typing* (chapter 24) for an example.

Classroom Activities

A Paternity Case

Mr. I. M. Megabucks, the wealthiest man in the world, recently died. Since his death, three women have come forward. Each woman claims to have a child by Megabucks and demands a substantial share of his estate for her child. Lawyers for the estate have insisted on DNA typing of each of the alleged heirs. Fortunately, Megabucks anticipated trouble like this before he died, and he arranged to have a sample of his blood frozen for DNA typing.

Laboratory technicians used the Southern hybridization method to look at three highly variable chromo-some regions. The results of the blots are shown in Fig. 26.1. Your job is to analyze the data and determine whether any of the children could be Megabucks' heir.

Remember that every person has two of each chromosome, one inherited from his mother and one inherited from his father. Half of every person's DNA comes from his mother, and half comes from his father, so some of the DNA bands showing in the Southern blots of the children will come from their moth-

Figure 26.1   Results of hybridization analysis.

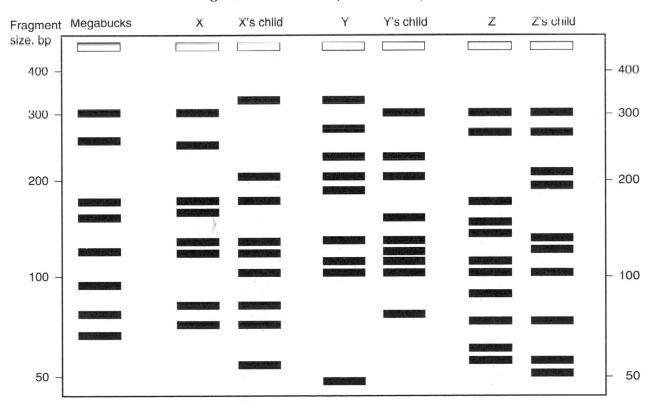

ers, and the rest will come from their fathers. The question is, could that father be Megabucks?

1. For the first child, identify the bands in the DNA profile that came from the mother. (Remember that not all of the mother's DNA is transmitted to the child; just one of each pair of chromosomes is transmitted.) Mark the bands that came from the mother with an M. Circle the remaining bands.

2. Compare the remaining bands with the DNA profile from Megabucks. If he is the father, then all of the circled bands in the child's profile should have a corresponding band in his profile. Use a straightedge to help you line up the bands accurately. (Remember that only half of the father's chromosomes are transmitted to a child, so not every band from the father would match the child's profile.)

3. Repeat the analysis for the other alleged heirs. Could any of them be Megabucks' children?

Classroom Activities

The Case of the Bloody Knife   27

About This Activity

In this activity, students must interpet DNA typing data from a murder scene. Students should have completed *Analyzing Genetic Variation: DNA Typing* (chapter 24) before attempting this activity. The activity also assumes that students are familiar with the polymerase chain reaction.

Class periods required: 1/2-1

Introduction

Since its introduction into U.S. courtrooms in 1987, DNA testing has been used in thousands of criminal investigations. Apart from paternity and parentage tests, DNA typing is used primarily in investigations of violent crimes. In rape cases, DNA is usually prepared from the semen. The semen DNA profile is compared to the profile generated from a blood sample taken from the suspect. DNA profiles can also be prepared from bloodstains, blood, saliva, or hair follicles

In a recent murder case, DNA testing linked a suspect to the crime scene. The suspect was arrested in Durham, N.C., after the DNA profile of blood from the murder scene was found to match the suspect's DNA profile. The suspect had been treated for a cut on his hand on the night of the murder (Durham "Herald-Sun," November 11, 1992, p. C1, C3).

The Federal Bureau of Investigation performs forensic DNA typing, and many states have their own DNA typing laboratories. Finally, private companies also perform forensic DNA analysis. When DNA typing evidence is entered as evidence in a trial, a scientist from the laboratory usually testifies as an expert witness. She would typically explain how DNA typing works and what its power and limitations are, and she would give the laboratory's interpretation of the test in question.

Objective

After completing this activity, students should be able to explain how DNA typing can be used to exonerate or further incriminate a suspect in a criminal case.

Materials

Photocopies of the *Student Activity* pages if students do not have manuals

Preparation

Make photocopies, if necessary.

Procedure

The activity is self-explanatory.

Related Activities

Have students collect newspaper and magazine clippings that refer to DNA typing for criminal or parentage cases. There will be quite a few!

Answer

The key to this case is the number of bands in the DNA profile from the knife. The chromosome regions and primers show that each chromosome should generate one band (unlike restriction fragment length polymorphism analysis, which can give varying numbers of bands per region analyzed depending on the restriction sites). Since people have two copies of every chromosome, two bands should be generated per set of primers. Two chromosomal regions are analyzed, and both Milhouse (the victim) and Smink (the suspect) show four bands in their profiles. The profile from the blood on Milhouse's clothes also shows four bands and clearly matches Milhouse's own DNA profile. The DNA profile from the knife, however, shows eight bands. Either something went wrong with the testing (such as contamination of the sample), or the blood on the knife is from more than one person.

A comparison of the banding pattern shows that four of the eight bands in the knife profile are exact matches to the victim's profile, while the remaining four are an exact match to Smink's. Given the circumstances, it seems highly likely that the blood on the knife is a mixture of blood from the two men. The knife was found under the bloody body, so it is possible that the victim's blood could be on it, even if the victim was not cut. Smink's hand wound could have been a cut from that knife, possibly as the victim tried to defend himself. On the other hand, poor laboratory practice could have resulted in mixing of the knife sample with Smink's sample during testing. The DNA typing should be repeated with special care to make sure that no contamination occurs.

Smink should definitely not be released. The current evidence suggests strongly that Smink's blood is on the knife. He should be questioned thoroughly. Additional DNA tests examining other highly variable chromosomes regions could make even stronger the conclusion that the "extra" blood on the knife is Smink's.

Optional Additional Question

Have students predict the lengths of the bands that could show up in a person's DNA profile judging from the chromosome regions and probes shown.

The Case of the Bloody Knife

Late one April night, government agents received an anonymous tip that the National Art Museum was about to be robbed of a priceless jewel collection. When they arrived at the museum, they saw they were too late: the jewels were gone. Lying facedown on the floor next to the empty jewel case was the body of a man the chief inspector recognized as the international jewel thief Heinrich Milhouse. Milhouse had been shot in the chest at close range; his clothes were saturated with blood. Underneath the body, the inspectors found a bloody knife.

At the airport the next day, police apprehended Englewood Smink, the murdered thief's occasional partner in crime. Smink denied all knowledge of the murder and the theft. When asked about the fresh cut on his hand, Smink said that he had had an accident in the kitchen that morning.

Suspicious, the chief inspector ordered DNA tests on the victim, the blood on the victim's clothes, the blood on the end of the knife found under the victim, and Smink. Police laboratory technicians used the polymerase chain reaction to look at two different chromosome regions that contain a variable number tandem repeat. They used one set of primers for each region. The chromosome regions, primers, and results of the tests are shown in Fig. 27.1.

What is your interpretation of the data? State your reasons. Should Smink be released? Should other tests be performed?

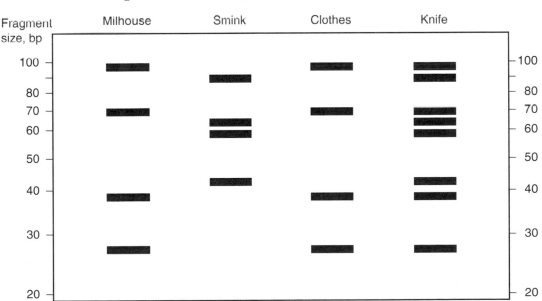

Figure 27.1 Results of polymerase chain reaction analysis.

Chromosome region 1:

5'-TCCGAGCTGGACGTGCAG...variable number of TAGA repeats...GTTACACGCCTGAGTTACGGT-3'

3'-AGGCTCGACCTGCACGTC...variable number of ATCT repeats...CAATGTGCGGACTCAATGCCA-5'

Primer set 1:5'-CCGAGCTGGACGTGCAG-3' + 3'-AATGTGCGGACTCAATG-5'

Chromosome region 2:

5'-CGACGCTTAGCATGTCCAG...variable number of CCAGT repeats...CGCTAGTCGACGCCATC-3'

3'-GCTGCGAATGCTACAGGTC...variable number of GGTCA repeats...GCGATCAGCTGCGGTAG-5'

Primer set 2:5'-GACGCTTACGATGTCC-3' + 3'-GCGATCAGCTGCGGTA-5'

The Molecular Basis of Genetic Diseases

28

About This Activity

The purpose of this activity is to get students to think about genetic diseases in terms of molecular biology. This chapter includes an extensive reading to provide teachers some background information, a student reading, and several questions that require thought. The questions could be used to extend classroom discussion and to review the relationship between protein structure and function, hybridization analysis, and meiosis. This activity assumes that students know what dominant and recessive traits are and have completed the DNA typing activity (chapter 24).

Class periods required: 1 2

Introduction

Our genes encode all the proteins present in our bodies: our enzymes, our structural proteins, our hormones, our immune proteins. Changes in our genes can result in changes in these proteins: absence of a protein, inappropriate expression of a protein, or expression of an altered form. Any of these aberrations in protein expression can cause problems. Some of the changes are fatal before birth, some result in improper development of physical or mental function, and others cause chronic metabolic deficiencies or impaired immune function. Any disorder caused by a change in the genes is a genetic disease. The characteristics of genetic diseases vary widely, depending on the type of change and the gene in which it occurs.

Genetic diseases can be divided into three categories: chromosomal defects, single-gene disorders, and multigenic traits. Chromosomal defects include missing or extra chromosomes and rearrangement or deletion of parts of chromosomes. Single-gene (Mendelian) disorders include dominant and recessive traits that are carried on either the autosomes or the sex chromosomes. Multigenic traits cover a range of conditions presumably influenced by many genes. The prevalence of genetic disorders in the human population is impossible to determine precisely. However, it

has been estimated that perhaps 2% of the population is affected by either a single-gene disorder or a chromosome defect.

Chromosome defects

Chromosome disorders include gain or loss of an entire chromosome (aneuploidy), loss of part of one or more chromosomes (deletion), transfer of one segment of a chromosome to another chromosome (translocation), and reversal of a segment of a chromosome (inversion). Considering the large number of genes involved, it is not surprising that chromosome disorders are usually fatal during early development. It is estimated that up to 50% of the fetuses in early miscarriages have chromosomal defects. However, certain chromosomal disorders permit a baby to develop to birth. Such babies display "syndromes," collections of symptoms and phenotypic abnormalities presumably caused by the abnormal number and/or expression of the large number of genes involved in the chromosomal abnormality.

The best known chromosomal disease is probably Down's syndrome, which is caused by an extra chromosome 21. Down's syndrome occurs in 1 of 600 to 800 live births and involves mental retardation, the characteristic "Mongoloid" appearance, and congenital heart defects. Why the presence of an extra copy of the genes of chromosome 21 produces this cluster of effects is not known.

Another relatively common chromosomal disorder is Turner's syndrome. Individuals with this disorder have only a single X chromosome and no Y (or second X). These individuals are phenotypic females but lack functional ovaries and do not develop secondary sexual characteristics, among other abnormalities. From 1 in 1,500 to 1 in about 5,000 female infants are thought to be affected by this condition.

A second chromosomal disorder involving the sex chromosomes is Klinefelter's syndrome. Klinefelter's syndrome is caused by the presence of two X chromosomes and a Y chromosome. These individuals are phenotypic males but are infertile because their testes fail to form properly. They also suffer other mental, behavioral, and physical abnormalities. This defect is present in 1 of 500 to 1,000 male infants.

Chromosomal deletion, the absence of part of a chromosome, is usually fatal, but some deletions cause disease instead of death. Certain deletions are associated with congenital tumors, such as retinoblastoma and Wilms' tumor. Deletions in the Y chromosome can result in XY individuals who are phenotypic females but infertile. A deletion in chromosome 5 results in the cri-du-chat syndrome, whose name refers to the peculiar catlike crying of babies with this defect. Other symptoms of this syndrome include mental retardation, a characteristic facial appearance, and motor and growth retardation.

Translocation, the relocation of one segment of a chromosome to a different chromosome, may or may not be harmful. In so-called balanced translocations, there is no net gain or loss of genetic material but simply a rearrangement. Some of these rearrangements cause no problems until the altered chromosome is passed without its balancing partner to an offspring (causing the offspring to have extra copies of genes or to be missing copies). Translocations can also be involved in disease, apparently depending on the site of the chromosome breakage and rejoining. Translocations are often observed in cancers that develop later in life. The "Philadelphia" chromosome associated with chronic myelogenous leukemia is the product of the translocation of a piece of chromosome 22 to chromosome 9. Burkitt's lymphoma is associated with the translocation of a piece of chromosome 8 to chromosome 14.

Translocations and deletions can give valuable clues to the location and function of particular genes. The translocation involved in Burkitt's lymphoma (a malignancy of antibody-producing cells) transfers an oncogene (see the reading on cancer) to a chromosomal region involved in antibody production. The XY "females" mentioned previously have helped scientists map the genes on the Y chromosome required for maleness. The retinoblastoma deletion led other scientists to the first discovery of a tumor suppressor gene (see the reading on cancer).

Multigene disorders

Multigenic disorders are often described as "an inherited tendency to develop. . . ." This sentence could be completed with "high blood pressure," "adult-onset diabetes," "mental illness," or the names of other dis-

eases. Multigene disorders have been difficult to analyze genetically because they do not show clear inheritance patterns. Environmental factors also contribute to whether or not disease actually develops.

Cancer is a multigene disease; evidence shows that disruption of the normal functions of several genes is usually required before cancer develops. In general, these alterations occur in individual cells during adulthood and are not inherited. Some families, however, have congenital defects in specific genes involved in cancer. These families inherit a predisposition to develop cancer, usually at an abnormally early age. Cancer is discussed in a separate reading at the end of this lesson plan.

Single-gene disorders

As the name implies, single-gene disorders result from the alteration of a single gene. Single-gene disorders are usually classified by whether they are dominant or recessive and by whether the mutation is carried on the sex chromosomes (X linked) or an autosome (all chromosomes except the sex chromosomes are called autosomes; they are the same in males and females). More than 3,000 single-gene disorders have now been characterized.

In general, single-gene mutations result in either the absence of a protein or the production of an altered form of the protein. Therefore, a dominant single-gene disease is one in which the lack of protein function or the production of an altered protein from one copy of a gene is sufficient to cause disease even in the presence of normal protein made from the second copy of the gene. Conversely, in a recessive genetic disease, the loss of a functional protein encoded by one gene (or production of an altered form) can be adequately compensated for by the production of normal protein from the other copy of the gene.

Consider an enzyme-encoding gene with a mutation that renders the enzyme nonfunctional. An individual who is heterozygous for this mutant gene will have half the normal amount of functional enzyme (the other copy of the gene makes functional protein). Whether this individual has a disease or not will be determined by whether or not half the normal amount of enzyme is sufficient.

Now think about a mutation in a gene encoding a structural protein. Suppose that this mutation causes the production of an altered form of the structural protein. Whether this mutation causes disease in the heterozygote depends on whether the single normal gene produces sufficient normal protein to meet the cell's needs and also on whether or not the presence of the altered mutant protein interferes with assembly of the normal structure.

Of the approximately 3,000 genetic diseases described in humans, about 1,000 are dominant. Two dominant disorders whose causes are not understood are polydactyly (extra fingers and toes) and achondroplasia (a type of dwarfism).

Familial hypercholesterolemia

An example of a dominant genetic disorder whose cause is well understood is familial hypercholesterolemia. Affected individuals have very high levels of cholesterol in their blood and usually die of heart attacks. Heterozygous individuals usually die before the age of 60, while homozygotes die much earlier. This disorder is caused by mutations in the gene encoding the low-density lipoprotein (LDL) receptor. Cholesterol is bound to LDL as the two compounds travel through the bloodstream. Cells capture cholesterol (which they need for a variety of functions, such as building cell membranes or synthesizing hormones) by binding cholesterol-LDL complexes with the LDL receptors in the cell membrane. An individual with one nonfunctional LDL receptor gene makes only half the normal number of receptors, leading to a high blood cholesterol concentration and an increased incidence of heart attack. Thus, the presence of a single mutant copy of the LDL receptor gene leads to disease.

Huntington's disease

Huntington's disease is another autosomal dominant inherited disease. It is characterized by involuntary spasmodic movements and a progressive loss of intellectual function. No symptoms are apparent until middle age, typically in the 40s, so an individual could have children before knowing that he (or even that his parent) was affected and might be transmitting the genetic defect. The gene involved has recently been identified on chromosome 4. Indications that the gene was on chromosome 4 came in 1983 from the studies mentioned below, but it took 10 more years of work to find the gene itself.

In 1983, a study of a very large Venezuelan family afflicted with Huntington's disease revealed a specific restriction fragment length polymorphism (RFLP) pattern that was consistently inherited with the disease gene. A smaller American family with the disease also showed coinheritance of the RFLP pattern and the disease. Since that time, it has been possible to perform DNA testing to look for the RFLP pattern as an indirect way of screening for the disease gene. Now that the gene has been identified, it will be possible to test directly for the disease-producing mutations.

Cystic fibrosis

Cystic fibrosis is the most common inherited disease of European-Americans. It is a recessive disorder. In cystic fibrosis patients, water and salt secretion is impaired, leading to production of abnormally thick mucus in the lungs, pancreas, and elsewhere. Digestion and breathing are impaired. Lung infections are very common, because the thick mucus prevents normal clearing of bacteria from the lungs. Affected individuals usually die before the age of 30.

The gene involved in cystic fibrosis was identified in 1989. It encodes a protein (called CFTR, for cystic fibrosis transmembrane conductance regulator) that allows chloride ions to cross the cell membrane. Exactly how the protein works and how the lack of its function results in cystic fibrosis are not yet understood. Heterozygous carriers of cystic fibrosis are completely healthy. Therefore, one copy of the CFTR gene produces enough protein to allow for normal cellular functioning.

The cystic fibrosis gene is located on chromosome 7 and spans 250,000 base pairs! The messenger RNA for CFTR, however, is only 6,500 base pairs long, and the protein has 1,480 amino acids. The most common mutation leading to cystic fibrosis is a 3-base-pair deletion in the 10th exon that results in a single missing phenylalanine in the CFTR protein (another good example of how the alteration of even one amino acid can dramatically affect protein function). To date, approximately 180 different cystic fibrosis-causing mutations in the CFTR gene have been characterized. Some mutations cause a more severe form of the disease than others, presumably because they result in different levels of deficiency in the protein's function.

Sickle-cell anemia

The most common single-gene disease among African-Americans is sickle-cell anemia, which occurs in 1 of every 600 African-Americans. This painful and eventually fatal blood disorder is caused by an altered form of hemoglobin, the protein that carries oxygen in the red blood cells. Hemoglobin is composed of four protein chains. The individual protein chains are called globins and are encoded by the globin genes. People normally have several different globin genes, which are denoted with the Greek letters alpha, beta, gamma, delta, and epsilon. The most common type of hemoglobin in adults is called hemoglobin A; it is composed of two alpha chains and two beta chains.

Sickle-cell anemia is caused by a mutation that leads to a valine rather than a glutamic acid at position 6 in the beta-globin chain. This change can be caused by a single base change from A to T, changing the glutamic acid DNA code GAA to the valine code GTA (see Table 3.1 in chapter 3). The mutant beta chain causes the hemoglobin to aggregate (see chapter 3), resulting in a change in the shape of the red blood cell from round to sickled. The sickle-shaped cells do not travel well through the capillaries, resulting in impaired circulation.

A heterozygous carrier of sickle-cell anemia does not experience clinical symptoms of sickle-cell disease. However, under special conditions such as lowered oxygen pressure, sickle-shaped red blood cells can be observed in their blood, since the mutant beta chain is produced in these individuals and is incorporated into hemoglobin along with normal beta-globin chains (produced from the healthy copy of the beta-globin gene). These heterozygous carriers are said to have *sickle-cell trait*.

Interestingly, people with sickle-cell trait are more resistant to malaria than are people with normal hemoglobin A. Most African-Americans are descended from slaves brought from West Africa, a region where malaria is endemic. For those original Africans, having one copy of the sickle-cell disease gene was probably an advantage.

Gaucher's disease

Gaucher's disease is a recessive autosomal disorder that affects 1 of every 2,500 American Jews. A patient with Gaucher's disease has an enlarged liver and spleen and can have painful bone lesions. Both the severity of the disease and the age of onset vary widely, with the severer forms appearing earlier in life. Gaucher's disease is caused by a lack of the enzyme glucocerebrosidase, which breaks down the glycolipid (a combination of carbohydrate and lipid) glucocerebroside. The gene encoding this enzyme has been identified and cloned. It is located on chromosome 1, consists of 11 exons, and is about 7,500 base pairs long. The spliced messenger RNA is about 2,000 base pairs long. One good copy of the gene can apparently produce enough enzyme to meet the body's needs, since heterozygotes are healthy.

Studies of Gaucher's disease patients revealed that these people have one of many different mutations in each glucocerebrosidase gene. The nature of the mu-

tation itself determines whether the form of the disease is severe or mild, although two individuals with the same mutations may still have disease symptoms that differ in severity (possibly because of effects exerted by other genes). A common mutation leading to mild disease is a base change from A to G at position 1226, resulting in an amino acid change from asparagine to serine. This mutation (1226G) accounts for 75% of the disease genes in the Jewish population. It is likely that this amino acid change leaves the protein able to function somewhat, accounting for the mildness of the disease in these cases. Two common mutations that cause severe disease are a single base change from T to C at position 1448, which causes an amino acid change from leucine to proline, and an insertion of a single G at position 84, which causes a reading frame disruption. The reading frame disruption early in the gene would likely result in a complete lack of protein function. Similarly, because the chemical characteristics of the amino acid proline are so different from those of leucine, substitution of a proline for a leucine within the protein probably severely disrupts the protein's structure and therefore its function. These two mutations (serine to proline and the reading frame disruption) would be predicted to impair the function of the enzyme much more than an asparagine-to-serine change. The fact that these two mutations result in a much severer form of the disease is consistent with this prediction.

Diagnosis of genetic disease and carrier status

In the last three decades, our ability to diagnose genetic diseases has improved dramatically. One factor that has been critical to these advances is our growing knowledge of the causes of various genetic diseases. Another factor is advances in analytical chemistry and biochemistry. When a genetic disease is caused by a known enzyme deficiency, it is now possible to test for that activity or for telltale metabolic products that indicate the presence of the disease. Many of these tests can be performed prenatally, using fetal cells from amniotic fluid. Chromosome spreads can be prepared from these cells to detect chromosome abnormalities.

Unfortunately, when the cause of a genetic disease is not known, there may be nothing known to test for. Although we have recently identified the protein whose alteration causes cystic fibrosis, it is not an enzyme that produces a measurable end product. As of now, there is no clinical assay for its function. Even when we can test for specific defects, there are sometimes problems with prenatal diagnosis. For example, to diagnose sickle-cell anemia by traditional methods, red blood cells must be examined. (Hemoglobin is not an enzyme, but the structural abnormality in sickle-cell disease can be detected through its effects on red blood cell shape.) Although it is possible to obtain a sample of fetal blood, the procedure involves significant risk to the fetus.

Finally, many of our best procedures cannot detect carrier status. It can be valuable for prospective parents to know whether they both carry a recessive disease gene, so that appropriate prenatal tests can be performed.

Now that we can analyze DNA directly, it is possible to bypass tests of protein function and look directly at the affected genes. By means of DNA analysis, any disorder for which a gene location is known can, theoretically, be detected. Prenatal DNA testing can be performed on amniotic cells. Even if the protein in question (such as hemoglobin in the case of sickle-cell anemia) is not produced in the amniotic-fluid cells, the genes are there and can reveal whether the defect is present.

Some genetic screening involves looking directly at the gene in question. Other approaches include looking for a particular RFLP pattern known to be consistently inherited with a disease gene (such as in Huntington's disease; see above). The major drawback of these tests is that most are designed to look for particular mutations or particular RFLP patterns. Although 95% of all cases of a particular disease may be caused by the three mutations that are tested for, the other 5% of cases will escape detection. Even if all known mutations (nearly 180 in the case of cystic fibrosis!) were tested for (probably a financial impossibility), any individual could carry a new, unique mutation. So DNA testing is not necessarily a 100% accurate method of diagnosis. However, in many cases, it is the safest and most accurate method available.

Many ethical issues need to be considered when we contemplate testing for genetic diseases. For example, should populations be screened for common disease genes? Should employers have access to knowledge about employees' genetic predispositions to disease? Should people with certain predispositions (for example, to cancer) be barred from working in occupations that might increase their risk of developing disease (such as one involving exposure to asbestos fibers)? Do people who bear a child knowing in advance that it has a genetic disease that will prevent it from ever being self-supporting (such as Down's syndrome) have the right to expect taxpayers (you and me) to pay for the child's lifelong care? Who should make these decisions?

A good approach to confronting ethical dilemmas is presented in chapters 32 through 34. This model forces students to analyze their thinking and channels discussion along ethical lines. We recommend that you refer to these chapter if you would like to raise ethical issues for your class to consider.

Objectives

After completing this activity, students should be able to

1. Explain some of the molecular considerations that determine whether a genetic disease is dominant or recessive;

2. Suggest an explanation based on protein function as to why some mutations in a gene cause mild disease while others cause severe disease.

Materials

- Photocopies of *Student Activity* pages, if students do not have manuals
- Transparency or photocopies of the genetic code for reference in question 1

Resource

In 1990, the Howard Hughes Medical Institute began to publish a series of reports on biomedical science. One report in the series, "Blood: Bearer of Life and Death," contains information on genetic disorders of the blood and refers to gene therapy. The Institute provides free copies of their reports to educators. Their address is

Howard Hughes Medical Institute
4000 Jones Bridge Road
Chevy Chase, MD 20815-6789
(301) 215-8500

Procedure

Use the reading and questions as best suits your class. The student reading is intended to get students thinking about the molecular basis of genetic diseases and to provide a starting point for discussion. The questions can be discussed in class, treated as small-group projects, or assigned to individuals.

Optional Follow-Up

- Invite a speaker from a laboratory that does DNA-based testing for genetic diseases (the director of your hospital laboratory should be able to tell you where the nearest one is).

Classroom Activities

- Invite someone from your hospital laboratory to talk about routine health screening for newborns.

- If you know anyone who has phenylketonuria, cystic fibrosis, or another inherited disease, ask that person if he or she would mind sharing information about their disease with your class.

- Invite a genetic counselor to talk to your class (an obstetrician or hospital laboratory director should be able to tell you where to find one).

Answers to Student Questions

1a. Normal sequence: valine, histidine, leucine, threonine, proline, glutamic acid, glutamic acid. Mutant sequence: valine, histidine, leucine, threonine, proline, valine, glutamic acid. The mutation causing sickle-cell disease is the change from glutamic acid to valine at position 6.

1b. The normal beta-globin gene sequence contains an *Mst*II site. The mutation that causes sickle-cell anemia also destroys the restriction site within the gene. Therefore, it is possible to test for the mutation by testing for the presence of the *Mst*II site.

To carry out the test, digest a DNA sample with *Mst*II, separate the fragments on a gel, and use the Southern blot procedure to transfer the fragments to a membrane. Probe the fragments with a probe that hybridizes to DNA on either side of the *Mst*II site in the beta-globin gene. Homozygous normal individuals will have two fragments that hybridize to the probe; one fragment will be approximately 1,150 base pairs long, and the other will be approximately 200 base pairs long. Homozygous affected individuals will have a single fragment of approximately 1,350 base pairs. Heterozygous carriers will have three bands: 1,350 (from the mutant gene), 1,150, and 200 base pairs long.

Students might design a procedure using *Mst*II and a second enzyme. Such a procedure could work as long as their logic is sound.

The hybridization step is essential. If students propose simply to cut the DNA with *Mst*II, separate the fragments on a gel, and examine the stained gel, remind them that all they would see is a smear. The human genome contains about 3 billion base pairs and would generate a vast number of *Mst*II fragments. Hybridization lets them look specifically at the genomic region of interest.

2a. Whether the offspring is healthy or not depends on which chromosomes he or she inherits. Assume that the parent has a balanced translocation in which part of one of his chromosome 3s is attached to one of his chromosome 5s. That parent also has a normal copy of chromosome 3 and one of chromosome 5. So when the parent's gametes are formed, there are four possibilities: 3 and 5 (both normal), 3* and 5* (both abnormal), 3* and 5, and 3 and 5*. Two of these combinations, 3 and 5 and 3* and 5*, would produce healthy offspring. The other two cases form a gamete with either missing genetic information or extra genetic information and would not produce a healthy offspring.

2b. The most likely reason has to do with the exact point at which the break occurs and where the segment reattaches. If either event disrupts an important gene, the consequences could be bad. Harmless translocations apparently do not interfere with normal gene expression.

3a. Use this question to make sure students realize that some mutations may have little effect on the function of the protein, while others can completely eliminate it. Mutations causing mild disease probably leave the enzyme able to function in a diminished capacity, so that disease symptoms arise but are not severe. Mutations causing severe disease may disrupt protein function altogether. Supply the specific mutation information to your class from the *Introduction*, if you wish.

3b. Again, the mild-disease mutation presumably results in an enzyme with a partial function. That partially functioning protein is present in the heterozygote who has two different disease genes. Thus, a patient with one mild mutation and one severe one has some enzyme activity, in contrast to a patient with two severe mutations, and his disease should be milder.

4. Discuss this question with your class if you like.

The Molecular Basis of Genetic Diseases

Our genes encode the proteins that orchestrate how we develop, what we look like, how we metabolize food, how we synthesize body components, and how we defend ourselves from disease. It is not surprising that changes (mutations) in our genes can cause many different kinds of problems. Any harmful condition caused by a change in genes is called a genetic disease. Genetic diseases are thought to affect approximately 2 of every 100 individuals.

Genetic diseases are usually divided into three categories: chromosome disorders, multigene disorders, and single-gene disorders. Just as it sounds, chromosome disorders are diseases caused by a problem with whole chromosomes: an extra or a missing chromosome, loss of a piece of a chromosome, or rearrangement of chromosome segments. Multigene disorders are conditions that clearly involve heredity but seem to involve many genes. Their genetic basis is not yet well understood. Multigene disorders are often talked about as "an inherited tendency to develop. . . ." Adult-onset diabetes, high blood pressure, and some mental illnesses are multigene disorders. Single-gene disorders are diseases caused by mutations in a single gene.

Chromosome Disorders

Think about a chromosome disorder for a moment. If a person were missing a chromosome, he or she would be missing one copy of many, many genes. Likewise, an extra chromosome would introduce an extra copy of thousands of genes. Even loss or gain of a piece of a chromosome could involve many genes. Would you expect a chromosome disorder to cause mild or severe problems? If you thought "severe," you were correct. In fact, most chromosome disorders are so harmful that fertilized eggs that have them cannot develop properly and thus die. Doctors estimate that as many as half of all early miscarriages involve chromosomal abnormalities in the embryo.

Some chromosomal abnormalities are not fatal. Individuals who have them usually show a range of problems and changes, which is not surprising when you consider how many genes are involved in a chromosome disorder. The most common chromosome disorder is Down's syndrome, which occurs in people with three copies of chromosome 21 instead of the normal two copies. People with Down's syndrome are mentally retarded, have a "Mongoloid" facial appearance, and have congenital heart problems and other abnormalities. At present, no one understands exactly which of the genes on chromosome 21 contribute to these conditions or why extra copies of them would do so. Do you know anyone with Down's syndrome?

Single-Gene Diseases

Now think about single-gene disorders. More than 3,000 of them, each caused by a change in one gene, have been described. These diseases are usually "discovered" by physicians who note that a particular illness or developmental problem is inherited in a specific way. By studying the pattern of inheritance, the physician can decide whether the disease is dominant or recessive. About 1,000 of the known human genetic diseases are dominant.

Let's think about what dominant and recessive actually mean on a molecular level. Since everyone has two copies of all their genes (except those on the X and Y chromosomes in males), the question is, "Is one copy of the mutant gene enough to cause the disease, or do both copies have to be mutant?" The answer to that question depends on the nature of the mutation and the gene it is in. If the mutation completely knocks out production of an enzyme, for example, the relevant question is, "Can the remaining good copy of the gene produce enough enzyme so that the individual will be healthy?"

If the answer to that question is "yes," then the disease is recessive: it takes two mutant genes for the disorder to manifest itself. If the answer is "no," then the disease is dominant.

If a different mutation doesn't knock out production but instead causes an altered, nonfunctional protein to be made, another relevant question (besides the one above) is, "Will the presence of the altered, nonfunctional protein interfere with the functioning of the normal protein produced by the other copy of the gene?" If the answer to this question is "yes," then the mutation is dominant. If the answer is "no," then the dominant-recessive question rests on the answer to the question above. Let's look at some single-gene disorders in the light of these questions.

Cystic fibrosis (CF) is the most common inherited disease of European-Americans. In CF patients, the secretion of water and salt from cells is impaired, resulting in production of a thick, sticky mucus in the lungs, digestive tract, and elsewhere. These patients have problems breathing and digesting their food. In addition, they are subject to frequent bacterial infections, because bacteria become trapped in the thick mucus rather than being swept away as in normal individuals.

CF is caused by mutations in a gene encoding a protein that allows chloride ions (part of salt) to cross the cell membrane. Nearly 180 different mutations in the gene have been described. Some of the mutations cause a severer form of CF than others cause. However, if a person has one mutant CF gene and one normal copy, that person is healthy. Apparently, one copy of the gene can produce all the protein needed to be healthy.

It also seems that altered forms of the protein do not keep the normal protein from performing its function. It is assumed that those mutations causing mild disease don't completely destroy the activity of the CF protein. About 1 in 2,500 European-American babies is born with CF. Do you know anyone who has the disease?

Sickle-cell anemia is the most common inherited disease of African-Americans, occurring in about 1 of every 600 babies. It is caused by a change in a single base in a gene encoding one of the proteins that makes up hemoglobin. Hemoglobin is the protein that carries oxygen in red blood cells. Sickle-cell anemia patients produce an altered form of hemoglobin that tends to stick to other similar hemoglobin molecules. Clumping of hemoglobin results in the red blood cells collapsing to a sickle shape. These sickled cells do not travel well through capillaries, causing poor blood circulation. As a result, sickle-cell anemia patients suffer painful circulation crises.

Does it take one or two copies of the mutant gene for sickle-cell anemia to result? Two copies. However, in people with one sickle-cell gene and one normal gene, the abnormal protein is produced. Although these people are healthy, it is possible to see some sickled red cells in their blood under some conditions. The abnormal hemoglobin is clumping, causing the shape change. These individuals are said to have sickle-cell trait. Do you know anyone who has sickle-cell trait or sickle-cell anemia?

Phenylketonuria is a genetic disease caused by the lack of the enzyme phenylalanine hydroxylase. Individuals with this disease are unable to break down the amino acid phenylalanine, and their urine contains high levels of the partial breakdown product, phenylketone. Immediately after birth, when these individuals begin to eat and consume phenylalanine, these breakdown products start to accumulate. Eventually, irreversible damage to the central nervous system occurs, resulting in mental retardation. One good copy of the gene for phenylalanine hydroxylase produces enough enzyme for healthy function.

Babies are now tested shortly after birth for phenylketonuria. If they have it, they are immediately put on a special diet that is very low in phenylalanine. On this diet, they can develop normally. Have you ever noticed any food labels with a warning to phenylketonurics (the name for people with phenylketonuria) that a particular food contains phenylalanine? Check cans of diet soda. The artificial sweetener aspartame (brand name NutraSweet) contains the amino acid phenylalanine.

Diagnosis of Genetic Diseases

Chromosome disorders are diagnosed after an examination of the chromosomes themselves. This is done by taking a sample of cells (usually white blood cells or amniotic cells) and culturing them to metaphase, the stage in which chromosomes are condensed and easy to see. The metaphase chromosomes are then spread on a microscope slide and stained with a special stain called Giemsa. This stain shows characteristic banding patterns on each chromosome, allowing all of them to be identified (from 1 to 22, X and Y). Any gross abnormalities can be detected visually.

Single-gene diseases like phenylketonuria are currently diagnosed by testing for activity of the enzyme involved. Sickle-cell anemia can be diagnosed by examining red blood cells under the microscope. Other diseases, like CF, are often diagnosed on the basis of clinical symptoms. Now that we can look directly at DNA and are learning about the mutations that cause these diseases, many of these diseases can be diagnosed by using Southern hybridization to detect restriction fragment length polymorphisms or by using the polymerase chain reaction.

The advantages of DNA-based diagnosis depend on the disease. Sometimes no other test predicts its occurrence (such as for Huntington's disease; see question 4 below). DNA tests can detect carrier status in the case of recessive diseases (by showing the presence of a single mutant gene), while most traditional tests cannot. Finally, any DNA-based test can be performed prenatally on cells from the amniotic fluid. This is a safer method of prenatal testing than those that require a sample of fetal blood (such as for sickle-cell anemia).

Questions for Thought and Discussion

1. Sickle-cell anemia is caused by a specific change in the sixth amino acid of the beta-globin protein, which forms part of hemoglobin. Shown below are a DNA sequence encoding the first seven amino acids of the normal beta-globin protein and a sequence encoding the first seven amino acids of the sickle-cell protein.

   Normal amino acid sequence:

   GTT CAT CTA ACC CCT GAG GAG . . .

   Sickle-cell sequence:

   GTT CAT CTA ACC CCT GTG GAG . . .

   a. Translate these sequences. What is the amino acid change that causes sickle-cell disease?

   b. The restriction enzyme *Mst*II cleaves the DNA sequence CCTGAGG. In human DNA, there is an *Mst*II site 1,150 base pairs 5' (to the left as shown) of the beta-globin coding region and 200 base pairs 3' (to the right) of the region shown. Design a DNA-based test for sickle-cell anemia using the enzyme *Mst*II. Draw a gel outline, and show the expected results of your test on a homozygous normal individual, a sickle-cell anemia patient, and a person with sickle-cell trait (a heterozygous carrier). You will have to supply a scale of base pair sizes to the left of the gel outline based on the expected results from your test.

2. A chromosomal translocation occurs when a segment of one chromosome breaks off and attaches to the end of a different chromosome. These events can occur during formation of the gametes, or they can happen in individual cells dur-

ing the life of the person. If the gamete carries a translocation, all the cells in the resulting person will have it.

   a. Occasionally, healthy adults who carry a translocation in all their cells are identified. The assumption we make about these people is that they did not gain or lose genetic information during the translocation and that the particular rearrangement (shifting of a segment of one chromosome to another) caused no harm. The translocations in these individuals are called balanced translocations. Would you expect the offspring of a person with a balanced translocation to be healthy?

   b. Translocations that happen in individual cells during adulthood can be harmless, or they can lead to cancer. For example, translocation of the end of chromosome 22 to chromosome 9 is associated with leukemia. Why do you think some translocations can be harmless while others can cause cancer or other problems?

3. Gaucher's disease is a recessive genetic disease caused by the lack of a particular enzyme. By looking at the gene for that enzyme in different Gaucher's patients, scientists have identified many different mutations that cause the disease. Interestingly, some mutations lead to a mild form of the disease (in homozygotes), while other mutations lead to a very severe form (again in homozygotes). Heterozygotes with one normal copy of the gene are healthy no matter what form of the mutant gene they have on the other chromosome.

   a. Why do you think some mutations in the gene could cause mild disease while others cause severe disease?

   b. Some patients with Gaucher's disease have two different mutant forms of the gene. If a person has one copy of the gene with a mutation leading to mild disease and the other copy of the gene with a mutation leading to severe disease, that patient will have mild disease. Propose an explanation for why the disease would be mild.

4. Huntington's disease is a dominant genetic disease. Symptoms of this disease do not appear until middle age, typically when the patient is in the 40s. However, when symptoms appear, they are devastating. The patient gradually loses the ability to think. At the same time, he begins to experience increasingly uncontrolled body movements, usually twitching and shaking. There is no cure. Since the disease is dominant, there is a 50%

Classroom Activities

chance that the child of an affected parent will also develop it. However, since the disease doesn't show until middle age, an individual can reproduce before he or she knows that the disease gene is present.

Suppose that when you are a teenager, one of your parents develops Huntington's disease. You watch the strong, intelligent adult you knew deteriorate into a nearly mindless, shaking body. You also know that there is a 50-50 chance that you will suffer the same fate and could transmit it to your children. There is now a DNA test that will tell you if you carry the Huntington's disease gene. Would you choose to take the test? What factors would enter into your decision?

The Molecular Genetics of Cancer

A Genetic Disease in a Single Cell

Most cancers start from a single cell. In order to become a cancer cell, a normal cell must accumulate mutations in several different genes. These mutations are then transmitted to the cell's cancerous descendants. Cancer can therefore be thought of as a genetic disease at the level of the cell.

Cancer in a Dish

What kind of genetic changes are involved in cancer? The first clues came from studies of certain animal viruses in cultured animal cells. Cultured cells behave in a preordained manner. They do not grow on top of one another but instead form a confluent layer in a dish. They usually live for a certain number of life cycles and then die. However, sometimes changes occur in these cells. The cells can become immortal, losing their predetermined life span. They can begin to grow over one another in an uncontrolled manner. Other changes can take place as well. The combination of several of these changes is called **transformation** (do not confuse this use of the word with the usage referring to the addition of new DNA). In the body, cancer cells lose their normal growth inhibitions and continue to multiply in an uncontrolled and inappropriate manner. Because of the similarities, transformation of cells in culture is considered the equivalent of cancer in a dish.

In the 1970s, it was discovered that infection by certain viruses could transform cultured cells. The ability to produce the cancerlike changes was traced to certain genes the viruses carried. These genes were christened **oncogenes** ("onco" is a prefix meaning tumor). Quite a number of oncogenes were identified; they were given three-letter names such as *ras, src,* and *myc.* Very shortly after the viral oncogenes were discovered, scientists found that copies of these genes (or nearly identical genes) resided in our chromosomes. The "viral" oncogenes were actually cellular genes that the viruses had acquired and transmitted by transduction.

Oncogenes

What do oncogenes do? An interesting and logical picture is emerging from a decade and a half of intense study in many laboratories. Oncogenes are normal, essential parts of our genetic material that appear to belong to the group of genes in charge of causing and regulating cell growth and division. These normal, necessary genes are called proto-oncogenes. It is actually changes in these genes that can lead to the development of cancer. Changed versions of the cellular proto-oncogenes were the oncogenes present in the transforming viruses.

Cell growth usually results from a signal, such as a hormone, received by a receptor on the cell membrane. After the receptor receives the "grow" signal, it transmits the message across the cytoplasm to the nucleus. There the signal promotes DNA replication. Mutations affecting any step of the signaling pathway could cause a cell to "think" it is constantly being instructed to divide. For example, many receptors respond to outside signals by adding phosphate groups to themselves and/or other proteins. If a receptor were mutated in such a way that it added phosphates whether or not it was being signaled, the cell would "believe" it was constantly being instructed to divide. Such a genetic change would promote uncontrolled cell growth, rather like stepping on the accelerator in a vehicle. The oncogenes identified in the transforming viruses had undergone mutations with this kind of effect.

Tumor Suppressor Genes

Evidence from genetic studies of cancer patients led scientists to believe that another type of gene might also be involved in the development of cancer. Subsequent research has proven their expectation to be correct. The product of this type of gene is like a brake on cell growth rather than an accelerator. Instead of being part of a signaling chain that tells a cell to divide, this type of protein represses cell growth and division. These repressor proteins apparently must be inactivated before full cancer can develop. The genes encoding these proteins are called **tumor suppressor genes.**

As might be expected, mutations in oncogenes that lead to cancer are usually different from mutations in tumor suppressor genes that lead to cancer. In oncogenes, the mutations must *activate* the protein to promote growth inappropriately. This kind of mutational change includes a change in a single amino acid that leads to an altered form of the protein, multiplication of the gene within the chromosome to provide greater activity, or alteration of the control regions of the gene to deregulate its expression or to cause it to be regulated inappropriately. These mutations also tend to be dominant; the presence of one normal copy of the oncogene cannot make up for the mutant, activated form. For example, the chromosome 9,22 translocation that brings about the inappropriate activation of the *abl* gene causes leukemia, even though a normal *abl* gene remains on the other copy of the chromosome. The cancer-causing viruses were able to transform cultured cells even though those cells had normal copies of the proto-oncogenes.

For tumor suppressor genes, however, *loss* of the protein function is required to promote cancer development. For this reason, mutations in suppressor genes tend to be recessive; one good copy of the gene can provide active protein. You might expect that individuals could inherit one copy of a recessive mutation in a tumor suppressor gene and that these individuals might be more likely to develop cancer during their lives. You would be right.

The first tumor suppressor gene to be identified was the retinoblastoma gene. Retinoblastoma is a tumor of the eye that can be hereditary. Analysis of patients with hereditary retinoblastoma revealed that one copy of a certain gene was inactivated in all of their cells and that both were inactivated in the tumors. A comparison with other patients whose retinoblastoma was not hereditary showed that the same gene, now called the retinoblastoma *(rb)* gene, was also inactivated in their tumor cells. Since loss of function of *rb* was apparently required for the tumor to develop, researchers concluded that the *rb* gene product must suppress tumor development. Further study of the *rb* gene suggests that its product somehow represses the expression of genes that stimulate cell growth.

We now know that children with hereditary retinoblastoma inherit one inactive copy of the *rb* gene from their parents. Every cell in their bodies has only one good copy of the *rb* gene. In these children, all that is needed is for one retinal cell to suffer a muta-

tion to the other copy of *rb,* and the retinoblastoma tumor can then develop. In genetically normal individuals, a single retinal cell would have to suffer two independent mutational events to knock out both copies of *rb,* an unlikely series of events. Retinoblastoma is a rare cancer. Most cases run in families, who apparently transmit the defective gene.

Mutation of another tumor suppressor gene, *p53,* is associated with a different inherited cancer syndrome. Li-Fraumeni syndrome is a rare inherited syndrome of cancers. Members of Li-Fraumeni families develop cancers (often multiple cancers) of the breast, brain, bone, and other tissues and leukemia, usually at early ages. Patients with Li-Fraumeni syndrome inherit one defective copy of the *p53* gene, so every cell in their bodies has only a single good copy. Apparently, inactivation of this good copy in any number of different cell types can lead to cancer.

The *p53* gene is also important in noninherited cancer. Scientists who analyze genetic alterations in cancerous cells have found that *p53* mutations are present in more different kinds of cancer than any other known cancer-related genetic alteration. Needless to say, this finding has sparked enormous interest in the biological role of *p53*. The p53 protein is a nuclear protein that appears to be involved with regulation of cell division. Recent research suggests that this protein acts as a gatekeeper in the cell division cycle. If cells with normal p53 suffer DNA damage, the p53 protein somehow prevents the cells from dividing, presumably until the damage is repaired. Cells that lack normal p53 go ahead and divide. The replication of damaged DNA leads to changes (mutations) in the DNA sequences of daughter chromosomes. So a cell that lacks p53 activity apparently can accumulate mutations at a much higher rate than normal cells do. According to this line of thinking, if one of these mutations leads to the activation of an oncogene, the result could be cancer. Exactly how p53 stops cell division is not known at this time.

Genetic analysis of families with hereditary cancer and of individual cancer patients is helping us identify additional genetic loci associated with cancer development. So far, the genetic picture of most common cancers looks very complicated, and no clear interpretation is available. For example, the following chromosomal locations have been identified as possible locations for tumor suppressor genes involved in breast cancer: 1p, 1q, 3p, 6q, 11p, 13q, 16q, 17p, 17q, 18, and 22 (p and q refer respectively to the short and long "arms" of the chromosomes as they appear on either side of the centromere). At this point, no one knows which genes might be at

Table 28.1 Some cancer genes and the physiological roles of their products

Gene	Role
Oncogenes	
sis	Growth factor
erbB, fms, neu	Cell surface growth factor receptor
ras, arc, abl	Signal transmission within the cell
myc, fos, myb	Regulators of DNA replication and transcription
Tumor suppressor	
rb, p53	Regulators of DNA replication and transcription; *p53* apparently stops cells from dividing if their DNA is damaged, allowing time for repair

these locations or what their roles in breast cancer development might be.

Some scientists expect that as many as 100 different oncogenes and tumor suppressor genes will eventually be identified. As the Human Genome Project progresses toward its goal of mapping and sequencing all human genes, answering some of these questions should become easier. Table 28.1 lists a few of the known oncogenes and tumor suppressor genes and gives their role in the control of cell growth.

Does our slowly increasing understanding of cancer genetics hold out any hope for improved therapies in the future? Many physicians and scientists believe so. Since cancer is the result of a malfunction of genes, the best way to fight it may be to restore proper genetic function. Possibly gene therapy could be used to introduce good copies of tumor suppressor genes into cells whose own copies are mutant. Antisense technology could be used to "turn off" the expression of tumor-promoting mutant genes. In fact, a Houston physician has already been granted approval to test an antisense gene therapy against an activated oncogene in 14 lung cancer patients. Scientists hope that the next few years will yield less toxic, more effective genetic therapies for one of our most dreaded genetic diseases.

Selected Readings

Cavenee, W. K., and R. L. White. 1995. Genes and cancer. *Scientific American* 272(3):72–79.

Ponder, B. A. J. 1992. Molecular genetics of cancer. *British Medical Journal* 304:1234–1236.

Classroom Activities

PART III

Societal Issues

*S*ocietal issues raised by biotechnology, or
any technology, are introduced by using background information and classroom activities
designed to provide tools for analyzing difficult issues rationally. The complex relationship between scientific understanding, technological development, and societal structure is
introduced and elaborated on in chapters on
assessing and debating risks and conducting
productive, thoughtful discussions of bioethical dilemmas. The final chapter provides
information on careers in biotechnology.

Societal Issues: Introduction

Parts I and II provide information on science and technology, or, more specifically, on the science of genetics and the technologies we have developed by using our deep and extensive understanding of molecular genetics. In this part, we introduce a related body of knowledge that is finding increasing relevance and importance in today's world: the nature of science and technology—their history, interrelationship, and societal effects.

The reasons for studying the sciences that spawned biotechnology and for practicing the hands-on techniques used in biotechnology research and product development are clear. Scientific knowledge and laboratory skills spell jobs. But why should one bother to learn *about* science and technology?

In short, because our lives have repeatedly been and will continue to be transformed by the omnipresent influences of science and technology. These influences are both obvious and subtle, both simple and profound. Science and technology permeate all aspects of our lives and our relationship to the natural world. As a result, not only do they shape our values, goals, and world view, but they also reflect these parts of ourselves.

In the past, we often embraced new technologies without making much of an attempt to analyze the potential societal effects, largely because we could not predict them. Imagine if we had been so prescient as to debate the introduction of automobiles on the grounds that they would lead to polluted air and fragmented family units. In contrast, modern biotechnology has generated a great deal of debate about its uses, from genetically engineered crops to recombinant microorganisms to gene therapy. This debate is healthy and smart.

Biotechnology will give us powers we never had before. These powers can be used for both good and evil. While this thought may be troubling, it is not new, nor is it unique to this technology. Societies have dealt with precisely the same issue in different

forms ever since humans began to fashion crude tools from stone. Democratic societies have given their citizens the power to determine how to use these technologies. As is clear from this quote by Thucydides, author of *The Peloponnesian War,* the public in Greek democracies viewed itself as the best judge of matters that affect the public:

. . . our ordinary citizens, though occupied with the pursuits of industry, are still fair judges of public matters; for, we alone regard the man who takes no part in public affairs not as one who minds his own business but as good for nothing. We Athenians are able to judge all events, and instead of looking on discussion as a stumbling block in the way of action, we think it an indispensable preliminary to any wise action at all.

Thucydides, 405 BC

Do you think a historian would describe democracy in today's America in the same way?

Like all technological applications, even the beneficial applications of biotechnology will probably be accompanied by some costs. To use the power of biotechnology wisely and to assess its costs and benefits correctly, we need to be conversant in both the science and the potential impacts of the science on society.

Dismissing decisions about the appropriate uses of biotechnology as someone else's concern guarantees dissatisfaction with the way the technology develops. Citizens who abdicate their oversight of technological development to a handful of scientists, engineers, and politicians are ignoring the power, privilege, and responsibility of a democratic way of life. They are also sacrificing a portion of their personal freedom.

The Activities in Part III

We begin this part with an overview of the interdependent relationship between science, technology, and society. Each of these ingredients influences the others and is in turn influenced by them. Analyzing this interdependence from a variety of perspectives will illustrate the extent to which they are interwoven.

Chapters 30 and 31 on risk assessment and weighing risks and benefits teach skills that will enable us to make a rational analysis and participate in a rational discussion of the risks and benefits of technological development in general. The examples used are drawn from applications of biotechnology.

Factual information plays a large role in formal risk assessments and risk-benefit analyses. These methodologies were developed in order to minimize the contribution of emotion to decision-making. In chapters 32 through 34, we move from these relatively unemotional topics to the highly charged contentious area of morals and values. Unless a clear, tight structure is provided, classroom discussions of ethical issues can quickly degenerate into heated exchanges in which facts take a backseat to strongly held views. In order to facilitate productive and informed discussion, we have provided a decision-making model for addressing the ethical dilemmas that arise in two scenarios relevant to biotechnology: gene therapy for the treatment of a genetic disease and genetic screening for inherited disorders. As will be evident, the model can be used to analyze thorny issues in areas other than biotechnology and biomedicine.

Finally, we end this part with a chapter on careers in biotechnology and the educational requirements for obtaining various jobs in this young and exciting field.

Selected Readings

In addition to the references found at the end of each chapter, here are other general references dealing with various aspects of this part.

Bishop, Jerry E., and Michael Waldholz. 1990. *Genome.* Simon & Schuster, New York.

Bronowski, J. 1956. *Science and Human Values.* Harper & Row, New York.

*Congressional Office of Technology Assessment. 1988. *Medical Testing and Health Insurance.* U.S. Government Printing Office, Washington, D.C.

*Congressional Office of Technology Assessment. 1990. *Genetic Monitoring and Screening in the Work Place.* OTA-BA-455. U.S. Government Printing Office, Washington, D.C.

Frankel, Mark S., and Albert H. Teich. 1995. *The Genetic Frontier: Ethics, Law and Policy.* AAAS Directorate for Science and Policy Program, American Association for the Advancement of Science, Washington, D.C. (Order from AAAS, 1333 H St., NW, Washington, DC 20005. The cost is $22.95.)

Holtzman, Neil. 1989. *Proceed with Caution: Predicting Genetic Risks in the Recombinant DNA Era.* John Hopkins University Press, Baltimore, Md.

National Research Council. 1989. *Field-Testing Genetically Modified Organisms: Framework for Decisions.* National Academy Press, Washington, D.C.

Nelkin, Dorothy, and Laurence Tancredi. 1989. *Dangerous Diagnostics: the Social Power of Biological Information.* Basic Books, Inc., New York.

Olson, Steve. 1989. *Shaping the Future.* National Academy Press, Washington, D.C.

Suzuki, David, and Peter Knudtson. 1989. *Genethics: the Clash Between the New Genetics and Human Values.* Harvard University Press, Cambridge, Mass.

*To order publications from the U.S. Government Printing Office, write to the Superintendent of Documents, Mail Stop: SSOP, Washington, DC 20402-9328.

Societal Issues

Science, Technology, and Society

29

Introduction

Today's world has been profoundly and unalterably shaped by advances in science and technology. Some of the ways in which science and technology have affected our lives are immediately obvious: airplanes, automobiles, television, computers, and microwave ovens. Others are so subtle that we fail to recognize them as benefits of science and technology. Instead, we take them for granted as a "natural" part of life. The clothes you're wearing, the chair you're sitting in, the pen you may be holding, the paper this book is printed on, the food you had for lunch are all products of science and technology.

If you doubt the extent to which technology has penetrated every tiny facet of your life, pay attention to your actions during the rest of the day and take note of all of the ways technology, both simple and sophisticated, has altered your world. As you are paying attention to the ways technology impinges on your life, also look at the flip side of the coin. How often do you come face to face with nature? In those encounters with nature, how many times do you experience a love for nature, and how many times do you wish it would go away and leave you alone? Thousands of years ago, we rejected the idea of living with nature. Our distaste for the adversities nature indifferently dispenses has driven our constant quest for new technologies. We use all of these technologies for a similar purpose: to change our environment so that the natural world suits us better.

Science and technology not only alter our environments—they also shape us. Our view of the world, the questions we ask, our sense of values, and the way we think are the products of life in a world shaped by scientific discoveries and technological innovations.

Has it always been this way? Yes and no. Technologies have always influenced human societies, our mental framework, and our relationship to the natural world.

Today, however, technological advances are more pervasive, and their effects are more powerful. Also, the changes they bring are occurring at an increasingly rapid rate. We can trace the increased presence, power, and rate of technological change to the role scientific understanding now plays in propelling technological developments.

Which Came First, Science or Technology?

Initially, technology drove science. People developed technologies on a trial-and-error basis, using the manual skills required in crafts such as clock making, lens grinding, and metal forging. Early scientists used those technologies to observe and question natural phenomena and thus to increase their understanding of the world around them. As their understanding of the natural world broadened and deepened, the relationship between science and technology shifted.

Equipped with a more sophisticated understanding of the world they sought to control, people utilized scientific understanding to drive technological change. They replaced the trial-and-error approach to technological advancement with the methodical, experimental approach that is characteristic of scientific investigation. Buttressed by sound science, the ensuing technological changes were more forceful, and the pace of technological change quickened. With ever more powerful technologies at their disposal, scientists were able to probe deeper and deeper into the causes and effects of natural phenomena.

Because the relationship between scientific understanding and technological innovation is circular and not linear, reciprocal and not hierarchical, we can only expect the pace of science and technological change to accelerate in the 21st century. A constantly accelerating rate of increase in scientific knowledge and technological achievement carries with it important implications for our species.

The Nature of Science and Technology

Many people equate science with technology, but the two differ in a number of fundamental ways. A few of the ways in which science differs from technology are provided in Table 29.1 in the *Student Activity* pages.

Despite their differences, science and technology converge in subtle ways that may not be apparent to people outside the scientific and engineering communities. Most people would readily agree that decisions to develop certain technologies and not others are often driven by societal factors such as politics, economics, and cultural values. In other words, the development of certain technologies instead of others is value laden. Fundamental, value-based assumptions drive the quest for more and better technologies spawned by ever-increasing scientific knowledge. One assumption is that science and technology result in progress. A second is that we should dominate nature.

What most people are not conscious of is that science is influenced by these same societal factors of politics, economics, and cultural values. People view science as an objective search for the truth, and in many ways it is. Adhering to the scientific method of systematic observation, hypothesis generation, and experimentation, today's researchers attempt to establish their hypotheses objectively and gather data through careful, methodical experimentation.

But scientists were raised in a society, and that society has shaped the way they view the world. The scientific method cannot shield scientists from the social context in which they conduct research. Society has molded their sense of values and goals since the day they were born. The narrower social context of the scientific world in which they function imposes pressures and erects powerful filters for selectively viewing the world. Only certain questions from the universe of available questions are asked; data are interpreted through these same filters (see the section on Dr. Barbara McClintock in chapter 2 for a real-world example of this phenomenon).

So, while the methodology scientists use to answer questions may be "objective," the questions asked and the interpretation of results are not value free.

Science, Technology, and Responsibility

Science and technology have altered the characteristics of societies throughout history. Some of the forces they have exerted have been revolutionary, and some have been minor. Different aspects of life—social, cultural, ecological, economical—have been affected, and different groups have been affected in different ways. Often, one group has benefited at another's expense.

Because of the extraordinary degree to which our environment and we ourselves are shaped by science and technology, individuals must understand the interrelationship between scientific understanding and technological change and must respect the impact technologies can have on the quality of life. While scientific investigation and technological advance may be the province of a few members of society, all citizens in a democracy should be prepared to consider the societal issues spawned by a science and its derivative technologies. In the absence of public understanding and interest, the stage is set for a handful of knowledgeable individuals to direct the course of scientific investigation and technological change and therefore, over time, society's future.

The effects of science and technology can be difficult to predict. It often seems that with every intended benefit comes an unintended cost. Daily we come face to face with the conflicts and contradictions brought about by technological innovation: running water may not be fit to drink, medical advances increase the average life span as well as the national debt. Many people question whether the benefits of technology are worth the price. They wonder whether the assumption that science and technology give rise to progress is valid.

Science and technology have given humans tremendous power over the natural world. With that power comes a responsibility that all of us must share: to make far-sighted and judicious choices that ensure we do not misuse our power.

Barry Commoner, a noted environmentalist, has applied the adage "There is no such thing as a free lunch" to the inevitable price we will pay if we continue to ignore the environmental costs of human activities. With every technological advance designed to control our environment, we attempt to remove ourselves from the natural order and live outside the boundaries Mother Nature has established. We want to live by our own rules and not nature's. Whether we can manage to have it our way, and at what cost, remains to be seen.

Societal Issues

About the Activities

Introduction

We cannot overestimate the importance of conducting classroom activities that encourage students to think about the complex relationship between science, technology, and society. While very few students will pursue biology as a career, *all* of them will be citizens in a society shaped, for better or worse, by science and technology. If they understand and respect the power of science and technology, perhaps they will direct these forces in ways that maximize benefits and minimize costs.

More than emotion and concern is required for guiding wise development of technologies. Instead, thoughtful development hinges on each individual's ability to identify the core issues, gather solid facts, subject the facts to thoughtful analysis, identify options, set priorities, be open-minded, detach from self-serving perspectives, and compromise with others who may have different priorities. It is not important that your students learn *what* to think about science and technology. It is critical that they learn *how* to think about them.

Objectives

After reading this chapter and completing the activities you select, students will

- Realize the impact technologies have on their daily life;

- Understand the interrelatedness of science, technology, and society;

- Understand that a technology has many first-order and second-order effects, of which some are intended and others are unintended.

Suggested formats

Activities on the interrelationship between science, technology, and society are by their very nature somewhat limited. There are no laboratory techniques to perform or basic science concepts to learn. Instead, your task is to teach essential mental skills: identification of issues, fact-finding, objective analysis, identification of options, priority setting, detachment, and compromise.

A variety of formats provide opportunities to practice these skills: case studies, research papers, debate, risk-benefit analysis, problem solving, decision making, role playing, brainstorming, team-based research, and classroom discussions. You will know which format is best for your class.

Methodology

Irrespective of the format chosen, it is critical that you help students learn how to

- Consider one issue at a time;
- Stay focused on the issue;
- Gather accurate information (too often, students believe that if something is in print, it must be true);
- Distinguish science from pseudoscience;
- Assess information critically and unemotionally;
- Draw conclusions;
- Discuss issues fairly and unemotionally, listening to each other so that they acquire the greatest understanding.

If the format you choose involves class discussions or debates, have the class establish ground rules. Examples of ground rules are provided in chapter 32, in which we describe the decision-making model for bioethical issues. Consider posting these ground rules somewhere in the classroom so that students can remind themselves and each other of the appropriate methodology.

Suggested activities

1. Have students make a list of *all* of the technologies that influence their activities during a day. Also have them list all the ways in which they interacted directly with the natural world on that day.

2. Have each student or student team pick a major technology—antibiotics, electricity, television—and thoroughly analyze its development.
   - What scientific discoveries permitted the development of this technology, and how did they come about?
   - What resources were required to develop the technology?
   - What sort of government bodies oversaw its development?
   - Did that oversight help or hinder development?

   Now ask the students to thoroughly analyze the effects (first order and second order; predicted and unintended; environmental, political, economic, cultural) of that technology in hindsight.
   - What were its intended effects?

   - Has it achieved that purpose?

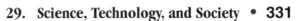

Societal Issues

- What were the broader changes in society caused by this technology that we could have predicted would occur?

- What were the unintended consequences we could not have predicted?

- How has the technology changed human activity, our socioeconomic structure, our environment, our relationship with each other, our government, our relationship with other nations?

- What would life be like without this technology?

- What other technologies did it give rise to?

3. Have students or student teams pick an emerging technology or one application of an all-encompassing technology like biotechnology (see chapter 1). Ask them

   A. To predict what some of the intended and unintended consequences of its development will be.

   - Who will be affected by it, and in what ways?

   - What will be the environmental, ethical, and socioeconomic impacts of this technological development?

   - What sort of policies, laws, and regulations should be formulated to oversee its developments?

   B. To explain what scientific knowledge, existing technology, or other resources are required in order for this technology to be fully developed.

   C. To prioritize the technologies that have been discussed (both individually and then working as a group).

   - Which should be developed first?
   - Which should be supported with the most federal funding?

4. Have students discuss the factors that should be used in evaluating technologies. Historically, technologies have been evaluated by three criteria:

- Can it be done?
- Will it sell?
- Is it safe?

Ask your students if they believe these criteria are sufficient. Would they add others? What would their criteria be? How would they implement these new criteria?

5. Assign each student a technology, and ask the student to determine whether that technology has increased our options.

Technologies are supposed to increase our options. You can now choose to be vaccinated against whooping cough or risk getting the disease. You can choose to fly or drive to Atlanta. But do technologies actually increase or decrease our options?

In a book by a wonderful children's writer, Edith Nesbit, a young boy awakens in a city in which every toy, building, or "city" he has ever built has become life-size. He lives surrounded by gigantic and tangible manifestations of his creative mind, in a fantasy world where every wish is granted. There is only one law. If you wish for a piece of machinery, you must use it for the rest of your life.

The parallel between Nesbit's world and ours is obvious. How many times has our "playing" resulted in enlarged "toys" on which we now depend and have no choice but to use? (What is particularly interesting is that Nesbit was born in the mid-1800s, years before we became immersed in technology the way we now are!)

6. Have students visit a senior citizen's center and ask the people there what life was like before antibiotics, the pill, washing machines, refrigeration.

7. Write a brief essay explaining the following quote by Edward Wenk:

"Those who control technology control the future."

Societal Issues

Science, Technology, and Society

Introduction

Today's world has been profoundly and unalterably shaped by advances in science and technology. Some of the ways in which science and technology have affected our lives are immediately obvious: airplanes, automobiles, television, computers, and microwave ovens. Others are so subtle that we fail to recognize them as benefits of science and technology. Instead, we take them for granted as a "natural" part of life. The clothes you're wearing, the chair you're sitting in, the pen you may be holding, the paper this book is printed on, the food you had for lunch are all products of science and technology.

If you doubt the extent to which technology has penetrated every tiny facet of your life, pay attention to your actions during the rest of the day and take note of all of the ways technology, both simple and sophisticated, has altered your world. As you are paying attention to the ways technology impinges on your life, also look at the flip side of the coin. How often do you come face to face with nature? In those encounters with nature, how many times do you experience a love for nature, and how many times do you wish it would go away and leave you alone? Thousands of years ago, we rejected the idea of living with nature. Our distaste for the adversities nature indifferently dispenses has driven our constant quest for new technologies. We use all of these technologies for a similar purpose: to change our environment so that the natural world suits us better.

Science and technology not only alter our environments—they also shape us. Our view of the world, the questions we ask, our sense of values, and the way we think are the products of life in a world shaped by scientific discoveries and technological innovations.

Has it always been this way? Yes and no. Technologies have always influenced human societies, our mental framework, and our relationship to the natural world. Today, however, technological advances are more per-vasive, and their effects are more powerful. Also, the changes they bring are occurring at an increasingly rapid rate. We can trace the increased presence, power, and rate of technological change to the role scientific understanding now plays in propelling technological developments.

Which Came First, Science or Technology?

Initially, technology drove science. People developed technologies on a trial-and-error basis, using the manual skills required in crafts such as clock making, lens grinding, and metal forging. Early scientists used those technologies to observe and question natural phenomena and thus to increase their understanding of the world around them. As their understanding of the natural world broadened and deepened, the relationship between science and technology shifted.

Equipped with a more sophisticated understanding of the world they sought to control, people utilized scientific understanding to drive technological change. They replaced the trial-and-error approach to technological advancement with the methodical, experimental approach that is characteristic of scientific investigation. Buttressed by sound science, the ensuing technological changes were more forceful, and the pace of technological change quickened. With ever more powerful technologies at their disposal, scientists were able to probe deeper and deeper into the causes and effects of natural phenomena.

Because the relationship between scientific understanding and technological innovation is circular and not linear, reciprocal and not hierarchical, we can only expect the pace of science and technological change to accelerate in the 21st century. A constantly accelerating rate of increase in scientific knowledge and technological achievement carries with it important implications for our species.

The Nature of Science and Technology

Many people equate science with technology, but the two differ in a number of fundamental ways. A few of the ways in which science differs from technology are provided in Table 29.1.

Despite their differences, science and technology converge in subtle ways that may not be apparent to people outside the scientific and engineering communities. Most people would readily agree that decisions to develop certain technologies and not others are often driven by societal factors such as politics, economics, and cultural values. In other words, the development of certain technologies instead of others is value laden. Fundamental, value-based assumptions drive the quest for more and better technologies spawned by ever-increasing scientific knowledge. One assumption is that science and technology result in progress. A second is that we should dominate nature.

What most people are not conscious of is that science is influenced by these same societal factors of politics, economics, and cultural values. People view science as an objective search for the truth, and in many ways it is. Adhering to the scientific method of systematic observation, hypothesis generation, and experimentation, today's researchers attempt to establish their hypotheses objectively and gather data through careful, methodical experimentation.

But scientists were raised in a society, and that society has shaped the way they view the world. The scientific method cannot shield scientists from the social context in which they conduct research. Society has molded their sense of values and goals since the day they were born. The narrower social context of the scientific world in which they function imposes pressures and erects powerful filters for selectively viewing the world. Only certain questions from the universe of available questions are asked; data are interpreted through these same filters (see the section on Dr. Barbara McClintock in chapter 2 for a real-world example of this phenomenon).

So, while the methodology scientists use to answer questions may be "objective," the questions asked and the interpretation of results are not value free.

Science, Technology, and Responsibility

Science and technology have altered the characteristics of societies throughout history. Some of the forces they have exerted have been revolutionary, and some have been minor. Different aspects of life—social, cultural, ecological, economical—have been affected, and different groups have been affected in different ways. Often, one group has benefited at another's expense.

Because of the extraordinary degree to which our environment and we ourselves are shaped by science and technology, individuals must understand the interrelationship between scientific understanding and technological change and must respect the impact technologies can have on the quality of life. While scientific investigation and technological advance may be the province of a few members of society, all citizens in a democracy should be prepared to consider the societal issues spawned by a science and its derivative technologies. In the absence of public understanding and interest, the stage is set for a handful of knowledgeable individuals to direct the course of scientific investigation and technological change and therefore, over time, society's future.

The effects of science and technology can be difficult to predict. It often seems that with every intended benefit comes an unintended cost. Daily we come face to face with the conflicts and contradictions brought about by technological innovation: running

Table 29.1 Fundamental differences in science and technology

Science	Technology
The search for knowledge	The practical application of knowledge
A way of understanding ourselves and the world	A way of adapting ourselves to the world
A process of asking questions and providing answers or broad, generalized explanations	A process of providing solutions to human problems to make our lives easier
Looks for order or patterns in nature	Looks for ways to control nature
Evaluated by how well the data support the answer, model, or theory	Evaluated by how well it works
Limited by our ability to collect relevant data	Limited by financial costs and risks
Discoveries give rise to technological advances	Advances give rise to scientific discoveries

Societal Issues

water may not be fit to drink, medical advances increase the average life span as well as the national debt. Many people question whether the benefits of technology are worth the price. They wonder whether the assumption that science and technology give rise to progress is valid.

Science and technology have given humans tremendous power over the natural world. With that power comes a responsibility that all of us must share: to make far-sighted and judicious choices that ensure we do not misuse our power.

Barry Commoner, a noted environmentalist, has applied the adage "There is no such thing as a free lunch" to the inevitable price we will pay if we continue to ignore the environmental costs of human activities. With every technological advance designed to control our environment, we attempt to remove ourselves from the natural order and live outside the boundaries Mother Nature has established. We want to live by our own rules and not nature's. Whether we can manage to have it our way, and at what cost, remains to be seen.

Suggested Activities

Your instructor will request that you complete one or more of the following assignments.

1. Make a list of *all* of the technologies that influence your activities during a day. On the same day, list all the ways in which you interact directly with the natural world.

2. Pick a major technology—antibiotics, electricity, television—and thoroughly analyze its development and its predicted and unpredicted effects on society.

3. Pick an emerging technology or one application of an all-encompassing technology like biotechnology (see chapter 1), and describe what scientific knowledge, existing technology, or other resources are required for this technology to be fully developed. In addition, try to predict what some of the intended and unintended consequences of its development will be.

4. What factors should be used in evaluating technologies. Historically, technologies have been evaluated by three criteria:
   • Can it be done?
   • Will it sell?
   • Is it safe?

   Are these criteria sufficient? If not, what others would you add?

5. Do you think technologies increase or decrease our options? Explain.

6. Write a brief essay explaining the following quote by Edward Wenk:

   "Those who control technology control the future."

Societal Issues

Weighing Technology's Risks and Benefits **30**

Introduction

Day in and day out, consciously or unconsciously, we take risks. We ride in cars, cross streets, and play sports without thinking twice about the risks involved, because we have "decided," usually subconsciously, that the risks are outweighed by the benefits.

Sometimes our subconscious risk-benefit analysis makes good sense. At other times, our decisions seem to involve no rational thought. We accept huge risks and worry about trivial ones—cigarette smokers protest the building of nuclear power plants; people risk pregnancy and AIDS by having unprotected sex and then worry about pesticide residues on their food—yet many more people are harmed by smoking and unprotected sex than by accidents at nuclear power plants or pesticides on food. Our choices are irrational because we sometimes make decisions about acceptable and unacceptable risks based not on facts but on emotions.

Emotions and Unconscious Risk Assessment

What are some of the factors that trigger emotions, interfering with our ability to make rational assessments of true risks?

Voluntary versus involuntary

If the risk is one we consciously decide to take (a voluntary risk, like smoking), we accept a much higher level of risk than if we feel we have had no choice in the matter (an involuntary risk, such as the location of a nuclear power plant).

Control versus no control

We are more fearful of risky situations over which we have no control. Many people choose to ride in their cars rather than fly in airplanes because they fear flying, even though more people are hurt or killed in

auto accidents than in plane crashes. In spite of the evidence, we perceive the risks of flying as greater than the risks of driving because we are not in control of the plane.

Familiar versus unfamiliar

Our view of risk also varies with our degree of familiarity with the risk. We have come to accept the large risks involved in driving a car, sunbathing, and drinking alcohol. A new risk elicits our concerns and avoidance not because the risk is greater but because it is unfamiliar.

Natural versus man-made

Another factor that affects our concept of the degree of risk is whether the risk is natural or man-made. We tend to view nature as more benevolent in spite of all the evidence to the contrary. Earthquakes, typhoons, hurricanes, and tornadoes take thousands of lives and cause billions of dollars in damage yearly. In underdeveloped countries, millions of children die every year from diarrhea resulting from microbial contamination of food and water, a product of nature. Man-made products and processes prevent this same tragedy from occurring in developed nations. How can we ignore this and other, similar examples and continue to remain skeptical of man-made products as we unthinkingly embrace natural products as safe?

Our confusion on the issue of natural and man-made risks deserves additional attention because an objective analysis of biotechnology requires a clear understanding of the pitfalls in equating nature with risk-free or man-made with risks.

Natural and Man-Made Risks

Some of the most instructive examples of problems caused by an irrational preference for natural products and an irrational fear of man-made products come from issues surrounding food safety. Even though food safety is very important to today's consumer, little has been done to place the issue of the relative risks of man-made versus natural compounds in perspective.

Natural toxins

Many people are frightened by the use of agricultural chemicals on our food crops. They have heard that these chemicals harm our health because they are "toxic" and "cancer causing." To avoid these chemicals, many consumers purchase "organically grown" food at much higher prices.

Some man-made chemicals, particularly those we first developed in the 1940s, are highly toxic. But are all man-made agricultural chemicals more toxic than substances we readily ingest like coffee and soft drinks? No. Even more to the point, are man-made chemicals we use in food production toxic while natural chemicals contained in these same food products are benign? No. In fact, some of the naturally occurring compounds in the plant foods we eagerly consume as "healthful" are more toxic than some of the agricultural chemicals we use for pest control.

How can it be that the fruits and vegetables we think of as healthy make their own toxic chemicals? Take a moment and think about the biology of plants. Plants did not evolve to serve as food for humans or any other organism. For millions of years they have done their very best *not* to be eaten. Stuck in one place, how can a plant defend itself? Chemicals. Plants make hundreds of different chemicals to ward off would-be predators. Biologists call these defense chemicals "secondary plant compounds."

For many years, crop scientists have directed much of their plant breeding to decreasing or eliminating these natural toxins. Without such selective breeding, many of the plants Mother Nature cooked up would not be fit for human consumption. In spite of years of work to strip our crop plants of secondary plant compounds, many of the plants we consume still contain toxins and carcinogens (cancer-causing agents), as defined by federal regulatory agencies such as the Food and Drug Administration and the Environmental Protection Agency. Does this mean we should stop consuming fruits and vegetables? Of course not. Most of these plant toxins occur in concentrations that are so low they have no effect on our health (see the first box in the *Student Activity* pages). The benefits of eating fruits and vegetables far outweigh the negligible to nonexistent risks posed by naturally occurring toxins. Does that mean all man-made chemicals are so safe they can be ignored as health risks? No. Our point is only that we must keep issues of chemical toxicology and food safety in perspective.

Natural carcinogens

According to Bruce Ames, the well-known biochemist who developed the most widely used test for measuring the cancer-causing potential of chemicals, in any given meal, an individual consumes about 100 to 150 natural carcinogens and 10,000 times more natural carcinogens than man-made carcinogens. Cabbage alone has over 40 natural carcinogens!

To shed more light on the relationship between food safety and natural versus man-made chemicals, here are the carcinogenic potentials of some familiar compounds. The higher the number, the greater the carcinogenic potential.

Substance	Carcinogenic potential
Wine	5.0
Beer	3.0
Mushrooms	0.1
Peanut butter	0.03
Chlorinated water	0.001
PCBs	0.0002

You may be surprised by these cancer-causing potentials. You probably did not expect peanut butter to be 100 times more carcinogenic than polychlorobiphenyls (PCBs), because you have heard that PCBs are very dangerous. Yet no one has warned you about eating peanut butter. How could you have gotten such misleading information on something as important to you as the food you eat? The answer is probably found in the information source you use.

Most of us get our information about science and technology from newspapers, magazines, and television. Most representatives of the media want to convey accurate information to the public, but a reporter writing a story on science and technology may not know where to get solid scientific information, how to assess the scientific merit of research, or how to interpret scientific data. These understandable shortcomings are sometimes made worse by other constraints. Newspapers, magazines, and television stay in business by selling their products. Fear sells; solid science doesn't.

Fear may sell newspapers, but it also ends up costing consumers a lot. Unnecessary fears may drive consumers to spend more money on products with no additional benefits. These same unfounded fears sometimes inspire Congress and regulatory agencies to pass certain regulations that do nothing to increase the protection of the public but much to increase the cost of goods and services (see the second box in the *Student Activity* pages). The relationship between in-

Societal Issues

creased fear and increased cost is best illustrated by an example.

The case of decaffeinated coffee

A few years ago, there was much ado about the use of methylene chloride to decaffeinate coffee. Members of the media reported that methylene chloride was a carcinogen in rats when inhaled. They neglected to add that it has a much lower carcinogenic potential when ingested. This distinction is very relevant, because in the case of decaffeinated coffee, the only variable coffee drinkers need be concerned with is the carcinogenic potential when ingested.

Many consumers changed their buying habits as a result of the media's attention to this issue. To avoid a perceived risk of getting cancer, coffee drinkers were more than willing to spend extra money for coffee decaffeinated by a different method, a process based on water. Those who couldn't afford the extra expense probably worried about what they were doing to their health.

Coffee itself has over 300 chemical compounds, many of which are more toxic and more carcinogenic than methylene chloride. In order to ingest enough methylene chloride to reach the level shown to cause cancer in rats, a person would have to drink 50,000 cups of coffee a day. In those 50,000 cups, natural coffee carcinogens would occur in far greater amounts than methylene chloride. More important, a person would die from drinking 50,000 cups of *water* a day. Long before the 50,000th cup was consumed, a person's kidneys would shut down.

Again, the message is not that man-made is safe and natural is unsafe. The situation is not that simple. The moral of the story is that all of us work from misconceptions derived from inaccurate or misleading information. Unfortunately, misinformation made worse by fuzzy thinking doesn't yield smart choices very often. How can we do a better job of making choices involving food safety risks or any other risks?

Facts and Conscious Risk Assessments

What drives your assessment of risks—emotions or facts—may or may not seem particularly important to you, yet it is. The risks you are willing to assume and the experiences or products you avoid because of faulty assumptions and misinformation affect the quality of your life and the lives of those around you. In today's world, people suffer from great anxiety

about the risks involved in simply living. Some of the fears that grip us are real; others are not. Wouldn't it be useful if we had a mechanism for helping us determine the difference? We do.

Toxicity versus hazards

First, to better understand and evaluate discussions of risks, particularly those involving food safety, we should differentiate between toxicity and hazard. Toxicity can be defined as the capacity of a substance to do harm or injury of any kind under any conditions. Given the definition of toxicity, anything can be toxic. In the decaffeinated coffee example above, water is toxic.

Contrast this to the definition of hazard. A hazard is the relative probability that harm will result when a particular quantity of a substance is used in a particular manner. Given these two definitions, it becomes clear that meaningful assessments of food safety, or the safety of any substance, must be based on the hazard a substance poses and not its toxicity. If we use this principle, we will be one step closer to assessing risks objectively.

Risk-benefit analysis

Using this concept of hazard, scientists have developed a process for assessing risks and benefits that attempts to maximize the contribution of factual information and minimize the contribution of emotion to the analysis: risk-benefit analysis. Here's how it works.

Assessing Risks

First of all, to analyze risks as thoroughly and unemotionally as possible, scientists conduct a risk assessment. The first step is defining what a risk actually is.

Risk is the probability of loss or injury, or

$$\text{risk} = \text{hazard} \times \text{exposure}$$

In other words, a hazard, which can be viewed as a source of danger, becomes a risk only if you are exposed to it. Your probability of loss or injury increases in relationship to the increase in hazard and/or exposure.

Once you have clearly defined the concept of risk, you are ready to begin a risk assessment. Assess the amount of risk involved in a particular situation by answering these three questions:

1. What could go wrong? (Identify the specific risk.)
2. How likely is it to happen? (Assess the probability of the risk occurring.)

3. How harmful is it if it does occur? (Assess the consequences.)

A simple formula for approximating the amount of risk is

$$\text{amount of risk} = \frac{\text{hazard} \times \text{exposure}}{\text{safeguards}}$$

If you are able to establish effective safeguards, the amount of risk decreases dramatically.

These conceptual formulas illustrate the first steps in using rational risk assessment methodology to get a handle on real, not perceived, risks. By no means are these steps perfect or devoid of subjectivity, but they provide a starting point for rational discussion of risk.

Why worry with all of this? Why not just overestimate risk to make sure we're safe? As the example of decaffeinated coffee demonstrates, responding to risks that don't exist can limit our options, increase the money we spend, and raise the costs of production of goods because of excessive regulatory burden. It is possible to be overly cautious. For a relative ranking of the risks of familiar activities and products, see Table 30.1 in the *Student Activity* pages.

Assessing Benefits

Assessing risks is a simple task compared to assessing benefits. Benefits are often very difficult to identify in the abstract. We need to experience them. Beyond the obvious and intended benefits of electric lights over gaslights, we would never have foretold all of the benefits we have derived from electrification of our homes. We would never have imagined other unintended, but direct, benefits of electrification: electric washing machines, irons, hot-water heaters, air conditioners, refrigerators, home computers, and VCRs. Nor could we have predicted the secondary, indirect effects of some of the unintended effects: homes in the desert and frozen north, the Internet, home shopping, more free time, the disappearance of household servants, more women in the workforce, and, as a result, a need for day-care centers.

In addition to being difficult to predict, many benefits are very subjective. They are based on emotions, values, and ethical principles, and they therefore vary from person to person. Some of the factors we use to measure benefits include the following:

- Saves lives
- Improves health
- Saves money
- Solves problems
- Makes life more pleasant or easier
- Increases emotional well-being

Putting Risks and Benefits Together

Defining and assessing the amount of risk and then comparing the risk to the perceived benefits and their probabilities of occurring are the components of a thoughtful, methodical analysis for assessing risks. Using this methodology, we are in a much better position to make judicious choices on issues that involve real or perceived risk. Yet that does not mean our problems are over.

Once we have conducted our own risk-benefit analysis of a certain issue and arrived at a decision regarding the acceptability of a risk, we may find that the person next to us has arrived at a different conclusion. This difference of opinion is to be expected, given the difficulties in assessing the amount of risk and the subjective and emotional nature of benefit analysis.

You may think that if the decision involves an individual's acceptance or rejection of personal risk, differences of opinion on the costs and benefits of individual behavior should present no problem. In other words, if I want to accept the risks when I skydive or ride a motorcycle without a helmet, it's my business. Once again, it's not that simple. When does someone's personal decision about the acceptability of a personal risk infringe upon the right of others?

People who drive drunk may feel comfortable with their own risk-benefit analysis of their behavior, but other drivers on the road may not. Given that example, most people would probably come to a similar conclusion regarding individual rights versus the rights of other members of society. Driving on public roads is a public matter made possible with public money. An individual's right to accept the risk of driving drunk is secondary compared to the public's right to use roads without fear of drunken drivers.

What about private matters like safe sex and AIDS? Surely the risks of having sex are relevant only to the people involved. If people risk contracting AIDS by having unprotected sex, isn't that their business and only their business? How can that choice infringe on the rights of others?

This may not be as simple and straightforward as it first appears or as we wish it were. As the cost of providing care for a growing population of AIDS victims increases, health insurance costs will increase. That cost increase will be borne by many people in ways that seem completely unrelated to AIDS and unprotected sex. For example, today, many employers pay for the health insurance of their employees. As health insurance costs increase, employers must pay more money to insure their workers. The company must then compensate for this increase either by decreasing costs in other areas or by increasing income. Cutting jobs is one way to cut costs. To increase income, companies usually raise the prices they charge for goods and services. So, an increase in health insurance costs could mean fewer jobs and/or higher prices. Should everyone pay the price for those who have chosen to have unprotected sex and, as a result, been infected with the human immunodeficiency virus?

Issues of individual rights are never as simple as we would like. As members of society, we have both rights and responsibilities to ourselves and each other. Finding the proper balance between the two can be very, very difficult.

About the Activities

As we said in the introduction to the activities in chapter 29, it is not important that your students learn *what* to think about assessing the risks and benefits of technological change, but it is critical that they learn *how* to think about this complex issue. Refer to that section for more comments on the value of these types of classroom activities as well as general statements on possible formats and methodologies applicable to activities that teach risk assessment skills.

Objectives

After reading this chapter and performing some of the activities listed below, students will

- Understand how emotions contribute to perceptions of risks;
- Be able to differentiate toxicity from hazard;
- Understand the basic steps of a risk-benefit analysis;
- Recognize that individual choices sometimes harm many people in ways that seem unrelated to the choices;
- Be better able to evaluate media coverage of the risks and benefits of technologies.

Suggested Activities

1. Select any familiar, pervasive technology we depend on, and have the students walk through the steps in a risk-benefit analysis: identifying the risks, assessing the probability of each risk occurring, assessing the consequences, analyzing the benefits.

2. Go back in time and evaluate that same technology before its widespread implementation. What risks and benefits would the students have pre-

dicted prior to that technology's development? What is the relationship between predicted versus actual risks and benefits?

3. Select a future application of biotechnology or another emerging technology and have students conduct a risk-benefit analysis. Pay particular attention to their assessment process, not the results. What information sources would they use to analyze a *future* development? How applicable is this information?

4. Select a risk that is considered a personal risk, such as skydiving, cigarette smoking, or using drugs, and have students analyze the societal ramifications of individual risk-taking. What is the appropriate role of the individual, society, or government in those cases?

5. Ask students to analyze the media coverage of a recent scientific discovery or technological advance that was reported in the newspaper, on radio, or on television. (They should record television and radio reports.) Compare and contrast the media coverage with the primary literature on the subject. Many scientific articles on biotechnology that receive popular coverage are published in the journals *Science* and *Nature,* which are available in many public libraries. A less formal, but still informative, assignment is to ask students to monitor media coverage of technology. Is it balanced? Is there mention of both the benefits and the risks of the technology in question?

6. Have students develop a case study similar to the one provided for decaffeinated coffee. For example, a few years ago the addition of sodium nitrite to processed food was scrutinized and criticized by the media because under certain conditions, ingestion of sodium nitrite causes the body to produce carcinogenic compounds called nitrosamines. The value of sodium nitrite is that it effectively inhibits the growth of *Clostridium botulinum,* the bacterium responsible for botulism, in processed foods. Students could conduct a risk-benefit analysis of the addition of sodium nitrite to processed food. This case study would entail conducting research on topics such as *C. botulinum* and the toxicity of the poison it secretes (botulin toxin) compared to the toxicity-carcinogenicity of nitrosamines; the rates of occurrence of botulism prior to adoption of the practice of adding sodium nitrite; alternative means food processors use to inhibit the growth of *C. botulinum;* the effectiveness and risks of these methods; the percentage of cancers that result from ni-

trosamines or food additives in general; the conditions under which sodium nitrite combines with secondary amines to produce nitrosamines; and the frequency of occurrence of these conditions. Other informative case studies would be the Alar-apples controversy, chlorinated water, and food irradiation. You might consider conducting these case studies in conjunction with the home economics teacher.

7. Have students write a short essay explaining either of the following quotes:

"To be alive is to be at risk."

Daniel Koshland, *Science* 1987

"Only the dose makes the poison."

Paracelsus, 16th century

Selected Readings

Ames, Bruce, et al. 1987. Ranking possible carcinogenic hazards. *Science* 236:271–279.

Covello, Vincent T. 1987. Risk comparisons, risk communication, and public perceptions of risk: issues and approaches. *Journal of Environmental Psychology,* July 1987.

Jones, Julie. 1992. *Food Safety.* Eagan Press, St. Paul, Minn.

Lave, Lester B. 1987. Health and safety risk analyses: information for better decisions. *Science* 236:291–295.

Schwig, R., and W. A. Albers (ed.). 1980. *Societal Risk Assessment: How Safe Is Safe Enough?* Plenum Press, New York.

Slovic, Paul. 1987. Perception of risk. *Science* 236:280–285.

Wilson, Richard, and E. A. C. Crouch. 1987. Risk assessment and comparisons: an introduction. *Science* 236:267–270.

Societal Issues

Weighing Technology's Risks and Benefits

Introduction

Day in and day out, consciously or unconsciously, we take risks. We ride in cars, cross streets, and play sports without thinking twice about the risks involved, because we have "decided," usually subconsciously, that the risks are outweighed by the benefits.

Sometimes our subconscious risk-benefit analysis makes good sense. At other times, our decisions seem to involve no rational thought. We accept huge risks and worry about trivial ones—cigarette smokers protest the building of nuclear power plants; people risk pregnancy and AIDS by having unprotected sex and then worry about pesticide residues on their food—yet many more people are harmed by smoking and unprotected sex than by accidents at nuclear power plants or pesticides on food. Our choices are irrational because we sometimes make decisions about acceptable and unacceptable risks based not on facts but on emotions.

Emotions and Unconscious Risk Assessment

What are some of the factors that trigger emotions, interfering with our ability to make rational assessments of true risks?

Voluntary versus involuntary

If the risk is one we consciously decide to take (a voluntary risk, like smoking), we accept a much higher level of risk than if we feel we have had no choice in the matter (an involuntary risk, such as the location of a nuclear power plant).

Control versus no control

We are more fearful of risky situations over which we have no control. Many people choose to ride in their cars rather than fly in airplanes because they fear flying, even though more people are hurt or killed in

auto accidents than in plane crashes. In spite of the evidence, we perceive the risks of flying as greater than the risks of driving because we are not in control of the plane.

Familiar versus unfamiliar

Our view of risk also varies with our degree of familiarity with the risk. We have come to accept the large risks involved in driving a car, sunbathing, and drinking alcohol. A new risk elicits our concerns and avoidance not because the risk is greater but because it is unfamiliar.

Natural versus man-made

Another factor that affects our concept of the degree of risk is whether the risk is natural or man-made. We tend to view nature as more benevolent in spite of all the evidence to the contrary. Earthquakes, typhoons, hurricanes, and tornadoes take thousands of lives and cause billions of dollars in damage yearly. In underdeveloped countries, millions of children die every year from diarrhea resulting from microbial contamination of food and water, a product of nature. Man-made products and processes prevent this same tragedy from occurring in developed nations. How can we ignore this and other, similar examples and continue to remain skeptical of man-made products as we unthinkingly embrace natural products as safe?

Our confusion on the issue of natural and man-made risks deserves additional attention because an objective analysis of biotechnology requires a clear understanding of the pitfalls in equating nature with risk-free or man-made with risks.

Natural and Man-Made Risks

Some of the most instructive examples of problems caused by an irrational preference for natural products and an irrational fear of man-made products come from issues surrounding food safety. Even though food safety is very important to today's consumer, little has been done to place the issue of the relative risks of man-made versus natural compounds in perspective.

Natural toxins

Many people are frightened by the use of agricultural chemicals on our food crops. They have heard that these chemicals harm our health because they are "toxic" and "cancer causing." To avoid these chemicals, many consumers purchase "organically grown" food at much higher prices.

Some man-made chemicals, particularly those we first developed in the 1940s, are highly toxic. But are all man-made chemicals more toxic than substances we readily ingest like coffee and soft drinks? No. Even more to the point, are man-made chemicals we use in food production toxic while natural chemicals contained in these same food products are benign? No. In fact, some of the naturally occurring compounds in the plant foods we eagerly consume as "healthful" are more toxic than some of the agricultural chemicals we use for pest control.

How can it be that the fruits and vegetables we think of as healthy make their own toxic chemicals? Take a moment and think about the biology of plants. Plants did not evolve to serve as food for humans or any other organism. For millions of years they have done their very best *not* to be eaten. Stuck in one place, how can a plant defend itself? Chemicals. Plants make hundreds of different chemicals to ward off would-be predators. Biologists call these defense chemicals "secondary plant compounds."

For many years, crop scientists have directed much of their plant breeding to decreasing or eliminating these natural toxins. Without such selective breeding, many of the plants Mother Nature cooked up would not be fit for human consumption. In spite of years of work to strip our crop plants of secondary plant compounds, many of the plants we consume still contain toxins and carcinogens (cancer-causing agents), as defined by federal regulatory agencies such as the Food and Drug Administration and the Environmental Protection Agency. Does this mean we should stop consuming fruits and vegetables? Of course not. Most of these plant toxins occur in concentrations that are so low they have no effect on our health (see the box). The benefits of eating fruits and vegetables far outweigh the negligible to nonexistent risks posed by naturally occurring toxins. Does that mean all man-made chemicals are so safe they can be ignored as health risks? No. Our point is only that we must keep issues of chemical toxicology and food safety in perspective.

Natural carcinogens

According to Bruce Ames, the well-known biochemist who developed the most widely used test for measuring the cancer-causing potential of chemicals, in any given meal, an individual consumes about 100 to 150 natural carcinogens and 10,000 times more natural carcinogens than man-made carcinogens. Cabbage alone has over 40 natural carcinogens!

To shed more light on the relationship between food safety and natural versus man-made chemicals, here are the carcinogenic potentials of some familiar com-

Societal Issues

Parts per Million and Parts per Billion

Part of the difficulty in accurately evaluating statements about the toxicity of substances and the safety of food stems from our extraordinary ability to measure chemical compounds. Our measuring techniques can detect quantities so small that they have no real significance for health and safety. We often hear the expressions "parts per million" and "parts per billion." We can now measure substances at the level of 1 part per trillion! What do these values really mean?

How much is a billion?
A billion seconds ago, World War II ended.
A billion minutes ago, St. Paul was writing the Epistles.

What is 1 part per million?
1 gram per ton
1 mouthful in a lifetime
1 drop in 1,000 quarts of water

How much is 1 part per billion?
1 inch in 16,000 miles
1 minute in 2,000 years
1 cent in 10 million dollars
1 drop in a million quarts of water

How much is 1 part per trillion?
1 inch in 16,000,000 miles

---

pounds. The higher the number, the greater the carcinogenic potential.

Substance	Carcinogenic potential
Wine	5.0
Beer	3.0
Mushrooms	0.1
Peanut butter	0.03
Chlorinated water	0.001
PCBs	0.0002

You may be surprised by these cancer-causing potentials. You probably did not expect peanut butter to be 100 times more carcinogenic than polychlorobiphenyls (PCBs) because you have heard that PCBs are very dangerous. Yet no one has warned you about eating peanut butter. How could you have gotten such misleading information on something as important to you as the food you eat? The answer is probably found in the information source you use.

Most of us get our information about science and technology from newspapers, magazines, and television. Most representatives of the media want to convey accurate information to the public, but a reporter writing a story on science and technology may not know where to get solid scientific information, how to assess the scientific merit of research, or how to interpret scientific data. These understandable shortcomings are sometimes made worse by other constraints. Newspapers, magazines, and television stay in business by selling their products. Fear sells; solid science doesn't.

Fear may sell newspapers, but it also ends up costing consumers a lot. Unnecessary fears may drive consumers to spend more money on products with no additional benefits. These same unfounded fears sometimes inspire Congress and regulatory agencies to pass certain regulations that do nothing to increase the protection of the public but much to increase the cost of goods and services (see the box). The relationship between increased fear and increased cost is best illustrated by an example.

The case of decaffeinated coffee

A few years ago, there was much ado about the use of methylene chloride to decaffeinate coffee. Members of the media reported that methylene chloride was a carcinogen in rats when inhaled. They neglected to add that it has a much lower carcinogenic potential when ingested. This distinction is very relevant, because in the case of decaffeinated coffee, the only variable coffee drinkers need be concerned with is the carcinogenic potential when ingested.

Many consumers changed their buying habits as a result of the media's attention to this issue. To avoid a perceived risk of getting cancer, coffee drinkers were more than willing to spend extra money for coffee decaffeinated by a different method, a process based on water. Those who couldn't afford the extra expense probably worried about what they were doing to their health.

Societal Issues

Thoughts on Public Perception of Risk

"A lot of our priorities are set by public opinion, and the public quite often is more worried about things that they perceive to cause greater risks than things that really cause risks. Our priorities often times are set through Congress. Their decisions may or may not reflect real risk; they may reflect people's opinions of risk or the Congressmen's opinions of risk."

Linda Fisher,
former Assistant Administrator of the
Environmental Protection Agency

"For many years there has been an argument among some technical experts that the public is all mixed up, that they just don't have their risk priorities right. But experimental results in psychology in recent years have suggested that that is not true. If you give the public a list of hazards and say "Sort these in terms of how many people die each year from each of them," they can do it. If instead you give them the same list and they are asked to sort them by how risky they are, you get a very different order. The point is that to most people risk does not equal expected numbers of deaths, it involves a lot of other things, things like equity, whether you can control the hazard, whether you understand it. And so partly the arguments derive from this difference between making judgments just on the basis of expected numbers of deaths versus considering all these other factors."

Granger Morgan,
Carnegie Mellon Institute

Coffee itself has over 300 chemical compounds, many of which are more toxic and more carcinogenic than methylene chloride. In order to ingest enough methylene chloride to reach the level shown to cause cancer in rats, a person would have to drink 50,000 cups of coffee a day. In those 50,000 cups, natural coffee carcinogens would occur in far greater amounts than methylene chloride. More important, perhaps, a person would die from drinking 50,000 cups of *water* a day. Long before the 50,000th cup was consumed, a person's kidneys would shut down.

Again, the message is not that man-made is safe and natural is unsafe. The situation is not that simple. The moral of the story is that all of us work from misconceptions derived from inaccurate or misleading information. Unfortunately, misinformation made worse by fuzzy thinking doesn't yield smart choices very often. How can we do a better job of making choices involving food safety risks or any other risks?

Facts and Conscious Risk Assessments

What drives your assessment of risks—emotions or facts—may or may not seem particularly important to you, yet it is. The risks you are willing to assume and the experiences or products you avoid because of faulty assumptions and misinformation affect the quality of your life and the lives of those around you. In today's world, people suffer from great anxiety about the risks involved in simply living. Some of the fears that grip us are real; others are not. Wouldn't it be useful if we had a mechanism for helping us determine the difference? We do.

Toxicity versus hazards

First, to better understand and evaluate discussions of risks, particularly those involving food safety, we should differentiate between toxicity and hazard. Toxicity can be defined as the capacity of a substance to do harm or injury of any kind under any conditions. Given the definition of toxicity, anything can be toxic. In the decaffeinated coffee example above, water is toxic.

Contrast this to the definition of hazard. A hazard is the relative probability that harm will result when a particular quantity of a substance is used in a particular manner. Given these two definitions, it becomes clear that meaningful assessments of food safety, or the safety of any substance, must be based on the hazard a substance poses and not its toxicity. If we use this principle, we will be one step closer to assessing risks objectively.

Risk-benefit analysis

Using this concept of hazard, scientists have developed a process for assessing risks and benefits that attempts to maximize the contribution of factual information and minimize the contribution of emotion to the analysis: risk-benefit analysis. Here's how it works.

Assessing Risks

First of all, to analyze risks as thoroughly and unemotionally as possible, scientists conduct a risk assessment. The first step is defining what a risk actually is.

Risk is the probability of loss or injury, or

$$\text{risk} = \text{hazard} \times \text{exposure}$$

In other words, a hazard, which can be viewed as a source of danger, becomes a risk only if you are exposed to it. Your probability of loss or injury increases in relationship to the increase in hazard and/or exposure.

Once you have clearly defined the concept of risk, you are ready to begin a risk assessment. Assess the amount of risk involved in a particular situation by answering these three questions:

1. What could go wrong? (Identify the specific risk.)
2. How likely is it to happen? (Assess the probability of the risk occurring.)
3. How harmful is it if it does occur? (Assess the consequences.)

A simple formula for approximating the amount of risk is

$$\text{amount of risk} = \frac{\text{hazard} \times \text{exposure}}{\text{safeguards}}$$

If you are able to establish effective safeguards, the amount of risk decreases dramatically.

These conceptual formulas illustrate the first steps in using rational risk assessment methodology to get a handle on real, not perceived, risks. By no means are these steps perfect or devoid of subjectivity, but they provide a starting point for rational discussion of risk.

Why worry with all of this? Why not just overestimate risk to make sure we're safe? As the example of decaffeinated coffee demonstrates, responding to risks that don't exist can limit our options, increase the money we spend, and raise the costs of production of goods because of excessive regulatory burden. It is possible to be overly cautious. For a relative ranking of the risks of familiar activities and products, see Table 30.1.

Assessing Benefits

Assessing risks is a simple task compared to assessing benefits. Benefits are often very difficult to identify in the abstract. We need to experience them. Beyond the obvious and intended benefits of electric lights over gaslights, we would never have foretold all of the benefits we have derived from electrification of our homes. We would never have imagined other unintended, but direct, benefits of electrification: electric washing machines, irons, hot-water heaters, air conditioners, refrigerators, home computers, and VCRs. Nor could we have predicted the secondary, indirect effects of some of the unintended effects: homes in the desert and frozen north, the Internet, home shopping, more free time, the disappearance of household servants, more women in the workforce, and, as a result, a need for day-care centers.

In addition to being difficult to predict, many benefits are very subjective. They are based on emotions, values, and ethical principles and they therefore vary from person to person. Some of the factors we use to measure benefits include the following:

- Saves lives
- Improves health
- Saves money
- Solves problems
- Makes life more pleasant or easier
- Increases emotional well-being

Putting Risks and Benefits Together

Defining and assessing the amount of risk and then comparing the risk to the perceived benefits and their probabilities of occurring are the components of a thoughtful, methodical analysis for assessing risks. Using this methodology, we are in a much better position to make judicious choices on issues that involve real or perceived risk. Yet that does not mean our problems are over.

Once we have conducted our own risk-benefit analysis of a certain issue and arrived at a decision regarding the acceptability of a risk, we may find that the person next to us has arrived at a different conclusion. This difference of opinion is to be expected, given the difficulties in assessing the amount of risk and the subjective and emotional nature of benefit analysis.

You may think that if the decision involves an individual's acceptance or rejection of personal risk, differences of opinion on the costs and benefits of individual behavior should present no problem. In other words, if I want to accept the risks when I skydive or ride a motorcycle without a helmet, it's my business.

Table 30.1   Relative amount of risk[a]

Risk	Source
0.2[b]	PCBs in diet
0.3[b]	DDT[c] in diet
1[b]	1 qt of city water a day
8[b]	Swimming 1 h/day in a chlorinated pool
18	Electricity—chance of dying of shock in a year
30[b]	2 tablespoons of peanut butter a day (from peanut mold)
60[b]	12 oz of diet cola a day
100[b]	3/4 teaspoon of basil a day
367	Accidents in the home
600[b]	Indoor air containing formaldehyde vapors from furniture
667	Contribution of air pollution in eastern United States to respiratory illness
800	Auto accident—chance of dying in 1 yr
2,800[b]	12 oz of beer a day
12,000[b]	One pack of cigarettes a day
16,000[b]	One tablet of phenobarbital a day

[a]The higher the number, the greater the risk. Drinking a quart of city water, which contains chloroform, a by-product of chlorination, serves as a reference point. For example, the risk in eating 3/4 teaspoon of basil a day is 100 times greater than the risk in drinking a quart of city water a day. The data are taken from articles by Bruce Ames and Richard Wilson in the April 17, 1987, issue of *Science*.

[b]These values indicate the risk of getting cancer.

[c]DDT, dichlorodiphenyltrichloroethane, a pesticide.

Societal Issues

Once again, it's not that simple. When does someone's personal decision about the acceptability of a personal risk infringe upon the right of others?

People who drive drunk may feel comfortable with their own risk-benefit analysis of their behavior, but other drivers on the road may not. Given that example, most people would probably come to a similar conclusion regarding individual rights versus the rights of other members of society. Driving on public roads is a public matter made possible with public money. An individual's right to accept the risk of driving drunk is secondary compared to the public's right to use roads without fear of drunken drivers.

What about private matters like safe sex and AIDS? Surely the risks of having sex are relevant only to the people involved. If people risk contracting AIDS by having unprotected sex, isn't that their business and only their business? How can that choice infringe on the rights of others?

This may not be as simple and straightforward as it first appears or as we wish it were. As the cost of providing care for a growing population of AIDS victims increases, health insurance costs will increase. That cost increase will be borne by many people in ways that seem completely unrelated to AIDS and unprotected sex. For example, today, many employers pay for the health insurance of their employees. As health insurance costs increase, employers must pay more money to insure their workers. The company must then compensate for this increase either by decreasing costs in other areas or by increasing income. Cutting jobs is one way to cut costs. To increase income, companies usually raise the prices they charge for goods and services. So, an increase in health insurance costs could mean fewer jobs and/or higher prices. Should everyone pay the price for those who have chosen to have unprotected sex and, as a result, been infected with the human immunodeficiency virus?

Issues of individual rights are never as simple as we would like. As members of society, we have both rights and responsibilities to ourselves and each other. Finding the proper balance between the two can be very, very difficult.

Activities

1. Select any familiar, pervasive technology we depend upon, and conduct a risk-benefit analysis.

2. Evaluate that same technology before its widespread implementation. What risks and benefits would you have predicted prior to that technology's development?

3. Select a future application of biotechnology or another emerging technology, and conduct a risk-benefit analysis.

4. Select a risk that is considered personal, such as skydiving, cigarette smoking, or using drugs, and describe the societal ramifications of taking this individual risk.

5. Compare and contrast the media coverage of a technological development or scientific discovery with the primary literature on the subject.

6. Select a food safety issue and develop a case study, focusing on real risks and benefits.

7. Write a short essay explaining either of the following quotes:

   "To be alive is to be at risk."
   Daniel Koshland, *Science* 1987
   "Only the dose makes the poison."
   Paracelsus, 16th century

Societal Issues

Debating the Risks of Biotechnology

31

Introduction

Ever since scientists produced the first recombinant DNA molecule in the early 1970s, we have heard much about the risks of biotechnology. Some claim that biotechnology's benefits far outweigh its risks, because biotechnology is the panacea for all of our ills, medical, environmental, and economic. Others say it is a Pandora's box from which will spring a new set of problems, most of which we can't predict, much less control. Which side is right, or does the truth lie somewhere in between?

When "authorities" espouse disparate views on the risks of biotechnology, or any technology, deciding which view to accept poses a problem for ordinary citizens not trained as scientists. Identifying specious arguments, superfluous facts intended to mislead, and intellectual dishonesty requires understanding some of the science underlying the issue. How can nonscientists both assess the arguments and engage in productive debates as society struggles to come to terms with the risks of biotechnology or any other technically complex and potentially volatile issue?

For a debate to be satisfying and productive, it must be driven by objective facts and not irrational fears. The participants should be as clear, specific, and as emotionally detached as possible. The primary goal should be not winning, but greater understanding by all parties. Remember, we usually learn the most from people whose ideas and opinions differ from our own.

Debate Methodology

What are some of the steps involved in preparing for and conducting a productive debate or discussion?

Step 1

Get clear on exactly what is being discussed. Delineate and define. Use precise language.

Ask people to define what they mean when they say the "risks of biotechnology," so that everyone will be talking about the same thing. The issues under debate are difficult in and of themselves. Adding the additional confusion of using sloppy language will only make the debate less productive. Having a productive discussion is possible only if the people involved are talking about the same issue.

To what type of risk are people referring? Do they mean environmental and/or public health risk? Are they including other types of risks, such as social, economic, and ethical risks? Do they equate the word "risk" with the probability of death or harm, or is their view of risk less specific and complicated by variables such as control, familiarity, and freedom of choice?

What does the word "biotechnology" mean to people discussing it? In chapter 1, you learned that biotechnology is a collection of technologies such as monoclonal antibody technology, cell culture, genetic engineering, and bioprocessing. You also learned that "biotechnology" is an ambiguous term used in a variety of ways by different people. To which biotechnology is someone referring? The risks of one technology differ greatly from those of another. Usually the focus is genetic engineering. The discussion should therefore be about the risks of genetic engineering, not biotechnology. Are people concerned with only one or two applications of genetic engineering or with genetic engineering in general?

Which facet of biotechnology development is the target of concern? Biotechnology development occurs along a continuum that begins in a research laboratory and extends through product development to widespread commercial use. The risk of biotechnology laboratory research, small-scale field testing, and large-scale commercial use differ from each other both qualitatively and quantitatively.

Finally, in discussing the risks of any technology, keep in mind that technologies may be used wisely or unwisely, for good or for bad. The risks of a technology per se must be distinguished from the risks of a technology placed in irresponsible hands. The flaws of a technology and the flaws of human nature are two very different considerations.

If your goal is greater understanding, the value of using discipline in the words you choose cannot be overestimated.

Step 2

For rational risk assessments as described in the previous chapter, get facts, not opinions. First, identify the risk; second, estimate the probability of the risk occurring; finally, assess the consequences.

In assessing the risks of biotechnology, obtain objective data that allow you to identify the risks clearly, specifically, and independently. If you want your assessment of risk to be grounded in sound science, go to primary sources such as scientific journals or scholarly articles or to review papers in publications like *Scientific American*. Talk to scientists conducting relevant research. Do not rely on news magazines, newspapers, television, or hearsay.

Often facts other than scientific information are important to the debate and should be clearly articulated. For example, farmers must keep their costs down if they are to make a profit. This is not a "scientific" fact, but it sure is true.

Once you have identified the risks, you must then estimate the probability of the risk occurring. Because much of biotechnology is similar to activities we have been engaged in for many years, we have abundant data on some products and applications that can be used to estimate the general probabilities of specific risks.

Then assess the consequences of a particular risk occurring. What is the significance of risk A occurring? If it occurs, will it matter? If it matters, under what circumstances? It may happen on a regular basis anyway without human interference (insects becoming resistant to chemicals produced by plants, bacteria transferring genes across species). It may be environmentally, ecologically, and economically insignificant (spraying a biopesticide in certain weather conditions causes a crop plant to lose a few leaves but suffer no yield loss).

Step 3

Compare the risk with the benefits, other risks, and other options for achieving the same goal.

Establishing that a risk of some type is involved is a first step in conducting a risk assessment but by itself is not particularly helpful. Every human activity carries some amount of risk with it. To weigh the value of taking a risk, we must place the risk in a meaningful context.

First, look at the benefits. An assessment of risk is meaningless unless it is compared with an equally objective analysis of the benefits. Only then can we begin to determine whether the risk is worth taking.

Second, compare risks. An assessment of the risks of one activity or product must be contrasted with the risks involved in achieving the same goal through different activities or products. In other words, if the benefits are substantial, then the goal is worthy. Is there another way of reaching the same goal that entails less risk?

This question is particularly pertinent in assessing the risks involved in biotechnology. Much of biotechnology is similar to activities we have been engaged in for centuries. Often the most appropriate question regarding the risks of biotechnology is not "Are there risks?" but "How do the risks of this new technology compare with those of existing technologies we have directed toward the same purpose?"

Suggested Activities

Introduction

Teaching students how to debate the risks of biotechnology is similar to teaching them how to analyze the relationship between science, technology, and society or how to conduct a risk-benefit analysis as described in chapters 29 and 30. Your task is to teach them not what to think, but how to think. Your goals should include helping students develop the following skills: fact finding, objective analysis, identifying the agendas of the various parties, decision making, developing options, and reaching consensus. The format you use should facilitate the development of these skills and may include case studies, debate, role playing, classroom discussion, and research papers. As an example, we have selected a case study on the use of herbicide-tolerant crops (HTC), and we suggest having the class debate the pros and cons of developing HTC. For your class, assignment of a research paper on the same topic or a classroom debate on a different topic might be more appropriate. At the end of the *Student Activity* section is a reading on genetically engineered organisms in the environment.

The suggested methodology, identical to that in the teacher's activities provided at the end of chapter 29,

Science, Technology, and Society, is worth reiterating here.

- Help students learn to consider one issue at a time.

- Show them how to stay focused on the issue.

- Help them learn how to gather accurate information. Too often, students believe that if something is in print, it must be true.

- Help them learn to distinguish science from pseudoscience.

- Show them how to assess information critically and unemotionally.

- Help them draw conclusions.

- Help them learn to discuss issues fairly and unemotionally, listening to each other so that they acquire the greatest understanding.

Although we have selected the development of HTC, many other areas of biotechnology can serve as topics for developing risk assessment and debate skills. Because these topics have been discussed for a number of years, abundant reference material is available. Topics include the following:

- Labeling of genetically engineered foods
- The use of bovine somatotropin in dairy cows
- Environmental releases of genetically engineered organisms
- Patenting of genes or genetically engineered organisms
- Biotechnology and sustainability (or choose one aspect of sustainability, such as genetic diversity)
- The forensic use of DNA typing

You might consider assigning research papers on different topics to different students or teams to illustrate the breadth of the societal issues associated with biotechnology.

Objectives

After reading the chapter and completing the assignment, students will

- Be better informed about an application(s) of biotechnology;

- Be able to discuss the pros and cons of at least one application of biotechnology;

- Understand that there are pros and cons to the use of any technology.

A Case Study: Herbicide-Tolerant Crops

Introduction

To illustrate how to conduct an objective, methodical analysis of the risks and benefits of a product of biotechnology, we have selected HTC for a number of reasons.

- A clear statement of the arguments against developing HTC is available in the report *Biotechnology's Bitter Harvest.*

- A clear statement of the arguments in favor of developing HTC is available in the report *Herbicide-Resistant Crops.*

- The issue is multifaceted.

- Farmers have used HTC for more than 40 years, so relevant data for conducting a risk assessment and weighing the validity of these arguments are readily available.

- The topic provides an opportunity for students to learn something about a technology they are totally dependent upon yet know little about: agriculture.

- The topic provides an opportunity for students to interview farmers, environmentalists, public health officials, and extension agents.

- Agricultural systems are complex ecosystems, so the students will be learning a lot of biology as they learn about agriculture.

Methodology

The development of HTC is one application of biotechnology that has received the negative attention we discussed in chapter 1.

A public interest group, the Biotechnology Working Group (BWG), published a report entitled *Biotechnology's Bitter Harvest* that condemns HTC as damaging to the environment and public health. The Council for Agricultural Science and Technology (CAST) counters BWG's assessment in its own report, *Herbicide-Resistant Crops.*

The primary issues the two groups disagree on are whether the use of HTC will

- Further our reliance on herbicides instead of other weed control methods;

Societal Issues

- Increase the amount of herbicides used;
- Cause the evolution of resistance to herbicides by weeds;
- Allow the transfer of genes for herbicide tolerance to weeds, making weeds no longer susceptible to herbicides.

These different viewpoints about the environmental effects of HTC can provide the key points for a classroom debate. The reports also discuss the economic impacts of using herbicides and HTC, if you would like to broaden the debate to include issues other than environmental ones.

Break your students into four-member teams, and assign the following tasks.

1. Write down the most important questions they need to ask and try to find answers for in order to determine which viewpoint appears to be the most accurate and objective. Each student should do this individually. Then, as a team, they will decide which questions are the most important. The group should then hone the questions so that they are phrased in clear, unambiguous terms that get to the heart of the issue.

2. Once each group has a final version of its questions, each student in a group should gather facts to answer one or two of the questions and then report the findings to the group. Remember not to rely on news magazines, newspapers, television, company brochures, newletters from environmental groups, or hearsay. Instead, students must use credible information from sources that have nothing invested in the topic. Your most objective agendaless sources are scientific articles and textbooks in agricultural science.

3. Each group should then determine the soundness of the key issues listed previously and conduct an informal risk assessment: weigh the risks and benefits of developing HTC, based only on those four points, unless you have elected to also include issues involving economics.

4. Each group should select a leader to report its findings to the class.

If the opinions of the teams regarding the development of HTC differ, then host a debate. After dividing the students into pro and con groups, have each group select four students to represent their side of the debate. Members of the group should share what they have learned in their research with their group representatives to prepare them for the debate.

During the debate, remind the students to follow these rules.

- Everyone on the team must participate.
- Listen to everyone's facts.
- Be specific.
- If necessary, ask the other team to define and clarify.
- Try to understand each other's point of view.
- Remember: the goal of this debate is not winning, but greater understanding.

Relevant facts, questions, and a suggested framework

Below you will find background information on herbicides and HTC, a general framework for assessing the risks and benefits of their use; a summary of the main points made by the BWG in its report *Biotechnology's Bitter Harvest* and the countering points made by the CAST in its report *Herbicide-Tolerant Crops;* some important questions for assessing the validity of these arguments, and some relevant information for answering questions. For additional information on weed control, herbicides, and HTC, contact the U.S. Department of Agriculture, your state Department of Agriculture, your county extension agent, or scientists in the crop science department at a state university with specialists in agriculture.

Background on HTC
Some of the first products of agricultural biotechnology will be crops genetically engineered to tolerate a variety of herbicides, the compounds farmers use to kill weeds in their fields.

HTC are not new. Virtually all of the acreage of our major crop plants such as corn, soybeans, cotton, and wheat is treated with herbicides. Therefore, all of our major crops are already tolerant to some herbicides.

Where did this tolerance originate? Genetically based herbicide tolerance exists in nature. Some of our crops were naturally resistant to the herbicides we have developed. These natural HTC serve as sources of herbicide tolerance genes for other crops.

Using existing, genetically based herbicide tolerance has been useful but, by its very nature, limited. Because of the constraints of crossbreeding, we are able to use only the herbicide tolerance genes found in crops or their close relatives. Genetic engineering now provides us with the ability to capitalize on genetic variation found anywhere in nature. Some bacteria naturally contain genes for herbicide tolerance.

We have moved those genes into both major and minor crops.

Framework and Questions
Establish the need for weed control.

Do farmers have to control weeds? If they don't, what happens? The average crop losses in the United States in the absence of weed control are as follows: corn, 70%; soybeans, 80%; cotton, 100%; peanuts, 100%.

Compare and contrast weed control options available to farmers.

If farmers must control weeds, are herbicides the only method available? No. The BWG states that other options available for controlling weeds are just as effective as herbicides and could replace herbicides as ways to control weeds reliably. Thus, there would be no need to develop HTCs. What questions would you need to ask to determine the validity of this assertion?

- What are the weed control options available to farmers?
- What are the pluses and minuses of each option?
- Are all of the options equally effective?
- Do they vary with the crop in question or the soil type?
- How much does each option cost per acre?
- What variables influence a farmer's choice?

Assess the environmental impacts of using herbicides.

The BWG assumes that herbicides as a group are damaging to the environment and public health. The CAST states that herbicides vary widely in toxicity and implies that some are relatively harmless because they have low toxicity and do not persist in the environment beyond a few days. How would you determine which opinion is most accurate scientifically?

Determine whether generalizing about herbicides is appropriate. Are all herbicides equal in toxicity? How do they vary? What variables affect whether a herbicide is damaging to public health or the environment?

- Toxicity-carcinogenicity
- Degradation rate
- Tendency to move into groundwater
  — Solubility in water
  — Binding to soil particles
- Residues on food
- Volatility
- Application rate

To assist students in evaluating toxicity, establish a frame of reference for placing toxicity in a meaning-

ful context. Have students compare the toxicities of herbicides with those of other herbicides, with familiar highly toxic substances, and with "toxic" substances they regularly consume: caffeine in sodas, aspirin, solanine in potato chips, etc.

Assess the environmental benefits of herbicides.

Now that students have assessed some of the environmental risks of herbicides and compared this method of weed control to other methods, the next important step is determining whether using herbicides provides any environmental benefits and, if so, what they are.

Are there environmental benefits to using herbicides? Yes. They permit the use of no- or minimum-tillage agriculture. This management practice significantly decreases soil erosion, a major environmental problem.

Are there other benefits?

Determine whether HTC will increase the amount of herbicides used.

The BWG asserts that the development of HTC will increase the amount of herbicides used in the United States, while CAST states that the development of HTC could have any of the following effects on the amount of herbicides used: no effect, decreased amount required, or increased amount required. The CAST report states that anyone trying to determine whether or not the use of HTC will increase the amount of herbicides used by U.S. farmers must examine the issue on a case-by-case basis, taking into account both the crop and the herbicide.

What are some of the features of the herbicide that affect the amount of herbicide we now use?

- Spectrum of the herbicide (broad or narrow)

- Unit activity (amount per acre)

- Preemergence or postemergence application (Preemergence herbicide is applied prophylactically; postemergence herbicide allows a "wait-and-see" approach to weed control.)

What are some of the features of the crop that affect the amount of herbicide used?

- The amount of weed-related damage the crops can sustain without the yield being affected
- The herbicides the crop tolerates and the effectiveness of these herbicides

How much do farmers use herbicides now?

We use herbicides on 98% of the acreage devoted to our major crops. Our top two crops, corn and soybeans, account for 67% of the total herbicide use in the United States, while the top 10 crops account for more than 91% of the total. Consequently, there is very little room for increasing the total number of acres treated with herbicides.

Will HTC increase the amount of herbicide per acre? If so, how?

In most cases, at least two applications of herbicides are used on these crops, and sometimes as many as four or five are used. If we can develop HTC that allow farmers to use one herbicide where they now commonly use two or three and to replace preemergence herbicides with postemergence herbicides, then the total amount of herbicide used on our major crop plants should decrease.

On the other hand, the amount of herbicides used on our minor crops could increase. (*Note:* Students will need to define what a "minor crop" is.) We do not currently use herbicides on many of these crops because they are not tolerant to herbicides. If we develop herbicide-tolerant minor crops, growers will probably use them, because HTC will allow farmers to control weeds by using herbicides instead of expensive manual labor.

What other factors might affect the amount of herbicide a farmer uses? Economics. Herbicides are expensive. Cost is the single most important factor in every decision a farmer makes.

Determine the relationship between the evolution of resistance to herbicides and the use of herbicides and HTC.

The BWG asserts that the use of HTC will cause the evolution of resistance to herbicides by weeds. Is this valid? Have weeds evolved resistance to the herbicides we have been using over the past 40 years? Yes, and weeds will continue to become resistant to the herbicides we use. The CAST report describes this phenomenon in detail. We will summarize here.

Populations of weeds have the genetic variation needed to evolve resistance. That's the nature of weeds. They have lots of genetically variable offspring every year, so their gene pools will contain the genetic flexibility they need to evolve resistance. Different weeds will become resistant to different herbicides. Some herbicides will be easy for weeds to side-

step in an evolutionary sense; others will be more difficult.

But weeds will evolve resistance to whatever means we use to kill them. They have evolved to look like crop plants so we won't pull them up by hand and to be very short, so they will escape lawn mower blades. The seeds of weeds have evolved to look like crop seeds so they will be saved and planted the next year. It doesn't matter how we try to beat them, they will do their best to escape. What does this imply for the suggestion that there are better options for weed control than herbicides?

The best thing we can do to slow the evolution of resistance to any control mechanism is to keep weeds guessing. To do this, we need a large bag of weed-killing tricks. Crops made tolerant to a wide variety of herbicides should increase our options for weed control. In other words, they should increase the size of our bag of tricks. If, however, we develop crops tolerant to one or two herbicides and then persist in using those one or two herbicides over and over, we will encourage the evolution of resistance to them.

Determine whether the development of HTC will lead to weeds resistant to herbicides because the genes for herbicide tolerance have moved from the crop to the weed.

The BWG maintains that gene transfer will occur between HTC and weeds, thus making weeds resistant to herbicides. What scientific facts do we know that support or refute this claim?

Some crop plants are capable of exchanging genes with some of their wild, weedy relatives. For this to occur, the crops and the weedy relatives must grow in very close proximity to one another. Very few of our major crop plants have wild relatives in the United States. Even fewer are able to transfer genes to these relatives and produce viable offspring able to reproduce. Examples of crops that can are sorghum, rice, and sunflowers. If we engineer herbicide tolerance into those crops and plant those HTC near their weedy relatives, we run the risk of creating herbicide-tolerant weeds.

Once we move outside of the boundaries of the United States, the issue of gene transfer between crops and wild relatives becomes more of a problem. Crops in Central and South America, Africa, and Asia grow surrounded by wild, weedy relatives. Little is known about gene transfer and the production of viable offspring by these weeds. So, what is true about the risks, or lack of risks, of developing HTC in the United States may not be true in foreign countries.

Is there any evidence of gene transfer between the HTC we have used for the last 40 years and their wild relatives in any part of the world? No. In every case in which weeds have evolved tolerance, the biochemical basis of herbicide tolerance has differed from that of the related crop plant. Therefore, the weed's herbicide tolerance did not come from the crop's herbicide-tolerant genes.

In making choices about developing HTC, which herbicide and crop traits would be important to consider?

We should first ask whether there is a need to develop that specific HTC. If there is, then here are some of the traits we should look for in developing new products:

• Tolerance to postemergence herbicide
• Tolerance to broad-spectrum herbicide
• Tolerance to nonpersistent herbicide
• Tolerance to low-toxicity herbicide
• Tolerance to herbicide with low water solubility
• Tolerance to inexpensive herbicide
• Low potential for gene transfer to weeds
• Tolerance to a herbicide difficult for plants to sidestep by evolving resistance

Selected Readings

*Congressional Office of Technology Assessment. 1981. *Impacts of Applied Genetics: Microorganisms, Plants, and Animals*. U.S. Government Printing Office, Washington, D.C.

*Congressional Office of Technology Assessment. 1987. *Field-Testing Engineered Organisms: Genetic and Ecological Issues*. U.S. Government Printing Office, Washington, D.C.

Davis, Bernard D. 1991. *The Genetic Revolution: Scientific Prospects and Public Perceptions*. The John Hopkins University Press, Baltimore, Md.

Dixon, Bernard. 1985. *Engineered Organisms in the Environment: Scientific Issues*. The Maple Press Co., York, Pa.

†Duke, Steve, A. Lawrence Christy, F. Dana Hess, and Jodie Holt. 1991. *Herbicide-Resistant Crops*. Council for Agricultural Science and Technology, Ames, Iowa.

‡Goldburg, Rebecca, Jane Rissler, Hope Shand, and Chuck Hassebrook. 1990. *Biotechnology's Bitter Harvest: Herbicide-Tolerant Crops and the Threat to Sustainable Agriculture*. Biotechnology Working Group, Washington, D.C.

Gould, Fred. 1991. The evolutionary potential of crop plants. *American Scientist* 79:496–507.

Heiser, Charles, B. 1981. *Seeds to Civilization: The Story of Food*. W. H. Freeman & Co., San Francisco.

§MacDonald, June F. (ed.) 1989. *Biotechnology and Sustainable Agriculture: Policy Alternatives*. National Agricultural Biotechnology Council report 1. National Agricultural Biotechnology Council, Ithaca, N.Y.

§MacDonald, June F. (ed.) 1991. *Agricultural Biotechnology at the Crossroads*. National Agricultural Biotechnology Council report 3. National Agricultural Biotechnology Council, Ithaca, N.Y.

Nash, Roderick. 1989. *The Rights of Nature: a History of Environmental Ethics*. The University of Wisconsin Press, Madison.

National Academy of Sciences. 1987. *Introduction of Recombinant DNA-Engineered Organisms into the Environment: Key Issues*. The National Academy of Science Press, Washington, D.C.

National Research Council. 1989. *Field-Testing Genetically Modified Organisms: Framework for Decisions*. National Academy Press, Washington, D.C.

Tiedje, J.M., R.K. Colwell, Y.L. Grossman, et al. 1989. The planned introduction of genetically engineered organisms: ecological considerations and recommendations. *Ecology* 70:298–315.

*To order publications from the U.S. Government Printing Office, write to the Superintendent of Documents, Mail Stop: SSOP, Washington, DC 20402-9328.

†To obtain a copy of the CAST report *Herbicide-Resistant Crops*, write CAST, 137 Lynn Ave., Ames, IA 50010-7197.

‡To obtain a copy of *Biotechnology's Bitter Harvest*, contact a member of the Biotechnology Working Group, for example, Dr. Rebecca Goldburg, Environmental Defense Fund, 257 Park Ave., New York, NY 10010, or Hope Shand, Rural Advancement Fund International, Pittsboro, NC 27312.

§To obtain a copy of any of the reports by the National Agricultural Biotechnology Council, write NABC, Boyce Thompson Research Institute, 159 Biotechnology Building, Cornell University, Ithaca, NY 14853-2703.

Societal Issues

Debating the Risks of Biotechnology

Introduction

Ever since scientists produced the first recombinant DNA molecule in the early 1970s, we have heard much about the risks of biotechnology. Some claim that biotechnology's benefits far outweigh its risks, because biotechnology is the panacea for all of our ills, medical, environmental, and economic. Others say it is a Pandora's box from which will spring a new set of problems, most of which we can't predict, much less control. Which side is right, or does the truth lie somewhere in between?

When "authorities" espouse disparate views on the risks of biotechnology, or any technology, deciding which view to accept poses a problem for ordinary citizens not trained as scientists. Identifying specious arguments, superfluous facts intended to mislead, and intellectual dishonesty requires understanding some of the science underlying the issue. How can nonscientists both assess the arguments and engage in productive debates as society struggles to come to terms with the risks of biotechnology or any other technically complex and potentially volatile issue?

For a debate to be satisfying and productive, it must be driven by objective facts and not irrational fears. The participants should be as clear, specific, and as emotionally detached as possible. The primary goal should be not winning, but greater understanding by all parties. Remember, we usually learn the most from people whose ideas and opinions differ from our own.

Debate Methodology

What are some of the steps involved in preparing for and conducting a productive debate or discussion?

Step 1

Get clear on exactly what is being discussed. Delineate and define. Use precise language.

Ask people to define what they mean when they say the "risks of biotechnology," so that everyone will be talking about the same thing. The issues under debate are difficult in and of themselves. Adding the additional confusion of using sloppy language will only make the debate less productive. Having a productive discussion is possible only if the people involved are talking about the same issue.

To what type of risk are people referring? Do they mean environmental and/or public health risk? Are they including other types of risks, such as social, economic, and ethical risks? Do they equate the word "risk" with the probability of death or harm, or is their view of risk less specific and complicated by variables such as control, familiarity, and freedom of choice?

What does the word "biotechnology" mean to people discussing it? In chapter 1, you learned that biotechnology is a collection of technologies such as monoclonal antibody technology, cell culture, genetic engineering, and bioprocessing. You also learned that "biotechnology" is an ambiguous term used in a variety of ways by different people. To which biotechnology is someone referring? The risks of one technology differ greatly from those of another. Usually the focus is genetic engineering. The discussion should therefore be about the risks of genetic engineering, not biotechnology. Are people concerned with only one or two applications of genetic engineering or with genetic engineering in general?

Which facet of biotechnology development is the target of concern? Biotechnology development occurs along a continuum that begins in a research laboratory and extends through product development to widespread commercial use. The risk of biotechnology laboratory research, small-scale field testing, and large-scale commercial use differ from each other both qualitatively and quantitatively.

Finally, in discussing the risks of any technology, keep in mind that technologies may be used wisely or unwisely, for good or for bad. The risks of a technology per se must be distinguished from the risks of a technology placed in irresponsible hands. The flaws of a technology and the flaws of human nature are two very different considerations.

If your goal is greater understanding, the value of using discipline in the words you choose cannot be overestimated.

Step 2

For rational risk assessments as described in the previous chapter, get facts, not opinions. First, identify the risk; second, estimate the probability of the risk occurring; finally, assess the consequences.

In assessing the risks of biotechnology, obtain objective data that allow you to identify the risks clearly, specifically, and independently. If you want your assessment of risk to be grounded in sound science, go to primary sources such as scientific journals or scholarly articles or to review papers in publications like *Scientific American*. Talk to scientists conducting relevant research. Do not rely on news magazines, newspapers, television, or hearsay.

Often facts other than scientific information are important to the debate and should be clearly articulated. For example, farmers must keep their costs down if they are to make a profit. This is not a "scientific" fact, but it sure is true.

Once you have identified the risks, you must then estimate the probability of the risk occurring. Because much of biotechnology is similar to activities we have been engaged in for many years, we have abundant data on some products and applications that can be used to estimate the general probabilities of specific risks.

Then assess the consequences of a particular risk occurring. What is the significance of risk A occurring? If it occurs, will it matter? If it matters, under what circumstances? It may happen on a regular basis anyway without human interference (insects becoming resistant to chemicals produced by plants, bacteria transferring genes across species). It may be environmentally, ecologically, and economically insignificant (spraying a biopesticide in certain weather conditions causes a crop plant to lose a few leaves but suffer no yield loss).

Step 3

Compare the risk with the benefits, other risks, and other options for achieving the same goal.

Establishing that a risk of some type is involved is a first step in conducting a risk assessment but by itself is not particularly helpful. Every human activity carries some amount of risk with it. To weigh the value of taking a risk, we must place the risk in a meaningful context.

First, look at the benefits. An assessment of risk is meaningless unless it is compared with an equally objective analysis of the benefits. Only then can we begin to determine whether the risk is worth taking.

Second, compare risks. An assessment of the risks of one activity or product must be contrasted with the risks involved in achieving the same goal through different activities or products. In other words, if the benefits are substantial, then the goal is worthy. Is there another way of reaching the same goal that entails less risk?

This question is particularly pertinent in assessing the risks involved in biotechnology. Much of biotechnology is similar to activities we have been engaged in for centuries. Often the most appropriate question regarding the risks of biotechnology is not "Are there risks?" but "How do the risks of this new technology compare with those of existing technologies we have directed toward the same purpose?"

Activities

Many areas in biotechnology can serve as topics for developing risk assessment and debate skills. Because these topics have been discussed for a number of years, abundant reference material is available. Topics include the following:

- Labeling of genetically engineered foods
- The use of bovine somatotropin in dairy cows
- Environmental releases of genetically engineered organisms
- Patenting of genes or genetically engineered organisms
- Biotechnology and sustainability (or choose one aspect of sustainability such as genetic diversity)
- The forensic use of DNA typing
- The development of herbicide-tolerant crops

A Case Study: Herbicide-Tolerant Crops

Introduction

To conduct an objective, methodical analysis of the risks and benefits of a product of biotechnology, we suggest using herbicide-tolerant crops (HTC) for a number of reasons.

- A clear statement of the arguments against developing HTC is available in the report *Biotechnology's Bitter Harvest.*

- A clear statement of the arguments in favor of developing HTC is available in the report *Herbicide-Resistant Crops.*

- The issue is multifaceted.

- Farmers have used HTC for more than 40 years, so relevant data for conducting a risk assessment and weighing the validity of these arguments are readily available.

- The topic provides an opportunity to learn something about a technology you are totally dependent upon yet may know little about: agriculture.

- The topic provides an opportunity to interview farmers, environmentalists, public health officials, and extension agents.

- Agricultural systems are complex ecosystems, so you will be learning a lot of biology as you learn about agriculture.

Methodology

The development of HTC is one application of biotechnology that has received the negative attention we discussed in chapter 1.

A public interest group, the Biotechnology Working Group (BWG), published a report entitled *Biotechnology's Bitter Harvest* that is a blanket condemnation of HTC as damaging to the environment and public health. The Council for Agricultural Science and Technology (CAST) counters BWG's assessment in its own report, *Herbicide-Resistant Crops.*

The primary issues the two groups disagree on are whether the use of HTC will

- Further our reliance on herbicides instead of other weed control methods;

- Increase the amount of herbicides used;

- Cause the evolution of resistance to herbicides by weeds;

- Allow the transfer of genes for herbicide tolerance to weeds, making weeds no longer susceptible to herbicides.

These different viewpoints about the environmental effects of HTC can provide the key points for a classroom debate.

Below, you will find background information on herbicides and HTC and a summary of the main points made by the BWG in *Biotechnology's Bitter Harvest* and, when appropriate, the countering points made by the CAST in *Herbicide-Tolerant Crops.* For additional information on weed control, herbicides, and HTC, contact the U.S. Department of Agriculture, your state Department of Agriculture, your county extension agent, or scientists in the crop science department at a state university with specialists in agriculture.

Background on HTC
Some of the first products of agricultural biotechnology will be crops genetically engineered to tolerate a variety of herbicides, the compounds farmers use to kill weeds in their fields. HTC are not new. Virtually all of the acreage of our major crop plants such as corn, soybeans, cotton, and wheat is treated with herbicides. Therefore, all of our major crops are already tolerant to some herbicides.

Where did this tolerance originate? Genetically based herbicide tolerance exists in nature. Some of our crops were naturally resistant to the herbicides we have developed. These natural HTC serve as sources of herbicide tolerance genes for other crops.

Using existing, genetically based herbicide tolerance has been useful but, by its very nature, limited. Because of the constraints of crossbreeding, we are able to use only the herbicide tolerance genes found in crops or their close relatives. Genetic engineering now provides us with the ability to capitalize on genetic variation found anywhere in nature. Some bacteria naturally contain genes for herbicide tolerance. We have moved those genes into both major and minor crops.

Major points in the two reports
Weed Control
The BWG states that other available options for controlling weeds are just as effective as herbicides and could replace herbicides as ways to control weeds reliably.

Herbicides and the Environment
The BWG report assumes that herbicides as a group are damaging to the environment and public health. The CAST report states that herbicides vary widely in their toxicities and implies that some are relatively

Societal Issues

harmless because they have low toxicity and do not persist in the environment beyond a few days. It also states that there are environmental benefits to using herbicides.

HTC and Herbicide Use
The BWG asserts that the development of HTC will increase the amount of herbicides used in the United States, while CAST states that the development of HTC could have any of the following effects on the amount of herbicides used: no effect, decreased amount required, or increased amount required. The CAST report states that anyone trying to determine whether or not the use of HTC will increase the amount of herbicides used by U.S. farmers must examine the issue on a case-by-case basis, taking into account both the crop and the herbicide.

HTC and the Evolution of Resistance to Herbicides
The BWG asserts that the use of HTC will cause the evolution of resistance to herbicides by weeds. The CAST report describes this phenomenon in detail. What questions do you need to ask and answer in order to assess this issue?

HTC and the Development of Weeds Resistant to Herbicides
The BWG maintains that gene transfer will occur between HTC and weeds, thus making weeds resistant to herbicides. What scientific facts do we know that support or refute this claim?

For Further Discussion
In making choices about developing HTC, which herbicide and crop traits would be important to consider for environmental reasons?

Societal Issues

Environmental Introductions of Genetically Engineered Organisms

Introduction

There has been considerable debate in the last 10 years about the risks of introducing genetically engineered organisms (GEOs) into the environment. As you might expect, some environmentalists urge caution in permitting these introductions. Those involved in developing GEOs for environmental uses, however, fear that excessive caution will deter the development of a promising industry.

In covering the issue, the press has tended to cast this debate in black-and-white terms. The media's treatment of the issue makes it appear as if the sides are totally polarized. Unfortunately, what has been ignored by the media is the general agreement among all parties on some of the key scientific issues germane to assessing the risks of environmental introduction of GEOs. The debate appears to have been driven primarily by factors outside of science—history, economics, politics, sociology, personalities, psychology—and not by issues of health and safety.

Before we address the specific concerns that have been raised regarding the environmental introductions of GEOs, let's look at three major points of agreement about the safety of environmental introductions of GEOs.

Points of Agreement

1. The level of risk of environmental introduction of GEOs

Perhaps the most important point of agreement is that most introductions of GEOs will pose minimal risk. As evidenced by the following quotes, ecologists and environmentalists share the view that most releases are low risk.

> *It is generally acknowledged that the overwhelming majority of introductions is likely to be environmentally benign.*
>
> Dr. Simon Levin, Past President of the Ecological Society of America

> *Most engineered organisms will pose minimal ecological risk.*
>
> Ecological Society of America

> *Like most novel organisms created or introduced by man, most genetically engineered organisms will be environmentally benign.*
>
> Dr. Rebecca Goldburg, Environmental Defense Fund

Everyone also agrees that a small number of introductions might present problems. Consequently, the federal government continues to regulate field tests of GEOs with which we have limited experience. At least one, and sometimes two, federal regulatory agencies have jurisdiction over these tests: the U.S. Department of Agriculture, the Environmental Protection Agency, and the Food and Drug Administration. As of 1995, thousands of field tests of GEOs have been done, and there have been no adverse consequences.

2. The basis of risk assessments of environmental introduction of GEOs

What virtually everyone also agrees upon is that the risk a given organism poses depends on *what* the organism is, not on *how* it was produced.

> *Risk should be assessed on the nature of the organism and the environment into which it is introduced, not on the method by which it was produced.*
>
> National Academy of Sciences, 1987

This view was espoused originally in a 1987 National Academy of Sciences report, *Introduction of Recombinant DNA-Engineered Organisms into the Environment: Key Issues,* and has been echoed by a wide variety of individuals and interest groups, such as the Ecological Society of America, the American Society for Microbiology, the Environmental Defense Fund, and the Audubon Society. This principle is often discussed as risk assessment based on "product not process."

Societal Issues

We have been releasing naturally occurring and genetically modified organisms—plants, animals, and microbes—into the environment for centuries. The methods used to genetically alter organisms have been selective breeding and mutagenesis, which, you will recall, are less predictable and less precise than genetic engineering. The "product-not-process" principle means that the risks posed by environmental introductions of GEOs are the same types of risks we have contended with in environmental introductions of naturally occurring organisms or of organisms genetically modified through mutagenesis and selective breeding.

3. The best model for risk assessments of environmental introductions of GEOs

The third point of consensus is the most appropriate model for assessing risks. You will hear different people focus on different portions of our historical record in order to support an opinion regarding the risks of releases of GEOs. Which of the following bodies of data provides us with the most appropriate information for assessing risks of field releases of GEOs?

- Introduced "exotic" species
- Laboratory experience with GEOs
- Genetic manipulation and introduction of agricultural species

Exotic-Species Model

Some people fear that introducing GEOs into the environment will lead to ecological disturbances similar to those resulting from the releases of foreign species into this country. The ecological problems most often cited are the disturbances caused by kudzu and starlings. In both of these situations, ecological havoc has resulted because the foreign species have exhibited uncontrolled population growth and, as a result, displaced native species.

While on the surface it may appear that the introduction of foreign species is analogous to environmental introductions of GEOs, scientists note a fundamental difference in the two situations.

The exotic-species model applies to situations in which species are accidentally or deliberately taken from their natural habitats and placed in a new environment. When nonnative species are introduced into a new environment, they sometimes cause ecological disturbances.

Most introductions of GEOs will involve reintroducing genetically modified organisms to their natural habitats. The transfer of species from one environ-

ment to another involves fundamentally different considerations than the ones involved in reintroduction of a slightly modified species into the environment from which it was taken.

If, however, an organism is taken from environment A, genetically modified, and then placed in environment B, the exotic-species model would apply. In that case, we can use our experience with exotic species to determine which negative outcomes could occur.

Laboratory Experience with GEOs

A second body of historical evidence that people may cite in predicting the effects of field releases of GEOs is our extensive and safe experience with GEOs in the laboratory. The reasoning goes something like this. We have been genetically engineering organisms by using recombinant DNA techniques for almost 20 years. Some of these organisms have undoubtedly escaped, and yet no health or environmental problems have resulted from these accidental releases. Therefore, the release of GEOs to the environment must involve no risk.

This argument is very misleading. Laboratory organisms—genetically engineered or not—are adapted to life in a laboratory. They cannot survive outside this sheltered environment. GEOs that are intentionally released into the environment must be able to survive if they are to serve their purpose. The lack of problems caused by accidental release of GEOs from the laboratory is not applicable to assessing the risks of intentional releases.

But just as our experience with exotic species is not wholly irrelevant, our experience with GEOs in the laboratory has taught us something useful regarding risk. It has provided us with data to support the assertion that genetic engineering techniques are not inherently dangerous. They alter the properties of organisms in highly predictable ways. From our laboratory experience, we have obtained data to substantiate the opinion that product, not process, is the relevant variable in risk assessment.

Genetic Manipulation and Introduction of Agricultural Species

Environmental, industrial, and academic scientists agree that genetic engineering of organisms and the subsequent release of these organisms into the environment correspond most closely to the genetic modification of agricultural species through human-imposed selection and mutation and their release into the environment. What has this experience with releases of plants, animals, and microbes taught us?

First, let's look at plants we have genetically modified through selective breeding to be disease resistant, insect resistant, herbicide resistant, and drought tolerant, the same traits we are trying to incorporate into crop plants through genetic engineering. A quote from a study by the National Research Council (1989) clearly supports the lack of problems we have encountered through our use of selectively bred crops.

Extensive experience has been gained from routine field introductions of plants modified through selective breeding. Hundreds of millions of field introductions of new plant genotypes have been made by American plant breeders this century. There have been no unmanageable problems from these introductions.

In addition to the experience we have gained from selective breeding of crop plants, a second body of experience we can draw upon includes the innumerable releases of microorganisms we have undertaken to date. Many of these organisms have been genetically manipulated through less precise techniques such as mutagenesis. Examples of these releases include

- Insect, bacteria, fungi, baculoviruses, and viruses for agricultural pest control
- Nitrogen-fixing bacteria for increased agricultural production
- Microorganisms applied in domestic waste treatment
- Microbes used by the food-processing industry

No environmental or public health problems have resulted from these millions of releases. In fact, the most important lesson we have learned from these releases is that getting introduced microbes to survive and persist is very difficult.

Examples of Concerns

What are some of the ecological issues regarding risk that have been raised, and what do we need to know in order to assess these issues objectively and knowledgeably?

Creating new pathogens and weeds

A common concern among the lay public is that we will create new pathogens or new weeds with genetic engineering. How valid is this concern? It depends on the organism the scientist is studying.

Known pathogenic species must be genetically modified with great caution. Pathogenic species often consist of virulent and nonvirulent strains, and small changes in the genetics of a nonvirulent strain can convert it into a virulent strain. That is how nature creates new strains of the influenza virus. Small genetic changes have also been important in the evolution of fungal plant pathogens such as black stem rust, a pathogen of wheat. If one is working with a nonpathogenic species with the intention of altering properties related to pathogenicity, similar caution should be exercised.

But can a scientist inadvertently convert a nonpathogen into a pathogen through genetic engineering? This is highly unlikely if the scientist is working with a nonpathogen and is engineering traits unrelated to pathogenicity. The likelihood of accidentally producing a pathogen is so negligible as to be nonexistent, because pathogenicity involves a very complicated set of characteristics. The pathogen must be able to recognize specific host cells, attach to specific host cells, overcome a wide range of host defense systems, synthesize toxic chemicals to kill cells, produce enzymes that degrade cellular components, and disperse readily to invade new hosts.

The same argument holds true for weedy plants. Weediness results from an equally complex set of traits, and therefore, it is highly unlikely that weediness will arise by accident when plants are engineered for characteristics unrelated to weediness.

Horizontal gene transfer

A much more substantial concern involves the possible exchange of genetic material between domesticated targets of genetic engineering and their relatives.

As we described in chapters 1 and 2, exchange of genetic material between domesticated plants and their wild relatives through sexual reproduction is one of a variety of mechanisms of exchange known to occur between species. Therefore, a gene in one plant species can be transferred to a different plant species. Such a transfer clearly has the potential to cause problems. To evaluate the potential risk of such an exchange, one needs to know something about the particular evolutionary and ecological relationships in a given community.

The accidental transfer of genes from crops to weeds is possible only if the two species are closely related, and even then, transfer may not occur. In the United States, a few of our major field crops, such as sorghum, sunflower, and clover, interbreed with their wild, weedy relatives. In the centers of origin of our major food crops, which are primarily locations in Central and South America, Africa, the Near East, and Asia, gene transfer between crops and wild, weedy relatives is more likely.

However, the potential for gene exchange—that is, the presence of a wild relative—does not necessarily imply that a risk is involved. To assess risk, first you must determine that the crop plant actually hybridizes with weedy relatives under field conditions. If it does, a second question must be asked in assessing risk: Are the offspring fertile? If not, the trait will not become established in the weed population. If the offspring are fertile, ask a third question. Does it matter if the gene moves from the crop plant into the weed? Will acquisition of this trait and its persistence in the weedy plant population pose a risk to public health or the environment or significantly alter existing ecological relationships or agricultural practices?

Bacterial populations can also exchange certain genes between species. The most familiar example of such exchange involves the proliferation of antibiotic resistance among bacterial species. Once again, however, the ability to transfer genes does not necessarily constitute a risk or danger. One must ask whether it matters that other bacterial species acquire this trait and, equally important, whether any selective pressure that will favor the spread of this trait through the population is operating.

Harm to nontarget species

The essence of the concern about harm to nontarget species is best illustrated by examples. A farmer could release genetically engineered parasitoids to control a certain species of caterpillar, but the parasitoid might also parasitize other, nonpest caterpillars. A bacterium genetically engineered to be pesticidal to one type of soil organism might also harm another soil organism.

Of course this could happen. It already does happen in current agricultural practices that rely on chemical pesticides, which are much less specific and precise than biological control mechanisms. How we view the harm done to the nontarget species seems to vary with the economic or esthetic value of the harmed species and our view of our role in the natural world and our relationship to other species.

Disruption of community structure

The concern about community structure deals with the unintended elimination of either wild or desirable

naturalized species through competition or predation. For example, the introduction of a highly competitive species of nitrogen-fixing bacteria into certain agricultural fields has made it difficult to introduce other more effective nitrogen fixers.

Disruption of community structure will become a risk when a GEO persists in the community for a long time. Therefore, most introductions into annual cropping systems should not result in any long-term shifts in the ecological community. Disruption of community structure is more likely when the GEO is being introduced into a nonmanaged system, such as a spruce-fir or oak-hickory forest, or when the plant in question is perennial, has broad dispersal abilities, or is sympatric with weedy relatives.

Disruption of ecosystem processes

The main players in fundamental ecosystem processes such the nutrient and mineral cycles are microorganisms. Some people have expressed concern that the introduction of genetically engineered microorganisms could result in the displacement of resident microbial species that play a vital role in these critical processes.

The generally accepted scientific opinion is that there is no basis for this concern. Natural communities have considerable redundancy; many species perform similar key roles. A recent paper by the Ecological Society of America states that, "because redundancy of function appears to be common in microbial communities, in many cases there would be little concern over microbial species displacement caused by an introduced transgenic organism."

Summary

In summary, while there may be environmental issues involved in releasing GEOs, none of the issues is new or unique to these organisms. Through our past experiences with releases of naturally occurring and genetically modified plants, animal, and microbes, we have acquired abundant experience that is useful in conducting a risk assessment for environmental introductions of GEOs: identifying the risk, assessing the probability of its occurrence, and determining the consequences.

Selected Readings

*Congressional Office of Technology Assessment. 1981. *Impacts of Applied Genetics: Micro-organisms, Plants, and Animals.* U.S. Government Printing Office, Washington, D.C.

*Congressional Office of Technology Assessment. 1987. *Field-Testing Engineered Organisms: Genetic and Ecological Issues.* U.S. Government Printing Office, Washington, D.C.

Davis, Bernard D. 1991. *The Genetic Revolution: Scientific Prospects and Public Perceptions.* The John Hopkins University Press, Baltimore, Md.

Dixon, Bernard. 1985. *Engineered Organisms in the Environment: Scientific Issues.* The Maple Press Co., York, Pa.

National Academy of Sciences. 1987. *Introduction of Recombinant DNA-Engineered Organisms into the Environment: Key Issues.* The National Academy of Science Press, Washington, D.C.

National Research Council. 1989. *Field-Testing Genetically Modified Organisms: Framework for Decisions.* National Academy Press, Washington, D.C.

Tiedje, J. M., R. K. Colwell, Y. L. Grossman, et al. 1989. The planned introduction of genetically engineered organisms: ecological considerations and recommendations. *Ecology* 70:298–315.

*To order publications from the U.S. Government Printing Office, write to the Superintendent of Documents, Mail Stop: SSOP, Washington, DC 20402-9328.

Societal Issues

Preface to Chapters 32 to 34

Introduction

As we have seen in chapters 29 to 31, no technology exists in a social vacuum. The use of any technology changes the society that uses it, often in unpredicted ways. Like the technologies that preceded it, biotechnology will provide us with new products, new processes, and new problems. But remember, the relationship between technology and society is circular. Not only do technologies shape societies, but societies shape technologies. If we make wise choices, we can minimize the problems and maximize the gains.

Biotechnology triggers strong emotional responses. Its remarkable capabilities raise both our hopes and our fears. Nowhere is this more apparent than in the area of medical biotechnology. The tremendous medical advances of the last decade give us reason to rejoice and also to reflect. As we said earlier, biotechnology gives us not only new knowledge but also new powers. As a result, we will be forced to wrestle with ethical issues we have never had to address. Our society must begin a discourse on these difficult issues before they become intractable problems.

The Bioethics Decision-Making Model

All too often, "debates" about controversial social issues are merely emotional exchanges of opinion. Having a highly structured framework for addressing these difficult issues helps us keep the discussion on track.

Teachers have found the bioethics decision-making model we describe in this section to be helpful in channeling productive thought and debate. The model is applied to specific cases that present ethical dilemmas. Ethical questions are raised and considered in the context of these cases. The decision-making model requires students to examine the facts of the case, consider alternatives, and choose the alternative they consider to be the best. They must then justify their choice in terms of fundamental ethical principles and defend their decision by citing relevant authorities. Finally, students present their decision and the justification for it in classroom presentations or formal reports.

In addition to increasing students' awareness about bioethical issues, this case study approach offers the following benefits.

- Critical thinking skills. Students must distinguish facts from inferences, and personal opinion and emotion from reasoned analysis.

- Experience gathering accurate information. To the extent a teacher requires it, students will read what authorities on ethics have to say about relevant issues. They will also do library research to answer factual questions relevant to the cases they are studying.

- Social skills. If rules are enforced by the teacher, students can learn how to discuss a potentially emotional topic, respect different answers to the same question, and agree to disagree.

As will be evident, the model can be used to analyze situations in areas other than biotechnology and biomedicine. The nature of the model makes it most readily applicable to situations involving individuals (as opposed to corporations or governments, for example). The case studies presented here therefore deal largely with biomedical issues.

Societal Issues

A Decision-Making Model for Bioethical Issues

32

Introduction to Bioethics

Since the middle of the 20th century, rapid improvements in technology have changed the practice of medicine profoundly. The technology often allows physicians to restore or supplant basic bodily functions. Mechanical ventilators, heart pacemakers, kidney dialysis machines, exotic drugs, organ transplantation, and artificial nutrition and hydration are some of the life-extending tools available to the modern practitioner.

Initially, the use of these techniques was seen as altogether positive, a must-do choice for the physician. It was not long, however, before difficult questions surfaced. Just because we *can* maintain a person with artificial ventilation and nutrition, *should* we? Who should answer such a question: the patient, physician, family, nurse, social worker, hospital administrator, insurance company, ethicist, politician?

Since the 1960s, questions like this one have provided us with an entirely new discipline or, more accurately, interdiscipline: bioethics. Bioethics, or biomedical ethics, has become an immensely important topic. Few major hospitals are without an ethics committee to assist patients or health care providers when they need help. Colleges offer courses of study in medical humanities or bioethics. High school students, too, must be aware of the questions bioethicists study.

Paralleling the medical developments described above were striking advances in our understanding of genetics and molecular biology. Genetic knowledge has only recently been applied to specific diseases or patients, but the ethical questions that plague us about any medical technology apply also to medical biotechnology.

How do we decide what is right or wrong, what is better or worse, as medical biotechnology offers us ever more choices? Should everyone be screened to determine who the carriers of recessive alleles for genetic diseases are? Should genetically engineered products be available to everyone, or should we sometimes screen requests for them (such as requests for human growth hormone)? Who should (and who actually will) have access to a person's DNA fingerprint? Are we on the road to employment and insurance discrimination based on genetic profiles? These are but a few of the ethical questions that can be posed.

Case Studies in Bioethics

Experience has demonstrated that while classroom discussions of bioethical questions are usually provocative and consciousness-raising, they often result in nothing more than a sharing of emotional opinions about what should or should not be done. One effective way to get students to evaluate ethical issues objectively is to use case studies and a structured method of analyzing each case. In the approach we present below, students are presented with specific cases that pose not just ethical issues but also ethical dilemmas. Students identify an aspect of the case that represents an ethical dilemma and then apply ethical principles in a systematic way, following the decision-making model, to formulate a suggested solution. Students are required to justify the decisions they make about the case by citing authoritative documents such as the Nuremberg Code or the Hippocratic Oath, two widely used documents that set forth standards of conduct for biomedical professionals.

How to identify a dilemma

Not every bioethical case study presents a dilemma; many times, the possible courses of action are clearly right or wrong. A dilemma exists when there is no "right" course of action in a certain situation but, instead, several options, none of which is wholly acceptable. Ethical dilemmas revolve around trying to find the best solution when no solution is completely good. Students must identify the dilemma they will address and then propose a course of action that is based on adherence to basic ethical principles. The following example may clarify what does and does not constitute a dilemma.

Assume that a patient with a certain condition would be an appropriate candidate for a drug research study. The patient's physician places her on this drug without getting her permission. This situation is not a dilemma. It is just plain wrong. Even if the doctor believes the drug will benefit the patient, by all modern medical standards, the physician has the obligation to get the patient's informed consent to include her in the study. (The Nuremberg Code and the Declaration of Helsinki specifically address the ethics of research using human beings.)

On the other hand, assume that the patient has been given all the information she needs for making a decision. She is told that the drug has the potential to help her but might also have harmful side effects. She sees benefits and costs regardless of which decision she makes. Now we have a dilemma. A dilemma exists when no choice is ideal but all options have benefits and risks that must be carefully assessed.

Once a dilemma has been identified, the next step is to pose the dilemma in the form of a question. Often, more than one question can be formulated. By working through one question at a time, everyone in the group can be sure that everyone else is talking about the same problem. It is also helpful to categorize the issue being discussed. In the example given previously, the issue might be human research, and the question posed could be, "Should the patient agree to be part of the study or not?"

Basic ethical principles

In order to make ethical decisions, we must agree on some basic guidelines about what constitutes moral conduct. In the field of biomedical ethics, certain guides are well established. The moral-action guides or principles can be divided into four major principles and several secondary ones. These principles are listed below (the terms in parentheses are those used in bioethical literature). Before simply listing these for your students, you may want to have the students list guidelines for behavior and decision making that are important to them. Most values can then be categorized into one of the general principles listed here.

Major Ethical Principles
1. Do no harm (nonmaleficence).
2. Do good (beneficence).
3. Do not violate individual freedom (autonomy).
4. Be fair (justice).

Secondary Ethical Principles
1. Tell the truth (truthtelling).
2. Keep your promise (fidelity and promise keeping).

3. Respect confidences (confidentiality).
4. Use the principle of proportionality: risk-benefit ratio (how much harm can be justifiably risked to effect good).
5. Attempt to avoid undesirable exceptions, also known as the wedge principle, the slippery slope, or the camel's nose.

Although these rules are simple, they represent fundamental values associated with respect for human dignity that most people agree to. These are the principles to which students should refer when making and justifying their decisions.

Using the Decision-Making Model

When faced with an ethical dilemma, how do we decide what to do, since there are no right answers? To assist students in learning how to follow a rational decision-making process based on ethical principles, you must provide a highly structured methodology. We suggest the decison-making model described below. Following the prescribed, step-by-step procedure laid out in the model greatly reduces the chance of getting off track. The model requires students to stay focused on the issue at hand and helps them develop critical-thinking skills.

Basic steps in the model, starting from a case study
1. Identify the question you want to address. Usually, for any given case, many questions could be considered. Choose the one you want to explore.
2. Identify the issue you are exploring (i.e., genetic screening, confidentiality, gene therapy, human research subjects, etc.). Naming the issue will help in the search for relevant literature.
3. State the facts in the case. Be sure to avoid inferences.
4. Think of as many possible decisions in the case as you can.
5. Gather additional information as needed.
6. Pick the decision you want to support.
7. State the ethical principle that supports your decision (your claim).
8. Identify an authority that supports your decision. Quote the authority, if possible.
9. Formulate a rebuttal. Under what circumstances would you abandon your claim?
10. How strongly do you believe your claim? What is your level of confidence, the qualifier?
11. "Box up" the case for reporting your decision.
12. Write a prose argument describing the case and your decision.

Two examples using this decision-making model are presented below. The first example is a typical school ethics dilemma, and the second is a case from medical ethics. Different teachers may find one or both effective in demonstrating the model.

Rules for classroom discussion

Since the issues discussed in bioethics are controversial, it is important that rules of etiquette be observed from the very beginning. The following list has proven effective in the classsroom.

1. Only one person at a time speaks after being recognized by the discussion leader.
2. Treat each other with respect; no name calling. Critique the argument, not the author of the argument.
3. Seek clarity by asking questions.
4. Look for gaps in the data.
5. Recognize your own biases.
6. Be true to your own position; do not jump on the bandwagon.
7. Keep emotions in check; use logic.
8. Do not follow authority blindly (the teacher doesn't know everything).
9. You must have a reason, not just an opinion. You can like pepperoni pizza better than anchovy pizza with no reason. You cannot decide bioethical issues based on opinions.
10. Be open minded and willing to be perplexed.

The case studies and "right" answers

In the two example cases that follow, as well as in several of the additional case studies, sample work-ups and decisions have been supplied to the teacher. These are included to show you how to use the model, not because they are the right answers. Remember that a dilemma is a dilemma because there is no perfect solution. The sample solutions may seem to be good resolutions of the situations, but it is critical to remember that your task as teacher is not to guide students to these particular resolutions or even to focus on the same questions. Rather, your task is to guide students to their own resolutions through use of the model.

For example, an alternative decision for case study 1 (Frank and Martin) would be for Martin to report what he saw to the judiciary council. If Frank is not guilty of cheating, he can explain himself to the council. The principles supporting this decision would be justice (being fair in that Martin is adhering to the agreed-upon honor code) and truthtelling.

To avoid creating the impression that there are "right" questions to focus on and "right" answers, the sample solutions are not provided in the student pages.

Example Case Study 1: Frank and Martin

This case shows an ethical dilemma but does not involve medicine or biotechnology. This case may be useful in introducing students to the method because it involves territory and issues that are very familiar to them. In addition, it serves as a useful example of how the decision-making model can be applied to many problems outside the bioethics arena.

Frank is an 11th-grade student at a small public high school that prides itself on its family atmosphere and strong academic reputation. The school has an honor code. Each student agrees to abide by a published list of rules. One part of the code obligates students to report observations of fellow students who may be cheating. A judiciary council composed primarily of students decides what should happen in each case. Frank works hard at school. He has a B+ average and hopes to go to a good college. He also works part-time, runs track, and helps at home with two younger brothers. Both of his parents work. Frank is a close friend of Martin. Martin is very bright and does well in school, although he does not have to put in as much time studying as Frank does. Martin does not have many extracurricular activities or responsibilities. During a history test, Martin notices that Frank appears to be cheating off a note card.

I. Identify the question

Often there are many questions. In this case, some might be as follows.

1. Should Martin report his observation to the judiciary council?
2. Should Martin tell Frank what he saw?
3. Should Frank report himself to the judiciary council?
4. Should other students be asked what they saw?

Other questions could be posed. It is important to encourage students to ask as many questions as possible. List them all, but finally choose one question for the group to focus on first. As students become more adept at using the model, they can be divided into small groups, with each group addressing a different question.

Societal Issues

At the beginning, it is better to start with the same question. Working as a large group the first few times is also helpful in modeling the kinds of behaviors you want to encourage. Next, break up into small groups, but work on different questions. Eventually, work in small groups on different cases. To demonstrate the model using this case, we will use question number 1: should Martin report what he saw to the judiciary council?

II. Identify the issue

What general problem does the case demonstrate? In this case, we might say cheating or school rules. This step is an attempt to categorize the case.

III. State the facts of the case

What are the facts in the case? It is important to help students differentiate between facts and inferences. This skill is one that students may already have been introduced to in the laboratory, when they have tried to draw conclusions based on data collected rather than on inferences. Many students will make unsupported assumptions or will jump to conclusions. There are "rights" and "wrongs" at this step of the decision-making model.

In this case, for instance, we are not positive that Frank's note card is actually a cheat sheet. There is a possibility, perhaps remote, that Frank was working on another assignment and had permission to use the card. Students might also propose that everyone cheats and ask why Frank should get punished when others don't. That may or may not be true at this particular school. Nor can we make assumptions about Frank and Martin's friendship. Maybe they had a fight and Martin just wants to get Frank in trouble. It is a possibility but not a fact in this case. Time spent discussing the facts is worthwhile, since it can prevent confusion later in the discussions.

Once the facts have been established, it is helpful to list them in an accurate but concise manner. Facts must be true, relevant, and sufficient.

- Frank was seen using a note card; the honor code requires Martin to report the incident.
- Frank is in 11th grade, works, is a B+ student, and runs track.
- School is small, with a good reputation.
- Frank plans to go to college.
- Martin does well in school with less effort than Frank.
- Martin does not have many extracurricular activities.

(The relevance of the last three facts is worth discussing.)

IV. List as many possible decisions as you can

What are the possible answers to the question of whether Martin should report what he saw? Encourage students to generate as many choices as possible. List them all. Promote creative and lateral thinking by asking leading questions and allowing adequate wait time. Here are some possible choices.

1. No, Martin should not report what he saw. It could harm Frank. Maybe Frank had permission to use the card. How does Martin know whether Frank was really cheating?

2. Yes, Martin should report what he saw. It is his duty under the code. He breaks his own promise to abide by the code if he does not tell. Frank may need help if he did indeed cheat. If he didn't cheat, he can explain to the council.

3. Martin should tell Frank's parents and let them decide what to do.

4. Martin should tell his own parents and let them decide what to do.

5. Martin should tell another student and let the other student decide what to do.

6. Martin should tell Frank what he saw and give Frank the opportunity to tell the judiciary council what happened. If Frank refuses to go to the judiciary council, then Martin should tell.

V. Gather additional information as needed

It is absolutely critical that students learn to seek relevant information and not to make decisions based on uninformed opinion. At the very least, students should recognize when there are gaps in the information they have. This sample case probably will not require gathering additional background information. However, other cases may. For example, in order to weigh the risks and benefits in a case involving use of an experimental drug to treat a patient, students may want to know just how severe the patient's disease is. It might also be appropriate for them to find out what sort of safety testing a drug must undergo before it is approved for experimental treatment of humans.

VI. Pick the decision you wish to support

These are six possible choices (your class may generate many more). Encourage solutions that represent compromise. Since ethical problems are usually complex, it is important to take time for thoughtful, honest reflection. The students must have a reason for the option they choose, and that reason should be related to one of the principles listed on page 367. However, students may make different decisions based on different principles.

For our example, we will choose option 6. With this choice, we respect Martin's obligation to tell what he saw, but at the same time, we might be able to avoid harming Frank.

VII. Identify the ethical principle that supports your decision

An ethically justifiable decision can be based on alternative principles. In a dilemma, adherence to one principle often results in the breach of another; dilemmas exist because principles often conflict. Application of different basic values can lead to different responses to a situation.

Students should learn what the basic principles are and recognize how they can be applied. One of the goals of using this model in the classroom is for students to develop an appreciation for the fact that two responsible, moral people can make very different decisions in a case because they are guided by two different basic principles.

The logic involved in the choice of principle(s) for these classroom cases should be explicitly discussed and/or written about by the students. They need to be able to justify why they have chosen one principle and why they are willing to breach the alternative principle(s). Although students can choose more than one principle to support their decision, it is helpful to force them to choose only one at the beginning to be certain they understand the differences between principles. In the sample case, the first principle we are adhering to is avoiding harm, and the second is truthtelling.

VIII. Identify a supporting authority

What experts or authorities would back up our position on this case? Normally, we would look to professional codes of ethics. In this case, the honor code itself would be the authority. In bioethical decisions, several codes of ethics are used; these codes include

the Hippocratic Oath, the American Medical Association Code of Ethics, the American Hospital Association's Bill of Rights for Patients, and the American Nursing Association Code of Ethics as well as the Nuremberg and Helsinki statements. Depending on the time available, library research is always desirable. Have your students find these documents in books and bring them to class.

IX. Formulate a rebuttal

Under what circumstances would we change our decision about what to do? Here again, it is important to encourage students to think creatively. This section is often difficult, because students are invested in their choice and may not be able to imagine a circumstance that would make them change. In this case, what if Frank were taking a different test, perhaps a more advanced one, and had permission to have the note card? What if the note card was actually an appointment slip that Frank was checking to be sure he was on time for an after-school doctor's appointment? If Martin knew that kind of information, then surely the problem would evaporate.

X. State your level of confidence in your decision

The student should formulate a one- or two-word statement to describe how strongly he or she believes his or her own argument. Until they have more experience, students tend to believe their own arguments are infallible. One way to assess the strength of the argument is to gauge the likelihood of rebuttal. If a rebuttal is highly unlikely and the rest of the argument makes sense, then the argument is a strong one. Also, if the principle ties the claim and the facts tightly together, then the argument is strong. Forcing students to qualify their arguments is one way to promote self-evaluation. As students work up more and more cases, they become better at constructing an argument and more realistic about its strength. Qualifiers might be "moderately confident," "absolutely confident," or "questionably confident." In this case, we will use "strongly confident."

XI. Box up the case for reporting

See Fig. 32.1 for a "box" of the sample case. Obviously, many other boxes could have been constructed. Students, usually working in groups of three or four, transfer this box to a transparency and present their case to the class, starting with the facts. Having students critique one another's arguments is a valuable lesson as well.

Issue:

Cheating

Question:

Should Martin report what he saw to the council?

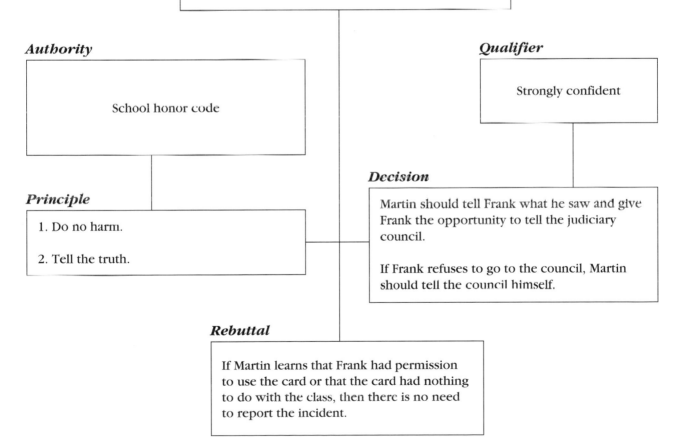

Facts

- Frank seen using note card during test
- Honor code requires Martin to report incident.
- Frank in 11th grade, works, B+ student, runs track
- Small school with good reputation
- Frank plans to go to college.
- Martin does well in school.
- Martin has few extracurricular activities.

Authority

School honor code

Qualifier

Strongly confident

Principle

1. Do no harm.

2. Tell the truth.

Decision

Martin should tell Frank what he saw and give Frank the opportunity to tell the judiciary council.

If Frank refuses to go to the council, Martin should tell the council himself.

Rebuttal

If Martin learns that Frank had permission to use the card or that the card had nothing to do with the class, then there is no need to report the incident.

Figure 32.1   "Box" of example case study 1: Frank and Martin.

Societal Issues

XII. Prepare a prose argument

Finally, if time permits, students can write up their arguments, using the box as an outline. They should produce a paper that can be understood by someone unfamiliar with the original case. Students who write well have little difficulty with this task. However, many need help using proper transitions. They should also explain more in the paper than the box can show. For instance, why is avoiding harm more important, at least initially, immediately than telling the truth? Students who are willing to apply this model to other classes often find it works very well in writing papers in English or history class. This reasoning model can be applied to almost any discipline.

Example Case Study 2: Mr. Johnson

Mr. Johnson, age 76, was admitted to a medical unit with pneumonia. He had a history of severe emphysema, for which he had been hospitalized twice in the past year because of secondary pneumonia. During the evening of the third day of intravenous antibiotic treatment, Mr. Johnson's complexion took on an increasingly bluish tint, and he was short of breath. Because the attending physician was out of town, the physician taking calls was notified by the evening nursing supervisor. The physician found the patient to be severely oxygen deficient. He tried unsuccessfully to reach Mrs. Johnson to notify her of his plan to transfer Mr. Johnson to the intensive care unit for ventilator support. During the process of arranging for the transfer, a nursing assistant said that she thought that Mr. Johnson and his wife had said that they would refuse life support measures if Mr. Johnson ever needed them. No documentation of these comments could be found.

When Mr. Johnson arrived in intensive care, he was anxious, and his breathing was so labored that he could not talk. He shook his head "no" when the physician explained the procedure for inserting a tube to connect him to the ventilator, and he attempted to push the physician away. Mr. Johnson's competency to make decisions about his treatment was questioned because his arterial blood oxygen was so low. Another unsuccessful attempt was made to reach Mrs. Johnson. Mr. Johnson's condition deteriorated as the physician and staff deliberated about proceeding with the intubation. Finally, the patient was sedated and placed on the ventilator.

Mrs. Johnson arrived the next morning and was shocked that her husband's condition had deteriorated so rapidly. She was upset that her husband was on the ventilator, explaining that they had agreed that life support measures would not be used for him. When the on-call physician asked, "What do you want me to do," Mrs. Johnson replied, "I don't know." The physician asked for assistance from the Ethics Committee.

I. Identify the question

Students will probably raise the most obvious question about what to do with Mr. Johnson: leave him on the ventilator or take him off. Other questions that they might raise include where Mrs. Johnson was all this time, where the attending physician was, whether Mr. Johnson is competent now, what will happen to Mr. Johnson if he stays on the ventilator, and whether Mr. Johnson has any children who disagree with Mrs. Johnson about her husband not wanting to be on a ventilator. These questions are interesting, and some of them could certainly add to the data in the case, but the central question about whether to take Mr. Johnson off the ventilator or leave him on would remain, so that is the question we will address here. Should Mr. Johnson remain on the ventilator?

II. Identify the issue

The general category of problem could be called withdrawal of life support or end-of-life decisions or extraordinary means. This step helps in grouping cases and may make it easier for students to find references in the literature.

III. State the facts of the case

What are the facts in this case? Students who have worked through this case are often angry at Mrs. Johnson and the regular doctor for not being there when they were needed, and students attribute all kinds of unworthy motives to both of them. It is an ideal opportunity to reiterate the differences between facts and inferences and to insist that only facts as they are known may be used. The facts could be summarized as shown in Fig. 32.2.

- 76-year-old Mr. Johnson
- Diagnosis: pneumonia secondary to severe emphysema
- Third hospitalization in last 12 months
- Treatment: intravenous antibiotics
- Day 3: short of breath, blue, oxygen deficient
- Doctor on call tried to reach Mrs. Johnson twice but could not locate her.
- Nursing assistant reported that she "thought patient and wife would refuse life support measures."
- No documentation of those wishes
- Patient unable to talk on admission to the intensive care unit

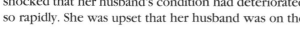

Issue:

Withdrawing life support

Question:

Should the physician disconnect
the ventilator?

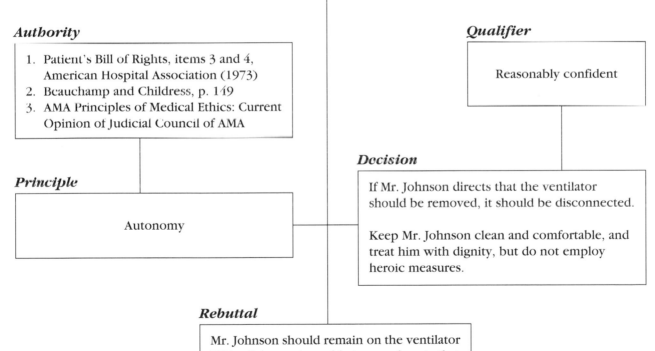

Facts

- Mr. Johnson is 76 years old.
- Diagnosis: pneumonia, secondary to emphysema
- Third hospitalization in last 12 months
- Intravenous antibiotics prescribed
- Day 3: short of breath, turning blue, oxygen deficient
- Doctor tried unsuccessfully to reach Mrs. Johnson.
- Nurse "thought Mr. J. and wife would refuse life support measures."
- No documentation of those wishes
- Mr. J. unable to talk on admission to the intensive care unit
- Blood O_2 level low; patient anxious
- Condition deteriorating
- Pushed doctor's hands, shook head "no"
- Sedated and intubated
- Next day, wife shocked at husband's condition
- Upset that Mr. J. was on ventilator
- Said she and Mr. J. had discussed not wanting life support

Authority

1. Patient's Bill of Rights, items 3 and 4, American Hospital Association (1973)
2. Beauchamp and Childress, p. 149
3. AMA Principles of Medical Ethics: Current Opinion of Judicial Council of AMA

Qualifier

Reasonably confident

Principle

Autonomy

Decision

If Mr. Johnson directs that the ventilator should be removed, it should be disconnected.

Keep Mr. Johnson clean and comfortable, and treat him with dignity, but do not employ heroic measures.

Rebuttal

Mr. Johnson should remain on the ventilator if Mrs. Johnson is unable to corroborate that he did not want heroic measures or if health care personnel have evidence that he did indeed want extraordinary measures.

Figure 32.2   "Box" of example case study 2: Mr. Johnson.

- Blood oxygen level low; patient anxious
- Condition deteriorating
- Pushed doctor's hands away, shook head "no"
- Sedated and intubated
- Next day, wife shocked at husband's condition
- Upset that Mr. Johnson was on ventilator
- Said that she and Mr. Johnson had discussed not wanting life support

IV. List the possible decisions

What are the possible courses of action in this case? Students should be encouraged to list as many as possible. In this case, for instance, here are some.

1. Mr. Johnson should remain on the ventilator.

2. Assuming that Mr. Johnson is competent, discuss the situation with him. If he directs the doctor to remove the ventilator, then remove it. Keep him clean and comfortable, and treat him with dignity, but do not employ heroic measures. If he wishes to be left on the ventilator, continue the present course.

3. If Mr. Johnson cannot communicate or is deemed to be incompetent or both, Mrs. Johnson, the attending physician, nurses, social workers, etc., should decide what Mr. Johnson would direct were he competent and able to communicate.

4. Assuming that Mrs. Johnson's position is corroborated (that Mr. Johnson would not want life support), then Mr. Johnson's ventilator should be removed as in option 2.

V. Gather needed background facts

Obviously, no further information about the Johnsons is available, but students will probably need to find out more about emphysema to understand Mr. Johnson's situation. They also may wish to learn more about breathing machines and their uses.

VI. Make a decision

For the purpose of this example, we will choose decision 2: if Mr. Johnson directs the doctor to remove the ventilator, then remove it, but keep him clean and comfortable, and treat him with dignity. Do not employ heroic measures.

VII. Identify the guiding principle

Since making the decision involves asking Mr. Johnson what he wants, the decision is based primarily on respect for his autonomy. We are trying to let him de-

cide for himself what he does or does not want. At the same time, we are trying to avoid harming him. Some would argue very convincingly, however, that removing him from the ventilator will result in his death, which is the ultimate harm. Others will argue that the ventilator is merely prolonging the moment of death and that in this situation, although removing the ventilator may hasten Mr. Johnson's death, that is not a harm. Herein lies the crux of these decisions. What principles do the individual students think are most important in each case? They need to be able to say why they think one principle takes precedence over another.

VIII. Cite a supporting authority

As time permits, students can read about similar cases, and they should examine the codes of ethics mentioned in the first sample case.

IX. Formulate a rebuttal

Under what circumstances would we abandon our decision? For instance, if Mr. Johnson is not competent and no one can corroborate that he did not want life support measures, then we may decide to leave him on the ventilator. Perhaps we discover that Mr. and Mrs. Johnson were having marital problems but that Mr. Johnson still has a $5 million insurance policy with Mrs. Johnson as the beneficiary. If he cannot speak for himself, we may be reluctant to discontinue the ventilator.

X. State your level of confidence

This case has quite a few unknowns, so we may not be very confident about our decision. We don't know whether Mr. Johnson is competent; we don't know what his prognosis is without the ventilator; we don't know how he felt about his quality of life before this episode. There are enough questions to make us somewhat unsure about our decision.

XI. Box

A box for this case is shown in Fig. 32.2.

XII. Prose argument

As time and interest dictate, students can write up their arguments in paragraph form.

In the next two chapters, the decision-making model will be applied to gene therapy and genetic screening cases. Blank "box-up" forms for those exercises are provided in Appendix K.

A Decision-Making Model for Bioethical Issues

Introduction to Bioethics

Since the middle of the 20th century, rapid improvements in technology have changed the practice of medicine profoundly. The technology often allows physicians to restore or supplant basic bodily functions. Mechanical ventilators, heart pacemakers, kidney dialysis machines, exotic drugs, organ transplantation, and artificial nutrition and hydration are some of the life-extending tools available to the modern practitioner.

Initially, the use of these techniques was seen as altogether positive, a must-do choice for the physician. It was not long, however, before difficult questions surfaced. Just because we *can* maintain a person with artificial ventilation and nutrition, *should* we? Who should answer such a question: the patient, physician, family, nurse, social worker, hospital administrator, insurance company, ethicist, politician?

Since the 1960s, questions like this one have provided us with an entirely new discipline or, more accurately, interdiscipline: bioethics. Bioethics, or biomedical ethics, has become an immensely important topic. Few major hospitals are without an ethics committee to assist patients or health care providers when they need help. Colleges offer courses of study in medical humanities or bioethics. High school students, too, must be aware of the questions bioethicists study.

Paralleling the medical developments described above were striking advances in our understanding of genetics and molecular biology. Genetic knowledge has only recently been applied to specific diseases or patients, but the ethical questions that plague us about any medical technology apply also to medical biotechnology.

How do we decide what is right or wrong, what is better or worse, as medical biotechnology offers us ever more choices? Should everyone be screened to determine who the carriers of recessive alleles for genetic diseases are? Should genetically engineered products be available to everyone, or should we sometimes screen requests for them (such as requests for human growth hormone)? Who should (and who actually will) have access to a person's DNA fingerprint? Are we on the road to employment and insurance discrimination based on genetic profiles? These are but a few of the ethical questions that can be posed.

The Bioethics Decision-Making Model

In attempting to answer difficult ethical questions, it is helpful to focus on specific cases and follow a step-by-step procedure. Below, you will be presented with specific case studies to analyze and a decision-making model to be used in these analyses. Part of the decision-making process will be to gather any additional background facts you need for evaluating the situation and to find out what sort of ethical standards that apply to your case have been established. Finally, you will make a decision as to the best course of action and justify it in terms of basic ethical principles.

Identifying a dilemma

The case studies we have provided raise bioethical dilemmas. A dilemma exists when there is no "right" course of action in a certain situation but, instead, several options, none of which is wholly acceptable. Ethical dilemmas revolve around trying to find the best solution when no solution is completely good. Not every situation presents a dilemma; many times, the possible courses of action are clearly right or wrong. The following example may clarify what does and does not constitute a dilemma.

Assume that a patient with a certain condition would be an appropriate candidate for a drug research study. The patient's physician places her on this drug without getting her permission. This situation is not a dilemma. It is just plain wrong. Even if the doctor believes the drug will benefit the patient, by all modern medical standards, the physician has the obligation to get the patient's informed consent to include her in the study. (The Nuremberg Code and the Declaration of Helsinki, two internationally recognized codes of ethics, specifically address the ethics of research using human beings.)

On the other hand, assume that the patient has been given all the information she needs for making a decision. She is told that the drug has the potential to help her but might also have harmful side effects. She sees benefits and costs regardless of which decision she makes. Now we have a dilemma. A dilemma exists when no choice is ideal but all options have benefits and risks that must be carefully assessed.

Once a dilemma has been identified, the next step is to pose the dilemma in the form of a question about a specific case. Often, more than one question can be formulated. You will need to choose one question at a time for analysis, although you may consider several questions, one after the other, about a single case. It is also helpful to categorize the kind of issue being discussed. In the example given previously, the issue might be human research and the question posed could be, "Should the patient agree to be part of the study or not?" Since these discussions are usually very interesting and emotional, it is easy to get off track. The decision-making model presented here reduces that risk.

Basic ethical principles

In order to make ethical decisions, we must agree on some basic guidelines about what constitutes moral conduct. In the field of biomedical ethics, certain guides are well established. The moral-action guides or principles can be divided into four major principles and several secondary ones. These principles are listed below. The terms in parentheses are those used in bioethical literature. You may encounter them in your library work.

Major Ethical Principles
1. Do no harm (nonmaleficence).
2. Do good (beneficence).
3. Do not violate individual freedom (autonomy).
4. Be fair (justice).

Secondary Ethical Principles
1. Tell the truth (truthtelling).
2. Keep your promise (fidelity and promise keeping).
3. Respect confidences (confidentiality).
4. Use the principle of proportionality: risk-benefit ratio (how much harm can be justifiably risked to effect good).

5. Attempt to avoid undesirable exceptions, also known as the wedge principle, the slippery slope, or the camel's nose.

Although these rules are simple, they represent fundamental values associated with respect for human dignity that most people agree to. These are the principles to which you should refer when making and justifying decisions.

Basic steps of the decision-making model

Here is a summary of the steps in the decision-making model. Each step will be explained more fully later on. Begin with a case study.

1. Identify the question you want to address. Usually, for any given case, many questions could be considered. Choose the one you want to explore.

2. Identify the issue you are exploring (i.e., genetic screening, confidentiality, gene therapy, human research subjects, etc.). Naming the issue will help in the search for relevant literature.

3. State the facts in the case. Be sure to avoid inferences.

4. Think of as many possible decisions in the case as you can.

5. Gather additional information as needed.

6. Pick the decision you want to support.

7. State the ethical principle that supports your decision (your claim).

8. Identify an authority that supports your decision. Quote the authority, if possible.

9. Formulate a rebuttal. Under what circumstances would you abandon your claim?

10. How strongly do you believe your claim? What is your level of confidence, the qualifier?

11. "Box up" the case for reporting your decision.

12. Write a prose argument describing the case and your decision.

Rules for classroom discussion

Since the issues discussed in bioethics are controversial, it is important that the following rules of etiquette be observed from the very beginning.

1. Only one person at a time speaks after being recognized by the discussion leader.
2. Treat each other with respect; no name calling. Critique the argument, not the author of the argument.
3. Seek clarity by asking questions.
4. Look for gaps in the data.
5. Recognize your own biases.
6. Be true to your own position; do not jump on the bandwagon.
7. Keep emotions in check; use logic.
8. Do not follow authority blindly (the teacher doesn't know everything).
9. You must have a reason, not just an opinion. You can like pepperoni pizza better than anchovy pizza with no reason. You cannot decide bioethical issues based on opinions.
10. Be open minded and willing to be perplexed.

Example Case Study 1: Frank and Martin

This case shows an ethical dilemma but does not involve medicine or biotechnology. It serves as a useful example of how the decision-making model can be applied to many problems outside the bioethics arena.

Frank is an 11th-grade student at a small public high school that prides itself on its family atmosphere and strong academic reputation. The school has an honor code. Each student agrees to abide by a published list of rules. One part of the code obligates students to report observations of fellow students who may be cheating. A judiciary council composed primarily of students decides what should happen in each case. Frank works hard at school. He has a B+ average and hopes to go to a good college. He also works part-time, runs track, and helps at home with two younger brothers. Both of his parents work. Frank is a close friend of Martin. Martin is very bright and does well in school, although he does not have to put in as much time studying as Frank does. Martin does not have many extracurricular activities or responsibilities. During a history test, Martin notices that Frank appears to be cheating off a note card.

I. Identify the question

Think of as many questions as you can. It may help you to think of questions that start with "Should. . . ."

For example, one question you might ask could be, "Should Martin ask other students if they saw anything?" List as many questions as you can. After you have done so, choose one question for analysis. Remember, you can consider other questions later.

II. Identify the issue

What general problem does the case demonstrate? In this case, it might be cheating or school rules. (Although the principal figures in the case are students, students per se are not the issue.) This step is an attempt to categorize the case in a way that will help you find additional information when you search the bioethical literature.

III. State the facts of the case

What are the facts in the case? Most of us have a tendency to draw conclusions based on some information and then believe that our conclusions are facts, too. Be sure to distinguish between the facts and your inferences. Once the facts have been established, list them in an accurate but concise manner.

IV. List as many possible decisions as you can

What are the possible answers to the the question you are addressing? Now is the time to be creative. Think of as many possible answers as you can. At this point, don't worry about which answer is best; that comes later.

V. Gather additional information as needed

Obviously, you don't have access to any additional information about Frank and Martin, but often, background information on the important issues involved in a case study is available. Frank and Martin's case will probably not require gathering additional background information. However, other cases may. For example, in order to weigh the risks and benefits in a case involving use of an experimental drug to treat a patient, you may want to know just how severe the patient's disease is. It might also be appropriate to find out what sort of safety testing a drug must undergo before it is approved for experimental treatment of humans.

The bottom line is that you should be as well informed as possible before you make an ethical decision.

Societal Issues

VI. Pick the decision you wish to support

Consider all the possible decisions listed in step IV. Since ethical problems are usually complex, it is important to take time for thoughtful, honest reflection. You must have a reason for the option you choose, and that reason should be related to one of the principles listed at the beginning of this reading.

VII. Identify the ethical principle that supports your decision

An ethically justifiable decision can be based on alternative principles. In a dilemma, adherence to one principle often results in the breach of another; dilemmas exist because principles often conflict. Application of different basic values can lead to different responses to a situation. Part of the decision-making process is realizing to which principle you are giving preference.

For example, someone who believes it is more important in this case to do no harm (principle 1) might make a different decision from someone who believes it is more important to uphold justice (principle 4). Neither person would be wrong. Both are adhering to high moral principles, but their principles are different.

Contrast these decisions to those of a person who decides to do nothing because someone might get angry or because he doesn't want to get involved. Keeping people from getting mad at you or keeping from getting involved are not high moral principles.

VIII. Identify a supporting authority

What experts or authorities would back up our position on this case? Normally, it is appropriate to look to professional codes of ethics. In this case, the honor code itself would be the authority. In bioethical decisions, several codes of ethics are used; these include the Hippocratic Oath, the American Medical Association Code of Ethics, the American Hospital Association's Bill of Rights for Patients, and the American Nursing Association Code of Ethics as well as the Nuremberg and Helsinki statements.

IX. Formulate a rebuttal

Under what circumstances would you change your decision about what to do? Try to imagine a circumstance or new information that could make you change your mind.

X. State your level of confidence in your decision

Use a one- or two-word statement to describe how strongly you believe in the argument you have made for your decision (your claim). One way to assess the strength of the argument is to gauge the likelihood of rebuttal. If a rebuttal is highly unlikely and the rest of the argument makes sense, then the argument is a strong one. Also, if the principle ties the claim and the facts tightly together, then the argument is strong.

Indicate the degree of confidence you have by using one of the following qualifiers or a similar one: "moderately confident," "absolutely confident," or "questionably confident."

XI. Box up the case for reporting

Your teacher will show you how to summarize your case in the box outline provided at the back of the book.

XII. Prepare a prose argument

Write up your argument, using the box as an outline. You should produce a paper that can be understood by someone who is completely unfamiliar with the original case. You should also explain more in the paper than the box can show. For instance, you should elaborate on why you selected the decision you chose and why the ethical principle justifying that decision was most important in this incident.

Example Case Study 2: Mr. Johnson

Mr. Johnson, age 76, was admitted to a medical unit with pneumonia. He had a history of severe emphysema for which he had been hospitalized twice in the past year because of secondary pneumonia. During the evening of the third day of intravenous antibiotic treatment, Mr. Johnson's complexion took on an increasingly bluish tint, and he was short of breath. Because the attending physician was out of town, the physician taking calls was notified by the evening nursing supervisor. The physician found the patient to be severely oxygen deficient. He tried unsuccessfully to reach Mrs. Johnson to notify her of his plan to transfer Mr. Johnson to the intensive care unit for ventilator support. During the process of arranging for the transfer, a nursing assistant said that she thought that Mr. Johnson and his wife had said that they would refuse life support measures if Mr. Johnson ever needed them. No documentation of these comments could be found.

When Mr. Johnson arrived in intensive care, he was anxious, and his breathing was so labored that he could not talk. He shook his head "no" when the physician explained the procedure for inserting a tube to connect him to the ventilator, and he attempted to push the physician away. Mr. Johnson's competency to make decisions about his treatment was questioned because his arterial blood oxygen was so low. Another unsuccessful attempt was made to reach Mrs. Johnson. Mr. Johnson's condition deteriorated as the physician and staff deliberated about proceeding with the intubation. Finally, the patient was sedated and placed on the ventilator.

Mrs. Johnson arrived the next morning and was shocked that her husband's condition had deteriorated so rapidly. She was upset that her husband was on the ventilator, explaining that they had agreed that life support measures would not be used for him. When the on-call physician asked, "What do you want me to do," Mrs. Johnson replied, "I don't know." The physician asked for assistance from the Ethics Committee.

I. Identify the question

What question will you address about this case? List possible questions ("Should . . . ?"), and then select one.

II. Identify the issue

The general category of problem could be called withdrawal of life support or end-of-life decisions or extraordinary means. This step helps in grouping cases and may make it easier to find references in the literature.

III. State the facts of the case

Only the facts as they are known may be used. List the facts in a concise manner, avoiding inferences or conclusions you may have drawn.

IV. List the possible decisions

What are the possible courses of action in this case? List as many as you can without worrying for now about which is best.

V. Gather needed background facts

Obviously, no further information is available about the Johnsons, but you may have other questions. What is emphysema? What is secondary pneumonia? What is arterial blood oxygen, and why would low oxygen levels make a doctor think Mr. Johnson wasn't competent to make decisions? What is a ventilator? Be sure you have a good understanding of the case before you attempt to make a decision.

VI. Make a decision

VII. Identify the guiding principle

VIII. Cite a supporting authority

As time permits, read about similar cases. What do recognized codes of ethics say about these situations? A list of recognized codes of ethics is given in the discussion of the Frank and Martin example.

IX. Formulate a rebuttal

Under what circumstances would you abandon your decision?

X. State your level of confidence

XI. Box up the case for reporting

Summarize the case and your claim in the box form provided at the back of the book.

XII. Prepare a prose argument

Write up your argument in paragraph form so that someone who has never seen the case before can understand it and the reasons for your decision.

Societal Issues

Bioethics Case Study: Gene Therapy

33

Background Information

Advances in genetic engineering technology offer hope to people afflicted with a number of genetic diseases. Although gene therapy is in its infancy, early studies are encouraging. In the first approved human gene therapy experiment, W. French Anderson is treating the genetic disease called severe combined immunodeficiency disease. Students are usually familiar with the dramatization (starring John Travolta) about the "bubble boy," a young man with this disorder. The disease is caused by the lack of the enzyme adenosine deaminase (ADA).

In 1990, Anderson inserted the correct sequence for ADA into the white blood cells of a young child who was not producing it. As of March 1992, Anderson's patient was producing the enzyme and interacting with the world freely for the first time in her life. In fact, she was one of the children in her preschool who did not get chicken pox during a recent outbreak. Other children have now received similar therapy and are also doing well.

Before this experimental treatment could begin, Anderson's proposal went through extensive review by the Recombinant DNA Advisory Committee at the National Institutes of Health (NIH), the government agency that funds most of the country's biomedical research. In fact, Anderson's proposal was reviewed 15 times. The need for a standard review procedure for proposed human gene therapy experiments was made clear in 1979–1980, when a UCLA researcher tried a human gene therapy experiment on two patients without getting approval from the appropriate review committee at his institution. He was eventually demoted, and his case demonstrated the need for a national policy on human gene therapy experiments.

A standard procedure is now in place, and with Anderson's work as a model, we can expect more and more trials. Several clinical trials for a number of different conditions are now under way.

What is gene therapy, and what ethical issues, if any, does it raise? At first glance, gene therapy, like any medical treatment that proposes to benefit the patient, seems free of any ethical implications. Closer examination raises a number of important questions regarding the acceptability of different kinds of gene therapy. Leroy Walters, of Georgetown University's Kennedy Institute of Bioethics, divided gene therapy into four possible categories.

1. Somatic-cell gene therapy for cure or prevention of disease. *Example:* Insertion of sequence in a person's cells to allow production of an enzyme like ADA.

2. Germ line gene therapy for cure or prevention of disease. *Example:* Insertion of ADA sequence into early embryo or reproductive cells; will affect not only the individual but all his or her offspring.

3. Somatic-cell enhancement. *Example:* Insertion of DNA sequence to improve memory, increase height, or increase intelligence; would affect only that individual.

4. Germ line enhancement. *Example:* Insertion of DNA sequence for enhancement into blastocyst, sperm, or egg; would affect future generations.

One of the ongoing issues concerning gene therapy is whether any or all of these four types of manipulation are ethically acceptable. At present, only somatic-cell gene therapy for cure or prevention of serious disease is considered ethically appropriate even by researchers like Dr. Anderson. Germ line therapy (altering disease genes so that the individual will not only be healthy but will pass on the healthy genes to his or her offspring) is considered desirable by some, but the techniques used to alter animal embryos have far too high a failure rate to consider their application to humans at this time. Human germ line therapy will probably be feasible in the future.

Enhancement therapy (whether somatic cell or germ line) is generally viewed as less acceptable. Anderson has stated, "I will argue that a line can and should be drawn to use gene transfer only for the treatment of serious disease, and not for any other purpose. Genetic transfer should never be undertaken in an attempt to enhance or 'improve' human beings." Anderson, a pioneer in this field, has maintained his position since 1980.

Guidelines for Gene Therapy

In considering any application of gene therapy, basic respect for human dignity is, as always, the underlying moral principle. Like any other experimental medical treatment, gene therapy should be used to benefit the patient. Harm should be avoided. Certain factors must be considered. An NIH committee suggested the following items in 1985.

1. What is the disease to be treated?

2. What alternative treatments for the disease exist?

3. What are the potential benefits of gene therapy for human patients?

4. How will patients be selected in a fair and equitable manner?

5. How will a patient's voluntary and informed consent be solicited?

6. How will the privacy of patients and the confidentiality of their medical information be protected?

These six questions address some ethical concerns that have been considered in evaluations of experimental proposals. Encourage students to consider extensions of these questions or completely new ones.

Sample Issues Related to Gene Therapy

Considering ethical issues related to gene therapy is like opening the proverbial can of worms. Using specific case studies will focus your students' discussion on one set of circumstances at a time. Students can consider broad issues in the context of one individual decision. The following paragraphs contain some of the issues and questions you may want to confront with your students.

What kind of diseases should be treated with gene therapy? How does one define a serious disease?

What is the difference between a serious genetic disease, a moderate genetic defect, and "normal" variation? Will there be subtle or even outright pressure on people with genetic diseases to be treated so they can be more nearly "perfect"? Are the procedures really safe? Could gene therapy have long-term consequences for the patient that cannot be predicted? Who will have access to this kind of treatment—the rich, the insured?

When germ line therapy becomes feasible, wouldn't it be much less costly to use such therapy to eliminate the bad genes from a whole family than to treat just one person at a time? Are we knowledgeable enough to consider changing the gene pool by doing germ line therapy? Are all alleles that we consider harmful really harmful in the long run? We only need to remember the sickle cell allele's effect on malaria infection to answer that question.

As our knowledge about genetic risk factors and our ability to manipulate the human genome increase, even more issues will be raised. For example, if we could lower a person's susceptibility to an environmental toxin like asbestos, should we? Who will decide? If such therapy were available, would insurance and employment discrimination follow?

And what about genetic enhancement? Presumably many people would want to be genetically altered to increase longevity or intelligence. In fact, students are usually quick to condemn the idea of genetic enhancement in general, but they frequently change their tune when asked if they themselves would like to receive treatment to become smarter or stronger, to have perfect teeth or a perfect complexion, or whatever. Similarly, adults view the issue differently when asked if they would like to ensure that their children will be intelligent, strong, and resistant to disease. Will standards be set for which traits are desirable and which are sufficiently undesirable to warrant therapy? When we consider enhancement gene therapy for height, for instance, how will we decide how short is too short? Many traits we might consider enhancing are multifactorial, a situation that increases the risks, since many genes may be involved.

In all of these discussions, the specter of eugenics cannot be ignored. Eugenics, the study of hereditary improvement through genetic control, was embraced by scientists in the United States in the 1920s. However, the horror of the way eugenics was practiced in Nazi Germany brought United States efforts to a halt. By 1939, to prevent racial deterioration, Germany had sterilized nearly 400,000 people who had genetic "defects," including alcoholism and feeblemindedness. It was not long before people who were costing the state too much money because of their genetic defects were killed. Seventy thousand such deaths are documented between 1939 and 1941. (Teaching about the Holocaust in this unit is an opportunity to bring in the history department to implement interdisciplinary study.)

Even if we are naive enough to believe that the eugenics movement as it occurred in Germany could never happen again, we still have numerous and difficult questions to answer about gene therapy. Obviously, we cannot provide our students with answers. First, we must teach them to ask as many questions as possible. We must then teach them to find the data they need to try to answer these questions by using critical-thinking skills.

Procedure for Using the Model

1. Review or present any necessary background material.

2. Have students read the case study.

3. Have students pose questions about the case that represent potential dilemmas ("Should . . . ?")

4. Select a specific question for analysis.

5. List the facts of the case. Seek additional background information.

6. Identify the relevant primary and secondary (if any) ethical principles.

7. List possible courses of action that could be taken.

8. Discuss how the possible solutions are related to the general ethical principles.

9. Have students select their recommended courses of action.

10. Discuss sources of authoritative opinion; assign students to look for authorities that support their chosen action. (This can be homework.)

11. Have students brainstorm about rebuttal conditions.

12. Have students decide their levels of confidence in their recommendation.

13. Prepare a summary. (If students have been assigned outside research, the summary should be written after that research is completed.)

14. Individual students or groups may present their summaries to the class. Be sure to examine how several different ethically defensible positions can result from one case.

Case Study 1: Anne B.

Students should have a basic understanding of enzyme activity and single-gene inheritance patterns. If they have that background and have used the decision-making model, they should be able to work through this case in two 50-min classes. All of these steps can be done as a whole class or in groups. The class may start out as a whole class and then break up into groups to decide courses of action. Alternatively, the decisions may be made by individual students.

Gaucher's disease, first identified in the late 19th century, is an autosomal recessive disorder. The disease results from the production of a structurally altered form(s) of the enzyme β-glucocerebrosidase, which breaks down a lipidlike substance, glucocerebroside. Failure to properly metabolize this substrate results in its accumulation in the spleen, liver, and bone. The disease varies widely in severity: some afflicted individuals die in early childhood, while others are diagnosed on autopsy after death from other causes at an old age. The severest form (the infantile form) follows a course somewhat similar to that seen in a related lipid disorder, Tay-Sachs disease.

Because of the variation in severity, Gaucher's disease is classified into forms I, II, and III: adult noncerebral, acute neuropathic, and subacute neuropathic, respectively. Studies suggest that these forms of the disease are allelic; that is, they are caused by different mutations within the same gene. The resulting structural defect in the protein is related to the severity of the disease. (See the lesson on genetic diseases for more information.)

The most common of the three types is form I (the adult form). It is known to occur frequently in Ashkenazic Jews, a population with a heterozygote frequency of 1 in 13. Patients with type I disorder have about 15% of the enzyme activity of normal individuals. There are estimated to be 20,000 cases in the United States. Form I Gaucher's disease varies in severity: some afflicted individuals die in early adulthood, while others live essentially symptom free to old age. Form I Gaucher's disease is therefore subdivided into acute, moderate, and mild courses.

Form I Gaucher's disease may first produce symptoms in a patient's late teens, with an enlarged spleen often being an early sign. Blood studies show reduced platelet, white blood cell, and red blood cell counts. Anemia and enlargement of the spleen may require removal of the spleen. Although this surgery decreases the anemia and relieves abdominal stress, the unmetabolized material is deposited in the bone more rapidly after splenectomy. Bone pain and pathologic fracture may occur.

In 1989, Joan B., age 32, died from the secondary effects of Gaucher's disease. Joan's 23-year-old sister, Anne B., also has the condition. However, she has been receiving enzyme replacement therapy each week since April 1992.

The biotechnology firm Genzyme marketed the β-glucocerebrosidase enzyme in June 1991. Anne receives the enzyme every week by intravenous infusion at the office of her physician. Her physician is considered one of the most knowledgeable practitioners with regard to this particular disorder. The infusion process takes about 4 h each week, including 2 h of travel time. The efficiency of the enzyme replacement therapy is not completely known, but early studies have been very encouraging. There have been no harmful side effects.

Anne recently read about a new experimental treatment that involves removing some of her own cells, treating them with a retrovirus that carries the correct sequence for glucocerebrosidase, and replacing the cells. The cells should reproduce, creating a colony of cells that make the proper enzyme. The hope is that this treatment could be done several times, resulting in enough cells to produce permanently adequate amounts of normal enzyme. The technology is experimental and there may be risks, some known and some unknown. For instance, since the sequence is added by using a retrovirus, some of Anne's normal DNA might be altered. This treatment is also more expensive than Anne's current treatment. Anne requests the experimental treatment from her physician.

Implementing the decision-making model

First of all, does an ethical dilemma exist? Anne has requested an experimental treatment that may harm her or may help her and improve her quality of life. Should her physician help her get the treatment she wants? Is it medically indicated? Will the treatment do more harm than good, since she is already receiving enzyme replacement therapy? Is her disease as serious as French Anderson says it should be in order for her to qualify for gene therapy? There are certainly

enough questions to indicate that this case does present bioethical issues.

I. Question
What question do we want to focus on to illustrate this case? Any question can be selected, depending on the interest of students and teacher. Let's choose the first one from above. Should Anne's physician help her get gene therapy?

II. Issue
We might choose to call this an issue of experimental treatment. (This is a good opportunity to discuss the ethics surrounding experimental treatments for human immunodeficiency virus.)

III. Facts
For the box, we want to restrict ourselves to the true, relevant, sufficient facts from the case itself. All the background material should not be included, since it will make the box very unwieldy.

- Anne, 23 years old, has Gaucher's disease.
- Anne's sister died from the disease.
- Anne has been on replacement therapy since April 1992.
- Anne is requesting experimental gene therapy.
- Anne thinks the therapy will improve the quality of her life.
- Anne has read extensively about the new treatment.
- Anne's doctor is considered an expert.

IV. Possible Decisions
Here are some options. There could be others that are more appropriate. None is intended to be the "right" answer, since there are no right answers in these issues, only more or less justifiable choices.

1. Anne's physician should arrange for her to have the treatment.

2. Anne's physician should refuse to help her, since she is already being treated with an effective, safe method.

3. Anne's physician should try to discourage her from having experimental treatment, but if she insists, he should help Anne.

4. Anne should not have either treatment, since both are experimental.

5. Anne should not have any treatment, since all treatments are very expensive.

6. Anne should have both kinds of treatment.

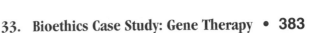

Societal Issues

Although the last three options may seem inappropriate, it is very important to let students suggest all alternatives and discuss them. Otherwise, students are misled into thinking that there are "right" answers to these questions. Furthermore, they will think that the teacher has the right answer! Be sure to have students consider the relevant ethical principles as they discuss the alternatives.

V. Additional Information

Do students need additional information to make a good decision? For example, what is a retrovirus, and what is known about the risk of cancer from retrovirus DNA insertion? Has the retrovirus approach been successful with other patients? Depending on the question the students address and the issues they focus on, these questions may be very relevant.

VI. Decision

In this example, the claim will be that the physician should try to discourage Anne because gene therapy is even more experimental than her current therapy. Gene therapy may pose unknown, unnecessary risks for her.

VII. Principle

Using option 3, the physician would be trying first to prevent harm from coming to Anne. However, if he cannot convince her to change her mind, then the physician, by helping Anne get what she wants, would be respecting her autonomy. Also, in order for Anne's autonomy to be respected, she would have to give informed consent for her treatment. Before she was treated, she would need to be able to understand her choices and the consequences of her choices.

VIII. Authority

What experts would back us up on our option? Again, we can turn to the various codes of medical ethics or to the extensive literature in bioethics. For this case, we can quote the Hippocratic Oath, the Nuremberg Statements, and the Patient Bill of Rights as preliminary sources.

IX. Rebuttal

Under what circumstances would we abandon our choice? What if Anne's physician had information unavailable to the public that the gene therapy was ineffective but not harmful or that it was harmful and ineffective? What if Anne's diagnosis of Gaucher's disease was incorrect to begin with? What if the cost of the gene therapy was beyond Anne's ability to pay? What if Anne was very depressed about her condition and her physician questioned her competence to make an informed choice? Again, encourage as much creative thinking as possible.

For this example, we chose that the physician had information that the treatment was harmful. If that were true, then the physician could try to talk Anne out of the treatment and should refuse to help her get it. If she were still insistent, Anne could fire her doctor and find another.

X. Level of Confidence

How strongly do you believe your position? In the example, we believe the rebuttal is so unlikely that we have strong confidence in the argument.

XI. Box

See the blank box-up forms in Appendix K.

XII. Prose Argument

Putting the argument in paragraph form gives students an opportunity to explain on paper why they chose one option rather than the other or why the principle supports the option. The written argument should reflect the thinking necessary to construct the box.

Additional case studies follow.

Case Study 2: Bobby K.

Bobby K. is a healthy 10-year-old boy. He is very agile and quick and loves sports. The coach of his city league basketball team has told Bobby's parents that their son's skills on the court are astounding for a child of his age.

Bobby's father, Mr. K., is a healthy man who is 5'3" tall. Bobby's mother is also short, only 5'0". Bobby's pediatrician has predicted that Bobby will attain an adult height of about 5'3" but has emphasized to his parents that he is a normal, healthy boy. Mr. K. remembers being teased constantly about his size and recalls that his lack of height kept him off all the varsity sports teams at his high school. He has often wondered if his shortness is a disadvantage to him in business dealings, too. Mr. and Mrs. K. both anticipate that their son will be the recipient of more and more pointed teasing as he reaches his teenage years. They also fear that he will not be selected for the basketball team when he reaches junior high or high school.

Bobby's coach has heard of a program at the local university in which gene therapy is being conducted on children who have a disease that results in the inadequate production of growth hormone. The children are being given working copies of the gene for human growth hormone, and the levels of growth hormone in their bodies have increased. These children's growth rates have also increased.

Bobby's coach tells the K. family about what he has heard. Bobby is thrilled, because he might be able to keep playing basketball, maybe even on a professional level. Bobby's parents are more cautious but would like their son to be spared the pain of being much smaller than his classmates.

The K.'s go to Bobby's pediatrician and request that Bobby be given the growth hormone gene therapy.

Using the model

As you use the model to discuss this case, you could head the discussion in any of several directions. For example, you could have the students do library work to try and learn whether any studies have determined whether it is safe to give additional growth hormone to normal children. Sample questions include the following.

- Should Bobby's doctor allow him to have the gene therapy?
- Should Bobby's parents let him have the therapy?
- Should Bobby's coach have told the K.'s about the program?

Case Study from Real Life

This case study also revolves around human growth hormone but concerns administration of the hormone as a drug, not gene therapy. It is a summary of the article "NIH stops enrolling children in growth hormone studies" in *Biotechnology Newswatch*, August 17, 1992.

In 1992, because of questions about the safety and appropriateness of the study, the NIH temporarily stopped doctors from enrolling more children in a study of the effects of growth hormone on healthy short children. Children already in this trial will continue. Without treatment, the boys in this study were expected to reach 5'6" or less and the girls were expected to reach 5'0" or less as adults. Half of the children are being given genetically engineered human growth hormone (the protein, not the gene), and the other half are receiving a placebo. These children do not know that their treatment is a placebo. They will serve as a control group for the children who are receiving the hormone. The physicians hope to learn how much additional growth the hormone produces in healthy short children.

Doctors normally prescribe growth hormone to children who fail to produce enough of it naturally and are expected to reach an adult height of 4'0" or less. The hormone has been on the market for 7 years, and Genentech (one of the two companies in this country

that make the hormone) is convinced that no safety problems are associated with long-term use of the hormone.

Scientists who defend the study present several points. One is that many parents obtain human growth hormone for their children on the black market. Another is that studies to determine the safety and efficacy of this treatment for short stature are needed. Another argument is that our society is preoccupied with height and that short stature can cause anxiety and poor self-image to the point that mental health is compromised. One of the scientists who defend the study is Dr. Arthur Levine, scientific director of the National Institute of Child Health and Human Development. At 5'4", Dr. Levine says he probably would have enrolled himself in the study if it had been available to him as a child.

Critics of the study argue that shortness is not a disease and should not be defined as requiring treatment. They question the use of a relatively experimental treatment in healthy children for apparently cosmetic reasons. They are concerned that virtually any cosmetic defect could eventually be redefined as illness.

Using the model

Apply the decision-making model as outlined previously. Some questions that could be asked include the following.

- Should this study be conducted?
- Should a parent enroll a short child in this study?
- Should doctors prescribe growth hormone to short but otherwise normal children?
- Should people be able to make their own decisions about taking growth hormone?
- Should any treatment be available for shortness?
- Should NIH prevent children who want to enroll from participating?

Sources of Additional Case Studies

Many bioethics texts contain case studies. News stories and magazine articles offer another rich source (the NIH case study discussed previously is based on a news article). Have your students watch for potential material, but remind them that factual information must come from primary sources, not newspapers and magazines.

No matter which case you are considering, make sure your students have access to real information about the issues. Library work is very important!

Societal Issues

Bioethics Case Study: Gene Therapy

Background Information

Advances in genetic engineering technology offer hope to people afflicted with a number of genetic diseases. Although gene therapy is in its infancy, early studies are encouraging. In the first approved human gene therapy experiment, W. French Anderson is treating the genetic disease called severe combined immunodeficiency disease. Students are usually familiar with the dramatization (starring John Travolta) about the "bubble boy," a young man with this disorder. The disease is caused by the lack of the enzyme adenosine deaminase (ADA).

In 1990, Anderson inserted the correct sequence for ADA into the white blood cells of a young child who was not producing it. As of March 1992, Anderson's patient was producing the enzyme and interacting with the world freely for the first time in her life. In fact, she was one of the children in her preschool who did not get chicken pox during a recent outbreak. Other children have now received similar therapy and are also doing well.

Before this experimental treatment could begin, Anderson's proposal went through extensive review by the Recombinant DNA Advisory Committee at the National Institutes of Health (NIH), the government agency that funds most of the country's biomedical research. In fact, Anderson's proposal was reviewed 15 times. The need for a standard review procedure for proposed human gene therapy experiments was made clear in 1979–1980, when a UCLA researcher tried a human gene therapy experiment on two patients without getting approval from the appropriate review committee at his institution. He was eventually demoted, and his case demonstrated the need for a national policy on human gene therapy experiments.

A standard procedure is now in place, and with Anderson's work as a model, we can expect more and more trials. Several clinical trials for a number of different conditions are now under way.

What is gene therapy, and what ethical issues, if any, does it raise? At first glance, gene therapy, like any medical treatment that proposes to benefit the patient, seems free of any ethical implications. Closer examination raises a number of important questions regarding the acceptability of different kinds of gene therapy. Leroy Walters, of Georgetown University's Kennedy Institute of Bioethics, divided gene therapy into four possible categories.

1. Somatic-cell gene therapy for cure or prevention of disease. *Example:* Insertion of sequence in a person's cells to allow production of an enzyme like ADA.

2. Germ line gene therapy for cure or prevention of disease. *Example:* Insertion of ADA sequence into early embryo or reproductive cells; will affect not only the individual but all his or her offspring.

3. Somatic cell enhancement. *Example:* Insertion of DNA sequence to improve memory, increase height, or increase intelligence; would affect only that individual.

4. Germ line enhancement. *Example:* Insertion of DNA sequence for enhancement into blastocyst, sperm, or egg; would affect future generations.

One of the ongoing issues concerning gene therapy is whether any or all of the four types of manipulation listed above are ethically acceptable. At present, only somatic-cell gene therapy for cure or prevention of serious disease is considered ethically appropriate even by researchers like Dr. Anderson. Germ line therapy (altering disease genes so that the individual will not only be healthy but will pass on the healthy genes to his or her offspring) is considered desirable by some, but the techniques used to alter animal embryos have far too high a failure rate to consider their application to humans at this time. Human germ line therapy will probably be feasible in the future.

Enhancement therapy (whether somatic cell or germ line) is generally viewed as less acceptable. Anderson has stated, "I will argue that a line can and should be drawn to use gene transfer only for the treatment of serious disease, and not for any other purpose. Genetic transfer should never be undertaken in an attempt to enhance or 'improve' human beings." Anderson, a pioneer in this field, has maintained his position since 1980.

Guidelines for Gene Therapy

In considering any application of gene therapy, basic respect for human dignity is, as always, the underlying moral principle. Like any other experimental medical treatment, gene therapy should be used to benefit the patient. Harm should be avoided. Certain factors must be considered. An NIH committee suggested the following items in 1985.

1. What is the disease to be treated?

2. What alternative treatments for the disease exist?

3. What are the potential benefits of gene therapy for human patients?

4. How will patients be selected in a fair and equitable manner?

5. How will a patient's voluntary and informed consent be solicited?

6. How will the privacy of patients and the confidentiality of their medical information be protected?

These six questions address some ethical concerns that have been considered in evaluations of experimental proposals. You may think of additional ones.

Case Study 1: Anne B.

Gaucher's disease, first identified in the late 19th century, is an autosomal recessive disorder. The disease results from the production of a structurally altered form(s) of the enzyme β-glucocerebrosidase, which breaks down a lipidlike substance, glucocerebroside. Failure to properly metabolize this substrate results in its accumulation in the spleen, liver, and bone. The disease varies widely in severity: some afflicted individuals die in early childhood, while others are diagnosed on autopsy after death from other causes at an old age. The severest form (the infantile form) follows a course somewhat similar to that seen in a related lipid disorder, Tay-Sachs disease.

Because of the variation in severity, Gaucher's disease is classified into forms I, II, and III: adult noncerebral, acute neuropathic, and subacute neuropathic, respectively. Studies suggest that these forms of the disease are allelic; that is, they are caused by different mutations within the same gene. The resulting structural defect in the protein is related to the severity of the disease.

The most common of the three types is form I (the least severe adult form). Form I Gaucher's disease is known to occur frequently in Ashkenazic Jews; approximately 1 in 13 members of this population is a carrier (heterozygotes). Patients with type I disorder have about 15% of the enzyme activity of normal individuals. There are estimated to be 20,000 cases in the United States. Although form I Gaucher's disease is the least severe of the three forms of the disease, it, too, varies in severity: some afflicted individuals die in early adulthood, while others live essentially symptom free to old age. Form I Gaucher's disease is therefore subdivided into acute, moderate, and mild courses.

Form I Gaucher's disease may first produce symptoms in a patient's late teens, with an enlarged spleen often being an early sign. Blood studies show reduced platelet, white blood cell, and red blood cell counts. Anemia and enlargement of the spleen may require removal of the spleen. Although this surgery decreases the anemia and relieves abdominal stress, the unmetabolized material is deposited in the bone more rapidly after splenectomy. Bone pain and pathologic fracture may occur.

In 1989, Joan B., age 32, died from the secondary effects of Gaucher's disease. Joan's 23-year-old sister, Anne B., also has the condition. However, she has been receiving enzyme replacement therapy each week since April 1992.

The biotechnology firm Genzyme marketed the β-glucocerebrosidase enzyme in June 1991. Anne receives the enzyme every week by intravenous infusion at the office of her physician. Her physician is considered one of the most knowledgeable practitioners with regard to this particular disorder. The infusion process takes about 4 h each week, including 2 h of travel time. The efficiency of the enzyme replacement therapy is not completely known, but early studies have been very encouraging. There have been no harmful side effects.

Anne recently read about a new experimental treatment that involves removing some of her own cells, treating them with a retrovirus that carries the cor-

Societal Issues

rect sequence for glucocerebrosidase, and replacing the cells. The cells should reproduce, creating a colony of cells that make the proper enzyme. The hope is that this treatment could be done several times, resulting in enough cells to produce permanently adequate amounts of normal enzyme. The technology is experimental and there may be risks, some known and some unknown. For instance, since the sequence is added by using a retrovirus, some of Anne's normal DNA might be altered. This treatment is also more expensive than Anne's current treatment. Anne requests the experimental treatment from her physician.

Implementing the decision-making model

First of all, does an ethical dilemma exist? Anne has requested an experimental treatment that may harm her or may help her and improve her quality of life. Should her physician help her get the treatment she wants? Is it medically indicated? Will the treatment do more harm than good, since she is already receiving enzyme replacement therapy? Is her disease as serious as French Anderson says it should be in order to qualify for gene therapy? There are certainly enough questions to indicate that this case does present bioethical issues.

I. Question

Several possible questions are listed above. List as many others as you can think of, and then choose one for consideration.

II. Issue

What is the issue you chose? Experimental treatment? Medical necessity? Something else?

III. Facts

There is a lot of background material in this case. For the box summary, try to select only the information that is necessary for presenting the dilemma. You could reasonably omit background information about Gaucher's disease from the box, for example.

IV. Possible Decisions

List as many alternatives as you can without trying to decide at this time which is best.

V. Additional Information

Do you need additional information to make a good decision? For example, if you are trying to decide whether the doctor should give Anne the treatment, do you need to know more about retroviruses, how they work, and how they might alter Anne's DNA? Do you need more information about Gaucher's disease

to decide whether it is serious enough to warrant gene therapy?

VI. Decision

Choose what you believe to be the best solution to your question.

VII. Principle

Which of the major ethical principles justifies your decision?

VIII. Authority

What experts would back up your option? Find relevant passages in documents such as the Hippocratic Oath, the Nuremberg Statements, and the Patient Bill of Rights.

IX. Rebuttal

Under what circumstances would you abandon your choice?

X. Level of Confidence

How strongly do you believe your position?

XI. Box

Use the outline at the back of the book.

XII. Prose Format

Explain on paper why you chose one option rather than the other and why the principle supports the option. Your written argument should reflect the thinking necessary to construct the box and be complete enough that someone who has never read the case study can understand the situation and your analysis.

Additional case studies follow.

Case Study 2: Bobby K.

Bobby K. is a healthy 10-year-old boy. He is very agile and quick and loves sports. The coach of his city league basketball team has told Bobby's parents that their son's skills on the court are astounding for a child of his age.

Bobby's father, Mr. K., is a healthy man who is 5'3" tall. Bobby's mother is also short, only 5'0". Bobby's pediatrician has predicted that Bobby will attain an adult height of about 5'3" but has emphasized to his parents that he is a normal, healthy boy. Mr. K. remembers being teased constantly about his size and recalls that his lack of height kept him off all the varsity sports teams at his high school. He has often wondered if his shortness is a disadvantage to him in business dealings, too. Mr. and Mrs. K. both anticipate that their son will be the recipient of more and more pointed teasing as he reaches his teenage years. They also fear that he will not be selected for the basketball team when he reaches junior high or high school.

Bobby's coach has heard of a program at the local university in which gene therapy is being conducted on children who have a disease that results in the inadequate production of growth hormone. The children are being given working copies of the gene for human growth hormone, and the levels of growth hormone in their bodies have increased. These children's growth rates have also increased.

Bobby's coach tells the K. family about what he has heard. Bobby is thrilled, because he might be able to keep playing basketball, maybe even on a professional level. Bobby's parents are more cautious but would like their son to be spared the pain of being much smaller than his classmates.

The K.'s go to Bobby's pediatrician and request that Bobby be given the growth hormone gene therapy.

Using the decision-making model

This case obviously presents many possible questions for consideration. For example,

- Should Bobby's doctor allow him to have the gene therapy?
- Should Bobby's parents let him have the therapy?
- Should Bobby's coach have told the K.'s about the program?

There are probably many background questions you will need to investigate. For example,

- Have there been any studies to determine whether it is safe to give additional growth hormone to normal children?

- What height range is considered normal? How can you tell whether a short person is normal?

Choose a dilemma question for analysis, and use the model to generate a solution.

Case Study from Real Life

This case study also revolves around human growth hormone but concerns administration of the hormone as a drug, not gene therapy. It is a summary of the article "NIH stops enrolling children in growth hormone studies" in *Biotechnology Newswatch*, August 17, 1992.

In 1992, because of questions about the safety and appropriateness of the study, the NIH temporarily stopped doctors from enrolling more children in a study of the effects of growth hormone on healthy short children. Children already in this trial will continue. Without treatment, the boys in this study were expected to reach 5'6" or less and the girls were expected to reach 5'0" or less as adults. Half of the children are being given genetically engineered human growth hormone (the protein, not the gene), and the other half are receiving a placebo. These children do not know that their treatment is a placebo. They will serve as a control group for the children who are receiving the hormone. The physicians hope to learn how much additional growth the hormone produces in healthy short children.

Doctors normally prescribe growth hormone to children who fail to produce enough of it naturally and are expected to reach an adult height of 4'0" or less. The hormone has been on the market for 7 years, and Genentech (one of the two companies in this country that make the hormone) is convinced that no safety problems are associated with long-term use of the hormone.

Scientists who defend the study present several points. One is that many parents obtain human growth hormone for their children on the black market. Another is that studies to determine the safety and efficacy of this treatment for short stature are needed. Another argument is that our society is preoccupied with height and that short stature can cause anxiety and poor self-image to the point that mental health is compromised. One of the scientists who defend the study is Dr. Arthur Levine, scientific director of the National Institute of Child Health and Human Development. At 5'4", Dr. Levine says he probably would have enrolled himself in the study if it had been available to him as a child.

Critics of the study argue that shortness is not a disease and should not be defined as requiring treatment. They question the use of a relatively experimental treatment in healthy children for apparently cosmetic reasons. They are concerned that virtually any cosmetic defect could eventually be redefined as illness.

Using the model

Apply the decision-making model as outlined previously. Some questions that could be asked include the following.

- Should this study be conducted?
- Should a parent enroll a short child in this study?
- Should doctors prescribe growth hormone to short but otherwise normal children?
- Should people be able to make their own decisions about taking growth hormone?

- Should any treatment be available for shortness?
- Should NIH prevent children who want to enroll from participating?

You may be able to think of more questions. Choose one for analysis, using the decision-making model.

Societal Issues

Bioethics Case Study: Genetic Screening

34

Background Information

Although gene therapy is a new development in health care, the capacity to screen for genetic diseases has been with us for more than 20 years. Biochemical assays or chromosomal analysis for sickle-cell anemia, Tay-Sachs disease, and Down's syndrome were under way in the 1970s. We have had genetic markers for Huntington's chorea (Huntington's disease [HD]) since 1983, for Duchenne muscular dystrophy since 1987, and for cystic fibrosis (CF) since 1989. The causative genes for all of those diseases have now been located. Genetic markers and the genes that contribute to multifactorial traits like colon cancer, breast cancer, Alzheimer's disease, and multiple sclerosis have also been identified.

The Human Genome Project proposes to map all 50,000 to 100,000 genes in the human genome. That should give us the power to screen for many more of the 3,000 to 4,000 known genetic defects. What ethical questions does genetic screening present? We have more experience with screening issues than with gene therapy. Some of that experience has made it clear that we must proceed with caution and with a clear policy.

In the 1970s, screening for sickle-cell anemia resulted in confusion and fear. People who merely carried the sickle-cell gene were threatened with loss of jobs and insurance. The airline industry and the armed forces argued that low air pressures would cause heterozygotes to experience symptoms. Some airlines eliminated these people as prospective employees. Although the charges were unsubstantiated, it took years to end the discrimination. The initial purpose of helping people understand their possible risk of bearing children with sickle-cell disease was lost in a mire of misunderstanding.

Genetic screening for late-onset diseases like HD is even more complex, and HD is particularly instructive because it demonstrates the situation in nearly all identifiable genetic conditions: we can identify them, but we cannot treat them. Therefore, should a 20-year-old be able to find out that he carries a dominant gene for a disease that will first debilitate and finally kill him in middle age? If people should be given this information, at what age is it best revealed? Does respect for human dignity in this situation mean that we try to prevent harm by not sharing the information on their genetic status or try to promote autonomy by sharing the information?

Because of the experiences with sickle-cell screening 20 years ago, screening for HD is being done under very strict guidelines at 14 centers in the United States. The centers require clients to undergo extensive psychological testing before screening and counseling after screening. If a cure or even a treatment were available, then these questions would not be nearly as difficult, but as mentioned earlier, neither treatment nor cure is available. In the case of sickle-cell anemia, the screening test has been available for 34 years, but there is still no cure and only palliative treatment.

The ethical questions surrounding screening are apparent. Three percent of the $3 billion budget for the Human Genome Project has been set aside to study the ethical considerations. The Working Group of Ethical Legal and Social Implications of the Human Genome Project has proposed nine areas that merit attention.

1. Fairness in the use of genetic information
2. Impact of knowledge of genetic variation on the individual
3. Privacy and confidentiality
4. Impact on genetic counseling
5. Impact on reproductive decisions
6. Issues raised by the introduction of genetics into mainstream medical issues
7. Uses and misuses of genetics in the past
8. Questions raised by commercialization
9. Conceptual and philosophical implications

Traditionally, early diagnosis is considered desirable. In the case of HD, that presumption cannot be applied to everyone. Hence the requirement for extensive psychological support. On the other hand, for conditions that are influenced by both the environment and genes, such as emphysema and high blood pressure, early detection of susceptibility may save many lives and significant amounts of money. If people do make appropriate lifestyle changes, good has been achieved. But who will be tested for these kinds of traits. Will the tests be voluntary? Who will pay for the test? Will we deny insurance or employment to people who refuse testing or to people who refuse to make lifestyle changes after tests indicate they have increased risks for certain conditions?

Requests for testing often come from pregnant women who want their fetuses tested. In the case of HD, some women want their fetuses tested but do not want themselves tested. The issue of abortion is impossible to ignore as a part of the effect of genetic screening. Again, what is the value of a diagnosis when no treatment can be offered?

Depending on the particular school system, classroom, and teacher, discussion of abortion can be handled well by using the same critical-thinking model introduced earlier. If you are willing to confront this issue, we suggest you include a brief presentation (or review) of human development, using filmstrips or the Nova tape "Miracle of Life," so that students connect trimesters to a meaningful picture of the embryo or fetus. None of the issues in bioethics have absolute answers, but abortion is so politically and emotionally charged that bringing logic and reason to it is beneficial. If a teacher has a good relationship with his or her students, the classroom is a good place to promote reasoned discussion.

As more and more screening tests become available to the public, it is incumbent on biology teachers to help students examine the pros and cons of screening. The students in your classrooms right now will have many choices to make. If you enable them to make a truly informed choice as they contemplate issues such as genetic screening, you will have served a worthy goal.

For additional information, see the lesson *The Molecular Basis of Genetic Diseases* (chapter 28), which contains background information on genetic diseases, including HD, sickle-cell anemia, and Gaucher's disease, and chapter 4, which has information on genetic markers.

Case Study 1: James and Carol H.

Refer to the step-by-step method (chapter 32) for using the model. In order to follow the methodology outlined in the model, students must have a basic knowledge of CF, including its genetic basis, symptoms, and treatments and the prognosis for patients who suffer from the disease.

Alice is the 6-year-old daughter of James and Carol H. Alice was diagnosed with CF at age 2. Alice has a moderately severe case of CF and receives postural drainage from her mother three times a day. In addition, she is on a regimen of vitamins and enzymes that requires her to take about 25 pills each day. Carol spends much of her day caring for Alice. At least twice each year, Alice requires hospitalization to fight respiratory tract infections.

James and Carol have been talking about having another child. Alice is cared for at a public clinic associated with a medical school that has a research project on CF. The clinic physicians suggest that James and Carol undergo screening for the CF gene as part of the project.

When the results come back to the clinic, Alice's physician is quite surprised to learn that Carol, but not James, carries the CF gene. Since James is not a carrier, it is virtually impossible for him to be Alice's biological father.

I. Identify the question

First of all, is there a dilemma, and if there is, what is it? Questions that might be posed include the following.

1. Was the physician morally justified in requesting the testing in the first place?

2. Should the physician divulge the test information to James and/or to Carol?

3. If the physician chooses not to divulge the information but James and/or Carol requests the results of the test, should the physician tell the truth or withhold it?

For purposes of this discussion, we select question 2.

II. Issue

What is the general topic in this case? Genetic testing.

III. Facts

What are the facts in the case? It is particularly important here to remind students about the difference between facts and inferences. Many students will make unsupported assumptions or jump to conclusions. For instance, in this case, we cannot be sure that Carol and James are not already aware of Alice's biological father. We simply want to list the facts in an accurate but concise manner.

- James does not carry the CF gene.
- Carol does carry the CF gene.
- Alice (6 years old) has moderately severe CF.
- James and Carol are considering another baby.
- Alice requires postural drainage three times each day.
- She is hospitalized twice a year for respiratory infections.
- She requires 25 pills a day.
- Mother is main caretaker.

The facts must be true, relevant, and sufficient. If a different question had been chosen, then the facts chosen from the case might be different.

IV. Possible decisions

What are some possible solutions to the problem? List as many as possible.

1. The physician should not tell James or Carol.
2. The physician should tell James only.
3. The physician should tell Carol only.
4. The physician should wait until Alice is 18 and tell Alice only.
5. The physician should tell only if asked.
6. The physician should tell Carol and direct her to tell James.
7. The physician should tell the minister of James and Carol.
8. If Carol will not tell James, then the physician should tell him.

These are eight of the possible decisions. Encourage solutions that represent compromise. Again, we need to make a decision that is supported by one of the major ethical principles listed earlier.

V. Additional information

Students may or may not need to gather additional information on CF, the accuracy of genetic tests, etc.

VI. Decision

For our example, we choose option 8: the physician should tell Carol and then tell James if she will not.

We need to know what principles we are abiding by and what principles we are breaching. First of all, we are telling Carol the results of James' test, and that is a breach of his right to confidentiality. We justify it by the principle of trying to avoid harm to Alice, Carol, and James. We fear the information may endanger their family structure and might even result in a divorce. Because Alice requires a lot of support and this family appeared to be happy prior to this event, threatening the stability of this family is seen as wrong in this argument.

On the other hand, we respect James' right to know. The information is about James, so he does have the right to know according to the principle of respect for individual freedom. Harm could be done to James by withholding this information. For example, if James were widowed and remarried, he might want more children but not be willing to take a fictional risk. By giving Carol the chance to tell James, we try to respect the rights of the parties involved. It is not a perfect solution, but there would be no dilemma if a perfect solution existed.

A student's (or student group's) decision should be stated in the form of an "I" or a "we" statement. For instance, "I believe the physician would be morally justified in telling Carol the results of the test first. But if she does not inform James, then the physician should reveal the results of the test to James."

VII. Principle

The principle we have chosen is to avoid harm.

VIII. Authority

Which authorities would back us up on this point? One of the first places to look is in professional codes of ethics. For instance, the Hippocratic Oath states that the physician should do no harm. Similar statements are found in the American Medical Association Code of Ethics. Students can also do library research to find similar cases to support their position.

IX. Rebuttal

Under what circumstances would we change our minds about what to do? Here again, it is important to encourage students to think freely. What if the physician found out that the test results were false? What if James came to the physician and said that he already knew he was not a carrier? In these two circumstances, the physician might change her mind.

Societal Issues

X. Level of confidence

In this case, the qualifier might be "moderately confident."

XI. Box

See Appendix K for blank box-up forms.

XII. Prose argument

See chapter 32.

Case Study 2: Angela

Angela is a healthy 32-year-old certified public accountant who is married with no children. She has been working for 8 years in a small accounting firm. She has decided she would like to move to a nearby large city and work for a major accounting firm.

Five years ago, Angela's mother was diagnosed with HD, an autosomal dominant disease. (Depending on student background, discussion of the disease may be as brief or as extensive as the teacher chooses.) Angela's mother is in a nursing home.

Angela's job interviews go well, and she is offered an excellent position that she readily accepts. However, as Angela is filling out the required paperwork in the personnel office, she notices that she must sign a consent form for a physical that includes a genetic screening test of her DNA.

When Angela questions whether she must have such a test done, she is told it is required by the company. The issue here could be called genetic screening.

A number of questions could be raised about this case. Here are a few examples.

1. Should the accounting firm require the testing?
2. Who should have access to Angela's results if she agrees to be tested?
3. Should Angela's employment be contingent on the results of her DNA?
4. If Angela has a sister who is also at risk, should she be told Angela's results?

Although this is a hypothetical case, it is interesting to note that at least 50 companies performed genetic screening in 1987. For the purposes of illustration, we will choose question 1. Should the accounting firm require the testing?

Question 1 provides a good opportunity to illustrate a major source of contention in all beginning ethics discussions: the difference between ethical behavior and legal behavior. Students must understand that their discussions are to focus on what is ethical, that is, what is right or wrong. An ethical decision may not necessarily be one that is supported by the law. Learning that what is legal may not necessarily be ethical is a lesson in itself for most students.

Even if the accounting firm is not legally prevented from DNA testing, that does not mean that such testing is right or ethical. In posing the questions listed above, the use of "should" rather than "can" implies that we are interested in the ethical rather than the legal questions. Although, the legal perspective cannot be ignored, it must be clearly separate from an ethical decision.

Helping students differentiate between what is an actual fact and what is an inference, a judgment from the facts that may or may not be true, is essential. To make this model work, we must use the available facts. Appropriate research should be encouraged. Students can be assigned library research, or they can interview local physicians, lawyers, insurance agents, employers, etc. The March of Dimes is also an excellent source of information on genetic disorders.

What are the facts in this case? Try to list them as briefly as possible without inferences.

- 32-year-old healthy female accountant
- mother has HD
- offered position with new firm
- firm requires physical including DNA screen

What are some possible solutions or choices in this case? Again here is a partial list.

1. The accounting firm is not morally justified in requiring the test.

2. The company should require the test only if it is trying to prevent possible harm to its employees (i.e., prevent exposure to a chemical to which a person may have a genetic susceptibility).

3. The company should request the screen but only for insurance purposes, not for employment decisions.

4. Angela should go somewhere else for a job.

5. Angela should go to an independent laboratory for the test and be sure that only she receives the results.

At this point, a class can be broken up into different groups to try to construct an argument for their choice, or they could be assigned to a position.

After choosing the option, the next step is to decide *why* that option is morally best. Each participant must decide which principle best supports the choice he or she has made. If our claim is that the accounting firm is not morally justified in requiring the DNA test, then we must be prepared to explain which moral-action guide is being violated by such a request.

In this case, more than one principle can be called on to support the choice. Next, locate backing for the position. Formulate a rebuttal, and state the level of confidence in the decision.

Where To Find More Case Studies

The literature in bioethics is extensive and growing quickly. Many texts on bioethics include case studies. The newspaper and weekly news magazines also provide material for case study analysis. Many of the sources listed in Appendix H will have more cases.

Societal Issues

Bioethics Case Study: Genetic Screening

Background Information

Genetic screening is examining a person's DNA to determine whether he or she carries a gene of interest. It is usually thought of in the context of determining whether someone has a genetic disease or is a heterozygous carrier of a recessive disease gene.

Although the ability to look directly at a person's DNA is relatively new, it has been possible for several years to screen for some genetic disorders by using other approaches. For example, Down's syndrome is caused by the presence of an extra chromosome 21, which can be detected by microscopic examination of an individual's chromosomes. Heterozygous carriers of diseases such as Tay-Sachs and sickle-cell anemia have also been identifiable for some time. Even though these individuals do not have the disease in question, their cells display characteristic differences from fully normal cells.

However, as we enter the new age of molecular genetics, it will be possible to screen for more and more genetic conditions. This change results from our new ability to analyze DNA directly and from research that is identifying more and more genes involved in human disease. With our current DNA technology, once a disease gene has been identified, it is usually possible to devise a test that will reveal whether a person carries a disease-causing mutation. In the past several years, genes responsible for cystic fibrosis (CF), muscular dystrophy, and many other diseases have been identified. The Human Genome Project, which proposes to map all 50,000 to 100,000 genes in the human genome, should give us the power to screen for many more of the 3,000 to 4,000 known genetic defects.

Many other diseases, such as colon cancer, Alzheimer's disease, multiple sclerosis, emphysema, and diabetes, are not simple genetic diseases but clearly result from an interaction between many genes, a person's genotype, and the environment. Research is currently focused on identifying genes that predispose individuals to these conditions. It is reasonable to assume that we will eventually be able to determine (through genetic screening) a person's genetic risk of developing these diseases as well.

What ethical questions does genetic screening present? We have more experience with screening issues than with gene therapy. Some of that experience has made it clear that we must proceed with caution and with a clear policy. In the 1970s, screening for sickle-cell anemia resulted in confusion and fear. It is evident that many people did not understand the difference between being a carrier and having the disease. People who merely carried the sickle-cell gene were threatened with loss of jobs and insurance. The initial purpose of helping people understand their possible risk of bearing children with sickle-cell disease was lost in a mire of misunderstanding.

Another problem with genetic screening is that we can identify many more diseases than we can treat. Huntington's disease (HD) is a good example. People who carry this dominant genetic-disease gene live their early lives normally and do not develop symptoms until middle age. At this point, however, the individual begins to lose the ability to think and develops uncontrollable body movements such as twitching and shaking. These conditions become worse and worse over a period of years until the victim dies. There is no treatment or cure. Unfortunately, since the disease is dominant, all of the children of a victim have a 50% chance of inheriting the disease gene. And since the disease does not reveal itself until middle age, most victims have already had children before they know whether or not they themselves have the gene.

Therefore, should a 20-year-old be able to find out that he carries a dominant gene for a disease that will first debilitate and finally kill him in middle age? If people should be given this information, at what age is it best revealed? Does respect for human dignity in this situation mean that we try to prevent harm by not telling or try to promote autonomy by telling the genetic status? Because of the experiences with sickle-cell screening 20 years ago, screening for HD is being done under very strict guidelines at 14 centers in the United States. The centers require clients to undergo extensive psychological testing before screening and counseling after screening. If a cure or even a treatment were available, then these questions would not be nearly as difficult, but as mentioned earlier, neither treatment nor cure is available. In the case of sickle-cell anemia, the screening test has been available for 34 years, but there is still no cure.

The ethical questions surrounding screening are apparent. Three percent of the $3 billion budget for the Human Genome Project has been set aside to study the ethical considerations. The Working Group of Ethical Legal and Social Implications of the Human Genome Project has proposed nine areas that merit attention.

1. Fairness in the use of genetic information
2. Impact of knowledge of genetic variation on the individual
3. Privacy and confidentiality
4. Impact on genetic counseling
5. Impact on reproductive decisions
6. Issues raised by the introduction of genetics into mainstream medical issues
7. Uses and misuses of genetics in the past
8. Questions raised by commercialization
9. Conceptual and philosophical implications

It is easy to see that different genetic conditions present different problems. With HD, there is presently no medical benefit to the individual in question in knowing that he or she has the disease gene (though such knowledge might affect the choices that person makes about how to live his or her early life). On the other hand, for conditions that have both environmental and genetic causes, such as high blood pressure and emphysema, early detection of susceptibility may save many lives and significant dollars. If people do make appropriate lifestyle changes, good has been achieved.

But who will be tested for these kinds of traits? Will the tests be voluntary? Who will pay for the test? (Remember that when anyone says, "The government pays," they really mean you pay, since the government gets its money from taxing people who work for a living. Middle class people pay about one-third of their incomes in various taxes.) Will we deny insurance or employment to people who refuse testing or to people who refuse to make lifestyle changes after tests indicate they have increased risks for certain conditions?

These are not easy questions. The medical and insurance professions are considering them now, and you will probably be affected by them during your lifetime.

Case Study 1: James and Carol H.

Refer to the step-by-step methodology for the decision-making model that is described in chapter 32.

Alice is the 6-year-old daughter of James and Carol H. Alice was diagnosed with CF at age 2. Alice has a moderately severe case of CF and receives postural drainage, a carefully specified, time-consuming program of thumping to loosen mucus in the lungs, from her mother three times a day. In addition, she is on a regimen of vitamins and enzymes that requires her to take about 25 pills each day. Carol spends much of her day caring for Alice. At least twice each year, Alice requires hospitalization to fight respiratory tract infections.

James and Carol have been talking about having another child. Alice is cared for at a public clinic associated with a medical school that has a research project on CF. The clinic physicians suggest that James and Carol undergo screening for the CF gene as part of the project.

When the results come back to the clinic, Alice's physician is quite surprised to learn that Carol, but not James, carries the CF gene. Since James is not a carrier, it is virtually impossible for him to be Alice's biological father.

I. Identify the question
First of all, is there a dilemma, and if there is, what is it? List some possible questions ("Should . . . ?").

II. Issue
What is the general issue in this case?

III. Facts
What are the facts in the case? It is easy to jump to some conclusions in this case, so try hard to stick to what you actually know (as opposed to what you have concluded or what you suspect). Depending on the question you are considering, different facts may be more relevant.

IV. Possible decisions
What are some possible solutions to the problem? Brainstorm as many as possible.

V. Additional information

Do you need additional background information to inform your decision?

VI. Decision

Formulate your decision (or your group's decision) in the form of an "I" or a "we" statement. For instance, "I believe the physician would be morally justified to. . . ."

VII. Principle

Which of the major ethical principles justifies your decision?

VIII. Authority

Which authorities would back you up on this point? One of the first places to look is in professional codes of ethics. For instance, the Hippocratic Oath states that the physician should do no harm. Similar statements are found in the American Medical Association Code of Ethics.

IX. Rebuttal

Under what circumstances would you change your minds about what to do? Think freely.

X. Level of confidence

Formulate a one- or two-word statement to describe how strongly you believe your argument.

XI. Box

Draw up a box as described previously (chapter 32).

XII. Prose argument

Write out an argument as described previously (chapter 32).

Case Study 2: Angela

Angela is a healthy 32-year-old certified public accountant who is married with no children. She has been working for 8 years in a small accounting firm. She has decided that she would like to move to a nearby large city and work for a major accounting firm.

Five years ago, Angela's mother was diagnosed with HD, an autosomal dominant disease. Angela's mother is in a nursing home.

Angela's job interviews go well, and she is offered an excellent position that she readily accepts. However, as Angela is filling out the required paperwork in the personnel office, she notices that she must sign a consent form for a physical that includes a genetic screening test of her DNA.

When Angela questions whether she must have such a test done, she is told it is required by the company. Although this is a hypothetical case, it is interesting to note that at least 50 companies performed genetic screening in 1987.

A number of questions could be raised about this case. Here are a few examples.

1. Should the accounting firm require the testing?
2. Who should have access to Angela's results if she agrees to be tested?
3. Should Angela's employment be contingent on the results of her DNA?
4. If Angela has a sister who is also at risk, should she be told Angela's results?

Think of more questions, and choose one for analysis. Apply the model as you have done before.

Societal Issues

Careers in Biotechnology

Introduction

Because of the relatively recent development of biotechnology, students may not be aware of the wealth of employment opportunities that are available in new biotechnology companies and in existing industries that are incorporating biotechnologies into their operations. Acquainting students with job types, educational requirements, and desired skills will not only inform them about these employment opportunities but might also encourage them to pursue careers in this exciting and expanding field.

The Biotechnology Industry

Because biotechnology is a collection of technologies that use cells and molecules isolated from cells, the phrase "biotechnology industry" is misleading. To be more accurate when describing possible careers in biotechnology, we should talk about "biotechnology companies (which are usually small, relatively young companies founded by scientists who have made a biological discovery with commercial potential) and the existing, long-established industries, such as the agricultural, chemical, and pharmaceutical industries, that are beginning to use biotechnologies in their research, development, and manufacturing processes." But since that phrase is long and clumsy, we are opting for the term "biotechnology industry" when describing career possibilities. The shorter phrase isn't perfectly accurate, but you will know what we really mean.

In 1995, there were approximately 1,500 biotechnology companies, defined above as small, relatively new companies, in the United States. Their total revenue was approximately $9 billion. Analysts predict that by the year 2000, this country's biotechnology industry will have sales of more than $50 billion, assets valued at $125 billion, and more than 500,000 people employed throughout the country. Figure 35.1 in the *Student Activity* pages shows the current distribution of biotechnology companies.

Companies involved in biotechnology range in size from small entrepreneurial start-ups to large multinational corporations employing thousands of people. Over time, new companies will emerge and existing companies will grow. Because the applications of biotechnology are so varied, the industry is growing rapidly, and the work environments are diverse, a career in biotechnology can be exciting and rewarding for students with an interest in science or engineering. If they acquire technical skills that are useful in one sector of the biotechnology industry, they will find that those skills are readily transferable to other sectors.

Types of Jobs in Biotechnology Companies

A well-established technology-based company usually has several divisions that carry out different functions. In each of these divisions, jobs are available for people with various skills and educational levels.

Research and development division

Scientific research is the basic foundation of any high-technology industry. New ideas represent the company's future, because they lead to a continuing line of new products. In some companies, research focuses on specific applications or products: how to apply scientific knowledge in new ways or how to improve an existing product. In companies with large budgets, some research teams carry out basic scientific research with no immediate application. These companies believe that simply acquiring new knowledge and understanding of how living systems work will pay off in the long run with new product ideas. Even though biotechnology companies have their own research teams and often contract with other companies for specialized work, much of the research that drives progress in biotechnology is carried out in universities by academic scientists.

Once a promising idea is generated, it is tested, refined, and made practical in a process known as product development. During the product development phase, scientists and engineers address issues such as safety, efficacy, and the most efficient and profitable way to manufacture the product.

The entire process of research and development is a substantial effort. The typical new pharmaceutical product can take as long as 10 years to develop and can cost up to $250 million.

In the research and development division of a biotechnology company, the research is usually directed by scientists with doctoral degrees. Research associates are typically college graduates with bachelor's degrees in science, although some graduates of technical schools (Associate of Applied Science [A.A.S.] degree) with specialized training are hired for these positions. In research and development divisions, individuals with high school diplomas might find employment as glassware washers, animal caretakers, or greenhouse assistants.

Production and quality control division

Workers in production actually make the products or deliver the services the company sells. Large-scale production and manufacturing often require not only people with scientific expertise but also people with a knowledge of engineering and industrial manufacturing technology and people with good mechanical skills.

Workers in quality control divisions make sure that the product meets specifications. Quality assurance personnel monitor the entire production process, and sometimes the research and development operations as well, to ensure that good manufacturing practices and standard operating procedures are followed at all times. Quality assurance is particularly important in the pharmaceutical industry, for which the Food and Drug Administration has established especially stringent guidelines for the testing and manufacture of medicines that will be used by humans.

There is less demand for advanced degrees in the manufacturing and production division of companies than in the research and development division. Conversely, there is more demand for people with A.A.S. degrees or high school diplomas in this aspect of company operations. Examples of entry-level jobs include chemical operator, manufacturing technician, instrument technician, and packaging operator. For those with a college degree in science or engineering, typical jobs would be quality assurance manager, process development scientist, or manufacturing en-

gineer. Even though high school graduates and holders of A.A.S. degrees will be hired in production and quality control, room for advancement is limited for those without a college degree.

Sales and marketing division

On the basis of its scientific research, a company may think it has a terrific product idea. But will it sell? Market researchers try to answer this question by assessing the need for the product, the number of people likely to buy it, and the price they might be willing to pay for it. Marketing personnel also try to find new markets for a product already being sold by the company and seek new ways to advertise and promote the product.

Salespeople are in the front lines, dealing directly with customers and selling the product. Sales personnel not only make sales but are also highly visible representatives of their company. They are often asked for technical advice about their products, and they relay feedback from customers to the company.

In the biotechnology industry, sales and marketing employees who have a scientific education have a competitive edge in securing jobs. To be good at sales and marketing, these employees must understand the nature of the highly technical products they sell and must know how to communicate with their customers, who often are scientists or medical professionals.

Regulatory affairs division

Various activities of companies of all types are regulated by federal and state agencies. This is particularly true of agricultural and pharmaceutical biotechnology companies, which must comply with intricate regulations imposed by the Food and Drug Administration, Environmental Protection Agency, and U.S. Department of Agriculture concerning the nature and manufacture of products. Many companies have teams of specialists, often with scientific backgrounds, who keep track of all federal and state regulations that apply to the company and make sure the company complies with them.

Legal affairs division

One of the most important tasks for a biotechnology company is to secure patent protection for its inventions. Without a patent, a new product idea may not be profitable for a company because competitors can make the same product. Having exclusive rights to market a new invention during the term of a patent is often the only way a company can be assured of sales

sufficient to pay for the high costs of research, development, and production of the product.

Consequently, companies may hire specialists to prepare and track patent applications, and larger corporations may have their own patent attorneys on staff. Many law firms now have attorneys who specialize in biotechnology patent law.

Public relations, communications, and training

Every company needs people who are effective communicators. This is particularly true in the biotechnology industry, where companies must be able to offer product information to the lay public in an easily understandable way. Technical writers may be employed to write internal or external scientific reports.

By their very nature, high-technology industries are involved in sciences that are breaking new ground rapidly. This rapid expansion of knowledge requires that employees be able to learn new technologies quickly. Large corporations may employ full-time staff development personnel who organize technical training within the company to keep their employees abreast of new developments.

Management and support functions

Managers at different levels in all divisions organize and supervise the activities described above. In most of the technical divisions of a company—research and development, production, and quality control, for example—people who become managers most often start out as scientists or engineers and work their way up. In entrepreneurial biotechnology companies, the chief executive officer or other high-level managers are often Ph.D.-holding research scientists whose ideas for new products provided the initial impetus to start the company. In other companies, these scientists remain in charge of research divisions, while managers with business training and experience assume other executive positions.

Like all companies, biotechnology companies need a variety of support personnel such as administrative assistants, accountants, information management specialists, and computer technicians.

Preparing for Careers in the Biotechnology Industry

People with all levels of education may be hired for various jobs in the research, development, or produc-

tion divisions in biotechnology companies. Types of jobs and job expectations related to different levels of education are discussed below.

The starting salaries for technical jobs vary widely with the type of job, the type of degree the applicant has, his or her prior job experience, and the geographical location of the job. Currently, typical salary ranges for technical positions in a biotechnology company are as follows.

Technical staff	$20,000–50,000
Technical middle management	$50,000–80,000
Vice president, research and development	$70,000–100,000+

High school diploma

Many people working in biotechnology-related manufacturing facilities have only a high school diploma. They are trained on the job to perform routine technical operations, or they work in more conventional jobs typical of those in any manufacturing facility (assembling, packaging, transporting, warehousing).

Less frequently, people with high school diplomas may be hired in research laboratories as laboratory assistants. They may perform routine experimental operations, prepare solutions and media, care for laboratory animals or greenhouse plants, or wash glassware.

Typical salary: $15,000–18,000

Typical job titles: media prep technician, greenhouse assistant I, laboratory assistant I

Opportunity for advancement: Limited. In industry or academia, advancement beyond a fairly low level is difficult or impossible without a college degree. However, a responsible, adaptable individual who can learn new skills quickly can earn job stability and pay raises.

Getting a job: Entry-level positions are advertised in local newspapers or through the local Employment Security Commission. The toughest competition is from the many other high school graduates seeking jobs. But since employers routinely find that only about 10 to 20% of applicants for entry-level positions have the basic skills to be considered after the first round of the selection process, job applicants are already way ahead of the crowd if they do nothing more than study hard enough to maintain a B average in high school. Good basic skills—reading, writing, speaking, and mathematics—are absolutely essential.

A.A.S. degree

People with a 2-year college degree in appropriate fields are very employable. Those graduating from biotechnology technician training programs at technical colleges readily find jobs in the industry, because such training programs focus on practical, hands-on skills that are in demand. The pragmatic training offered by technical colleges gives students more laboratory experience and "how-to" knowledge than many graduates with bachelor of science (B.S.) degrees acquire.

Two-year college graduates are most frequently employed in industrial production facilities. Some also find jobs as technicians in research laboratories in industry, universities, and medical centers. Since technical college graduates with specific training in biotechnology are few, industries do not usually specify a 2-year degree as a job requirement. Nonetheless, technical college graduates with training in biotechnology compete successfully for many jobs in the biotechnology industry that are usually held by people with a bachelor's degree.

Typical starting salary: $16,000–20,000, advancing to $22,000–26,000

Typical job titles: Same as for high school diploma (given above); may also include quality control inspector, instrumentation technician, laboratory technician, research assistant

Opportunities for advancement: Somewhat limited with only a 2-year degree but better than opportunities for those with no postsecondary training. However, people with an A.A.S. degree who work awhile and then go back to school and get a bachelor's degree have a competitive advantage over new college graduates because of their hands-on training and industrial experience.

Getting a job: Same as for high school diploma (described above). Most jobs at this level are advertised in local newspapers. For associate-degree graduates, the chances for employment as an academic research technician are better than the chances for high school graduates. Job seekers should visit the employment offices of universities and research centers and send résumés and letters to corporations with production facilities.

B.S. degree

For a career in the biotechnology industry, receiving a 4-year degree in the biological sciences is a logical beginning. The biological sciences are organized differently in different colleges and universities, so depending on the institution, students might major in general biology, genetics, biochemistry, microbiology, botany, or zoology. Students who major in general biology, botany, or zoology should choose as many courses as possible in molecular and cellular biology. A chemistry degree is also an excellent foundation, since so much of modern molecular biology is based on chemistry. Chemical engineering is a very marketable specialty. Other specialties useful in the industry are environmental science, toxicology, pharmacology, computer science, and agricultural science.

Employers value the B.S. degree because it gives the graduate a well-rounded background in general principles of science, provides a good knowledge base in a particular area (for example, biology or chemistry), and enhances basic communication and problem-solving skills. A B.S. in science or engineering is not only essential preparation for a scientific career but is increasingly in demand by biotechnology companies as a basic credential for those who want to work in sales, marketing, or other business areas not traditionally considered scientific.

Typical starting salary: $23,000–35,000, advancing to $30,000–45,000

Typical job titles: research associate, technician, technologist, manufacturing or production associate, product development engineer, quality control analyst or engineer, microbiologist, chemist

Opportunities for advancement: Good. In industry, an enterprising, talented employee with a bachelor's degree in science or engineering can advance to supervisory positions with higher pay. Such an employee may also be able to move from one division to another within a company, thus acquiring a new set of skills and improving the prospects for advancement.

Although the emphasis in this discussion is on research and production jobs, a B.S. in science or engineering is good preparation for jobs in sales, marketing, regulatory affairs, and other aspects of a company's business operations.

Getting a job: Jobs at this level are often advertised in local newspapers as well as in scientific journals. One of the main sources of scientific want ads is the weekly journal *Science,* published by the American Association for the Advancement of Science. *Science* runs ads for all levels of positions, from technicians to research directors, in universities as well as industry. Other sources for national ads are *The Scientist,*

Societal Issues

Chemical & Engineering News, and other trade journals. Entry-level jobs for which a B.S. is sufficient qualification are frequently filled by local applicants. However, jobs for people with industrial experience or postgraduate training are often advertised nationally.

For positions at the B.S. level, job seekers should answer want ads, identify companies that might need their skills, and send out many résumés (100 is not too many). If there are only a few companies in the area, job seekers may have to relocate.

Industrial employers place a premium on prior industrial experience and often prefer an applicant with industrial work experience over a college graduate fresh out of school. Therefore, students should seek out cooperative education opportunities, internships, or other summer jobs in industry during college. Or they can identify a mentor in their academic department and carry out an undergraduate research project or get a part-time job in a university research laboratory. These kinds of experiences add great value to a résumé.

Grades matter. Some companies award a better starting salary to an applicant with better grades, and some will not even consider applicants who have less than a B average. Good grades are important even for obtaining summer internships.

M.S. degree

In general, people with master of science (M.S.) degrees hold jobs similar to those held by people with bachelor's degrees. Pay is about $2,000 to 5,000 more, and advancement to supervisory positions may be more likely. The increased value of the M.S. degree is that it gives students more hands-on laboratory experience and, through work on a thesis project, teaches them how scientific research is actually carried out. Also, someone with a B.S. degree who did not include much specific biotechnology training in the undergraduate program can acquire advanced skills in an M.S. degree program. Obtaining both a B.S. and an M.S. is also a way to combine two specialties, such as biology and business.

Doctoral degree

A Ph.D. is the primary qualification for a scientific research career. Ph.D.'s design the research process, directing the activities of B.S. and M.S. technicians.

Typical starting salary: $40,000–55,000, advancing to $60,000–100,000+

Typical job titles: research scientist, senior scientist, principal scientist, research or scientific director, vice president for research (listed in order of increasing seniority)

Opportunities for advancement: Excellent. Companies have different promotion pathways for Ph.D.'s. In many, promotion to higher levels traditionally involves more management and less science. But since many scientists want to continue practicing science and not become involved in company management, some companies are finding ways to reward their top-notch researchers other than by promotion to management positions.

Getting a job: Students who aspire to become research scientists need only keep in mind that they should take all the science and mathematics they can in high school and seek out the best undergraduate education possible. If they excel, faculty at their college will counsel them in selecting a graduate school and research area. Taking summer jobs in industry and working on high school or undergraduate research projects are important preparation.

Academic Careers

Much of the basic research in biotechnology is carried out in universities and government research laboratories, not by companies. Another career choice open to students interested in science is that of academic research scientist. Traditionally, pay across all levels is substantially lower in universities than in industry, and for technicians with B.S. or M.S. degrees, the opportunities for promotion are limited.

In the typical university or government research laboratory, there are three classes of scientists: technicians, scientists-in-training, and professors. Technicians may be hired with B.S. or M.S. degrees. Frequently, undergraduates working on their B.S. degree can get part-time positions. Getting your first job as a technician is sometimes easier at a university than in industry, particularly if you have worked in that laboratory on a part-time basis while still in school. Even if your ultimate goal is to work in industry, acquiring technical skills and work experience in a university laboratory can provide a solid foundation and a competitive edge for seeking a job in industry.

Scientists-in-training may be graduate students working on Ph.D. thesis research or postdoctoral fellows. The postdoctoral fellowship is a temporary research position in which new Ph.D.'s get more extensive research experience. A "postdoc" is now considered a

Societal Issues

standard part of the career path for Ph.D. scientists. Some companies offer postdoctoral fellowships.

Professors at research universities and most 4-year colleges are Ph.D.'s. Smaller 4-year colleges and community colleges may hire scientists with M.S. degrees as faculty members or, less often, those with B.S. degrees as laboratory instructors.

Essential Skills of a Biotechnician

Although some students might be interested in pursuing a Ph.D., the great majority of those interested in a career in biotechnology will have a B.S. or A.A.S. degree. What are the specific skills these students should acquire to increase their competitiveness in the job market?

A number of nationwide studies have addressed this question for an entry-level job as a biotechnician. Although each study describes some skills that are industry specific, other sets of skills appear to be common to all industries that utilize some aspect of biotechnology. The skills that seem to be in universal demand by very diverse companies are listed below.

Technical skills

Basic Laboratory Skills

Be familiar with standard laboratory equipment (balances, pH meters, glassware, centrifuges) and procedures (filtration, distillation, weighing, measuring). Maintain and calibrate laboratory equipment and instruments. Prepare buffers, reagents, and solutions. Prepare dilutions of solutions correctly. Perform basic separation techniques. Use aseptic techniques when appropriate. Follow protocols and standard operating procedures.

Biological Laboratory Techniques

Basic microbiological techniques such as identifying, screening, and quantifying microorganism; isolating and maintaining pure cultures; analyzing products; and harvesting microorganisms. Basic cell biology techniques such as using microscopes, preparing cells for microscopic analysis, and propagating cells.

Safety Skills

Monitor, use, store, and dispose of hazardous materials properly. Use protective equipment and hoods. Maintain, understand, and follow Material Safety Data Sheets. Know and comply with current federal, state, and local regulations. Maintain safe work area.

Quality Control Skills

Perform validation testing. Document product specifications. Use analytical equipment. Compare results to government and/or company standards. Collate large volumes of data. Perform statistical tests and data analysis.

Instrumental Analysis Skills

Obtain representative samples. Prepare samples for analysis. Use and calibrate standard analytical instruments such as gas chromatographs, high-pressure liquid chromatographs, and spectrophotometers.

General employability skills

Communication Skills

Create, follow, and record protocols and standard operating procedures. Maintain accurate and clear records in laboratory notebook. Summarize experiment results for reports. Write business letters, memos, and technical reports. Proofread and edit written materials for correct spelling and grammar. Organize and present oral summaries. Locate and review scientific reference material. Comprehend and follow verbal instructions. Conduct literature searches.

Mathematical and Statistical Skills

Read, construct, and comprehend data in graphical form. Calculate and interpret ratios. Perform basic statistical tests (mean, standard deviation). Recognize anomalies in data collection. Perform calculations using exponents, roots, and logarithms. Solve simple algebraic equations. Convert units between metric and English systems. Express a number in scientific and standard notation. Convert word problems to mathematical expressions. Use linear regression to forecast data.

Computer Skills

Use basic word processing systems. Enter, store, and retrieve data. Create and use a spreadsheet. Manipulate data electronically with graphical software. Use electronic communication techniques and systems. Develop and maintain a database. Transfer data to and from remote databases and instruments.

Interpersonal Skills

Develop and use listening skills. Participate as a team member. Demonstrate an understanding of team planning, problem-solving, and the value, roles, and responsibilities of individuals. Develop conflict resolution and consensus-building techniques. Recognize and respect organizational structure and goals. Be open and adaptable to new technology and applications. Develop initiative taking and observation skills. Work independently.

In 1994, the U.S. Department of Labor and the U.S. Department of Education initiated a number of national skills standards projects, three of which relate to biotechnology: agricultural biotechnology, chemical processing, and biomedical sciences. If you are interested in obtaining detailed information on the skills employers expect entry-level biotechnicians to have, you may want to contact the organizations that conducted studies and get a copy of their reports. They are as follows.

Agricultural biotechnology: National FFA Foundation, 5632 Mt. Vernon Highway, Box 15160, Alexandria, VA 22309; (703) 360-3600. No charge.

Chemical processing: American Chemical Society, Education Division, 1155 16th Street, NW, Washington, DC 20036; (202) 872-8734. No charge.

Biomedical sciences: Education Development Center, 55 Chapel Street, Newton, MA 02160; (617) 969-7100. This report costs $15 for educational institutions.

For additional information on careers in biotechnology, contact the Biotechnology Industry Organization (BIO) at their mailing address (1625 K Street, NW, Washington, DC 20006-1604) or their Internet address (comments@bio.com).

BIO also has a biotechnology information site on the World Wide Web (http://www.biocom).

Careers in Biotechnology

Introduction

Because of the relatively recent development of biotechnology, many people are not aware of the wealth of employment opportunities that are available in new biotechnology companies and in existing industries that are incorporating biotechnologies into their operations. Becoming acquainted with the biotechnology industry, job types, educational requirements, and desired skills will put you in an excellent position for finding employment in this exciting and expanding field.

The Biotechnology Industry

Because biotechnology is a collection of technologies that use cells and molecules isolated from cells, the phrase "biotechnology industry" is misleading. To be more accurate when describing possible careers in biotechnology, we should talk about "biotechnology companies (which are usually small, relatively young companies founded by scientists who have made a biological discovery with commercial potential) and the existing, long-established industries, such as the agricultural, chemical, and pharmaceutical industries, that are beginning to use biotechnologies in their research, development, and manufacturing processes". But rather than being perfectly accurate, we are opting for the term "biotechnology industry" when describing career possibilities, and you will know what we really mean.

In 1995, there were approximately 1,500 biotechnology companies, defined above as small, relatively new companies, in the United States. Their total revenue was approximately $9 billion. Analysts predict that by the year 2000, this country's biotechnology industry will have sales of more than $50 billion, assets valued at $125 billion, and more than 500,000 people employed throughout the country. Figure 35.1 shows the current distribution of biotechnology companies.

Companies involved in biotechnology range in size from small entrepreneurial start-ups to large multina-

tional corporations employing thousands of people. Over time, new companies will emerge and existing companies will grow. Because the applications of biotechnology are so varied, the industry is growing rapidly, and the work environments are diverse, a career in biotechnology can be exciting and rewarding for those with an interest in science or engineering. If you acquire technical skills that are useful in one sector of the biotechnology industry, you will find that those skills are readily transferable to other sectors.

Types of Jobs in Biotechnology Companies

A well-established technology-based company usually has several divisions that carry out different functions. In each of these divisions, jobs are available for people with various skills and educational levels.

Research and development division

Scientific research is the basic foundation of any high-technology industry. New ideas represent the company's future, because they lead to a continuing line of new products. In some companies, research focuses on specific applications or products: how to apply scientific knowledge in new ways or how to improve an existing product. In companies with large budgets, some research teams carry out basic scientific research with no immediate application. These companies believe that simply acquiring new knowledge and understanding of how living systems work will pay off in the long run with new product ideas. Even though biotechnology companies have their own research teams and often contract with other companies for specialized work, much of the research that drives progress in biotechnology is carried out in universities by academic scientists.

Once a promising idea is generated, it is tested, refined, and made practical in a process known as product development. During the product development phase, scientists and engineers address issues such as

Figure 35.1 Distribution of biotechnology companies in the United States (from G. S. Burrill and K. B. Lee, *Biotech 91: a Changing Environment,* Ernst and Young, San Francisco).

safety, efficacy, and the most efficient and profitable way to manufacture the product.

The entire process of research and development is a substantial effort. The typical new pharmaceutical product can take as long as 10 years to develop and can cost up to $250 million.

In the research and development division of a biotechnology company, the research is usually directed by scientists with doctoral degrees. Research associates are typically college graduates with bachelor's degrees in science, although some graduates of technical schools (Associate of Applied Science [A.A.S.] degree) with specialized training are hired for these positions. In research and development divisions, individuals with high school diplomas might find employment as glassware washers, animal caretakers, or greenhouse assistants.

Production and quality control division

Workers in production actually make the products or deliver the services the company sells. Large-scale production and manufacturing often require not only people with scientific expertise but also people with a knowledge of engineering and industrial manufacturing technology and people with good mechanical skills.

Workers in quality control divisions make sure that the product meets specifications. Quality assurance personnel monitor the entire production process, and sometimes the research and development operations as well, to ensure that good manufacturing practices and standard operating procedures are followed at all times. Quality assurance is particularly important in the pharmaceutical industry, for which the Food and Drug Administration has established especially stringent guidelines for the testing and manufacture of medicines that will be used by humans.

There is less demand for advanced degrees in the manufacturing and production division of companies than in the research and development division. Conversely, there is more demand for people with A.A.S. degrees or high school diplomas in this aspect of company operations than in research and development. Examples of entry-level jobs include chemical operator, manufacturing technician, instrument technician, and packaging operator. If you have a college degree in science or engineering, typical jobs would be quality assurance manager, process development scientist, or manufacturing engineer. Even though high school graduates and holders of A.A.S. degrees may be hired in production and quality control, room for advancement is limited for those without a college degree.

Sales and marketing division

On the basis of its scientific research, a company may think it has a terrific product idea. But will it sell? Market researchers try to answer this question by assessing the need for the product, the number of people likely to buy it, and the price they might be willing to pay for it. Marketing personnel also try to find new markets for a product already being sold by the company and seek new ways to advertise and promote the product.

Salespeople are in the front lines, dealing directly with customers and selling the product. Sales personnel not only make sales but are also highly visible representatives of their company. They are often asked for technical advice about their products, and they relay feedback from customers to the company.

In the biotechnology industry, sales and marketing employees who have a scientific education have a competitive edge in securing jobs. To be good at sales and marketing, these employees must understand the nature of the highly technical products they sell and must know how to communicate with their customers, who often are scientists or medical professionals.

Regulatory affairs division

Various activities of companies of all types are regulated by federal and state agencies. This is particularly true of agricultural and pharmaceutical biotechnology companies, which must comply with intricate regulations imposed by the Food and Drug Administration, Environmental Protection Agency, and U.S. Department of Agriculture concerning the nature and manufacture of products. Many companies have teams of specialists, often with scientific backgrounds, who keep track of all federal and state regulations that apply to the company and make sure the company complies with them.

Legal affairs division

One of the most important tasks for a biotechnology company is to secure patent protection for its inventions. Without a patent, a new product idea may not be profitable for a company because competitors can make the same product. Having exclusive rights to market a new invention during the term of a patent is often the only way a company can be assured of sales

sufficient to pay for the high costs of research, development, and production of the product.

Consequently, companies may hire specialists to prepare and track patent applications, and larger corporations may have their own patent attorneys on staff. Many law firms now have attorneys who specialize in biotechnology patent law.

Public relations, communications, and training

Every company needs people who are effective communicators. This is particularly true in the biotechnology industry, where companies must be able to offer technical product information to the lay public in an easily understandable way. Technical writers may be employed to write internal or external scientific reports.

By their very nature, high-technology industries are involved in sciences that are breaking new ground rapidly. This rapid expansion of knowledge requires that employees be able to learn new technologies quickly. Large corporations may employ full-time staff development personnel who organize technical training within the company to keep their employees abreast of new developments.

Management and support functions

Managers at different levels in all divisions organize and supervise the activities described above. In most of the technical divisions of a company—research and development, production, and quality control, for example—people who become managers most often start out as scientists or engineers and work their way up. In entrepreneurial biotechnology companies, the chief executive officer or other high-level managers are often Ph.D.-holding research scientists whose ideas for new products provided the initial impetus to start the company. In other companies, these scientists remain in charge of research divisions, while managers with business training and experience assume other executive positions.

Like all companies, biotechnology companies need a variety of support personnel such as administrative assistants, accountants, information management specialists, and computer technicians.

Preparing for Careers in the Biotechnology Industry

People with all levels of education may be hired for various jobs in the research, development, or production divisions in biotechnology companies. Types of jobs and job expectations related to different levels of education are discussed below.

The starting salaries for technical jobs vary widely with the type of job, the type of degree the applicant has, his or her prior job experience, and the geographical location of the job. Currently, typical salary ranges for technical positions in a biotechnology company are as follows.

Technical staff	$20,000–50,000
Technical middle management	$50,000–80,000
Vice president, research and development	$70,000–100,000+

High school diploma

Many people working in biotechnology-related manufacturing facilities have only a high school diploma. They are trained on the job to perform routine technical operations, or they work in more conventional jobs typical of those in any manufacturing facility (assembling, packaging, transporting, warehousing).

Less frequently, people with high school diplomas may be hired in research laboratories as laboratory assistants. They may perform routine experimental operations, prepare solutions and media, care for laboratory animals or greenhouse plants, or wash glassware.

Typical salary: $15,000–18,000

Typical job titles: media prep technician, greenhouse assistant I, laboratory assistant I

Opportunity for advancement: Limited. In industry or academia, advancement beyond a fairly low level is difficult or impossible without a college degree. However, a responsible, adaptable individual who can learn new skills quickly can earn job stability and pay raises.

Getting a job: Entry-level positions are advertised in local newspapers or through the local Employment Security Commission. The toughest competition is from the many other high school graduates seeking jobs. But since employers routinely find that only about 10 to 20% of applicants for entry-level positions have the basic skills to be considered after the first round of the selection process, job applicants are already way ahead of the crowd if they do nothing more than study hard enough to maintain a B average in high school. Good basic skills—reading, writing, speaking, and mathematics—are absolutely essential.

A.A.S. degree

People with a 2-year-college degree in appropriate fields are very employable. Those graduating from biotechnology technician training programs at technical colleges readily find jobs in the industry, because such training programs focus on practical, hands-on skills that are in demand. The pragmatic training offered by technical colleges gives A.A.S. degree holders more laboratory experience and "how-to" knowledge than many graduates with bachelor's of science (B.S.) degrees acquire.

Two-year-college graduates are most frequently employed in industrial production facilities. Some also find jobs as technicians in research laboratories in industry, universities, and medical centers. Since technical college graduates with specific training in biotechnology are few, industries do not usually specify a 2-year degree as a job requirement. Nonetheless, technical college graduates with training in biotechnology compete successfully for many jobs in the biotechnology industry that are usually held by people with a bachelor's degree.

Typical starting salary: $16,000–20,000, advancing to $22,000–26,000

Typical job titles: Same as for high school diploma (given above); may also include quality control inspector, instrumentation technician, laboratory technician, research assistant.

Opportunities for advancement: Somewhat limited with only a 2-year degree but better than opportunities for those with no postsecondary training. However, people with an associate degree who work awhile and then go back to school and get a bachelor's degree have a competitive advantage over new college graduates because of their hands-on training and industrial experience.

Getting a job: Same as for high school diploma (described above). Most jobs at this level are advertised in local newspapers. For A.A.S. graduates, the chances for employment as an academic research technician are somewhat better than the chances for high school graduates. Job seekers should visit the employment offices of universities and research centers and send résumés and letters to companies, particularly those with production facilities.

B.S. degree

For a career in the biotechnology industry, receiving a 4-year degree in the biological sciences is a logical beginning. The biological sciences are organized differently in different colleges and universities, so depending on the institution, you might major in general biology, genetics, biochemistry, microbiology, botany, or zoology. If you major in general biology, botany, or zoology, choose as many courses as possible in molecular and cellular biology. A chemistry degree is also an excellent foundation, since so much of modern molecular biology is based on chemistry. Chemical engineering is a very marketable specialty. Other specialties useful in the industry are environmental science, toxicology, pharmacology, computer science, and agricultural science.

Employers value the B.S. degree because it gives the graduate a well-rounded background in general principles of science, provides a good knowledge base in a particular area (for example, biology or chemistry), and enhances basic communication and problem-solving skills. A B.S. in science or engineering is not only essential preparation for a scientific career but is increasingly in demand by biotechnology companies as a basic credential for those who want to work in sales, marketing, or other business areas not traditionally considered scientific.

Typical starting salary: $23,000–35,000, advancing to $30,000–45,000

Typical job titles: research associate, technician, technologist, manufacturing or production associate, product development engineer, quality control analyst or engineer, microbiologist, chemist

Opportunities for advancement: Good. In industry, if you are an enterprising, talented employee with a B.S. in science or engineering, you can advance to supervisory positions with higher pay. You may also be able to move from one division to another within a company, thus acquiring a new set of skills and improving your prospects for advancement.

Although the emphasis in this discussion is on research and production jobs, a B.S. in science or engineering is good preparation for jobs in sales, marketing, regulatory affairs, and other aspects of a company's business operations.

Getting a job: Jobs at this level are often advertised in local newspapers as well as in scientific journals. One of the main sources of scientific want ads is the weekly journal *Science,* published by the American Association for the Advancement of Science. *Science* runs ads for all levels of positions, from technicians to research directors, in universities as well as industry. Other sources for national ads are *The Scientist, Chemical & Engineering News,* and other trade journals. Entry-level jobs for which a B.S. is sufficient qualification are frequently filled by local applicants. However, jobs for people with specialized industrial experience or postgraduate training are often advertised nationally.

For positions at the B.S. level, job seekers should answer want ads, identify companies that might need their skills, and send out many résumés (100 is not too many). If there are only a few companies in your area, you may have to relocate.

Industrial employers place a premium on prior industrial experience and often prefer an applicant with industrial work experience over a college graduate fresh out of school. Therefore, seek out cooperative education opportunities, internships, or other summer jobs in industry during college. Or identify a mentor in an academic department and carry out an undergraduate research project or get a part-time job in a university research laboratory. These experiences will add great value to your résumé.

Grades matter. Some companies award a better starting salary to an applicant with better grades, and some will not even consider applicants who have less than a B average. Good grades are important even for obtaining summer internships.

M.S. degree

In general, people with master of science (M.S.) degrees hold jobs similar to those held by people with bachelor's degrees. Pay is about $2,000 to 5,000 more, and advancement to supervisory positions may be more likely. The increased value of the M.S. degree is that it gives students more hands-on laboratory experience and, through work on a thesis project, teaches them how scientific research is actually conducted. Also, someone with a B.S. degree who did not include much specific biotechnology training in the undergraduate program can acquire advanced skills in an M.S. degree program. Obtaining both a B.S. and an M.S. is also a way to combine two specialties, such as biology and business.

Doctoral degree

A Ph.D. is the primary qualification for a scientific research career. Ph.D.'s design the research process, directing the activities of B.S. and M.S. technicians.

Typical starting salary: $40,000–55,000, advancing to $60,000–100,000+

Typical job titles: research scientist, senior scientist, principal scientist, research or scientific director, vice president for research (listed in order of increasing seniority)

Societal Issues

Opportunities for advancement: Excellent. Companies have different promotion pathways for Ph.D.'s. In many, promotion to higher levels traditionally involves more management and less science. But since many scientists want to continue practicing science and not become involved in company management, some companies are finding ways to reward their top-notch researchers other than by promotion to management positions.

Getting a job: If you aspire to become a research scientist, take all the science and mathematics you can in high school, and seek out the best undergraduate education possible. If you excel, faculty at your college or university will counsel you in selecting a graduate school and research area. Taking summer jobs in industry and working on high school or undergraduate research projects are also important preparation for a career as a research scientist.

Academic Careers

Much of the basic research in biotechnology is carried out in universities and government research laboratories, not by companies. Another career choice open to students interested in science is that of academic research scientist. Traditionally, pay across all levels is substantially lower in universities than in industry, and for technicians with B.S. or M.S. degrees, the opportunities for promotion are limited.

In the typical university or government research laboratory, there are three classes of scientists: technicians, scientists-in-training, and professors. Technicians may be hired with B.S. or M.S. degrees. Frequently, undergraduates working on their B.S. degree can get part-time positions. Getting your first job as a technician is sometimes easier at a university than in industry, particularly if you have worked in that laboratory on a part-time basis while still in school. Even if your ultimate goal is to work in industry, acquiring technical skills and work experience in a university laboratory can provide a solid foundation and a competitive edge for seeking a job in industry.

Scientists-in-training may be graduate students working on Ph.D. thesis research or postdoctoral fellows. The postdoctoral fellowship is a temporary research position in which new Ph.D.'s get more extensive research experience. A "postdoc" is now considered a standard part of the career path for Ph.D. scientists. Some companies offer postdoctoral fellowships.

Professors at research universities and most 4-year colleges are Ph.D.'s. Smaller 4-year colleges and community colleges may hire scientists with M.S. degrees as faculty members or, less often, those with B.S. degrees as laboratory instructors.

Essential Skills of a Biotechnician

Although some students might be interested in pursuing a Ph.D., the great majority who are interested in a career in biotechnology will have a B.S. or A.A.S. degree. What are the specific skills you should acquire to increase your competitiveness in the job market?

A number of nationwide studies have addressed this question for an entry-level job as a biotechnician. Although each report describes some skills that are industry specific, other sets of skills appear to be common to all industries utilizing some aspect of biotechnology. The skills that seem to be in universal demand by very diverse companies are listed below.

Technical skills
Basic Laboratory Skills
Be familiar with standard laboratory equipment (balances, pH meters, glassware, centrifuges) and procedures (filtration, distillation, weighing, measuring). Maintain and calibrate laboratory equipment and instruments. Prepare buffers, reagents, and solutions. Prepare dilutions of solutions correctly. Perform basic separation techniques. Use aseptic techniques when appropriate. Follow protocols and standard operating procedures.

Biological Laboratory Skills
Basic microbiological techniques such as identifying, screening, and quantifying microorganism; isolating and maintaining pure cultures; analyzing products; and harvesting microorganisms. Basic cell biology techniques such as using microscopes, preparing cells for microscopic analysis, and propagating cells.

Safety Skills
Monitor, use, store, and dispose of hazardous materials properly. Use protective equipment and hoods. Maintain, understand, and follow Material Safety Data Sheets. Know and comply with current federal, state, and local regulations. Maintain safe work area.

Quality Control Skills
Perform validation testing. Document product specifications. Use analytical equipment. Compare results to government and/or company standards. Collate large volumes of data. Perform statistical tests and data analysis.

Societal Issues

Instrumental Analysis Skills

Obtain representative samples. Prepare samples for analysis. Use and calibrate standard analytical instruments such as gas chromatographs, high-pressure liquid chromatographs, and spectrophotometers.

General employability skills

Communication Skills

Create, follow, and record protocols and standard operating procedures. Maintain accurate and clear records in laboratory notebook. Summarize experiment results for reports. Write business letters, memos, and technical reports. Proofread and edit written materials for correct spelling and grammar. Organize and present oral summaries. Locate and review scientific reference material. Comprehend and follow verbal instructions. Conduct literature searches.

Mathematical and Statistical Skills

Read, construct, and comprehend data in graphical form. Calculate and interpret ratios. Perform basic statistical tests (mean, standard deviation). Recognize anomalies in data collection. Perform calculations using exponents, roots, and logarithms. Solve simple algebraic equations. Convert units between metric and English systems. Express a number in scientific and standard notation. Convert word problems to mathematical expressions. Use linear regression to forecast data.

Computer Skills

Use basic word processing systems. Enter, store, and retrieve data. Create and use a spreadsheet. Manipulate data electronically with graphical software. Use electronic communication techniques and systems. Develop and maintain a database. Transfer data to and from remote databases and instruments.

Interpersonal Skills

Develop and use listening skills. Participate as a team member. Demonstrate an understanding of team planning, problem solving, and the value, roles, and responsibilities of individuals. Develop conflict resolution and consensus-building techniques. Recognize and respect organizational structure and goals. Be open and adaptable to new technology and applications. Develop initiative taking and observation skills. Work independently.

In 1994, the U.S. Department of Labor and the U.S. Department of Education initiated a number of national skills standards projects, three of which relate to biotechnology: agricultural biotechnology, chemical processing, and biomedical sciences. If you are interested in obtaining detailed information on the skills employers expect entry-level biotechnicians to have, you may want to contact the organizations that conducted the studies and request a copy of their reports. They are as follows.

> Agricultural biotechnology: National FFA Foundation, 5632 Mt. Vernon Highway, Box 15160, Alexandria, VA 22309; (703) 360-3600. No charge.

> Chemical processing: American Chemical Society, Education Division, 1155 16th Street, NW, Washington, DC 20036; (202) 872-8734. No charge.

> Biomedical sciences: Education Development Center, 55 Chapel Street, Newton, MA 02160; (617) 969-7100. This report costs $15 for educational institutions and $22 for others.

For additional information on careers in biotechnology, contact the Biotechnology Industry Organization (BIO) at their mailing address (1625 K Street, NW, Washington, DC 20006-1604) or their Internet address (comments@bio.com).

BIO also has a biotechnology information site on the World Wide Web (http://www.biocom).

PART IV

Appendixes

These appendixes contain information to assist you in incorporating the science and techniques of DNA-based technologies into your curriculum. We review basic laboratory techniques, equipment use, and biosafety information. Recipes for solutions used in the protocols, lists of recommended reading, and sources of further information are provided.

Appendixes

Appendix A: Laboratory Biosafety

Handling Microorganisms in the Laboratory

Escherichia coli is a normal inhabitant of the digestive tract. Of the many strains of *E. coli,* some inhabit the human gut, and others reside in animals. A few strains of *E. coli* cause significant disease in humans; one such strain was involved in the contaminated-hamburger incidents reported in the media in autumn 1993.

The laboratory strains of *E. coli* discussed in this book, MM294, cI, cII, CR63, and B$_E$, have been used in the laboratory for years and do not normally cause disease. MM294 is reported to be ineffective at colonizing the human digestive tract and so is especially harmless in that respect. However, all of these strains could cause infection if introduced into an open wound or into the eye. It is therefore very important to use aseptic technique (see Appendix C) when handling the organisms. Students should never eat, drink, smoke, or apply cosmetics in the laboratory. Many instructors require that students wear protective goggles while working in the laboratory.

When transferring cultures of *E. coli,* keep pipette tips away from the face to avoid inhaling any aerosol that might be created. If you or any student contaminates his or her hands with a culture, wash immediately. Avoid contaminating any cuts with bacterial culture, and keep all bacteria away from the eyes. If a student believes he or she may have contaminated an area of broken skin, wash that area immediately. If a student gets bacteria in his or her eyes, use the eyewash fountain to rinse the eyes, and call a physician's office for further advice.

Agrobacterium tumefaciens does not cause disease in humans. However, the same precautions should be taken with it as with *E. coli.*

Bacteriophage T4 is harmless to humans. It cannot infect human cells. It should be handled in the same manner as *E. coli,* mostly to avoid inadvertent contamination of laboratory *E. coli* cultures.

Disinfect all plates and cultures before disposing of them.

Disinfecting

Keep disinfectant solutions available in the laboratory in squeeze bottles. These solutions can be 2% Lysol, 70% ethanol, rubbing alcohol, or other special disinfectants.

Before carrying out any experiments involving bacteria or phage, wipe down the laboratory bench with disinfectant solution. At the end of the laboratory period, wipe down the benchtop with disinfectant again. Clean up any spills involving organisms immediately, and disinfect the area thoroughly.

After the laboratory period, disinfect all materials that have come in contact with bacteria or phage (micropipette tips, pipettes, agar plates, culture tubes, flasks, etc.) either with pressurized steam or by soaking them in concentrated disinfectant. To use steam, place all biological waste in an autoclave bag, and sterilize it in an autoclave or pressure cooker. To use disinfectant, soak all contaminated materials for at least 15 min in either 10% Lysol or 15 to 20% chlorine bleach (note that these solutions are not the same as the benchtop solutions described above; see Recipes [Appendix E]). Drain the liquid. Place trash in a plastic bag, and put the bag in the trash. Thoroughly rinse any containers to be reused. Clean the containers as usual, and then sterilize as needed.

It is not necessary to disinfect materials that have come in contact only with DNA and restriction enzymes (for example, from the gel electrophoresis laboratories).

Regulations for Recombinant DNA Work

The National Institutes of Health (NIH) oversees and regulates all research involving transfer of DNA between species. Rules for conducting recombinant DNA research are published as the NIH *Guidelines for Research Involving Recombinant DNA Molecules*. Certain kinds of recombinant DNA research are designated as exempt from these guidelines. Provision III-D-3 states that "The following molecules are exempt from these guidelines . . . those that consist entirely of DNA from a prokaryotic host, including its indigenous plasmids or viruses when propagated only in the host (or a closely related strain of the same species) or when transferred to another host by well-established physiological means." *Under this guideline, all the experiments and DNA molecules used in this book are exempt and may be conducted and used in a high school setting.*

If students wish to pursue further research involving recombinant DNA, instructors should make sure that the work involves only exempt molecules and procedures or should make arrangements for the students to work in an NIH-approved laboratory.

Selected Reading

Horn, T. M. 1992. *Working with DNA and Bacteria in Precollege Science Classrooms*. National Association of Biology Teachers, Reston, Va.

Appendix B: Basic Microbiological Methods

The Care and Feeding of *Escherichia coli*

Escherichia coli is not a fussy microbe and will grow on a variety of media at a variety of temperatures. Although *E. coli* cultures grow faster if aerated, they will also grow without aeration. All of this means that you have many options for growing the organism and can choose methods that suit your circumstances.

Three medium recipes are given in Appendix E (Recipes): Luria broth (and agar), nutrient broth (and agar), and tryptic soy broth (and agar). Any of these will do, and they can be freely substituted for each other in the procedures in this book.

Liquid cultures

Use an inoculating needle or loop to touch an isolated colony on an agar plate. Introduce the cells into sterile broth. The cells will multiply fastest at 37°C with constant shaking (for aeration). If you have an incubator but no shaker, grow the cells at 37°C, and shake the flask yourself whenever it's convenient. If you can't shake the flask at all, the cells will still grow. If you don't have an incubator, they will grow at room temperature, but don't expect heavy growth overnight. It's a good idea to test how long it will take *E. coli* to grow under your conditions if you need to be certain of having the cells ready at a particular time.

When microbiologists need to grow large volumes of cells, they usually grow a small overnight culture first and then use that culture to inoculate the larger volume of medium. You don't have to follow this procedure, but if you do, you'll be able to tell much sooner that your cells are actually growing.

If you have a magnetic stirrer, you can use it to aerate your cultures. Sterilize your medium in a loosely capped or cotton-plugged flask with a stir bar in it. Introduce the cells, replace the cap or plug, put the flask on the stirrer, and turn the stirrer on. The stirring action will keep the culture aerated.

Dispose of liquid cultures as soon as you reasonably can after using them. Contaminants can grow in cultures that have been opened and used.

Growing E. coli *on agar plates*

The factors that apply for growth in liquid also apply for growth on agar but without the question of stirring. The warmer you keep the plates (up to 37°C), the faster the cells will grow. Once the colonies have grown, keep the plates in the refrigerator. Chilling the plates prevents the growth of bacterial and fungal contaminants. Plates can be stored in the refrigerator for a month or so, and colonies taken from them will grow in fresh media. When you start a series of experiments using *E. coli,* it is a good idea to streak a "master plate" to use as a source of isolated colonies for inoculating liquid cultures. Keep this plate in the cold.

Long-Term Storage of *E. coli*

The best long-term storage method is to add glycerol to 20% (vol/vol) to a fresh liquid culture and keep the culture in a –80°C freezer. Given the frequency with which we find –80°C freezers in schools, it is good that other methods will work.

Frozen cultures

Frozen cultures can be stored in a regular freezer, preferably *not* a frost-free one. If your freezer is frost-free, put the cells in a place away from the heating coils. Even better, place a small Styrofoam box inside the frost-free freezer, and keep the frozen cultures inside the box to protect them from defrosting cycles.

To make a frozen culture, grow up the desired cells in any medium. With a sterile pipette, remove a specific volume of cells. Add sterile glycerol so that it will be 20% of the final volume. For example, to 2 ml of culture, add 0.5 ml of sterile glycerol. Label (all frozen cultures look alike), shake well to mix in the glycerol, and freeze.

Appendixes

Glycerol is very viscous and therefore extremely difficult to deliver from pipettes with accuracy (because so much clings to the pipette walls). It is more accurate to measure glycerol by weight. The density of glycerol is 1.25 g/ml. The method we use is to add 0.625 g (0.5 ml) of glycerol to several autoclavable screw-cap plastic tubes, autoclave them, and then add 2 ml of culture to one tube whenever we need to make a frozen stock. The extra tubes with sterile glycerol can be kept tightly closed at room temperature.

Stab cultures

E. coli cultures can be kept for a year or longer in stab cultures at room temperature. A stab culture is a screw-cap test tube containing agar medium. To make these cultures, mix up one of the solid media listed in Appendix E (Recipes), and heat it to boiling to melt the agar. Let the medium cool a little, and then fill clean glass screw-cap test tubes or vials about one-third full. The size of the test tubes is not important. Autoclave the test tubes, and let the medium harden. Use an inoculating needle to stab some *E. coli* cells into the agar. Incubate the stab tubes with the caps loose until you see growth. Then screw the caps on tight, label the tubes, and put them in a drawer or other convenient place.

Working cultures

The laboratory strains of *E. coli* recommended in this book are safe for student use with reasonable precautions. Because no one can say the same thing for random contaminants of cultures, it is important for safety (as well as for the success of your experiments) that you not contaminate your cultures. It is especially important to protect your long-term storage cultures. The best way to do this is to use them as little as possible and as carefully as possible.

As directed above, when you begin a series of *E. coli* experiments, use your long-term storage culture once to streak an agar plate for isolated colonies. Store the plate in the refrigerator after the culture grows up, and use this plate as a source of cells for 2 to 4 weeks. This approach minimizes risks of contamination to your long-term storage cultures.

Streaking Agar Plates To Obtain Isolated Colonies

Using isolated colonies is another method to protect yourself from contaminated cultures. Microbiological procedures recommend starting with an isolated colony because an isolated colony is considered a pure culture. When an isolated single cell on an agar

plate multiplies repeatedly, it forms a nice, round, isolated colony. This colony is a pure culture of descendants of that founder cell. If you touch it to inoculate another culture, you can be assured that you are introducing only one type of organism into that culture. On the other hand, if you scrape up cells from a region in which cells are growing in large streaks, you are introducing descendants of many cells and may even be introducing contaminants that you couldn't detect in the thick growth.

So isolated colonies are best for starting cultures. If you don't have any such colonies, however, you can start your cultures from streaks of growth or directly from long-term storage cultures and hope for the best.

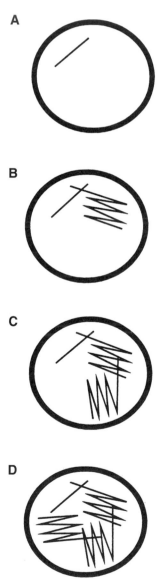

Figure B.1 Schematic diagram showing how to streak an agar plate to obtain isolated bacterial colonies (refer to text).

A

B

C

D

Procedure

1. Flame your inoculating loop. Allow it to cool (you may touch it to the surface of the sterile agar to speed this process), and then touch the source of cells (liquid or solid culture). Do not try to get a large volume of culture into the loop; it will only make the rest of the job harder. The goal is to isolate single *E. coli* cells on the agar plate.

2. On the fresh plate, make a single streak with the loop (Fig. B.1A).

3. Reflame the loop, and allow it to cool.

4. With the loop lightly touching the surface of the agar plate, drag the loop once across the streak of cells you just made. This line should be at somewhat more than a 90° angle to the streak. You are dragging some of those cells farther out across the plate.

5. Without lifting the loop from the agar surface, zigzag the loop back and forth in a tight pattern across about a quarter of the plate. The idea is to spread those cells out (Fig. B.1B).

6. Reflame the loop.

7. Drag the cooled loop through the area that you last streaked (one drag). Repeat the zigzag pattern in a new area of the plate (Fig. B.1C).

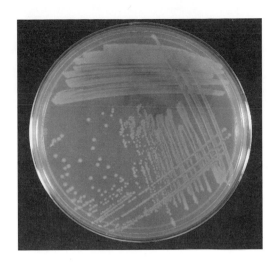

Figure B.2 A streaked agar plate.

8. Reflame the loop.

9. Repeat the "drag and zigzag" (Fig. B.1D).

10. Incubate the plates until visible colonies form. Figure B.2 is a photograph of a streaked plate. You can see the zigzag pattern and some isolated colonies.

You may want to practice this technique a few times.

Appendix C: Aseptic Technique

General Information

In experiments with microorganisms, it is essential to avoid contamination of the experiment by other microbes. Contamination could cause the procedure to fail. More important, contaminating organisms cannot be guaranteed to be harmless. Avoiding contamination is most important when cultures are being inoculated. If a contaminating microbe should find its way into the growth medium at the beginning of the growth period, the contaminant could grow along with the intended experimental organism.

To prevent microbiological contamination, a system of laboratory practice called "sterile technique" or "aseptic technique" is used. Proper aseptic technique in the laboratory minimizes the risk of contamination.

There are a few principles to remember when learning aseptic technique.

- An object or solution is sterile only if it contains *no* living thing.

- In general, objects and solutions are sterile only if they have been treated (autoclaved, irradiated, etc.) to kill contaminating microorganisms.

- Any sterile surface or object that comes in contact with a nonsterile thing is no longer considered sterile.

- Air is not sterile (unless it was sterilized inside a closed container).

Acting on those principles, we first sterilize all containers and solutions to be used when culturing microorganisms. Thereafter, material is transferred between sterile containers with sterile tools (sterile pipettes, micropipettes with sterile tips, sterile inoculating loops, etc.) in such a way as to minimize exposure to outside air and avoid contact with nonsterile surfaces and items.

The keys to good aseptic technique are as follows.

- Keep lids off sterile containers for the shortest time possible.

- Pass the mouths of open containers through a flame. Flaming warms the air at the opening, creating positive pressure and preventing contaminants from falling into the tube. Even plastic items can be flamed briefly.

- Hold open containers at an angle whenever possible to prevent contaminants from falling in.

Specific Techniques

Use of glass or plastic pipettes

Glass pipettes are put into containers or wrapped and then autoclaved. Plastic pipettes are purchased presterilized in individual wrappers. To use a pipette, remove it from its wrapper or container by the end opposite the tip. Do not touch the lower two-thirds of the pipette. Do not allow the pipette to touch any laboratory surface. Draw the lower length of the pipette through a Bunsen burner flame. Insert only the untouched lower portion of the pipette into a sterile container.

Using test or culture tubes

Sterilize test tubes with lids or caps on. When you open a sterile tube, touch only the outside of the cap, and do not set the cap on any laboratory surface. Instead, hold the cap with one or two of your fingers while you complete the operation, and then replace it on the tube. This technique usually requires some practice, especially if you are simultaneously opening tubes and operating a sterile pipette. If you are working with a laboratory partner, one person can operate the test tube and the other can operate the pipette.

After you remove the cap from the test tube, pass the mouth of the tube through a flame. If possible, hold the open tube at an angle. Put only sterile objects into the tube. Complete the operation as quickly as you reasonably can, and then flame the mouth of the tube again. Replace the lid.

Inoculating loops and needles

Inoculating loops and needles are the primary tools for transferring microbial cultures. Loops and needles are sterilized by flaming. Put the business end of the tool directly into a Bunsen or alcohol burner flame, and hold it there until the end glows bright red. Withdraw the tool from the flame. The tool is now hot and sterile. Count to 5 or 10 to let it cool, and then transfer the organisms.

If you are moving organisms from an agar plate, touch an isolated colony with the transfer loop. Be sure your inoculating loop is cool before you do this. Replace the plate lid. Open and flame the culture tube, and inoculate the medium in it by stirring the end of the transfer tool in the medium. If you are removing cells from a liquid culture, insert the loop into the culture. The loop may hiss. If so, wait until the hissing stops, move the loop a little in the culture, and then withdraw it. Even if you cannot see any liquid in the loop, there will be enough cells there to inoculate a plate or a new liquid culture.

Transferring large volumes

If you don't have to be careful about the volume you transfer, a pure culture or sterile solution can be transferred to a sterile container or new sterile medium by pouring. Remove the cap or lid from the solution to be transferred. Thoroughly flame the mouth of the container, holding it at an angle as you do so. Remove the lid from the target container. Hold the container at an angle and flame it, if possible (you may not be able to if you are holding many items in your hands). Quickly and neatly pour the contents from the first container into the second. Flame the mouth of the second container. Replace the lid.

If you must transfer an exact volume of liquid, use a sterile pipette or a sterile graduated cylinder. When using a sterile graduated cylinder, complete the transfer as quickly as you reasonably can to minimize the time the sterile liquid is exposed to the air.

disposable plates may be used directly from the package.

- Make and sterilize the agar (or melt presterilized agar). You will be able to pour 30 to 40 standard plates from 1 liter of medium.

- Allow the agar to cool to 55 to 60°C. At this temperature, the container will feel very warm but not painful when you hold it against your cheek. Add antibiotics at this time, and swirl the flask.

- Wipe down your working surface with disinfectant. Arrange the empty dishes into a convenient pattern. Some people like to make stacks of two to three plates; others prefer not to stack them for pouring. If circumstances permit, light a burner.

- Holding the flask of molten agar at an angle, remove the stopper and flame the mouth. Open the lid of the first plate enough so that you can pour in the agar solution. Pour in enough to cover the bottom of the plate, and then pour a little more. A depth of 1/4 in. is sufficient. Close the lid. Go to the next plate.

If you touch something nonsterile with the mouth of the flask, flame the flask immediately and thoroughly. Otherwise, flame it occasionally. There is no need to flame it after pouring each plate. If you are pouring plates at home, just be careful, and do the best you can. If any plates are contaminated, you will be able to see growth in a few days. Disinfect and discard those plates.

Storing agar plates

Let the newly poured plates sit at room temperature for 2 or 3 days before moving them to the refrigerator for longer-term storage. They will dry out a little, and you will be able to see if any of them are contaminated and need to be discarded. After the 2 or 3 days, put the plates into plastic containers (food storage containers with lids, resealable plastic bags, or the plastic sleeves that presterilized plastic petri plates are packaged in), and refrigerate them.

Remember that antibiotics will lose their activity after 2 months even in the refrigerator.

Appendix E: Recipes

Media

The three media whose recipes are given here can be freely substituted for each other in all of the exercises in this book. All of these media can be purchased as premixed powders that need only be suspended in water (these mixes come with and without agar). Medium mixes and individual components can be purchased from biological supply houses. Of the three media listed below, tryptic soy agar and broth tend to be the least expensive. Some supply houses sell ready-to-use sterile media.

For instructions on sterilizing media, see *Sterilization of Equipment and Media* (Appendix D).

Luria broth (L broth)

Per liter of medium:

tryptone	10 g
yeast extract	5 g
NaCl	10 g
water	1 liter

Or you can buy premixed dry powder and follow the instructions on the bottle.

Nutrient broth

Per liter of medium:

peptone	5 g
beef extract	3 g
water	1 liter

Or you can buy premixed dry powder.

Tryptic soy broth (or Trypticase soy broth)

Buy premixed powder.

Adding agar to medium

For normal agar plates, add 15 g of agar per liter of medium. For soft (top) agar, add 7 g of agar per liter of medium.

Antibiotics

Sterile antibiotic solutions can be purchased from some molecular biology supply houses; this is the easiest way to get them. It is, however, more common to find antibiotics for sale as dried powders. Making sterile antibiotic solutions can present a challenge, because antibiotics are inactivated by prolonged heating and so cannot be autoclaved. Filter sterilization is the recommended procedure.

As an alternative to filter sterilization, you can make the solutions aseptically. The powdered antibiotics can be dissolved in sterile, preferably distilled water (buy it in the grocery store, and sterilize it yourself). The resulting solutions can be considered sterile if you have used good sterile technique in making them.

To make the solutions aseptically, buy as close to the desired amount of antibiotic as you can, and use it all to avoid a weighing step. Use Table E.1 to calculate the amount of water you need for the entire amount of antibiotic.

Table E.1   Guide for making and using 100× antibiotic solutions

Antibiotic[a]	Weight of powder	Dissolve in this amount of water	Use this much per liter of medium	Final concentration in medium
Ampicillin, sodium salt form	1.0 g (1,000 mg)	100 ml	10 ml	100 mg/liter
Kanamycin sulfate	1.0 g (1,000 mg)	100 ml	10 ml	100 mg/liter
Streptomycin sulfate	0.5 g (500 mg)	100 ml	10 ml	50 mg/liter

[a]Purchase the chemical form of the antibiotic listed.

Carefully remove the lid from the container, and aseptically transfer all the powder to a sterile container. Add the proper amount of sterile water, and swirl the mixture until the antibiotic dissolves. Using a sterile pipette, transfer convenient volumes (perhaps the amount for 0.5 or 1 liter of medium) to sterile plastic screw-cap tubes. Store the tubes in the freezer. They will keep for years in a frost-free freezer away from the heating coils.

Antibiotic concentrations

A check of the molecular biology literature will show that there is a range of recommended working concentrations for antibiotics. The table below uses the working concentrations of ampicillin and kanamycin recommended in *DNA Science* (see Appendix H), although these concentrations are higher than those recommended in many other sources. The standard, lower concentrations will work in the procedures in this book.

Table E.1 gives instructions for making 100× solutions. If you have been successfully using other stock and working concentrations of antibiotics, there is no need to change them. If you purchase prepared solutions from a supply house, follow their instructions for using the solutions.

Calculating the amount of water needed for different antibiotic amounts

To find out how much water you need to add to different amounts of antibiotic, use this formula:

ml of H_2O you need =

$$\frac{\text{weight of your antibiotic in grams} \times 100 \text{ ml}}{\text{weight of antibiotic in Table E.1 in grams}}$$

For example, if you buy 0.3 g of streptomycin sulfate (300 mg), you will need

$$\frac{0.3 \text{ g} \times 100 \text{ ml}}{0.5 \text{ g}} = 60 \text{ ml of water}$$

You will still use 10 ml of the solution for 1 liter of medium.

Solutions

Nearly all of these solutions can be purchased ready to use from molecular biology and chemical supply houses. Ingredients for these solutions can be purchased from chemical and biological supply companies (some companies are listed in Appendix I). Methylene blue is a common microbiological dye. You may need to look under "dyes" or "stains" to find it. Tris comes in acid and base form. Buy the base for Tris-borate-EDTA (ethylenediaminetetraacetic acid) (TBE).

Calcium chloride, 0.1 M

$CaCl_2 \cdot 2H_2O$	14.7 g
water	100 ml

Sterilize by autoclaving. If you have anhydrous $CaCl_2$ (no H_2O), use 11.0 g instead.

10% sodium dodecyl sulfate (SDS)

SDS	10 g
water	90 ml

Dissolve SDS in water; heating helps. After the SDS is dissolved, adjust the final volume to 100 ml by adding a little more water. There is no need to sterilize the solution. SDS is also called sodium lauryl sulfate or lauryl sulfate, sodium salt.

Wear a dust mask when you weigh out the SDS. It is a very fine powder that can irritate your nose.

Note that a 50% solution of dishwashing detergent can be substituted for 10% SDS in the activity *Extraction of Bacterial DNA* (shampoo and Woolite also work).

10× TBE electrophoresis buffer

Tris base	108 g
boric acid	55 g
0.5 M EDTA solution	20 ml
water	to 1 liter

First, dissolve everything in slightly less than 1 liter of water, and then adjust the volume to 1 liter by adding more water. There is no need to sterilize the solution. If you notice that white clumps begin to precipitate in your 10× TBE, place the bottle in hot water until the clumps dissolve. The solution can be stored at room temperature. The 1× form is also stored at room temperature. The 10× TBE may be autoclaved, if you like. Some people report that autoclaving and storing it in the refrigerator helps prevent precipitation.

To use as a buffer, dilute the 10× stock 10-fold. Commercially available concentrated TBE may be 5× or 20×. Dilute accordingly for 1× stock.

1× TBE for electrophoresis

10× TBE	100 ml		20× TBE	50 ml
water	900 ml	*or*	water	950 ml

0.5 M EDTA

EDTA, disodium salt · 2H$_2$O	93.05 g
NaOH	as described below
water	to 500 ml

Add 400 ml of water to 93.05 g of disodium EDTA · 2H$_2$O. While stirring the mixture, adjust the pH to 8.0 by adding NaOH (about 10 g of solid NaOH). EDTA will not dissolve until the pH reaches 8.0. When the EDTA has finished dissolving, adjust the volume to 500 ml by adding water. This solution may be autoclaved, but autoclaving is not necessary unless there is a particular reason for the solution to be sterile.

10× Tris-EDTA (TE) buffer for DNA solutions

1 M Tris solution, pH 7.6 to 8.0	10 ml
0.5 M EDTA	2 ml
water	88 ml

Autoclave the mixture. A 10× TE buffer is 0.1 M Tris–0.01 M EDTA.

1× TE buffer

10× TE	10 ml
water	90 ml

Autoclave the mixture. A 1× TE buffer is 0.01 M Tris–0.001 M EDTA.

1 M Tris, pH 7.6 to 8.0

Tris base	12.1 g
concentrated HCl	as described
water	to 100 ml

Dissolve the Tris base in about 80 ml of water in a beaker. Add concentrated HCl and stir until the pH is in the 7.6-to-8.0 range. It is best to add the HCl and stir it in a chemical hood because of fumes from the acid. The solution will become hot. After it cools, put the solution in a graduated cylinder, and bring the volume to 100 ml with water.

Methylene blue

For 0.25% solution:

methylene blue	0.25 g
water	100 ml

Stir until the methylene blue is dissolved. Avoid getting any of it on your clothes, because it stains.

For 0.025% solution (1 liter):

0.25% methylene blue	100 ml
water	900 ml

Restriction enzyme digestion buffers

Nearly every company that sells restriction enzymes now ships 10× or 5× concentrated digestion buffer with the enzyme. The best strategy is to purchase your enzymes from such a company and use their buffers. Different restriction enzymes work best with different buffers, and the company will supply the correct buffer for each enzyme you purchase.

Disinfectant solutions

10% Lysol:

concentrated Lysol	100 ml
water	900 ml

15 to 20% bleach:

chlorine bleach	150 to 200 ml
water	800 to 850 ml

final solution volume = 1 liter

Buy chlorine bleach and concentrated Lysol in your supermarket or drugstore. Chlorine bleach is found with the laundry products; Lysol is among the cleaners. Concentrated Lysol usually comes in fairly small, dark brown bottles. It is a thick, syrupy brown liquid (as opposed to the already diluted sprays and solutions that can also be found; don't get these). Concentrated disinfectant solutions can also be purchased from scientific supply houses; dilute them as directed.

Appendix F: Biotechnology Laboratory Equipment

Equipment Called For by Lessons in This Book

Many scientific supply companies sell biotechnology equipment. It pays to compare prices and features of different brands and suppliers. The following equipment is referred to in the lesson plans.

Micropipettors

The micropipettes specified in the experiments in this book are low range (capacity from about 1.0 to 10 or 20 µl) and high range (capacity from about 100 to 1,000 µl). Midrange micropipettes are also available (top volume, about 200 µl). As an alternative, the larger volumes called for in the procedures could be measured and delivered with plastic or glass 1-ml pipettes. Several brands of micropipettors are available; talk to colleagues who have used them, or ask a salesman to show them to you so that you can compare features.

Micropipettors are expensive, precise instruments. Some brands are especially vulnerable to jamming when used by students who don't know what they are doing. Make sure that you understand how to use your instruments and that your students also understand how to use them. We have seen instruments abused by students from middle school to college.

Alternatives to micropipettors

Micropipettors are expensive and require continuing purchases of tips. Glass microcapillary tubes are one less expensive alternative. These tubes are available in a variety of sizes and have volume increments on the sides. They come with rubber tubing and mouthpieces to use for filling and emptying the tube (you attach the rubber tubing to the glass capillary and put the mouthpiece in the other end of the tubing). The capillary tubes are designed to be thrown away after use, so you must plan to purchase new capillary tubes periodically.

You can also buy tiny wire plungers to use with microcapillary tubes in the manner of a small syringe. One brand of these devices is Wiretrol. The plungers are reused and the glass tubes are thrown away after use. Have students practice with these devices before using them in laboratory procedures.

The least expensive alternative, which works for some applications, is small, graduated, disposable plastic dropping pipettes. These can be used for loading gels and making measurements when precise volume control is not imperative. They take a little practice to use effectively (for us, more practice than the Wiretrol and glass capillaries); students are often better at it than their teachers.

Gel electrophoresis chambers (gel boxes)

The minigel-size gel box is sufficient for the experiments in this book. Minigels vary somewhat in size depending on the manufacturer. The wider the gel, the more samples you will be able to run in it, depending on the comb (the number of wells determines the number of samples you can load). Gels larger than the minigel also work fine and give excellent results: you can achieve better resolution of bands by using a longer gel and running the samples further. Gel boxes differ in convenience features. Some have built-in casting trays; others require the use of tape. Some have a built-in holder for lifting the gel out of the box; this holder may be transparent to ultraviolet light so that it can be placed directly on a transilluminator for viewing ethidium bromide-stained DNA. Choose a gel box that suits your needs.

Gel electrophoresis power supplies

There is a lot of variety in the power supplies available and the price you can pay for them. For agarose gel electrophoresis, a maximum voltage of 120 to 150 V is all you need, and less will do. Some of the least expensive models run at one low voltage, such as 15 V (you cannot adjust it). With this kind of power supply, a minigel must be run for about 24 h to get good separation of the DNA fragments generated in the procedures in this book. This type of power supply is not necessarily a bad choice. It is the least expensive available. The chief drawback is that you are committed to overnight gel runs and must be able to leave your gels on. Although many teachers prefer to run gels faster, you get sharper bands with a long, slow run.

Other power supplies (more expensive than the one-setting, low-voltage types) have two or three preset voltages. It is nice if one of these settings is low enough to let you run the gel overnight (12 to 15 V). A 40- to 50-V setting gives you a good separation in about 3 to 4 h. Higher voltages run gels faster, but you lose band sharpness. More expensive models have a continuously adjustable voltage control. These give you the most control and flexibility.

Safety

All gel electrophoresis boxes designed for classroom use have built-in safety features.

A student would have to *try* to shock himself; it is virtually impossible to do so by accident if the equipment is used properly. The low-voltage power supplies are the safest, but the safety features of all the models make them acceptable for ordinary classroom use.

Additional Equipment (Nice If You Can Get It)

Microcentrifuge

Although none of the procedures in this book call for a microcentrifuge, some of the experiments in *DNA Science* (by Micklos and Freyer; see Appendix H) do. Prices on microcentrifuges vary widely, from less than $500 to $2,000 or more. If you purchase a microcentrifuge, get one that is capable of pelleting DNA; this function requires that it be able to spin at 5,600 g or more. Some less expensive models can do this, so shop around.

Incubators

Incubators are helpful for growing cells and incubating restriction digests. Water bath incubators and dry incubators are available. Whether you buy a water bath or a dry incubator, you are not likely to require one that gets significantly hotter than 37°C; lower-cost models usually reach 65 to 70°C and are perfectly adequate.

- *Escherichia coli* in liquid medium grows fastest at 37°C with shaking. Wet and dry incubators with shaking platforms are available.

- It is convenient to incubate restriction digests in small water baths set at 37°C.

- Petri plates can be incubated only in a dry incubator (resembling a small refrigerator but warm).

If you are setting up a laboratory and have plenty of money, provide yourself with a shaking incubator for liquid cultures (make sure it is large enough to accommodate a fairly large flask). In addition, get a small stationary water bath for digests and a dry incubator for plates.

If your finances aren't up to that, a dry incubator is probably the most versatile item. Get one that has at least one shelf in it besides the bottom. For restriction digests, put a rack that will hold microcentrifuge tubes in a pan of water inside the incubator, and let the water equilibrate at 37°C. Reactions in microcentrifuge tubes will be warmed more efficiently in the warm water than in air. For liquid cultures of *E. coli*, set the containers inside the incubator, and shake them when you can. Even if you don't shake them at all, the cells will grow up. Small dry incubators are available for a few hundred dollars.

Refrigerator

Get a refrigerator for storing solutions and agar plates with cultures. Small refrigerators can be purchased for around $200 at department stores. Their freezer compartments are usually not frost-free—just the ticket for laboratory needs (see *Freezer*, below). These small refrigerators might be too small if you need to share the space with colleagues.

Freezer

If you can get access to a *non*-frost-free freezer, by all means do so. Restriction enzymes and frozen cultures will last much longer if they are not subjected to repeated heating cycles in a frost-free freezer. The good news is that "frosty" freezers are often less expensive than the more modern frost-free ones. You may even be able to find a small, old refrigerator with a frosty freezer compartment that your department could purchase very cheaply. Check used-appliance stores.

This equipment can be purchased from a variety of sources, both companies that cater to educational institutions and companies that sell to research laboratories, not to mention retail stores. Be sure to shop around. Many models of each item are available, and they vary greatly in price.

Appendixes

Appendix G: Using the Equipment

Micropipettes

Different brands of micropipettors vary in the volume range they will measure, the type of tips they fit, and the type of device used to set the volume. Be sure that you understand how to operate the micropipettes you have and that you obtain the right tips.

A few features and rules are common to all micropipettes.

Rules

1. Never try to force the volume-setting device beyond its stated range. You can break the pipette.

2. Never use a micropipettor without a tip. Getting solutions on the plunger can ruin it.

3. Always pipette gently, releasing the plunger button slowly. Letting the button pop up causes liquid to splash into the tip and can contaminate the plunger.

4. Never force the volume control dial, even if the volume indicator shows that the pipette "should" adjust up or down. When a pipette has been damaged by improper handling, its volume indicator can be very inaccurate. Forcing the dial could make things worse.

5. Use good sense, and be kind to the instruments. Don't drop, throw, etc.

Use

Setting the Volume
All micropipettors have a volume control dial. Determine whether yours shows tenths of microliters or whole microliters in the smallest place, so that you can read the scale correctly. In general, low-range (10- to 20-μl) devices show tenths, while high-range devices don't. This can be confusing, since different sizes of pipettor by the same maker can have different scales on the volume dials.

Drawing Up and Expelling Liquid
Most micropipettors have two stops as you depress the plunger to expel liquid. The first stop is the correct stop; the second stop puffs a little air to squeeze out any extra drops. When you draw liquid into the pipette tip, depress the plunger control *only* to the *first* stop. If you go to the second stop, you will draw too much liquid into the tip. The most common pipetting error with micropipettes is missing the first stop and thus drawing too much liquid into the tip. It is worthwhile to check each student for correct technique before beginning laboratory procedures requiring use of the pipettes. In the past, we assumed that because we had explained the correct use of the pipettor and demonstrated the first and second stops, students would use the pipettors correctly. We were quickly disabused of that notion.

Practice with the micropipette until you are comfortable. Colored water makes a convenient practice solution.

Electrophoresis Equipment

Directions for casting gels are given in the gel electrophoresis experiments in the book.

1. Cast the gel, and place it in the chamber. The comb end of the gel should be by the black electrode connectors (some gel boxes are built so that this is the only way you can put the gel in).

2. Cover the gel with 1× Tris-borate-EDTA buffer (for recipe, see Appendix E). Remove the comb.

3. Put the lid on the gel box, connecting the black terminal on the lid to the black terminal on the box and the red terminal to the red terminal.

4. Plug the black lead from the lid into the black socket on the power supply, and plug the red lead

from the lid into the red socket. Turn on the power. Look for tiny bubbles rising from the wires in the gel box to verify that the power is reaching the box.

Gel boxes have built-in safety features. When the lid is removed from the box, the electrical circuit is broken. Students cannot shock themselves by sticking their fingers into the buffer, because they can't get their fingers in there when the lid is on the box. Do not allow students to stick any objects into the sockets on the power supply or the lid of the gel box. If they tried hard enough, they could shock themselves this way.

When the electrophoresis run is finished and the gel has been removed from the box, pour the buffer out (it can be reused a few times but not forever), and rinse the box with water. Do not wash inside the box, because you might break the thin electrode wires that are typically exposed in the bottom. Do not dry the box for the same reason. Turn it upside down, and let it drain instead.

Microcentrifuges

Read the directions that come with your microcentrifuge for details of its operation. Microcentrifuges will not run unless their lids are firmly closed. Either they will not allow you to open the lid until the rotor has quit moving, or they will automatically stop if you open the lid.

The most important thing to remember about using a microcentrifuge is that the tubes in the rotor must be balanced before the equipment is run. Before you close the microcentrifuge lid for a run, make sure that every tube in the rotor has a tube directly across from it. Empty tubes can be used to balance tubes that contain 20 μl of fluid or less, but if a tube contains more than 20 μl, the balance tube needs to have about the same amount of fluid as the tube of medium has in it. If you don't have an even number of similarly filled tubes, make a "water blank" by adding the correct amount of water to an additional tube.

Signs of an unbalanced microcentrifuge:

- A very noisy run
- Bad vibrations during the run
- (Worst case) shaking and moving of the microcentrifuge

If you notice any of these things, stop the microcentrifuge immediately, and investigate. If every tube has another one opposite it, remove each pair of tubes and check to see that the fluid levels are the same. Fix the balancing problem, and restart the run for the full time requirement.

Appendix H: Recommended Reading

Books

Additional background and experiments

Micklos, D., and G. Freyer. 1990. *DNA Science: a First Course in Recombinant DNA Technology*. Carolina Biological Supply and Cold Spring Harbor Press, Cold Spring Harbor, N.Y.

This very readable text tells the story of the development of molecular biology, explaining the science along the way. It covers all the major techniques of DNA science plus a variety of different applications. This book could be used as a text for a molecular biology course for high-achieving high school seniors or for college or community college students. It also makes an easy-to-read reference book for teachers.

DNA Science contains a laboratory experiment section with several well-explained procedures. These could be used to supplement the wet laboratories in this book. It also contains a great two-page set of photographs of "bad gels," with explanations of what caused the problems.

Glick, B. R., and J. J. Pasternak. 1994. *Molecular Biotechnology: Principles and Applications of Recombinant DNA*. American Society for Microbiology, Washington, D.C.

Bloom, M. 1995. *Laboratory DNA Science: an Instructor's Prep Guide*. Benjamin/Cummings Publishing Co., Redwood City, Calif.

Bloom, M., G. Freyer, and D. Micklos. 1995. *DNA Lab Manual*. Benjamin/Cummings Publishing Co., Redwood City, Calif.

This book is the student manual to accompany *Laboratory DNA Science*.

The last two books are teacher and student versions of a laboratory manual for a more advanced molecular biology course. They contain wet-laboratory protocols for procedures such as Southern blotting, constructing a genomic library of lambda, colony hybridization, etc.

General reference

Watson, J., M. Gilman, J. Witkowski, and M. Zoller. 1992. *Recombinant DNA*, 2nd ed. Scientific American Books, W. H. Freeman Co., New York.

This shorter, easier-to-read general reference covers the science and applications of molecular biology. It is an excellent reference for teachers and would be a good text for a specialized college-level course in molecular biology. It includes references to relevant articles in the scientific literature at the end of each chapter.

Watson, J., N. Hopkins, J. Roberts, J. Steitz, and A. Weiner. 1987. *The Molecular Biology of the Gene*, 4th ed. Benjamin/Cummings Publishing Co., Redwood City, Calif.

Alberts, B., D. Bray, J. Lewis, M. Raff, K. Roberts, and J. Watson. 1994. *The Molecular Biology of the Cell*, 3rd ed. Garland Publishing Co., New York.

These are "heavy-duty" molecular biology texts that give thorough yet readable treatments to every scientific topic touched on in this text and more. They were used as references in preparing this text. If you can, get one of these or a similar book for yourself.

Special topics

Ptashne, M. 1992. *A Genetic Switch*, 2nd ed. Blackwell Scientific Publishing, Boston.

This small book describes one of the best-understood systems of gene regulation, the lambda repressor, in molecular detail. Ptashne lays out the biology clearly and without wasted words and then describes some of the experiments that led to the knowledge. References to the original scientific articles are included.

Branden, J., and C. Tooze. 1991. *Introduction to Protein Structure*. Garland Publishing Co., New York.

If you are intrigued by protein structure and function, this is the book for you. Not too long, clearly written, and amply illustrated.

Human genome project

The Human Genome Project. 1992. Los Alamos Science series no. 20. Los Alamos National Laboratory, Los Alamos, N. Mex.

Excellent explanations of the broad outline, goals, techniques, and science of the Human Genome Project. Thorough treatment, helpful illustrations. Contains interviews with scientists active in the Human Genome Project. This would be a good text for a human genetics course.

Mapping and Sequencing the Human Genome: Science, Ethics, and Public Policy. 1992. Biological Sciences Curriculum Study, 830 N. Tejon St., Suite 405, Colorado Springs, CO 80903-4720.

This free monograph was developed by the Biological Sciences Curriculum Study and the American Medical Association. It focuses on the social issues of the human genome project. Contains classroom activities and case studies.

Biosafety

Horn, T. M. 1992. *Working with DNA and Bacteria in Precollege Science Classrooms*. National Association of Biology Teachers, Reston, Va.

A 22-page booklet covering the topics treated in these appendices but in somewhat more detail and with helpful illustrations and photographs.

Periodicals

Periodicals are the best way to keep up with recent scientific developments. Everyone who teaches science should read at least one regularly. The easiest ones to use for general purposes are the science news and summary types rather than the research journals. Some good ones are as follows.

Scientific American. Address: *Scientific American*, Dept. SAF, 415 Madison Ave., New York, NY 10017.

This monthly publication is probably the best for thoroughly explaining important new concepts in science. The articles, written by leading scientists, are rather long for most high school students but are excellent for providing background to teachers. In general, they do not present the latest news in science but instead present a new research area and discuss applications. There is a special rate for educators ($19.97 per year in 1993).

Science. Address: AAAS, 1333 H St., NW, Washington, DC 20005; fax, (292) 842-1065.

This weekly publication is a combination of research articles and science news written for scientists. The front one-third of each issue contains summaries of leading-edge research, science policy analysis, science education, and review articles. The research articles are written for specialized professionals; don't be discouraged by them. *Science* is the best magazine for news from the world of professional scientists. It also features scores of ads for biotechnology supplies and equipment (with pictures), announcements of meetings and fellowships, and an extensive want ad section in the back. The want ads are a nice way to make connections to the world of work for your students. (These positions are usually at the Ph.D. level.) A subscription to *Science* includes membership dues for the American Association for the Advancement of Science. Student subscription rates are available and are reasonably priced. Ask about special rates for high school teachers.

Science News. Address: Science Service, Inc., 1719 N St., NW, Washington, DC 20036. For new subscriptions, call 1-800-247-2160.

A thin, weekly science newsmagazine that typically features two short articles and a number of news summaries. The style and length are suitable for high school students. College students and teachers may feel frustrated with some of the news summaries because they are so brief and lack background. This would make a good classroom magazine for high schools.

Note that all of these magazines cover all of science, from physics to biotechnology to paleontology. Perhaps your science department could share subscriptions.

Biotechnology periodicals

Bio/Technology. Address: Nature Publishing Co., 65 Bleecker St., New York, NY 10012-2467; phone, (212) 477-9600; fax, (212) 505-1364.

A monthly periodical that contains news of interest to scientists and the biotechnology industry as well as a few research articles.

Genetic Engineering News. Address: GEN Publishing, Inc., 1651 Third Ave., New York, NY 10128; phone, (212) 289-2300.

A twice-monthly newspaper-format publication that covers the biotechnology industry. Articles feature research developments and new products as well as business and relevant government news.

Publications from the Howard Hughes Medical Institute (HHMI)

HHMI is publishing a series of reports on biomedical science. The reports are very readable and beautifully illustrated, and each includes interviews with scientists conducting groundbreaking research. The series includes the following volumes.

- *From Egg to Adult* describes what worms, flies, and other creatures can teach us about the switches that control human development.

- *Blood: Bearer of Life and Death* reports on new ways to fight diseases caused by faults in the bloodstream.

Copies are free for educators. Contact HHMI at 4000 Jones Bridge Rd., Chevy Chase, MD 20815-6789; phone, (301) 215-8500.

Appendixes

Appendix I: Teaching Resources

Suppliers

The following companies are listed in this book as potential sources of supplies.

Carolina Biological Supply Co.
2700 York Rd.
Burlington, NC 27215-3398
Phone: (800) 334-5551

Connecticut Valley Biological Supply
P.O. Box 326, 82 Valley Rd.
Southampton, MA 01073
Phone: (800) 628-7748

Edvotek, Inc.
P.O. Box 1232
West Bethesda, MD 20827-1232
Phone: (301) 251-5990

Fotodyne
950 Walnut Ridge Dr.
Hartland, WI 53029
Phone: (414) 369-7000

Sigma Chemical Co.
P.O. Box 14508
St. Louis, MO 63178
Phone: (800) 325-3010

The following is a partial list of other companies that sell biotechnology supplies.

Boehringer Mannheim
P.O. Box 50414
Indianapolis, IN 46250-0414
Phone: (800) 262-1640

Modern Biology, Inc.
111 N. 500 West
West Lafayette, IN 47906
Phone: (800) 733-6544

New England Biolabs
32 Tozer Rd.
Beverly, MA 01915-5599
Phone: (800) 632-5227

Perkin Elmer
761 Main Ave.
Norwalk, CT 06859-0012
Phone: (203) 762-1000

Promega
2800 Woods Hollow Rd.
Madison, WI 53711-5399
Phone: (800) 356-9526

Stratagene
11011 North Torrey Pines Rd.
La Jolla, CA 92037
Phone: (800) 424-5444

Sources of Information

Professional societies and trade organizations

As we mentioned in chapter 1, biotechnology has developed from findings in many different scientific disciplines and can be applied in many industries. Consequently, many scientific professional societies and industrial trade organizations promote biotechnology education. Some offer educational materials; others have programs pairing scientists with classroom teachers or offer fellowships for educators.

American Association for the Advancement of Science
Office of Opportunities in Science
1776 Massachusetts Ave., NW
Washington, DC 20036
Phone: (202) 467-5438

American Chemical Society
Education Division
1155 16th St., NW
Washington, DC 20036
Phone: (202) 872-8734

American Institute of Biological Sciences
1401 Wilson Blvd.
Arlington, VA 22209
Phone: (703) 527-6776

American Institute of Chemical Engineers
345 East 47th St.
New York, NY 10017
Phone: (212) 705-7338

American Petroleum Institute
1220 L St., NW
Washington, DC 20005
Phone: (202) 820-8000

American Society for Cell Biology
9650 Rockville Pike
Bethesda, MD 20814
Phone: (301) 530-7153

American Society for Microbiology
1325 Massachusetts Ave., NW
Washington, DC 20005-4171
Phone: (202) 737-3600

American Society of Agricultural Engineers
2950 Niles Rd.
St. Joseph, MI 49085
Phone: (616) 429-0300

American Society of Agronomy
Soil Science Society of America
677 South Segoe Rd.
Madison, WI 53711
Phone: (608) 273-8080

American Society of Animal Science
309 West Clark St.
Champaign, IL 61820
Phone: (217) 356-3182

American Society of Biological Chemists
9650 Rockville Pike
Bethesda, MD 20814
Phone: (301) 530-7145

American Society of Cytology
130 South 9th St., Suite 810
Philadelphia, PA 19107
Phone: (215) 922-3880

American Society of Human Genetics
P.O. Box 6015
Rockville, MD 20850
Phone: (301) 424-4120

American Veterinary Medical Association
930 North Meacham Rd.
Schaumburg, IL 60196
Phone: (312) 885-8070

Association for Women in Science
1346 Connecticut Ave., NW, Suite 1122
Washington, DC 20036
Phone: (202) 833-1998

Biomedical Engineering Society
P.O. Box 2399
Culver City, CA 90230
Phone: (213) 206-6443

Biotechnology Industry Organization
1625 K St., NW, Suite 1100
Washington, DC 20006
Phone: (202) 857-0244

Genetics Society of America
9650 Rockville Pike
Bethesda, MD 20814-3998
Phone: (301) 571-1825

Institute of Food Technologists
Suite 2120, 221 North LaSalle St.
Chicago, IL 60601
Phone: (312) 782-8424

Society for Economic Botany
University of Illinois
College of Pharmacy
P.O. Box 6998
Chicago, IL 60680

Society for Industrial Microbiology
1401 Wilson Blvd.
Arlington, VA 22209
Phone: (703) 256-0337

Society of Women Engineers
345 East 47th St.
New York, NY 10017
Phone: (212) 705-7855

Appendixes

Educational associations

American Association of Teacher Educators in
 Agriculture
Department of Agricultural Education
North Dakota State University
Fargo, ND 58105
Phone: (701) 237-7436

American Society for Engineering Education
11 Dupont Circle, NW, Suite 200
Washington, DC 20036
Phone: (202) 293-7080

National Association of Biology Teachers
11250 Roger Bacon Drive, #19
Reston, VA 22090
Phone: (703) 471-1134

National Science Teachers Association
1840 Wilson Blvd.
Arlington, VA 22201-3000
Phone: (703) 243-7100

Society for College Science Teachers
Dr. William J. McIntosh
Delaware State College
Dover, DE 19901
Phone: (302) 739-5206

State biotechnology centers

State-supported biotechnology centers often provide educational materials, coordinate education programs, or employ individual scientists who are interested in becoming involved in science education. Below is a partial list of biotechnology centers listed by state. The one in your state may offer programs and provide materials. All will have scientists able to assist you by offering technical advice on the laboratory activities or directing you to references. Because these laboratory scientists order laboratory supplies in bulk and often have a microbial culture "on the stove," they may be more than willing to provide you with a small amount of chemicals, enzymes, DNA, or a starter culture of *Escherichia coli*.

Arizona

Biotechnology Division
University of Arizona
Arizona Research Labs
Gould-Simpson, 1011
Tucson, AZ 85721
Phone: (602) 621-4064
Fax: (602) 621-1364
Contact: Ms. Caroline Garcia, Assistant Director

Arkansas

UAMS Biomedical Biotechnology Center
4301 W. Markham, Slot 718
Little Rock, AR 72205
Phone: (501) 686-6696
Fax: (501) 686-8501
Contact: Ms. Alice Rumph Smith, Associate Director

California

Bay Area Bioscience Center

1300 Clay St., Suite 320
Oakland, CA 94612
Phone: (510) 874-1464
Fax: (510) 874-1466
Contact: Mr. Frederick Dorey, President

California

Biotechnology Program
University of California-Davis
College of Agricultural & Environmental Sciences
Davis, CA 95616
Phone: (916) 752-3260
Fax: (916) 752-4125
E-mail: mmmcgloughlin@ucdavis.cdu
Contact: Ms. Martina McGloughlin, Associate
 Director

California

California State University Program for Education
 and Research in Biotechnology
San Diego State University
Molecular Biology Institute
San Diego, CA 92182-0328
Phone: (619) 594-5578
Fax: (619) 594-1613
E-mail: cdahms@sciences.sdsu.edu
Contact: Dr. Stephen Dahms, Director

California

California Interagency Task Force on
 Biotechnology
801 K St., Suite 1700
Sacramento, CA 95814
Phone: (916) 322-5665
Fax: (916) 322-3524
Contact: Mr. Wes Ervin, Director

Appendixes

Appendix I: Teaching Resources • 437

California

Systemwide Biotechnology Research and Education Program
University of California
345 Giannini Hall
Berkeley, CA 94720-3100
Phone: (510) 643-0725
Fax: (510) 643-1450
E-mail: biotech@violet.berkeley.edu

Colorado

Colorado Advanced Technology Institute
1625 Broadway, Suite 700
Denver, CO 80202
Phone: (303) 620-4777
Fax: (303) 620-4789
E-mail: fcp@cati.org
Contact: Dr. Fred Pearson, Biotechnology Programs Director

Connecticut

Biotechnology Center
University of Connecticut
184 Auditorium Rd.
Storrs, CT 06269-3149
Phone: (203) 486-5011
Fax: (203) 486-5005
E-mail: biotctr1@uconnvm.uconn.edu
Contact: Ms. Mary Tokes, Program Assistant

Florida

Institute for Biomolecular Science
University of South Florida
University of South Florida, LIFI36
Tampa, FL 33620-5150
Phone: (813) 974-2392
Fax: (813) 974-1614
E-mail: ibs@chuma.cas.usf.edu
Contact: Dr. Mary Jane Saunders, Director

Florida

The Biotechnology Program
University of Florida
Newell Dr., South, P.O. Box 110580
Gainesville, FL 32611-0580
Phone: (904) 392-8408
Fax: (904) 392-8598
E-mail: jani@nervm.nerdc.ufl.edu
Contact: Ms. Jani Sherrard, Associate Director

Georgia

Georgia Biomedical Partnership, Inc.
P.O. Box 54151
Atlanta, GA 30808-4151
Phone: (404) 817-5919
Fax: (404) 817-4345
Contact: Ms. Lauren Owenby, Program Manager

Georgia

University of Georgia Biotechnology Program
University of Georgia
Athens, GA 30602
Phone: (706) 542-6512
Fax: (706) 542-5638
Contact: Dr. John Ingle, Director

Illinois

Center for Biotechnology
Northwestern University
2153 Sheridan Rd.
Evanston, IL 60208-3500
Phone: (708) 467-1453
Fax: (708) 467-2180
Contact: Mr. Richard Loerzel, Program Assistant

Indiana

Institute for Molecular and Cellular Biology
Indiana University
Jordan Hall 322A
Bloomington, IN 47405
Phone: (812) 855-4183
Fax: (812) 855-6082
E-mail: mgossard@indiana.edu
Contact: Ms. Mary Gossard, Industrial Liaison Office

Indiana

Purdue University Biotechnology Institute
Purdue University
1057 Agricultural Research Building
West Lafayette, IN 47907
Phone: (317) 494-4596
Fax: (317) 496-1219
Contact: Dr. Peter Dunn, Director

Iowa

Office of Biotechnology
Iowa State University
1210 Molecular Biology Building
Ames, IA 50011
Phone: (515) 294-9818
Fax: (515) 294-4629
E-mail: w_fehr@molebio.iastate.edu
Contact: Dr. Walter Fehr, Director of Biotechnology

Kansas

Higuchi Biosciences Center
University of Kansas
2099 Constant Ave.
Lawrence, KS 66047-2535
Phone: (913) 864-5183

Fax: (913) 749-7393
E-mail: decedue@smissman.hbc.ukans.edu
Contact: Dr. Charles Decedue, Executive Director

Maine

Center for Innovation in Biomedical Technology
412 State St., Wing Park
Bangor, ME 04401
Phone: (207) 941-9855
Fax: (207) 941-0873
Contact: Dr. Donald Colbert, Executive Director

Maryland

Center for Agricultural Biotechnology
University of Maryland
2111 Agriculture/Life Sciences Surge Building
College Park, MD 20742
Phone: (301) 405-1582
Fax: (301) 314-9075
Contact: Mr. Greg Silsbee, Assistant Director

Maryland

Center of Marine Biotechnology
Columbus Center Building, Suite 236
701 E. Pratt St.
Baltimore, MD 21202
Phone: (410) 234-8800
Fax: (410) 234-4896
E-mail: @mbimail.umd.edu
Contact: Mr. William Cooper, Assistant Director

Maryland

Columbus Center
111 Market St.
Baltimore, MD 21202
Phone: (410) 547-8727
Fax: (410) 547-9089
Contact: Ms. Cheryl Hudgins, Marketing and Public
Relations Director

Maryland

Suburban Maryland Montgomery County High
Technology Council, Inc.
2092 Gaither Rd.
Rockville, MD 20850
Phone: (301) 258-5005
Fax: (301) 258-9148
Contact: Ms. Kathleen Manning, Communications
Director

Maryland

University of Maryland Biotechnology Institute
University of Maryland
4321 Hartwick Rd., Suite 500
College Park, MD 20740
Phone: (301) 403-0501

Fax: (301) 454-8123
Contact: Mr. Robert Hando, Vice President

Massachusetts

Massachusetts Biotechnology Research Institute
3 Biotech Park, One Innovation Dr.
Worcester, MA 01605
Phone: (508) 797-4200
Fax: (508) 799-4039
Contact: Mr. Marc Goldberg, President/CEO

Michigan

Michigan Biotechnology Institute
P.O. Box 27609
Lansing, MI 48909-0609
Phone: (517) 337-3181
Fax: (517) 337-2122
Contact: Ms. Gretchen Smith, Public Relations
Manager

Minnesota

Biological Process Technology Institute
University of Minnesota
1479 Gortner Ave., Suite 240
St. Paul, MN 55108
Phone: (612) 624-6774
Fax: (612) 625-1700
E-mail: bpti@biosci.cbs.umn.edu
Contact: Dr. Jeffrey Tate, Assistant to Director

Nebraska

Center for Biotechnology
University of Nebraska-Lincoln
P.O. Box 880665
Lincoln, NE 68588-0665
Phone: (402) 472-2635
Fax. (402) 472-3139
E-mail: btec001@unlvm.unl.edu
Contact: Ms. Karen Henricksen, Administrative
Coordinator

New Jersey

Agricultural Molecular Biology Center
Rutgers University
P.O. Box 231, Cook College
New Brunswick, NJ 08903
Phone: (908) 932-8165
Fax: (908) 932-6535
Contact: Dr. Geetha Ghai, University-Industry
Liaison

New Mexico

Plant Genetic Engineering Lab—Desert Adaptation
New Mexico State University
Box 3GL
Las Cruces, NM 88003

Appendixes

Phone: (505) 646-5453
Fax: (505) 646-5975
E-mail: phavstad@nmsu.edu
Contact: Ms. Patti Havstad, Administrative Manager

New York

Center for Biotechnology
SUNY-Stony Brook
130 Life Sciences
Stony Brook, NY 11794-5208
Phone: (516) 632-8521
Fax: (516) 632-8577
E-mail: glenn.prestwich@sunysb.edu
Contact: Ms. Diane Fabel, Associate Director

New York

Cornell Center for Advanced Technology
   (Biotechnology Program)
Cornell University
130 Biotechnology Building
Ithaca, NY 14853-2703
Phone: (607) 255-2300
Fax: (607) 255-6249
Contact: Dr. Richard Holsten, Research Director

North Carolina

North Carolina Biotechnology Center
P.O. Box 13547
15 TW Alexander Dr.
Research Triangle Park, NC 27709
Phone: (919) 541-9366
Fax: (919) 990-9544
E-mail: barry_teater@ncbiotech.org
Contact: Mr. Barry Teater, Public Affairs Manager

Ohio

Edison Biotechnology Center
University Circle Research Center
11000 Cedar Ave.
Cleveland, OH 44106
Phone: (216) 229-0400
Fax: (216) 229-7323
Contact: Dr. Cinda Herndon-King, Vice President

Oregon

Advanced Science and Technology Institute
University of Oregon
1239 Hendricks Hall
Eugene, OR 97403
Phone: (503) 346-3189
Fax: (503) 346-1352
E-mail: robert_mcquate@ccmail.uoregon.edu
Contact: Dr. Robert McQuate, Executive Director

Oregon

Oregon Biotechnology Association

2611 SW 3rd Ave., Suite 200
Portland, OR 97201
Phone: (503) 241-7802
Fax: (503) 241-0827
E-mail: orbio@ortel.org
Contact: Dr. Nanette Newell, Executive Director

Pennsylvania

Center for Biotechnology and Bioengineering
University of Pittsburgh
300 Technology Dr.
Pittsburgh, PA 15219
Phone: (412) 383-9700
Fax: (412) 383-9710
Contact: Ms. Eleanor Cadugan, Administrative
   Manager

Rhode Island

Rhode Island Partnership for Science and
   Technology
7 Jackson Walkway
Providence, RI 02903
Phone: (401) 277-2601
Fax: (401) 277-2102
Contact: Ms. Claudia Terra, Executive Director

South Carolina

Institute for Biological Research and Technology
University of South Carolina
Coker Life Sciences Building, Room 701
Columbia, SC 29208
Phone: (803) 777-2468
Fax: (803) 777-4002
E-mail: ibrt@biol.scarolina.edu
Contact: Dr. Bert Ely, Director

Tennessee

Center for Environmental Biotechnology
University of Tennessee
10515 Research Dr., Suite 100
Knoxville, TN 37932-2567
Phone: (615) 974-8080
Fax: (615) 974-8086
Contact: Ms. Kimberly Davis, Senior Research
   Associate

Texas

State Industry-University Cooperative Research
   Center
University of Texas
Health Science Center-San Antonio
7703 Floyd Curl Dr.
San Antonio, TX 78284-7823
Phone: (210) 567-2023
Fax: (210) 567-2052
Contact: Dr. Barbara Boyan, Director

Texas

University of Texas Institute of Biotechnology
Texas Research Park
San Antonio, TX 78245
Phone: (210) 567-7200
Fax: (210) 567-7277
E-mail: hottle@thorin.uthscsa.edu
Contact: Ms. Judy Wolfe, Public Affairs Director

Utah

Biotechnology Center
Utah State University
Logan, UT 84322-4700
Phone: (801) 797-2753
Fax: (801) 797-2766
E-mail: wscouten@csces1.usu.edu
Contact: Dr. William Scouten, Director

Virginia

CIT Institute of Biotechnology
Virginia Commonwealth University
P.O. Box 980126
Richmond, VA 23298-0126
Phone: (804) 828-8565
Fax: (804) 828-8566
E-mail: woodworth@gems.vcu.edu
Contact: Dr. Terry Woodworth, Associate Director

Virginia

Center for Biotechnology

Virginia Tech
Blacksburg, VA 24061-0346
Phone: (703) 231-6933
Fax: (703) 231-7126
E-mail: biotech@vtvm1.cc.vt.edu
Contact: Mr. Donald Ball, Administrative Assistant

Washington

Department of Trade and Economic Development
2001 6th Ave., Suite 2600
Seattle, WA 98121
Phone: (206) 464-7143
Fax: (206) 464-7222
E-mail: 74674.3005@compuserve.com
Contact: Mr. Philip Ness, Biotechnology Program
Manager

Wisconsin

University of Wisconsin Biotechnology Center
University of Wisconsin-Madison
1710 University Ave.
Madison, WI 53705
Phone: (608) 262-8606
Fax: (608) 262-6748
E-mail: fitzmaur@macc.wisc.edu
Contact: Dr. Leona Fitzmaurice, Assistant Director
for Outreach

Videotapes

The following high-quality videotapes deal with DNA technology and applications.

The Infinite Voyage: the Geometry of Life. 55 min. DNA and related biotechnologies with emphasis on human health applications. Produced by the National Academy of Science and WQED Pittsburgh.

The Web of Life. 50 min. DNA and related biotechnologies. Features the perspectives of diverse scientists. Produced by WETA-TV and the Smithsonian Institution.

The Secret of Life. This eight-part series with geneticist David Suzuki aired in fall 1993 on public television. Each of the episodes listed below is 1 h long. Suzuki brings sensitive, thoughtful insights to a consideration of molecular biology and the impact of biotechnology on society. Produced by WGBH Boston and BBC-TV.

1. *The Immortal Thread* introduces the DNA molecule, heredity, the human genome, and other themes of the series.

2. *Accidents of Creation* traces the small errors in DNA that have, over millennia, allowed adaptation to a changing environment.

3. *Birth, Sex and Death* looks at gene differentiation and cases in which it has gone awry, leading to a better understanding of gene function.

4. *Conquering Cancer* looks at current knowledge of cancer genetics and explores what treatments may lie ahead.

5. *Cell Wars* highlights the immune system.

6. *The Mouse That Laid the Golden Egg* weighs the pros and cons of genetically engineering everything from sheep to tomatoes.

7. *Children by Design* looks at issues arising from advances in gene therapy and genetic engineering.

8. *Who Are You?* highlights investigations of identical twins and what studies have revealed about the relationship between genes and identity.

Appendixes

Appendix J: National Science Education Standards and the Content of This Book

What constitutes a good education in science has recently been under serious discussion. Two well-respected scientific organizations, the American Association for the Advancement of Science (AAAS) and the National Research Council (NRC), organized a concentrated series of discussions and debates between scientists, teachers, and others about this issue.

AAAS has published two documents, *Science for All Americans* and *Benchmarks for Science Literacy*, that describe the consensus reached in its endeavor to define what the substance and character of science education should be. The NRC is still working with scientific and educational organizations to elaborate standards for science education. It has published several draft versions of the standards, which are complementary to the outline set forth in *Benchmarks*.

The concepts and approaches of this book are consistent with many of the objectives set forth by AAAS in *Benchmarks for Science Literacy*. Given below are specific objectives for high school and middle school students, taken from *Benchmarks*, that are addressed in this book.

12th Grade

By the end of the 12th grade, students should know the following.

Diversity of life
- The variation of organisms within a species increases the likelihood that at least some members of the species will survive under changed environmental conditions.

- The degree of kinship between organisms or species can be estimated from the similarity of their DNA sequences, which often closely matches their classification based on anatomical similarities.

Heredity
- Some new gene combinations make little difference, some can produce organisms with new and perhaps enhanced capabilities, and some can be deleterious.

- The sorting and recombination of genes in sexual reproduction result in a great variety of possible gene combinations for the offspring of any two parents.

- The information passed from parents to offspring is encoded in DNA molecules.

- Genes are segments of DNA molecules. Inserting, deleting, or substituting DNA segments can alter genes. The resulting features may help, harm, or have little or no effect on the offspring's success in its environment.

- Gene mutations can be caused by such things as radiation and chemicals.

Cells
- The work of the cell is carried out by the many different types of molecules it assembles, mostly proteins. Protein molecules are long, usually folded chains made from 20 different kinds of amino acid molecules. The function of each protein molecule depends on its specific sequence of amino acids, and the shape the chain takes is a consequence of attractions between the chain's parts.

- The genetic information in DNA molecules provides instructions for assembling protein molecules. The code used is virtually the same for all life forms.

- Gene mutation in a cell can result in uncontrolled cell division, called cancer.

Evolution of life

- Molecular evidence substantiates the anatomical evidence for evolution and provides additional detail about the sequence in which various lines of descent branched off from one another.

- Heritable characteristics of structure, chemistry, and behavior can be observed at the molecular and whole-organism levels.

- New heritable characteristics can result from new combinations of existing genes or from mutations of genes in reproductive cells.

Human identity

- The similarity of human DNA sequences and the resulting similarity in cell chemistry and anatomy identify human beings as a single species.

Physical health

- Faulty genes can cause body parts or systems to work poorly.

Health technology

- Knowledge of genetics is opening whole new fields of health care. In diagnosis, mapping of genetic instructions in cells makes it possible to detect defective genes that may lead to poor health. In treatment, substances from genetically engineered organisms may reduce the cost and side effects of replacing missing body chemicals.

- Knowledge of molecular structure and interactions aids in synthesizing new drugs and predicting their effects.

- Biotechnology has contributed to health improvement in many ways, but its cost and application have led to a variety of controversial social and ethical issues.

Agriculture

- New varieties of farm plants and animals have been engineered by manipulating their genetic instructions to produce new characteristics.

Social trade-offs

- In deciding among alternatives, a major question is who will receive the benefits and who (not necessarily the same people) will bear the costs.

Human development

- The development and use of technologies to maintain, prolong, sustain, or terminate life raise social, moral, ethical, and legal issues.

8th Grade

Relevant objectives for students who are completing 8th grade include the following.

- In some kinds of organisms, all the genes come from a single parent, whereas in organisms that have sexes, half of the genes typically come from each parent.

- In sexual reproduction, a single specialized cell from a female merges with a specialized cell from a male. As the fertilized egg, carrying genetic information from each parent, multiplies to form the complete organism with about a trillion cells, the same genetic information is copied in each cell.

- New varieties of cultivated plants and domestic animals have resulted from selective breeding for particular traits.

- Small differences between parents and offspring can accumulate (through selective breeding) in successive generations so that descendants are very different from their ancestors.

- It is becoming increasingly possible to manufacture chemical substances such as insulin and hormones that are normally found in the body. These manmade substances can be used by individuals whose own bodies cannot produce the amounts required for good health.

- Trade-offs are not always between desirable possibilities. Sometimes social and personal trade-offs require accepting an unwanted outcome to avoid some other unwanted one.

Appendix K: Templates

This appendix contains 48 templates that can be cut out from this section or photocopied as needed for use with the *Student Activity* pages.

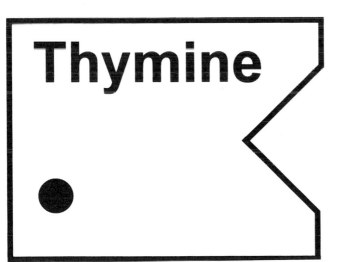

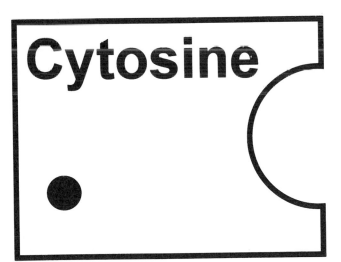

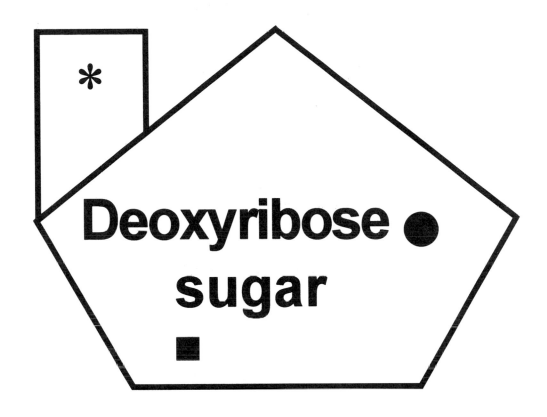

* Deoxyribose ● sugar ■

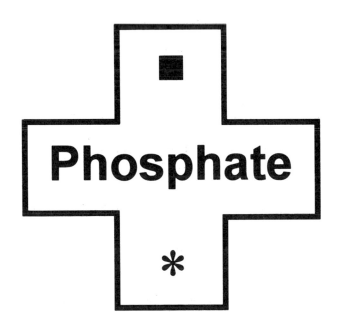

■ Phosphate *

449

DNA, mRNA, tRNA, and amino acid cards for *From Genes to Proteins*

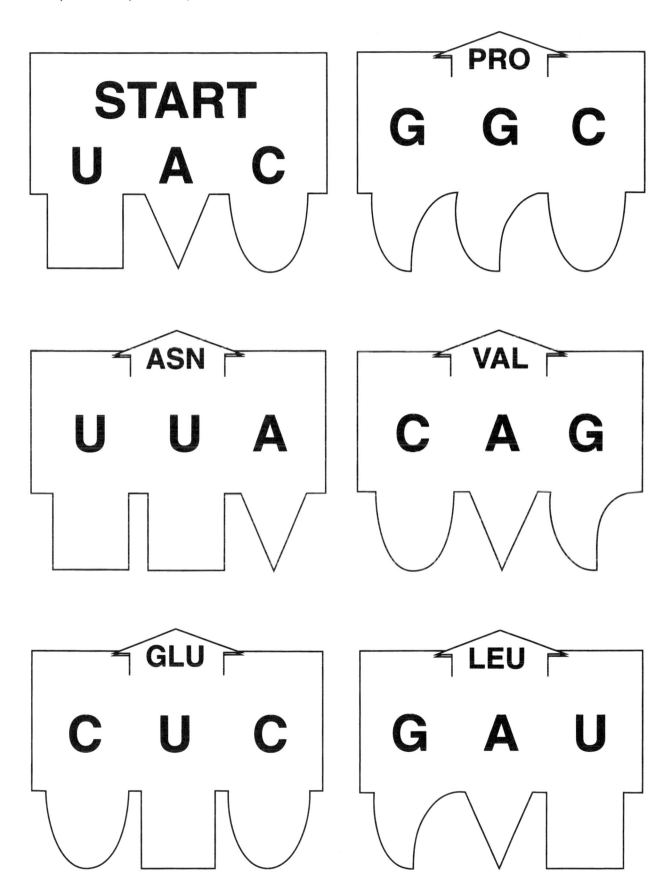

tRNA

tRNA

tRNA

tRNA

tRNA

tRNA

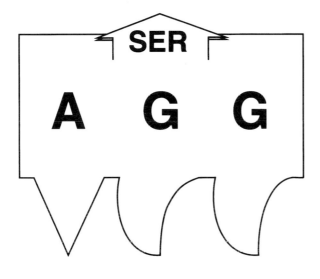

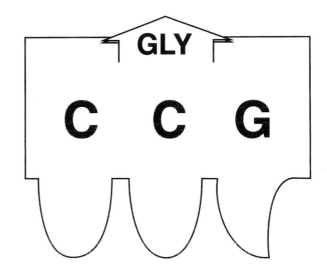

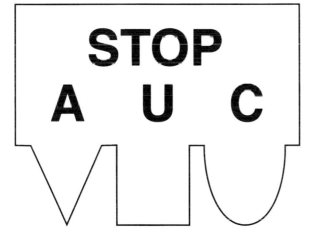

tRNA

tRNA

tRNA

454

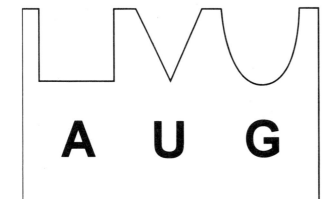

A U G

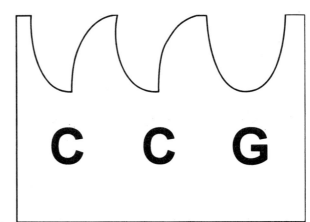

C C G

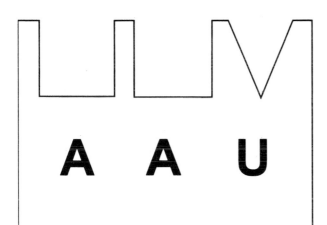

A A U

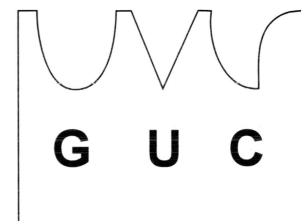

G U C

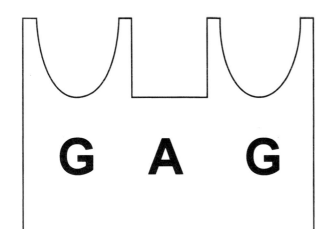

G A G

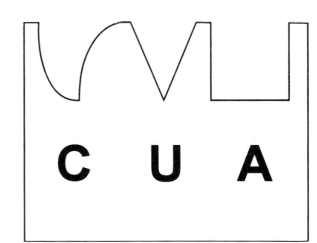

C U A

mRNA

mRNA

mRNA

mRNA

mRNA

mRNA

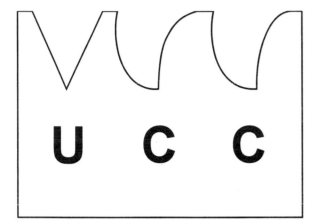 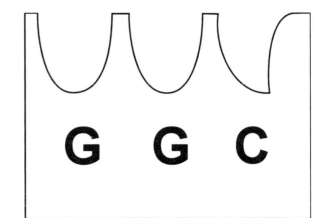

mRNA

mRNA

mRNA

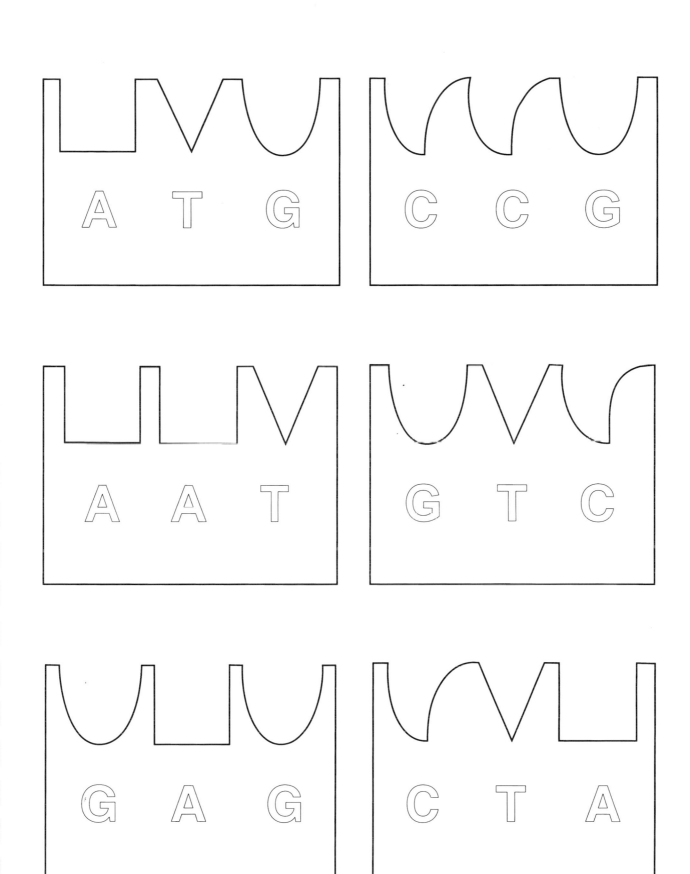

DNA

DNA

DNA

DNA

DNA

DNA

460

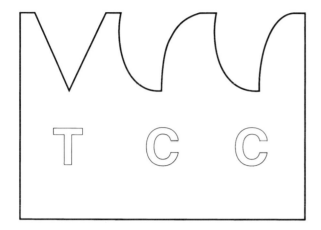

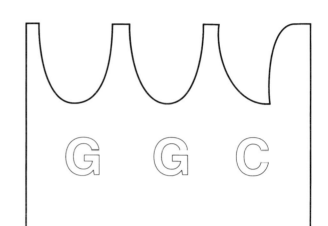

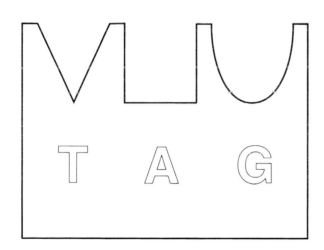

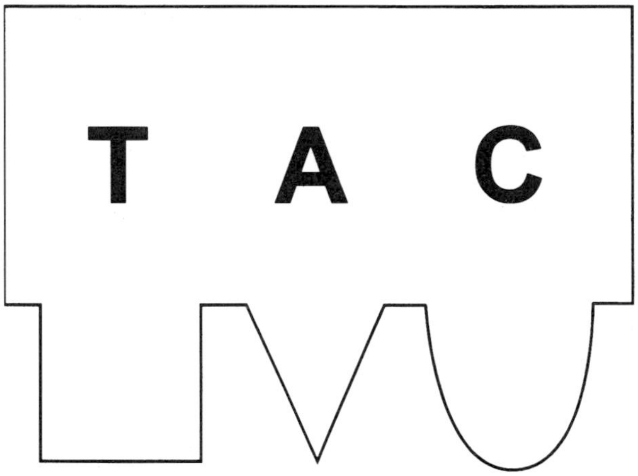

T A C

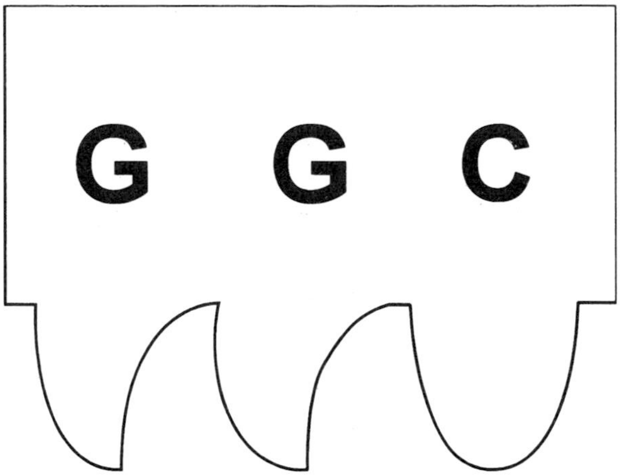

G G C

T T A

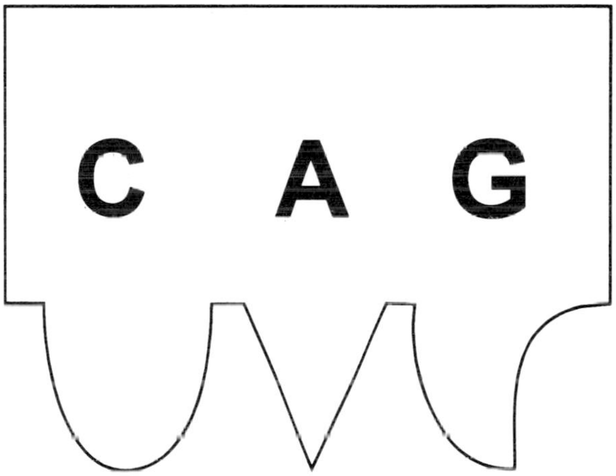

C A G

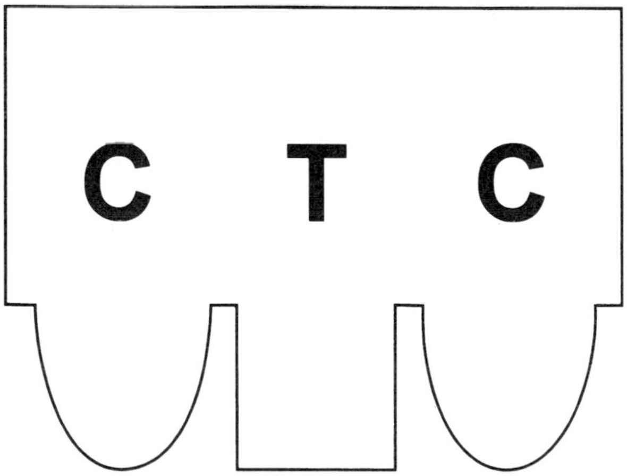

C T C

G A T

463

DNA

DNA

DNA

DNA

DNA

DNA

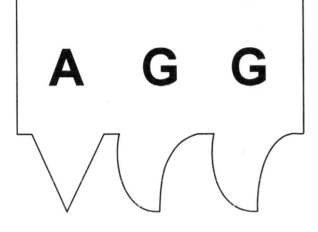

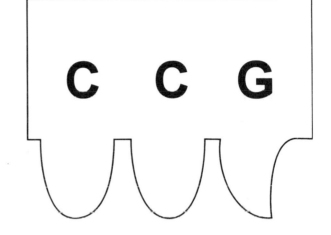

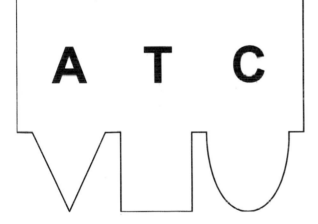

465

DNA

DNA

DNA

466

PROLINE

ASPARAGINE

VALINE

GLUTAMIC ACID

LEUCINE

SERINE

GLYCINE

Amino Acid

Amino Acid

Amino Acid

Amino Acid

Amino Acid

Amino Acid

Amino Acid

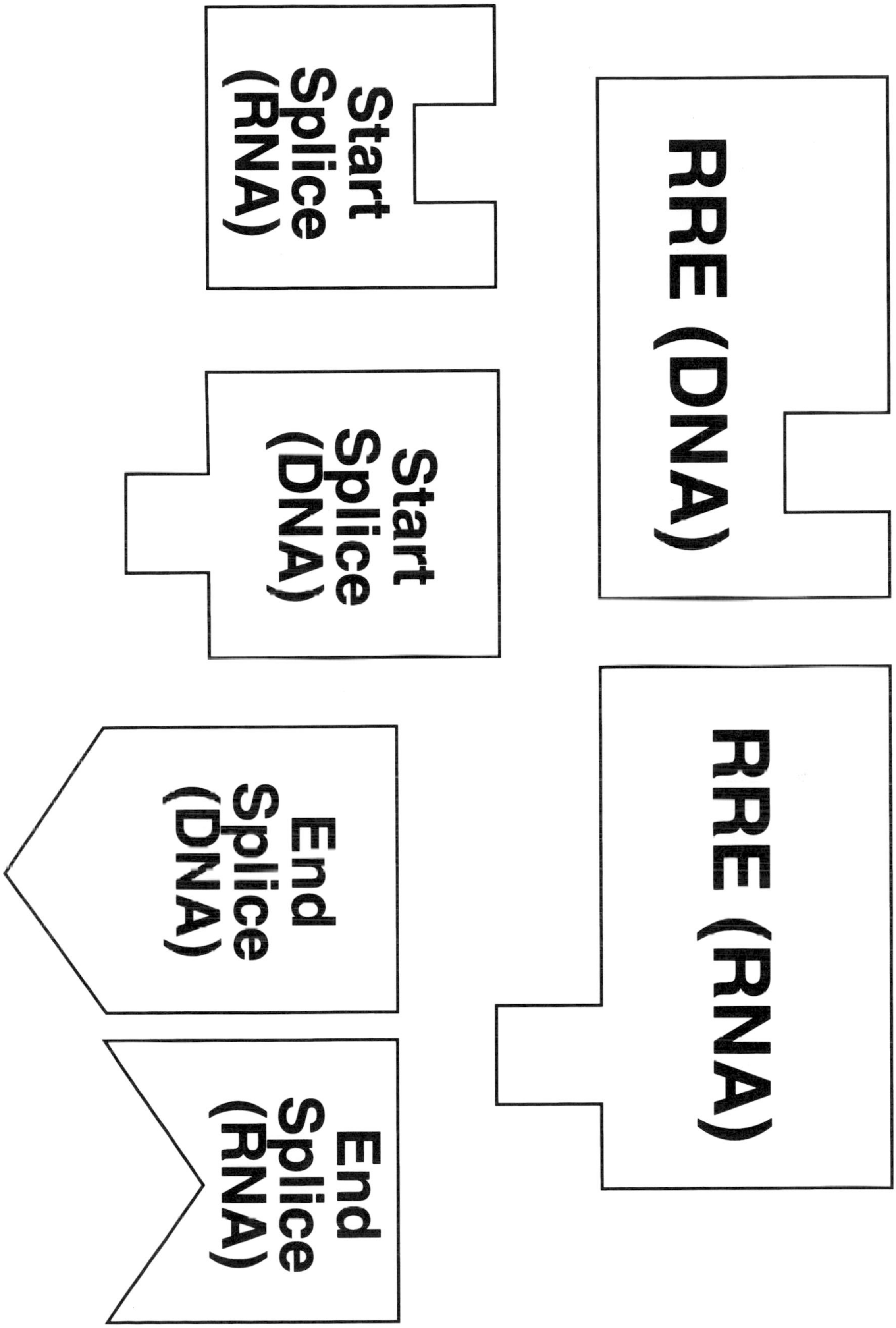

Start
Splice
(RNA)

RRE (DNA)

Start
Splice
(DNA)

End
Splice
(DNA)

RRE (RNA)

End
Splice
(RNA)

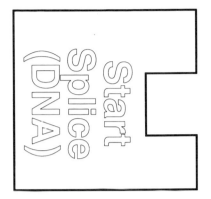

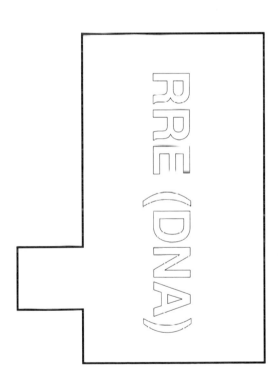

Promoter
(no RNA transcript)

Terminator
(no RNA transcript)

Promoter
(no RNA transcript)

Terminator
(no RNA transcript)

DNA sequence strips for *DNA Scissors*

```
1                                                    1

     5'-TAGACTGAATTCAAGTCA-3'
        ||||||||||||||||||
     3'-ATCTGACTTAAGTTCAGT-5'


2                                                    2

     5'-ATACGCCCGGGTTCTAAA-3'
        ||||||||||||||||||
     3'-TATGCGGGCCCAAGATTT-5'


3                                                    3

     5'-CAGGATCGAAGCTTATGC-3'
        ||||||||||||||||||
     3'-GTCCTAGCTTCGAATACG-5'


4                                                    4

     5'-AATAGAATTCCGATCCGA-3'
        ||||||||||||||||||
     3'-TTATCTTAAGGCTAGGCT-5'
```

Restriction maps for *DNA Goes to the Races*

Below are three representations of a 15,000-base-pair DNA molecule. Each representation shows the locations of different types of restriction site, with vertical lines representing the cut site. The numbers between the cut sites show the sizes (in base pairs) of the fragments that would be generated by digesting the DNA with that enzyme.

EcoRI sites

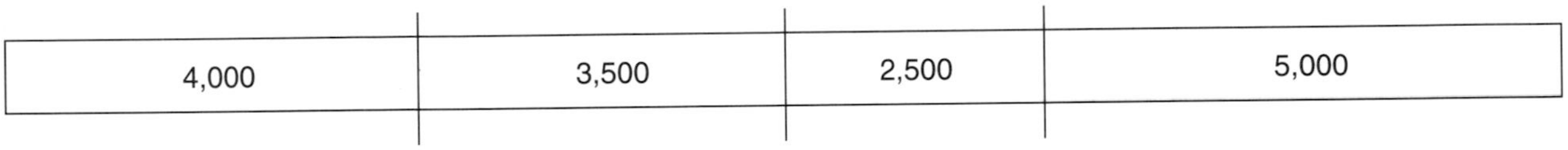

| 4,000 | 3,500 | 2,500 | 5,000 |

BamHI sites

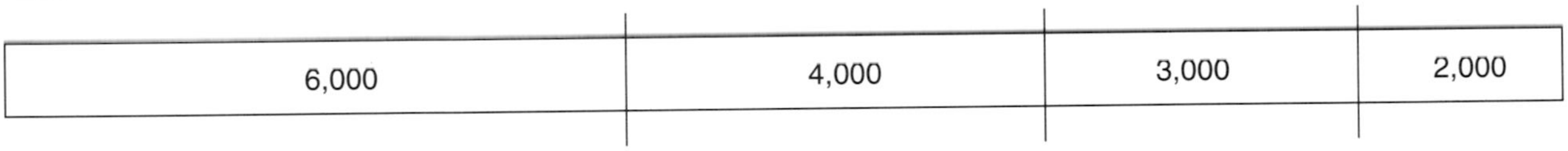

| 6,000 | 4,000 | 3,000 | 2,000 |

HindIII sites

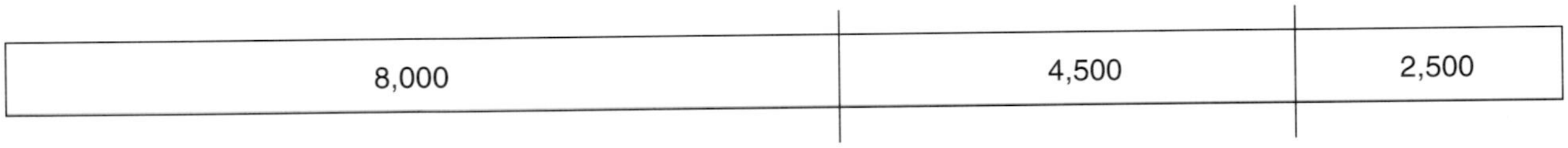

| 8,000 | 4,500 | 2,500 |

Gel outline for *DNA Goes to the Races*

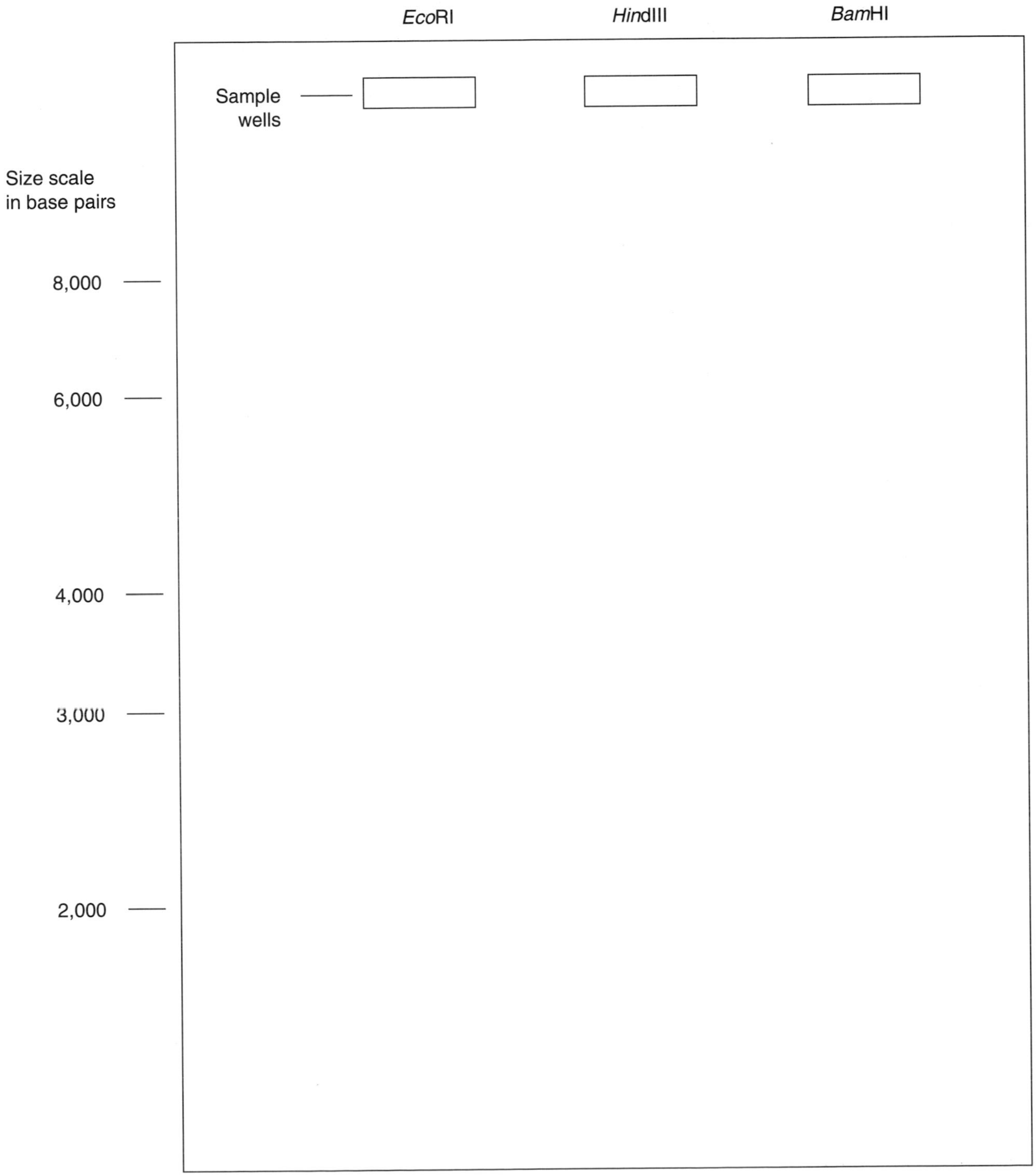

	EcoRI	HindIII	BamHI
Sample wells			

Size scale
in base pairs

8,000 ——

6,000 ——

4,000 ——

3,000 ——

2,000 ——

Paper pAMP plasmid model for *Recombinant Paper Plasmids*

	1 Ampicillin resistance gene	paste 2
5'	AATTCGATGAATTCXXXXXXXXXXXXXXXXXXXXXXXXXXXGAATTCTGAAGGTTCGAAGCGCTAT	
3'	TTAAGCTACTTAAGXXXXXXXXXXXXXXXXXXXXXXXXXXXCTTAAGACTTCCAAGCTTCGCGATA	

	2	paste 3
5'	GTCGGATCCAGATCCGAAGTCTCTCTAGGACCTTGCGAAGCCACGTAGTTCAGATTAATGCCTGAT	
3'	CAGCCTAGGTCTAGGCTTCAGAGAGATCCTGGAACGCTTCGGTGCATCAAGTCTAATTACGGACTA	

	3 Origin of replication	paste 1
5'	CGCTACAAGCTTATAGCGGCCXXXXXXXXXXXXXXXXXXXXXAATATTGCGCAGTCTTAGCACTCC	
3'	GCGATGTTCGAATATCGCCGGXXXXXXXXXXXXXXXXXXXXXTTATAACGCGTCAGAATCGTGAGG	

Paper pKAN plasmid model for *Recombinant Paper Plasmids*

1 — Origin of replication

5' TACTCGATGAAATCXXXXXXXXXXXXXXXXXXXXXXXXXXXXAGCTATGTTCTGAAGGATCCATATAGCGC

3' ATGAGCTACTTAGXXXXXXXXXXXXXXXXXXTCGATACAAGACTTCCTAGGTATATCGGG

paste 2

2 — Kanamycin resistance gene

3' TACTGGCAGTCTACGAAGXXXXXXXXXXXXXXXXXXXXXXXXXXXXXXXXXXXXXXXXAGCTTGCATGCCAGGCT

5' ATGACCGTCAGATCCGATGCTTCXXXXXXXXXXXXXXXXXXXXXXTCGAACGTACGGTCCGA

paste 3

3

5' GATCACATGCTTATAAATATTGCGAAGCTTCAGTCAGGCGGGTAGCACTCCTTAACGGATGCATTAA

3' CTAGTGTACGAATATTTATAACGCTTCGAAGTCAGTCGGGCCATCGTGAGGAATTGCGCTACGTAATT

paste 1

Worksheet 16.I: single-stranded DNA sample sequence and probe for *Detection of Specific DNA Sequences: Hybridization Analysis*

Probe:

3' GGATGCTACCATAGC 5'

3' GGATGCTACCATAGC 5'

Sample DNA, sequence written 5' to 3':

1
GGATCAGACTTCTAGCAGGCTCTTGACCAATGATCACAGCTTCCGATCTAG

2
AGCTCGATCTCTTGATCTCGATCTCGTGTCGGAATCTAGCCCGGGTCAGC

3
TATCGCTAAGATAGACCGGAATCGAGAATTCCGGATCGATCGATTGTGCGA

4
CCGCGATTATTCGATCGTTTGCCCGGGATCTAGCTTTCCGATCTAGCTGTG

5
TCAAGCTAGTGGAATCGAATTCGGAACTCGGCCCGATCTTGATCTCCCGGG

6
ACAATTGTCGATCTGATGCTAGCTGAATCGATGTGCCTAAGTGCTAGCCCGG

7
GCGATCTGGGATCGATTCCCGGGATCTAGGCCTACGATGGTATCGTTAGC

8
TAGCTCTCTAGCTTAGCTCTCAAGTGATCTACCCGGGTAGATCTAGTATATTG

9
TATCGATATTTTCGCTTAGCTAGCCCGGGCTAGCTCTCTCTAGCTAATAGATAG

10
TCTAGCTAGCTAGCTAGCTGTGTCTAGTCGATCGTTTGTCGATCTTCGATC

487

Worksheet 16.II: outlines for gel and results of hybridization analysis for *Detection of Specific DNA Sequences: Hybridization Analysis*

Stained electrophoresis gel

Fragment size, bp

Sample well

100 —
90 —
80 —
70 —
60 —
50 —
40 —
30 —
20 —

Results of hybridization analysis

100 —
90 —
80 —
70 —
60 —
50 —
40 —
30 —
20 —

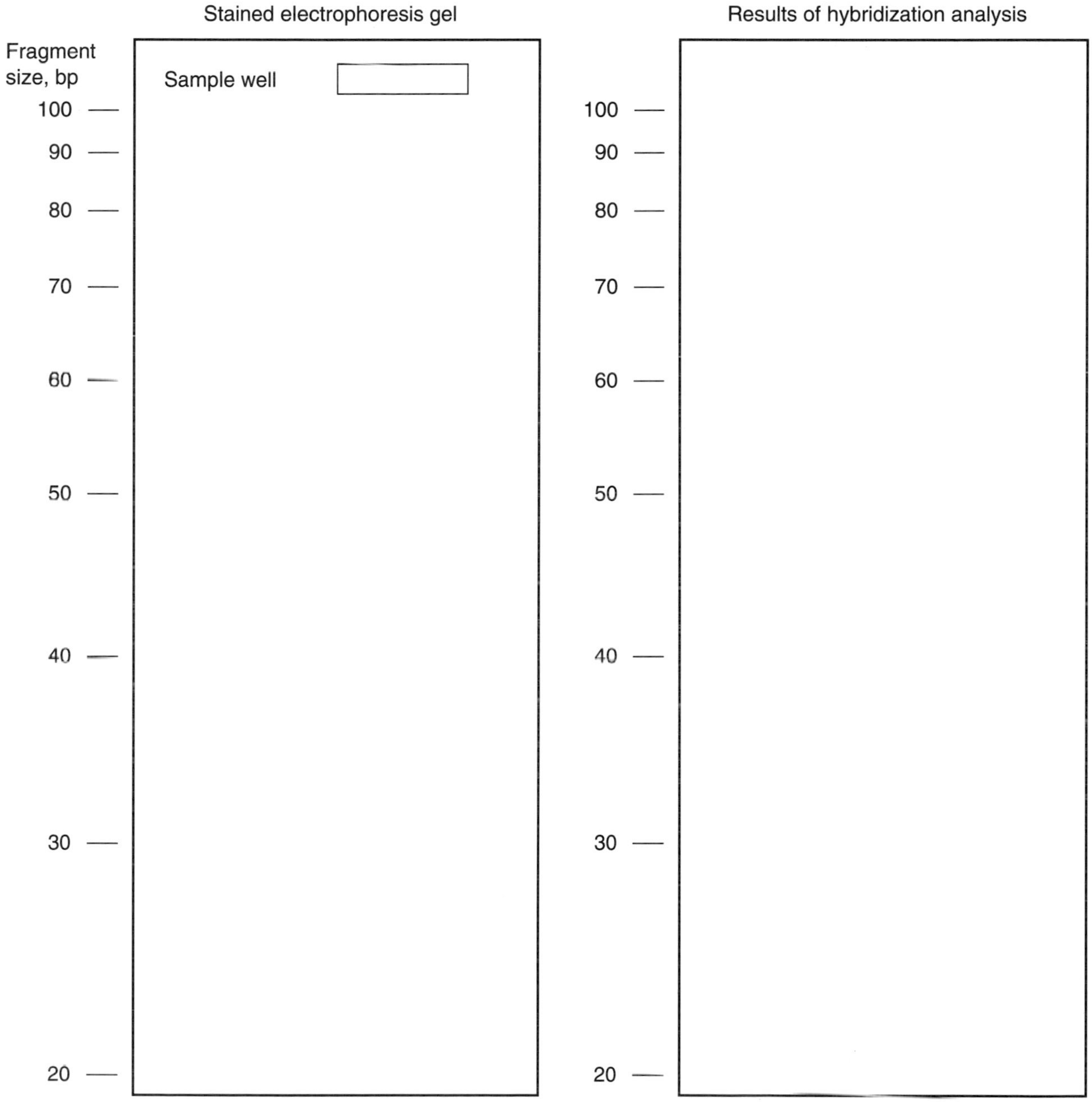

Worksheet 16.IIIA: restriction maps and hybridization analysis of virus X for *Detection of Specific DNA Sequences: Hybridization Analysis*

The map at the top shows the *Eco*RI restriction sites. The bottom line shows the *Bam*HI restriction sites. Fragment sizes are in base pairs. The stained elec- trophoresis gel shows *Eco*RI and *Bam*HI fragments of the virus. The hybridization analysis shows which of the bands in the stained gel hybridized to the probe.

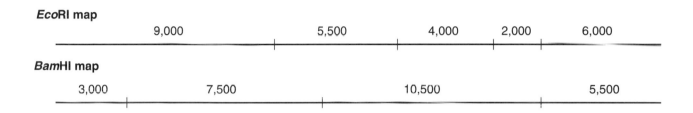

EcoRI map

9,000 5,500 4,000 2,000 6,000

BamHI map

3,000 7,500 10,500 5,500

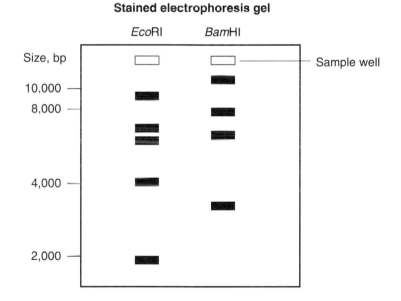

Stained electrophoresis gel

*Eco*RI   *Bam*HI

Sample well

Size, bp

10,000 —
8,000 —

4,000 —

2,000 —

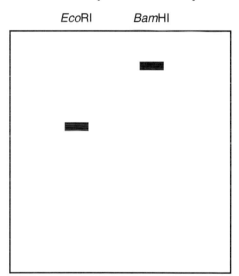

Results of hybridization analysis

*Eco*RI   *Bam*HI

Worksheet 16.IIIB for *Detection of Specific DNA Sequences: Hybridization Analysis*

Shown below are restriction maps for bacteriophage lambda, showing the *Bam*HI, *Hin*dIII, and *Eco*RI sites. Fragment sizes are in base pairs. The gel and hybridization analysis outlines are provided for your use.

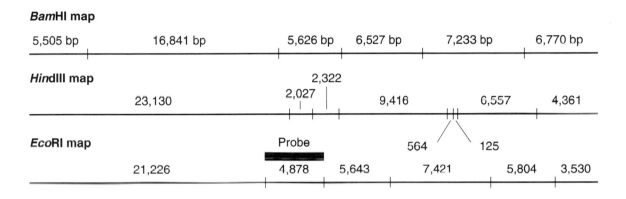

Bam*HI map

| 5,505 bp | 16,841 bp | 5,626 bp | 6,527 bp | 7,233 bp | 6,770 bp |

Hin*dIII map

23,130   2,027   2,322   9,416   6,557   4,361

Eco*RI map

Probe

21,226   4,878   5,643   564   125   7,421   5,804   3,530

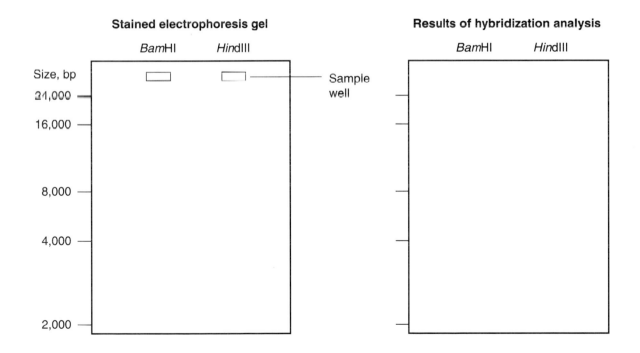

Stained electrophoresis gel

Results of hybridization analysis

*Bam*HI   *Hin*dIII   *Bam*HI   *Hin*dIII

Size, bp
21,000
16,000
8,000
4,000
2,000

Sample well

5' CGAACATC
3' GCTTGTAGAAGCTGCATTGACGCT 5'
1 2 3 4 5 6 7 8 9 10 11 12 13 14 15 16

5' CGAACATC
3' GCTTGTAGAAGCTGCATTGACGCT 5'
1 2 3 4 5 6 7 8 9 10 11 12 13 14 15 16

5' CGAACATC
3' GCTTGTAGAAGCTGCATTGACGCT 5'
1 2 3 4 5 6 7 8 9 10 11 12 13 14 15 16

5' CGAACATC
3' GCTTGTAGAAGCTGCATTGACGCT 5'
1 2 3 4 5 6 7 8 9 10 11 12 13 14 15 16

Parental DNA molecule and primers for
The Polymerase Chain Reaction

5' **TACGACCCGGTGTCAAAGTTAGCTTAGTCA** 3'

5' **TACGACCCGGTGTCAAAGTTAGCTTAGTCA** 3'

5' **TACGACCCGGTGTCAAAGTTAGCTTAGTCA** 3'

5' **TACGACCCGGTGTCAAAGTTAGCTTAGTCA** 3'

5' **CCCGG** 3'   5' **CCCGG** 3'

5' **CCCGG** 3'   5' **CCCGG** 3'

Sample DNAs and primers for PCR-based diagnosis for *The Polymerase Chain Reaction*

3' ATGCTGGGCCACAGTTTCAATCGAATCAGT 5'

3' ATGCTGGGCCACAGTTTCAATCGAATCAGT 5'

3' ATGCTGGGCCACAGTTTCAATCGAATCAGT 5'

3' ATGCTGGGCCACAGTTTCAATCGAATCAGT 5'

3' TCGAA 5'   3' TCGAA 5'

3' TCGAA 5'   3' TCGAA 5'

Patterns for cutting out chromosomes (activity I) for *Generating Genetic Variation: the Meiosis Game*

1	2	3	4	5	6	7	8
9	10	11	12	13	14	15	16
17	18	19	20	21	22	X	

1	2	3	4	5	6	7	8
9	10	11	12	13	14	15	16
17	18	19	20	21	22	Y	

1	2	3	4	5	6	7	8
9	10	11	12	13	14	15	16
17	18	19	20	21	22	X	

1	2	3	4	5	6	7	8
9	10	11	12	13	14	15	16
17	18	19	20	21	22	Y	

Patterns for maternal chromosomes (activity II) for
Generating Genetic Variation: the Meiosis Game

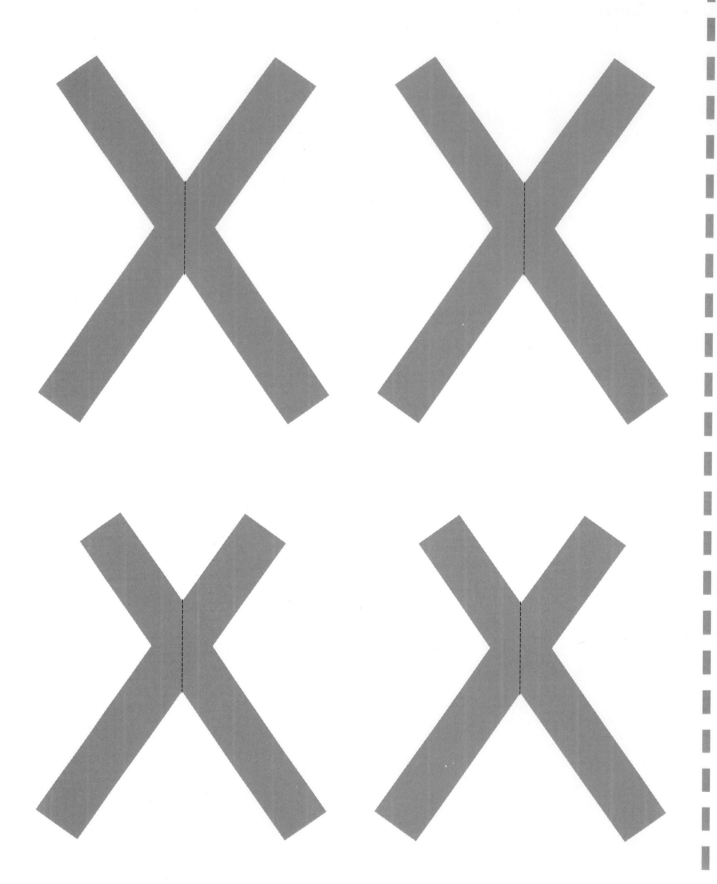

504

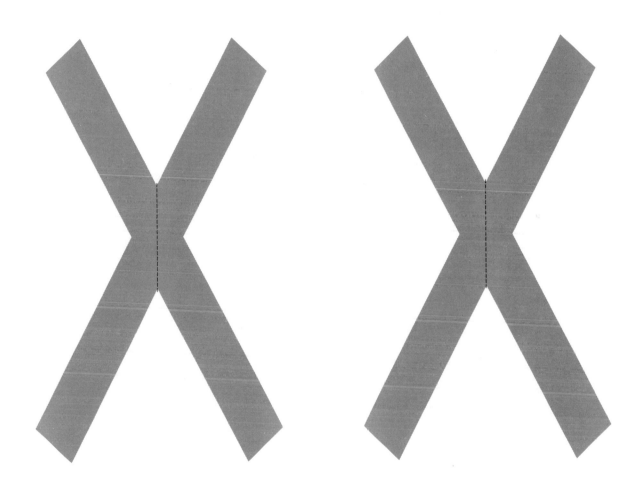

505

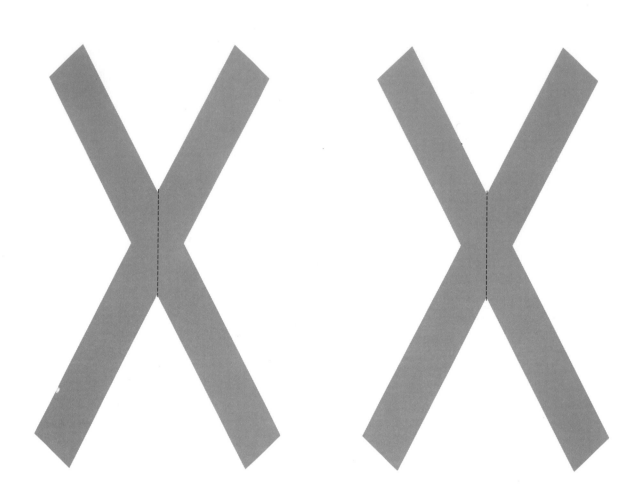

506

Patterns for paternal chromosomes (activity II) for
Generating Genetic Variation: the Meiosis Game

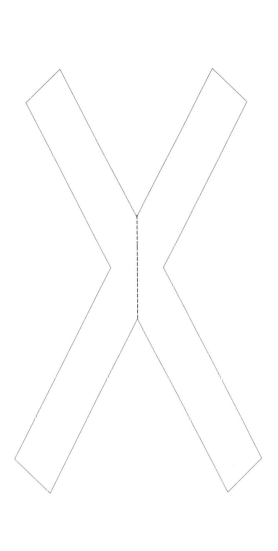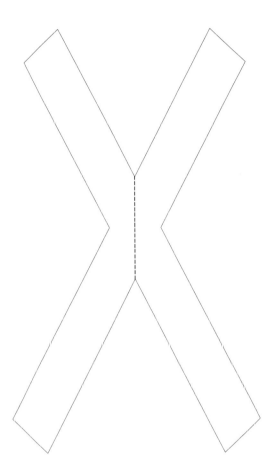

"Box-up" forms for *A Decision-Making Model for Bioethical Issues*

Issue: *Question:*

Facts

Authority

Qualifier

Decision

Principle

Rebuttal

Issue: *Question:*

Facts

Authority

Qualifier

Decision

Principle

Rebuttal

Issue: *Question:*

Facts

Authority

Qualifier

Decision

Principle

Rebuttal

Issue: *Question:*

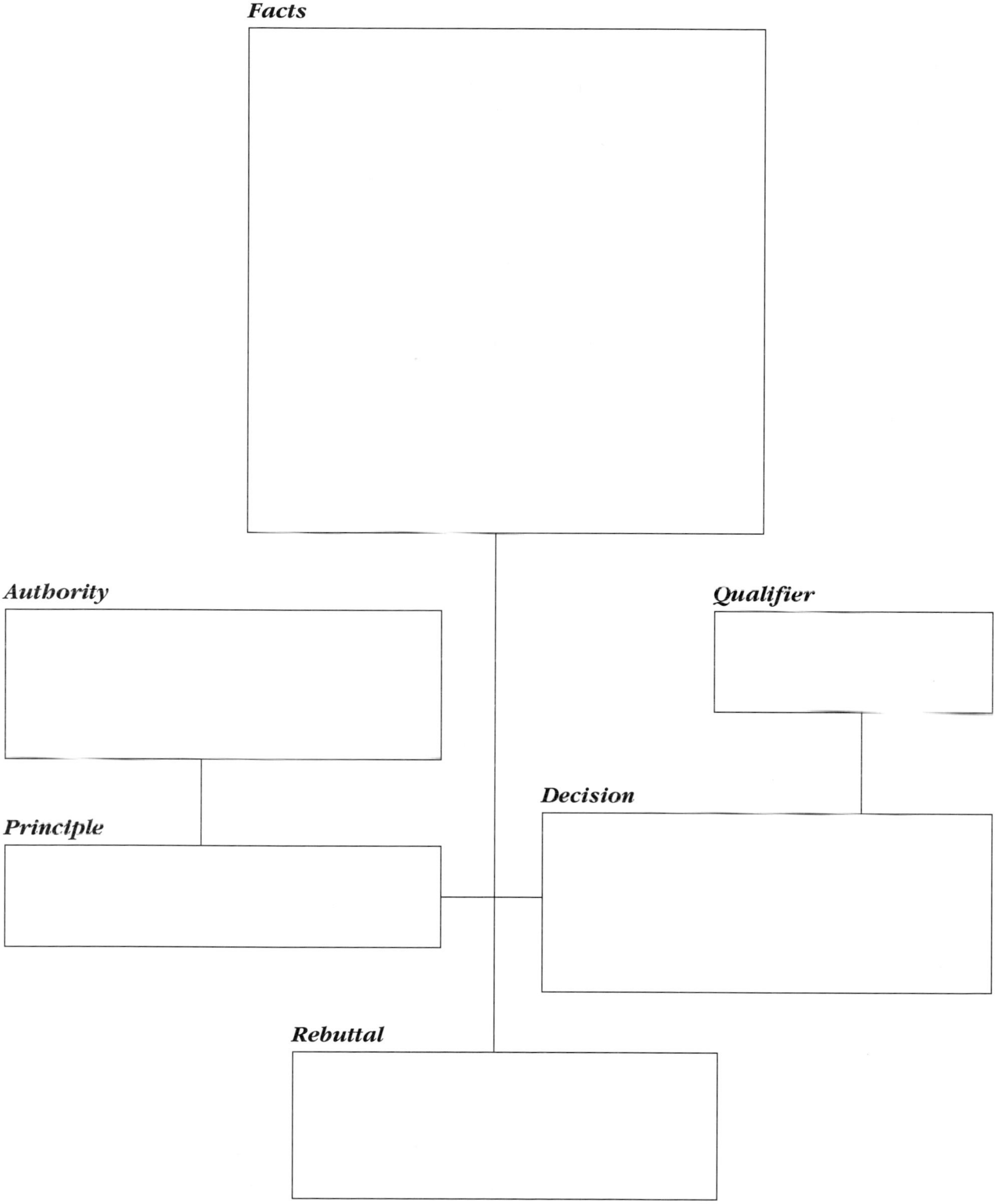

Facts

Authority

Qualifier

Principle

Decision

Rebuttal

517

Issue: *Question:*

Facts

```
┌─────────────────────────┐
│                         │
│                         │
│                         │
│                         │
│                         │
│                         │
│                         │
│                         │
└─────────────────────────┘
```

Authority   *Qualifier*

```
┌─────────────────────┐        ┌──────────────────┐
│                     │        │                  │
│                     │        │                  │
│                     │        │                  │
└─────────────────────┘        └──────────────────┘
```

Principle   *Decision*

```
┌─────────────────────┐        ┌──────────────────────┐
│                     │        │                      │
│                     │        │                      │
│                     │        │                      │
└─────────────────────┘        └──────────────────────┘
```

Rebuttal

```
┌──────────────────────┐
│                      │
│                      │
│                      │
└──────────────────────┘
```

519

Issue: *Question:*

Facts

Authority   *Qualifier*

Principle   *Decision*

Rebuttal

521

Appendix L: Overhead Masters

Mendel's principle of segregation

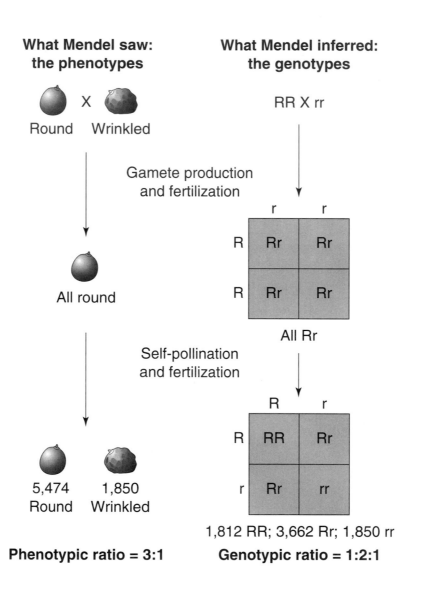

What Mendel saw: the phenotypes

Round X Wrinkled

↓ *Gamete production and fertilization*

All round

↓ *Self-pollination and fertilization*

5,474 Round   1,850 Wrinkled

Phenotypic ratio = 3:1

What Mendel inferred: the genotypes

RR X rr

↓

	r	r
R	Rr	Rr
R	Rr	Rr

All Rr

↓

	R	r
R	RR	Rr
r	Rr	rr

1,812 RR; 3,662 Rr; 1,850 rr

Genotypic ratio = 1:2:1

Principle of independent assortment

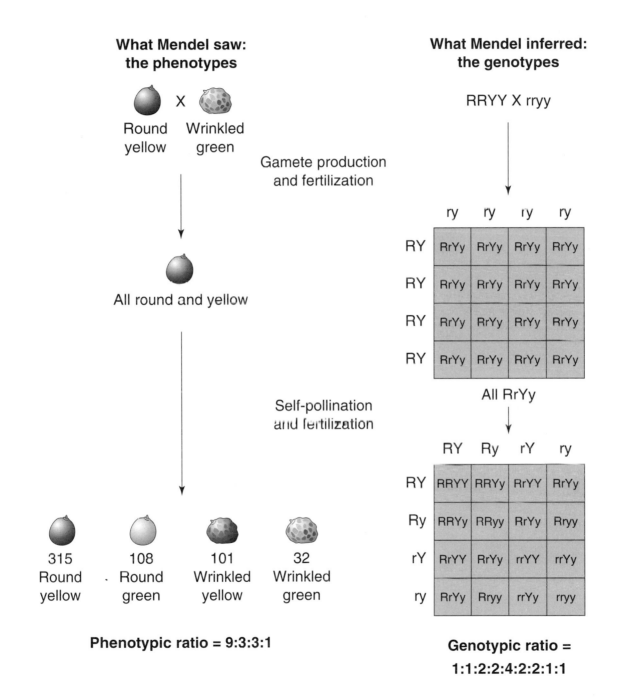

What Mendel saw: the phenotypes

Round yellow X Wrinkled green

Gamete production and fertilization

All round and yellow

Self-pollination and fertilization

315	108	101	32
Round yellow	Round green	Wrinkled yellow	Wrinkled green

Phenotypic ratio = 9:3:3:1

What Mendel inferred: the genotypes

RRYY X rryy

	ry	ry	ry	ry
RY	RrYy	RrYy	RrYy	RrYy
RY	RrYy	RrYy	RrYy	RrYy
RY	RrYy	RrYy	RrYy	RrYy
RY	RrYy	RrYy	RrYy	RrYy

All RrYy

	RY	Ry	rY	ry
RY	RRYY	RRYy	RrYY	RrYy
Ry	RRYy	RRyy	RrYy	Rryy
rY	RrYY	RrYy	rrYY	rrYy
ry	RrYy	Rryy	rrYy	rryy

Genotypic ratio = 1:1:2:2:4:2:2:1:1

Ribbon model of DNA

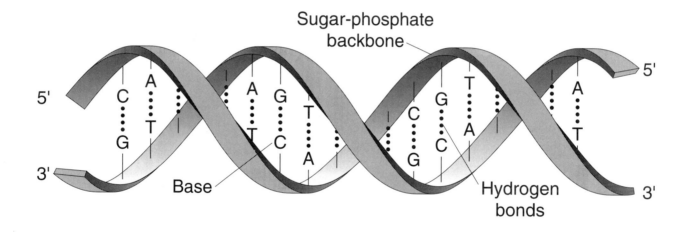

Transcription

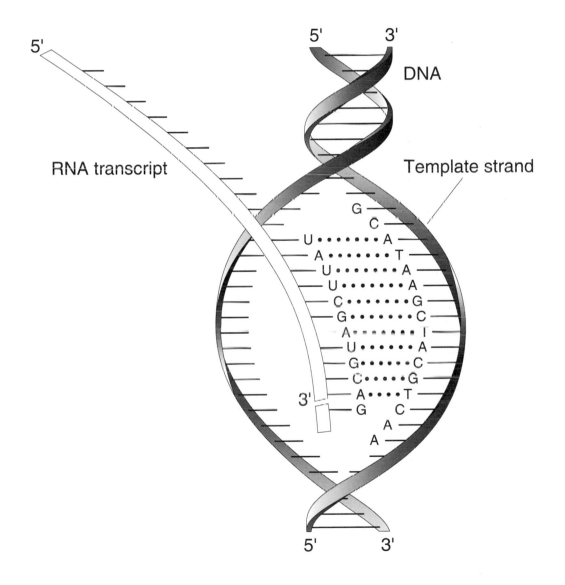

RNA transcript

5'

DNA

5'   3'

Template strand

G
  C
U • • • • • A
A • • • • • • T
U • • • • • • A
U • • • • • • A
C • • • • • • G
G • • • • • C
A • • • • • • I
U • • • • • A
G • • • • C
C • • • • • G
A • • • • T
3'   G   C
         A
       A

3'

5'   3'

tRNA molecule

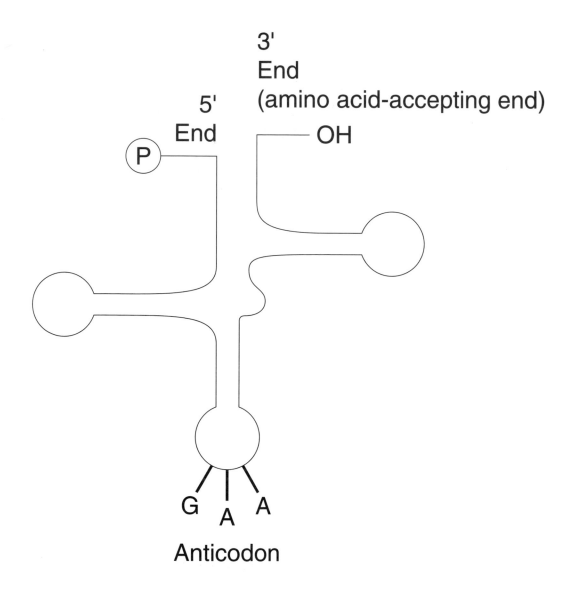

Translation

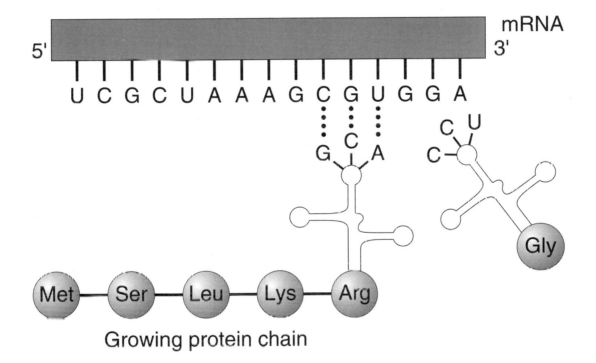

Splicing of precursor

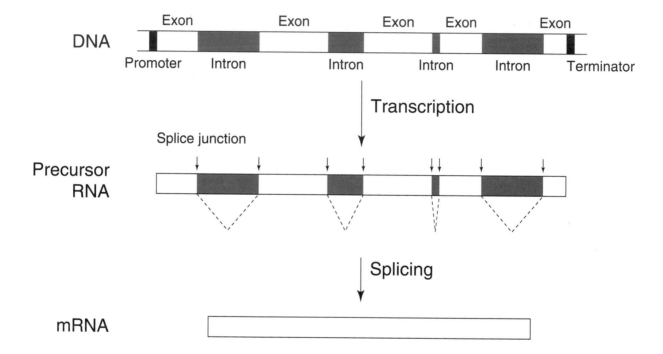

Transcriptional regulation of the *lac* operon

Active *lac* repressor protein + Lactose → Inactive lactose-repressor complex

A. No lactose in cells: active repressor prevents transcription.

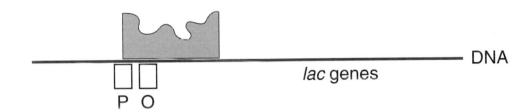

DNA

lac genes

P O

B. Lactose in cells: inactive lactose-repressor complex allows transcription.

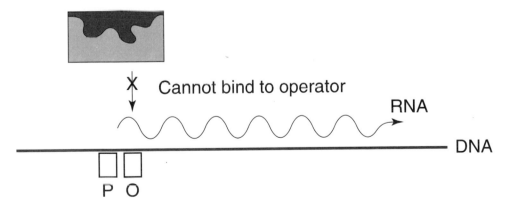

Cannot bind to operator

RNA

DNA

P O

Transcriptional regulation of the *trp* operon

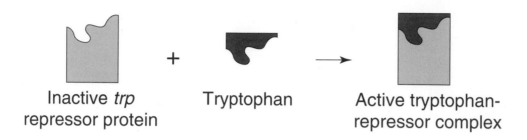

Inactive *trp* repressor protein + Tryptophan → Active tryptophan-repressor complex

A. Low tryptophan concentration: inactive repressor allows transcription.

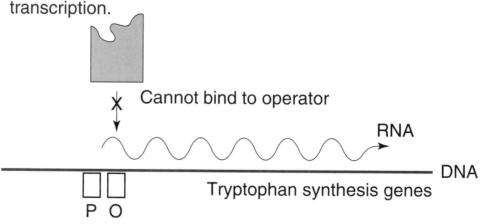

Cannot bind to operator

RNA

DNA

Tryptophan synthesis genes

P O

B. High tryptophan concentration: active repressor complex binds to operator and prevents transcription.

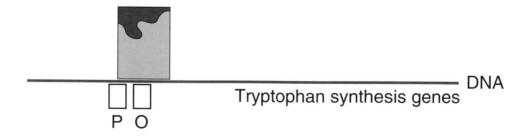

DNA

Tryptophan synthesis genes

P O

View of chromosomes at the molecular level during meiosis

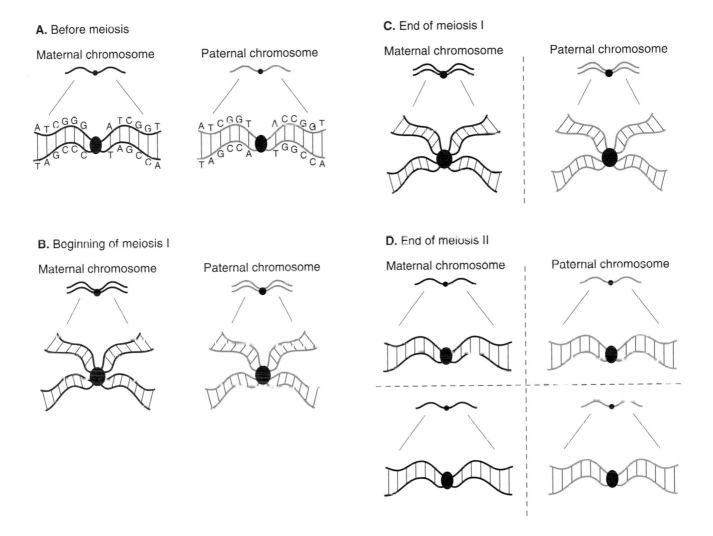

A. Before meiosis

Maternal chromosome   Paternal chromosome

B. Beginning of meiosis I

Maternal chromosome   Paternal chromosome

C. End of meiosis I

Maternal chromosome   Paternal chromosome

D. End of meiosis II

Maternal chromosome   Paternal chromosome

Chromosome behavior during meiosis

A. Before meiosis begins

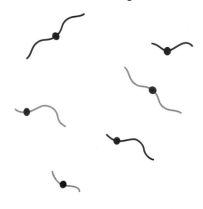

C. Middle of meiosis I

D. End of meiosis I

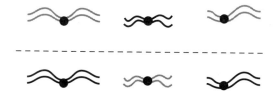

B. Beginning of meiosis I

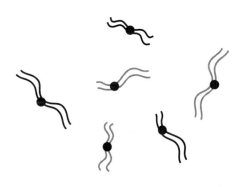

E. End of meiosis II

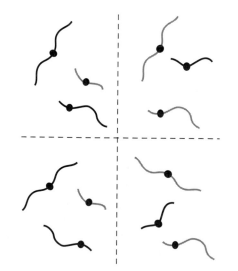

Glossary

Abzyme   An antibody designed to catalyze a specific chemical reaction. Also called a catalytic antibody.

Adenine   A nitrogen-containing base found in DNA and RNA.

Agarose   A substance purified from seaweed that dissolves in boiling water and then solidifies into a gel as the solution cools. Agarose gels are used to separate DNA fragments during electrophoresis.

Agrobacterium tumefaciens   A common soil bacterium that causes crown gall disease by transferring some of its DNA to the plant host. Scientists alter *A. tumefaciens* so that it no longer causes the disease but is still able to transfer DNA. They then use this altered organism to ferry desirable genes into plants.

Allele   One of several alternative forms of a specific gene that occupies a certain locus on a chromosome.

Alpha helix   One of a small number of stable arrangements of the peptide backbone within proteins.

Amino acids   The fundamental building blocks of a protein molecule. A protein is a chain of hundreds or thousands of amino acids. Our bodies can synthesize most of the amino acids from their component parts (carbon, nitrogen, oxygen, hydrogen, and sometimes sulfur). However, eight amino acids (called essential amino acids) must be obtained from food.

Annealing   The formation of a double-stranded nucleic acid molecule from two complementary single strands. (See **DNA hybridization.**)

Antibiotic resistance marker   A gene encoding a protein that renders a cell resistant to an antibiotic. The organisms, usually bacteria, that have this gene are not susceptible to that antibiotic. Often used to identify organisms that have been successfully transformed.

Antibody   A protein produced in response to the presence of a specific antigen.

Anticodon   A triplet of nucleotides in a transfer RNA molecule that is complementary to a codon in a messenger RNA molecule.

Antigen   A foreign substance that elicits the production of antibodies.

Antisense RNA   An RNA molecule that is complementary to the messenger RNA transcribed from a gene. It blocks the expression of that gene by interfering with protein production.

Assay   A method of determining the presence or quantity of a component.

Autosome   A chromosome that is the same in males and females of the species (as opposed to the sex chromosomes). Humans have 22 pairs of autosomes and 1 pair of sex chromosomes. Mutations in or traits encoded by genes on these chromosomes can be described as autosomal. For example, cystic fibrosis is an autosomal recessive trait.

Bacillus thuringiensis   A naturally occurring bacterium with pesticidal properties. *B. thuringiensis* produces a protein (Bt toxin) that is toxic only to certain insect larvae that consume it.

Bacteriophage   A virus that infects bacteria. Also called a phage.

Baculovirus   A class of insect viruses used as cloning vectors for eukaryotic cells.

Beta sheet   One of a small number of stable arrangements of the peptide backbone within proteins.

Bioassay   A method of determining the effect of a compound by quantifying its effect on living organisms or their component parts.

Biolistics   A method of getting DNA into cells by using small metal particles coated with DNA. These particles are fired into a cell at very high speed.

Biological control   The use of one organism to control the population size of another organism.

Biological molecules   Large, complex molecules, such as proteins, nucleic acids, lipids, and carbohydrates, that are produced only by living organisms. Biological molecules are often referred to as macromolecules or biopolymers.

Bioprocess   A process in which microorganisms, living cells, or their components are used to produce a desired end product.

Bioreactor   A container used for bioprocessing.

Bioremediation   The use of organisms, usually microorganisms, to clean up contamination.

Biosensor   An electronic system that uses cells or biological molecules to detect specific substances. Consists of a biological sensing agent coupled to a microelectronic circuit.

Biosynthesis   Production of a chemical by a living organism.

Biotechnology   (Broad definition) The use of living organisms to solve problems and make useful products. (Modern definition) A collection of technologies that use living cells and/or biological molecules to solve problems and make useful products.

B lymphocyte   A type of immune system cell that is responsible for the production of antibodies.

Bovine somatotropin (BST)   The proteinaceous growth hormone found naturally in cattle; very similar in structure and function to human growth hormone. BST is chemically similar to human growth hormone and is being used commercially to increase growth rate, the protein/fat ratio, and milk production in cows.

Callus   A cluster of undifferentiated plant cells that have the capacity to regenerate a whole plant in some species.

Capsid   Protein coat of a virus. Plants can be genetically engineered to be resistant to a virus by giving them the gene that encodes the capsid.

Catalyst   A substance that speeds up a chemical reaction but is not itself changed during the reaction.

cDNA library   A collection of genetic clones that contains all of the cDNA derived from a source organism. The cDNA is "housed" in the library by splicing portions of the entire complement of cDNA into suitable vectors. (see **Complementary DNA**.)

Cell culture   A technique for growing cells under laboratory conditions.

Cell fusion   The formation of a hybrid cell by the fusing of two different cells.

Chromatin   The DNA-protein complex found in the nucleus.

Chromosomal aberration   An abnormality in chromosome number or structure.

Chromosomes   Components in a cell that contain genetic information. Each chromosome contains numerous genes. Chromosomes occur in pairs: one is obtained from the mother, and the other is obtained from the father. Chromosomes of different pairs are often visibly different from each other (see also **DNA**).

Chromosome walking   The step-by-step analysis of a long stretch of DNA by the sequential isolation of clones that carry overlapping sequences of this DNA. Used to locate unknown genes. Using a gene library and starting from a known sequence (usually the site of a restriction fragment length polymorphism), scientists isolate clones containing DNA that hybridizes to DNA probes taken from the ends of the known sequence. The ends of these clones are then used to screen the library for clones that hybridize to the ends of the first clone. This screening and isolation is repeated again and again until the unknown gene of interest is reached. If each clone covers a long stretch of DNA, the researcher can "walk" the chromosome quickly, because each "step" is like a "giant step." So researchers prefer to use clones or vectors that carry a large amount of foreign DNA. (See **Clone, Cloning,** and **Genetic library.**)

Clone   A cell, collection of cells, or collection of individuals containing genetic material identical to that of the parent cell and of each other. Clones are produced from a single parent cell and thus show little if any variation compared to that in similar organisms produced through sexual reproduction. The word "clone" also refers to the identical piece of DNA a collection of cells (usually bacterial) contains.

Cloning   Isolating DNA sequences and incorporating them into plasmids or other vectors so that they can be inserted into a suitable organism (bacterium, yeast cell) for copying. Cloning also refers to the production of genetically identical cells from a single parent cell. These genetically identical cells are referred to as a clonal population.

Codon   A triplet of bases that specifies an amino acid.

Colony hybridization   A technique that uses hybridization to identify bacteria containing DNA that is complementary to a certain sequence.

Competent   Able to take in DNA molecules from the environment, as in competent cells.

Complementary base pairs   Pairs of bases that form hydrogen-bonded pairs in nucleic acid molecules (in DNA, adenine-thymine and cytosine-guanine).

Complementary DNA (cDNA)   A single-stranded DNA that is synthesized in vitro from an RNA template by reverse transcriptase.

Conditional lethal mutations   Mutations that are fatal to the organism under some environmental conditions but not under others. Usually refers to laboratory conditions that can be manipulated.

Conjugation   The transfer of genetic material from one bacterium to another through physical contact between the cells. In *Escherichia coli*, the contact occurs through a special structure called a pilus.

Cosmid   A plasmid that is packaged in a phage coat. Scientists use cosmids to transfer a relatively long stretch of DNA into host organisms.

Covalent   A type of chemical bond that consists of two electrons shared by two atomic nuclei.

Crossing over   A natural process that occurs during meiosis in which pieces of homologous chromosomes are exchanged.

Culture   To grow living organisms in prepared medium.

Culture medium   A nutrient system for growing bacteria or other cells in the laboratory.

Cytosine   A nitrogen-containing base found in DNA and RNA.

Deletion   A form of chromosomal aberration in which a portion of a chromosome is lost.

Denaturation   The complete unfolding of a protein or the separation of the two complementary strands of a DNA double helix.

Deoxynucleotide   A compound made up of the sugar deoxyribose, phosphate, and a nitrogen-containing base. Found in DNA.

Deoxyribose   The five-carbon sugar found in DNA.

Dimer   A molecule made up of two identical subcomponents. The subcomponents are called monomers.

DNA (deoxyribonucleic acid)   The chemical molecule that is the basic genetic material found in all cells. DNA is the carrier of genetic information from one generation to the next. Because DNA is a very long, thin molecule, it is packaged into units called chromosomes. DNA belongs to a class of biological molecules called nucleic acids.

DNA hybridization   The formation of a double-stranded nucleic acid molecule from two separate but complementary single strands. The single strands can be two DNA strands or one RNA and one DNA strand. The term also refers to a molecular technique that uses one nucleic acid strand to locate another.

DNA library   A collection of cloned DNA fragments that collectively represent the genome of an organism. A complementary DNA library collectively represents the messenger RNA species that were present in the cells when the library was made.

DNA ligase   An enzyme that rejoins cut pieces of DNA.

DNA polymerase   An enzyme that replicates DNA. DNA polymerases synthesize new DNA complementary to a template strand. Synthesis occurs in the 5′-to-3′ direction only and requires a primer (see **Primer**).

DNA probe   A relatively short single strand of DNA that is used to detect a specific sequence of nucleotides through hybridization.

DNA repair enzymes   Proteins that recognize and repair abnormalities in DNA.

DNA sequence   The order of nucleotide bases in the DNA molecule.

Dominant allele   An allele that is phenotypically expressed in the same way in individuals who are either homozygous or heterozygous for that allele.

Duplication   A chromosomal aberration in which part of the chromosome is present in duplicate form in a cell.

EDTA   Ethylenediaminetetraacetic acid.

Electronegativity   The ability of an atom to attract electrons.

Electroporation   A technique that uses an electrical current to create temporary pores in a cell membrane through which DNA can enter.

Embryo transplant   A technique used in animal biotechnology. After in vitro fertilization, the zygote is cultured for a few days and then implanted into a female. The developing embryo is sometimes separated into individual cells at the four- to eight-cell stage, and each cell is implanted in a female.

Endonuclease   An enzyme that cleaves a nucleic acid at nonterminal phosphodiester bonds. (See **Exonuclease** for comparison; exonucleases cleave at terminal sites.) One class of endonucleases, the restriction endonucleases, recognize specific sequences of bases along a DNA molecule and cleave the molecule following recognition.

Enhancers   DNA-binding sites of certain transcription activator proteins that are important for maximal transcription of associated promoters.

Enzyme   A protein that accelerates the rate of chemical reactions. Enzymes are catalysts that promote reactions repeatedly without being changed by the reactions.

Enzyme-linked immunosorbent assay (ELISA)   A technique for detecting specific proteins by using antibodies linked to enzymes.

Erythropoietin   A growth factor that stimulates the cells that give rise to red blood cells.

Escherichia coli   A bacterium commonly found in the intestinal tracts of most vertebrates. *E. coli* is used extensively in recombinant DNA research because it has been genetically well characterized.

Eukaryote   An organism whose genetic material is located within a nucleus. Yeast cells, fungi, protozoans, plants, and animals are eukaryotes.

Evolution   Changes in the gene pool of a population over time. These changes in the frequencies of certain genotypes result primarily from natural selection. Other factors that may contribute to changes in the genetic composition of a population are genetic drift and migration.

Exons   The regions of a gene that determine the amino acid sequence of a protein.

Exonuclease   An enzyme that cleaves the terminal phosphodiester bond of a nucleic acid molecule, releasing a single nucleotide. Exonucleases must have access to the end of a molecule for activity; they will not cleave circular nucleic acid molecules.

Expression   The physical manifestation (protein production) of the information contained in a gene.

Factor VIII   One of many compounds involved in blood clotting in humans. If this protein is missing or defective, the resulting condition is hemophilia. Because factor VIII is a small protein, it can be produced in large quantities by bacteria genetically engineered for its production.

Fermentation   A process of growing microorganisms to produce various chemical or pharmaceutical compounds. Microbes are usually incubated under specific conditions in large tanks called fermenters. Fermentation is a specific type of bioprocessing.

Gel electrophoresis   A process for separating molecules by forcing them to migrate through a semisolid material (gel) under the influence of an electric field.

Gene   A unit of hereditary information. A gene is a section of a DNA molecule that specifies the production of a particular protein.

Gene amplification   The increase, within a cell, of the number of copies of a given gene.

Gene mapping   Determining the relative locations of genes on a chromosome.

Gene therapy   The addition of genetic material to an individual so that a defect or disease can be corrected. To date, human gene therapy has involved changing the genetic makeups of somatic cells only. Genetic changes to germ cells are prohibited in humans and have been restricted to animals.

Genetic code   The system of nucleotide triplets in genes that encode the amino acids in proteins. All living organisms on Earth use the same genetic code.

Genetic engineering   The technique of removing, modifying, or adding genes to a DNA molecule in order to change the information it contains. By changing this information, genetic engineering changes the type or amount of proteins an organism is capable of producing.

Genetic library   A collection of DNA that, taken collectively, represents all of an organism's genome. The DNA molecules are "housed" in microorganisms as recombinant DNA molecules and are copied when the microorganism replicates.

Genome   The total hereditary material of a cell.

Genotype   The specific genetic makeup of an organism, as opposed to the actual characteristics of an organism (see **Phenotype**).

Growth factors   Naturally occurring proteins that stimulate growth and reproduction of specific cell types. For example, epidermal growth factor stimulates the production and differentiation of cells in the upper skin layer. Fibroblast growth factor stimulates the growth of cells in connective tissue. Growth factors are being studied as possible therapeutic compounds to be used in the treatment of diseases or injuries. For example, the two growth factors just mentioned could be useful for treating burn victims.

Growth hormones   Hormones that stimulate growth in plants and animals. The growth hormones in plants bear no chemical resemblance to the growth hormones in animals. In the vertebrates, growth hormone is a protein hormone secreted by the anterior pituitary. It stimulates protein production in its target organs. Also known as somatotropin.

Guanine   A nitrogen-containing base found in DNA and RNA.

Homologous   Two chromosomes are said to be homologous if they carry alleles for the same traits. In each cell containing homologous chromosomes, each member of a homologous pair is derived from a different parent. Nonhomologous chromosomes carry genes for different traits.

Hybridization   Production of offspring, or hybrids, from genetically dissimilar parents. In selective breeding, it usually refers to the offspring of two different species (See also **DNA hybridization.**)

Hybridoma   A type of hybrid cell produced by fusing a normal cell with a tumor cell. When lymphocytes (antibody-producing cells) are fused to the tumor cells, the resulting hybridomas produce antibodies and maintain rapid, sustained growth, producing large amounts of an antibody. Hybridomas are the source of monoclonal antibodies.

Hydrogen bonds   Weak electrostatic attractions. Hydrogen bonds exist between paired bases in DNA and are important in determining protein structure.

Hydrophilic   Favoring chemical associations with water molecules.

Hydrophobic   Disfavoring chemical associations with water molecules.

Immunoassay   A technique for identifying substances that is based on the use of antibodies.

Immunotoxin   A molecule that is toxic to the cell and is attached to an antibody.

Initiation codon The codon in messenger RNA that tells the ribosomes to start synthesizing a protein. Usually 5′ AUG 3′.

Initiation factor A protein necessary to begin translation. Initiation factors are not parts of the ribosomes and do not participate in translation once the process has begun.

Insulin A protein hormone that lowers blood glucose levels. The first commercial product derived from genetically engineered bacteria.

Interferon A protein produced naturally by the cells in our bodies. It increases the resistance of surrounding cells to attacks by viruses. One type of interferon, alpha interferon, is effective against certain types of cancer. Others may prove effective in treating autoimmune diseases.

Interleukin-2 A protein produced naturally by our bodies to stimulate our immune systems. There are at least six kinds of interleukins.

Introns Noncoding regions within a gene. They are transcribed into RNA but removed by splicing prior to protein synthesis.

Inversion A chromosomal aberration in which a section of chromosome is reversed.

In vitro Performed in a test tube or other laboratory apparatus.

In vitro selection Selection at the cellular or callus stage of individuals possessing certain traits, such as herbicide resistance.

In vivo In the living organism.

Keratins The family of structural proteins that make up hair, wool, feathers, claws, hooves, etc.

Ligation The joining of the ends of two DNA molecules.

Linkage The tendency of pairs or groups of genes to be inherited together because they occur close together on the same chromosome.

Locus The position a gene occupies on a chromosome.

Lysate The mixture of cellular components obtained after cells have been broken open.

Lysis The breaking open of cells.

Macrolesion A genetic change in which a large amount of DNA is altered by changing the total amount of DNA or changing the relative position of genes on a chromosome.

Macromutation See **Macrolesion.**

Marker Restriction fragment(s) seen only when a particular genetic disease is present.

Marker gene Genes that identify which plants, bacteria, or other organisms have been successfully transformed.

Melting temperature The temperature required to denature a DNA or protein molecule.

Messenger RNA (mRNA) The RNA molecules that carry genetic information from the chromosomes to the ribosomes.

Metabolic engineering Changing cellular activities by manipulating the enzymatic, transport, and regulatory functions of the cell.

Microinjection Method of delivering DNA, primarily to animal cells, by using a microscopic needle to pierce the nucleus.

Microlesion A genetic change, sometimes called a micromutation, that involves a small amount of DNA. (See also **Point mutation.**)

Micromutation See **Microlesion.**

Molecular genetics The study of the molecular structures and functions of genes.

Monoclonal antibody Highly specific, purified antibody that is derived from only one clone of cells and recognizes only one antigen. Also see **Hybridoma.**

Monomer The simplest subcomponent that is repeated to form a multicomponent molecule. For example, the nucleotide is the monomer found in DNA and RNA. See **Dimer.**

Multigenic A multigenic, or polygenic, trait is one whose expression is governed by many genes.

Mutagen A substance that induces mutations.

Mutant A cell or organism that manifests new characteristics because of a change in its genetic material.

Mutation Any change in the base sequence of a DNA molecule.

Mycorrhiza A symbiotic association between certain fungi and the roots of vascular plants.

Natural selection The differential rate of reproduction of certain phenotypes in a population. If those phenotypes have a genetic basis, natural selection can lead to a change in gene frequencies in a population.

Noncoding DNA DNA that does not encode any product (RNA or protein). The majority of DNA in plants and animals is noncoding.

Nonhomologous Chromosomes are described as nonhomologous if they carry genes for different traits. Contrast with the definition for **Homologous.**

Northern blotting A technique for identifying an RNA sequence by transferring it from a gel to a filter and hybridizing it to a DNA probe. Useful for measuring gene expression.

Nuclease   An enzyme that cleaves the phosphodiester bonds of a nucleic acid molecule. See **Endonuclease** and **Exonuclease.**

Nucleic acid   A biological molecule composed of a long chain of nucleotides. DNA is made of thousands of molecules of four different nucleotides repeated randomly.

Nucleoside   A nucleotide-like molecule containing only the sugar and a base.

Nucleotide   A compound made up of three components: a sugar (either ribose or deoxyribose), phosphate, and a nitrogen-containing base. Found as individual molecules (e.g., adenine triphosphate, the "energy molecule") or as many nucleotides linked together in a chain (nucleic acid such as DNA).

Oligonucleotide   A polymer consisting of a small number of nucleotides. Oligonucleotides can be synthesized by automated machines and so are widely used as probes and primers.

Oncogene   A gene thought to be capable of producing cancer.

Oncology   The study of tumors.

Operator   A sequence of bases near a bacterial gene promoter where a repressor protein binds and shuts off transcription.

Operon   A collection of adjacent genes that are transcribed together whose products usually have related functions.

Origin of replication   A sequence of DNA bases that tell DNA polymerase and its helper proteins where to begin duplicating a DNA molecule.

Pathogenic   Disease causing.

Peptide bond   The chemical bond that links adjacent amino acids within proteins.

Peptides   Molecules composed of a few or several amino acids linked by peptide bonds. The difference between peptides and proteins is that proteins are much longer chains of amino acids and may even consist of an association between two or several peptide chains.

Phage   See **Bacteriophage.**

Phenotype   The observable characteristics of an organism as opposed to the set of genes it possesses (its genotype). The phenotype that an organism manifests is a result of both genetic and environmental factors. Therefore, organisms with the same genotype may display different phenotypes because of environmental factors. Conversely, organisms with the same phenotypes may have different genotypes.

Phosphodiester bonds   The chemical bonds that connect nucleotides in the backbones of DNA and RNA.

Plasmid   A small, self-replicating piece of DNA found outside the chromosome. Plasmids are the principal tools for inserting new genetic information into microorganisms or plants.

Point mutation   A change in the DNA sequence in a single gene. Most often, this term refers to a change in a single base or a single base pair in a gene.

Polar   Referring to a covalent bond or an entire molecule in which the electrons are unevenly distributed, leading to regions of partial positive and partial negative electrical charge.

Polyhydroxybutyrate (PHB)   A naturally occurring compound produced by bacteria. PHB may be used in the production of biodegradable plastic.

Polymerase chain reaction (PCR)   A method of making millions of copies of a single DNA molecule by using a heat-stable DNA polymerase.

Polypeptides   Molecules composed of many amino acids linked by peptide bonds. Another term for a single protein chain. See **Protein** and **Peptides.**

Porcine somatotropin (PST)   The growth hormone found in pigs. See **Growth hormones.**

Primary structure   The linear sequence of amino acids within a protein molecule.

Primer   A single-stranded nucleic acid molecule (DNA or RNA) hybridized to a template strand in such a way that the primer's 3′ end is available to serve as the starting point for synthesis of a new DNA strand complementary to the template. Required for DNA synthesis by DNA polymerase enzymes.

Probe   A single-stranded DNA or RNA molecule used to detect the presence of a complementary nucleic acid.

Prokaryotes   Organisms whose genetic material is not enclosed by a nucleus. The most common examples are bacteria.

Promoter   A special sequence of bases in DNA that is recognized by RNA polymerase enzymes. The promoter signals RNA polymerase to begin transcription of a gene.

Protein   A complex biological molecule composed of a chain of units called amino acids. Proteins have many different functions: structure (collagen), movement (actin and myosin), catalysis (enzymes), transport (hemoglobin), regulation of cellular processes (insulin), and response to stimuli (receptor proteins on surface of all cells). Protein function is dependent on the protein's three-dimensional structure (tertiary structure), which depends on the linear sequence of amino acids in the protein (secondary structure). The information for making proteins is stored in the sequence of nucleotides in the DNA molecule.

Proteinase An enzyme that cleaves the peptide bonds of protein backbones.

Protoplast A plant or bacterial cell whose wall has been removed by artificial treatment.

Quaternary structure The arrangement of multiple protein subunits in a larger complex.

Recessive allele An allele whose expression is masked in the heterozygous state by a dominant allele.

Recombinant DNA (rDNA) DNA that is formed by combining DNA from two different sources.

Recombinant DNA (rDNA) technology The laboratory manipulation of DNA in which DNA or fragments of DNA from different sources are cut and recombined by using enzymes. This rDNA is then inserted into a living organism. rDNA technology is usually synonymous with genetic engineering.

Recombination The formation of new combinations of genes. Recombination occurs naturally in plants and animals during the production of sex cells (sperm, eggs, pollen) and their subsequent joining in fertilization. In microbes, genetic material is recombined naturally during conjugation, transformation, and transduction.

Regeneration The process of growing an entire plant from a single cell or group of cells.

Renaturation The re-formation of normal molecular structure after denaturation. In DNA, the re-formation of bonds between the two separated strands of a molecule. In protein, the refolding of the unfolded polypeptide chain into the normal three-dimensional structure. See **Denaturation.**

Repressors Proteins that bind to DNA and block transcription.

Restriction endonuclease See **Restriction enzyme.**

Restriction enzyme Also called restriction endonuclease. An enzyme that recognizes a specific sequence of bases in a DNA molecule and cleaves the molecule at or near that sequence. The recognition sequence is called a restriction site. Different restriction enzymes recognize and cleave at different restriction sites.

Restriction fragment A short length of DNA that results from the cleavage of a large DNA molecule by a restriction enzyme.

Restriction fragment length polymorphism (RFLP; pronounced "riflip") A difference in restriction fragment lengths between very similar DNA molecules (such as homologous chromosomes from two different individuals). RFLPs are caused by relatively minor differences in the base sequences of the molecules. RFLP analysis is used to detect differences in DNA molecules that are, on a large scale, quite similar. Applications of RFLP analysis include DNA typing and prediction of genetic disease through DNA testing.

Restriction map A diagram of the sites on a DNA molecule that are cleaved by different restriction enzymes.

Reverse transcriptase The enzyme that uses an RNA molecule as a template for synthesizing a complementary DNA molecule.

Ribonucleotide A nucleotide made with the sugar ribose. Found in RNA. See **Nucleotide.**

Ribose The 5-carbon sugar found in RNA.

Ribosomes The protein-RNA complexes that form the site of protein synthesis.

Ribozymes RNA molecules that catalyze reactions, often the breakdown of RNA molecules. Also called catalytic RNA.

RNA (ribonucleic acid) Like DNA, a type of nucleic acid. RNA differs from DNA in three ways: RNA nucleotides contain the sugar ribose instead of deoxyribose; RNA contains the base uracil instead of thymine; and RNA is primarily a single-stranded molecule rather than a double-stranded helix. The three major types are messenger RNA, transfer RNA, and ribosomal RNA. All are involved in the synthesis of proteins from the information contained in the DNA molecule.

RNA polymerase The enzyme that synthesizes RNA by using a DNA template.

Secondary structure Local regions of alpha helixes, beta sheets, and unstructured loops within a protein molecule.

Sex-linked inheritance A trait that is determined by a gene on a sex chromosome, most often the X chromosome. As a result, the trait shows a different pattern of inheritance in males and females. In humans, the ability to discriminate color is a sex-linked trait.

Somaclonal variant selection A form of plant genetic manipulation that is analogous to selective breeding at the plantlet and not the reproductive stage.

Somatotropin A synonym for growth hormone.

Southern blotting A technique for identifying a specific DNA sequence by transferring single-stranded DNA from a gel to a filter and then hybridizing the DNA with a complementary nucleic acid probe.

Splicing   The process of removing introns from mRNA.

Stop codon   A codon in messenger RNA that causes protein synthesis to stop because it cannot be translated into an amino acid. There are three codons that can terminate protein synthesis, UAA, UAG, and UGA.

Structural motif   Simple combination of a few secondary-structure elements frequently found in protein molecules.

Subcloning   Breaking a large cloned gene into smaller parts and making a new clone from each of the DNA pieces.

Taxol   A chemotherapeutic compound found in yew trees.

T cells   Lymphocytic cells of the immune system involved in cell-mediated immunity and interactions with B cells.

Terminator   Sequence of DNA bases that tells the RNA polymerase to stop synthesizing RNA.

Tertiary structure   The total three-dimensional structure of a protein.

Thymine   A nitrogen-containing base found in DNA.

Tissue culture   A procedure for growing or cloning cells or tissue by in vitro techniques.

Tissue plasminogen activator   A naturally occurring protein that dissolves blood clots; currently being produced for commercial use by genetically engineered bacteria. Also known as tPA.

Transcription   The process of using a DNA template to make a complementary RNA molecule.

Transcription factor   A protein that helps RNA polymerase begin transcription at many promoters.

Transcriptional activator   A protein that helps RNA polymerase begin transcription at one or more promoters.

Transduction   The transfer of DNA from one bacterium to another via a bacteriophage.

Transfection   Using a virus particle as a vector to deliver a gene or genes of interest into eukaryotic cells. The word is a combination of the words transformation and infection.

Transfer RNA (tRNA)   The RNA molecules that match codons and amino acids at the ribosome.

Transformation   A change in the genetic structure of an organism as a result of the uptake and incorporation of foreign DNA.

Transgenic   A transgenic organism is one that has been altered to contain a gene from an organism that belongs to a different species.

Translation   The process of using a messenger RNA template to make a protein.

Translocation   A chromosomal aberration in which a segment of one chromosome breaks off and joins a nonhomologous chromosome.

Transposon   A mobile genetic element that can move from one location in a plasmid or chromosome to another location.

Tumor suppressor gene   A gene whose normal cellular product acts to block tumor formation.

Uracil   A nitrogen-containing base found in RNA.

Vector   The agent used to carry new DNA into a cell. Viruses or plasmids are often used as vectors.

Virus   An infectious agent composed of a single type of nucleic acid (DNA or RNA) enclosed in a coat of protein. Viruses can multiply only within living cells.

Western blot   A technique for identifying a protein by transferring it from a gel to a membrane and then probing it with a labeled antibody.

Index

Page numbers in italics indicate pages in the *Student Guide* corresponding to the preceding page numbers in this volume. Index entries from chapters 1 to 4 are listed only once, as this material is common to both volumes.

Crop plant
  genetically engineered, 361, *232*
  wild relatives of, 361–362, *232–233*
Crossbreeding, 26
Crossing over, 8, 38, 46–47, 266–267, 271, 273, 275–277, *184–186*
Crown gall disease, 258, 261–262, *179*
Culture tube, 420–421
Cystic fibrosis, 88, 112, 208, *152*, 273, 313, 318, *206*, 391, 396, *245*
  gene therapy in, 249
  genetic screening for, 392–394, 397–398, *246–247*
Cytochrome *c*, 103–104
Cytosine, 41, 53–54, 118–119, 123, *118*

D

Darwin, Charles, 50
ddC, 220, *160*
ddI, 220, *160*
Decision-making model, 349, 365–379, *235–239*
Deletion, 44, 311–312
Denaturation
  of DNA, 201
  of proteins, 78, 155
Deoxynucleotide, 53–54, 118–119
Deoxyribose, 41, 53–54, 58, 118, 123, *118*
Designer drugs, 13–14
Diagnostics, 12–13, 18, 22–23
  DNA-based, 106
  PCR-based, 223, 229, *164*
  using monoclonal antibodies, 4–5
Dideoxy sequencing method, 213, 217–219, *157–159*
Dideoxycytidine, 220, *160*
Dideoxyinosine, 220, *160*
Dideoxynucleotide, 213
Dilemma, identification of, 366–367, 375–376, *235–236*
Dilution mathematics, 252
Dimer, 76
Diphtheria, 249
Diplococcus pneumoniae, 39
Discrete particle model of inheritance, 35–37
Disinfectant solution, 427
Disinfecting, 415
Disulfide bridge, 78, 80–81, 112–113
DNA, 3–4
  A-form, 120
  amplification of, *see* Polymerase chain reaction
  ancient, 103, 223
  automated synthesis of, 95
  base sequence of, 57–58, 60
    changes in, 102–103
    comparisons of sequences, 102
    evolutionary studies, 103–105
  B-form, 120
  cellular content of, 149–154, *127–129*
  coding strand of, 136
  complementary, *see* Complementary DNA
  denaturation of, 201
  extraction of
    from animal tissue, 158
    from bacteria, 155–160, *130–131*
    from yeast, 157–158
  genomic organization, 65–68
  hybridization analysis of, *see* Hybridization analysis

"junk," *see* DNA, noncoding
lambda, 182–188, *140–145*
melting temperature of, 102, 201
mitochondrial, 102, 282–284, 303
noncoding, 42, 68–69, 267, 276, *185*
proof that it is genetic material, 38–40
purification of, 155–160, *130–131*
restriction digestion of, 162–166, *134–136*, 172, 183
staining of, 168, 172, *136*, 184, 188, *145*
structure of, 40–41, 53–55, 58
  constructing paper helix, 118–124, *118–119*
  major groove, 120
  minor groove, 120
transcription of, 59–60
in transformation, *see* Transformation
transmission of information by, 56–58
Z-form, 120
DNA fingerprinting, *see* DNA typing
DNA fragments
  blunt-ended, 163, 165, *134*
  cloning of, 100–101, 163
  gel electrophoresis of, *see* Gel electrophoresis
  ligation of, 196, 198–199, *149–150*
  with sticky ends, 163, 165, *134*
  transfer of, 97–98
DNA gun, *see* Gene gun
DNA library
  production of, 101–102
  screening of, 202–203
DNA ligase, 92–93, 100–101, 129, 166, *135*, 189–190, 193, *146*, 196, 198–199, *149–150*
DNA palindrome, 162, 165, *134*
DNA polymerase, 56, 68, 91–92, 98–99, 125–132, *120–121*, *see also* Polymerase chain reaction
  chain terminator, *see* Chain terminator
  proofreading function of, 126
DNA profiling, *see* DNA typing
DNA repair enzymes, 68
DNA replication, 55–56, 60, 125–132, *120–121*, 218, *158*
  direction of, 126–127, 129, 131–132, *120–121*
  errors in, 126
  initiation of, 127
  lagging strand in, 127–129
  leading strand in, 127–129
  trombone model of, 128
DNA sequencing, 99–100, 204
  chain terminators, 213–221, *157–161*
DNA typing, 69, 105–106, 281–298, *190–198*
  accuracy of, 282
  applications of, 283–284, 292, *192*
  case of the bloody knife, 307–310, *203–204*
  hospital mix-up of babies, 299–302, *199–200*
  mitochondrial DNA, 282–283
  paternity case, 303–306, *201–202*
  by PCR, 291–292, *191–192*, 294–295, *194–195*, 298, *198*, 309–310, *203–204*
  by Southern hybridization analysis, 291–294, *191–194*, 305–306, *201–202*
  wet laboratories, 287
DNA virus, 66
DNA-binding protein, 76, 81–83, 85
DNase, 155
Domain, of proteins, 76–80

Information sources, 435–441
Inheritance
    chromosomal nature of, 37–38
    discrete particle model of, 35–37
    fluid blending model of, 35
    sex-linked, 38
Inherited disease, *see* Genetic disease
Initiation codon, 61, 134
Initiation factors, 134
Inoculating loop/needle, 421
Insulin, 14, 111
Interferon, 18
Interleukin-2, 13, 18
Interpersonal skills, 404, 412, *254*
Introductions, *see* Environment, introduction of GEOs into
Intron, 42, 62, 68, 79, 134–135
Inversion, 44–45, 311

J
Jumping gene, *see* Transposon
"Junk" DNA, *see* Noncoding DNA

K
Kalanchoe plant, 260
Kanamycin resistance, 190–195, *146–148*, 233
Keratin, 79–81, 85
Kinship determination, 283–284, 292, *192*
Klinefelter's syndrome, 311–312
Kringle domain, 78

L
L broth, 425
Laboratory
    biosafety, 415–416
    microbiological methods, 417–419
    release of GEOs from, 360, *231*
Laboratory equipment, 428–431
lac operon, 63–64, 140
β-Lactamase, 233
Lambda, *see* Bacteriophage lambda
Latent viral infection, 248
Legal behavior, 394
Leukemia, 312, 322, *210*
Life support measures, 372–374, 378–379, *238–239*
Li-Fraumeni syndrome, 322, *210*
Linkage, 38, 273
Liquid culture, 417
Livestock productivity, 18–19
Loggerhead turtle, 106
Luria broth, 425
Lysate, 155, 159, *130*, 249
Lysis, 155, 159, *130*
    by bacteriophage T4, 251–252, 256, *177*
Lysogen, 248
Lysogenic viral infection, 249
Lysozyme, 112–113
Lytic viral infection, 248

M
Macrolesion, 44
Macromutation, 44
Maentyranta, Eero, 86
Malaria, 314
Man-made risk, 336–338, 342–345, *220–223*

Marker, 108–109
Marker gene, 237, *168*
McClintock, Barbara, 51
Media
    presterilized, 423
    recipes for, 425
    sterilization of, 423
Medical biotechnology, 12–16, 27–30
Medical research tools, 16
Meiosis, 36, 46–47, 266–280, *184–189*
Meiosis I, 271–272, 275, *184*
Meiosis II, 272, 279–280, *188–189*
Melting temperature
    of DNA, 102, 201
    of protein, 78
Mendel, Gregor, 34–35
Mendelian principles, 35–36, 273
Messenger RNA (mRNA), 58, 133–148, *122–126*
    splicing of precursor RNA, 62
    synthesis of, 59–60
    synthesis of protein from, 60–61
Metabolic engineering, 12
Metal objects, sterilization of, 422
Methylene blue, 168, 427
Methylene chloride, 338, 344, *222*
Microbial culture, transferring of, 421
Microbiological methods, 417–419
Microcentrifuge, 429, 431
Microcentrifuge tube, 422
Microinjection of DNA, 263–264, *181–182*
Microlesion, 44
Micromutation, 44
Micropipette, 430
Micropipette tip, 422–423
Micropipettor, 428
    alternatives to, 428
Milk production, 19
Minigel, 428
Mitochondrial DNA, 102, 282–284, 303
Modular protein, 78–80
Molecular biology, 89
Monoclonal antibody, 4–5, 13
Moral-action guides, 367, 376, *236*
Morgan, Thomas Hunt, 38
Motif, in proteins, 75–77
mRNA, *see* Messenger RNA
Muller, Herman, 38
Multigene disorder, 311–312, 317, *205*, 396, *245*
Multigenic trait, 17
Multiplication rule, 282, 287–289
Mutagenesis, 360, *231*
Mutation, 35, 43, 68–69, 250–251, 266–267, 274–275, *184*, 312–313, 317, *205*
    beneficial, 69, 86–87
    cancer and, 321, *209*
    effect of, 68–69, 85–87
    harmful, 68–69, 85–86
    "silent," 68
    as source of genetic variation, 44–45
    types of, 44
Mycorrhiza, 17

N
NAD-binding domain, 78